DAVID I. MARX
*Southern Illinois University at Carbondale*

*to accompany*

# PHYSICS

*FOURTH EDITION*

**JOHN D. CUTNELL**
**KENNETH W. JOHNSON**
*Southern Illinois University at Carbondale*

**JOHN WILEY & SONS, INC.**
NEW YORK • CHICHESTER • WEINHEIM • BRISBANE • SINGAPORE • TORONTO

Copyright © 1998 by John Wiley & Sons, Inc.

Excerpts from this work may be reproduced by instructors for distribution on a not-for-profit basis for testing or instructional purposes only to students enrolled in courses for which the textbook has been adopted. *Any other reproduction or translation of this work beyond that permitted by Sections 107 or 108 of the 1976 United States Copyright Act without the permission of the copyright owner is unlawful. Requests for permission or further information should be addressed to the Permissions Department, John Wiley & Sons, Inc., 605 Third Avenue, New York, NY 10158-0012.*

ISBN 0-471-16415-1

Printed in the United States of America

10 9 8 7 6 5 4 3 2 1

Printed and bound by Victor Graphics, Inc.

# PREFACE

This volume of *Test Bank and Homework Choices* has undergone complete revision since the previous edition. The *Test Bank* contains 2006 questions of varying difficulty. The *Homework Choices* section lists all of the even-numbered problems in the text and provides multiple-choice answers for them. The notation and language used throughout this volume match the usage in the text; and the questions have been placed into the section of each chapter from which its subject originates. The sections are grouped in the same manner as the homework problems are at the end of each chapter in the text.

The test bank has been designed for use in test preparation software for both Windows and Macintosh platforms by Brownstone Research Group. The software is very easy to use and provides full-editing capability. I highly recommend trying the program. Copies are available from your Wiley representative.

Questions are designed to test conceptual understanding of the material as well as problem-solving ability. The questions have also been labeled with an difficulty indicator:

- ○ – less than average difficulty
- ◓ – average difficulty
- ● – greater than average difficulty

Instructors may choose to use the test bank questions in either open- or closed-book test conditions. Some problems include the necessary constants, but, in general, students will require access to the constants that are listed on the inside covers of the text. Each question has five answers labeled a, b, c, d, and e for use in a multiple choice format. An instructor who prefers to not use the multiple choice test format may still use the questions by simply omitting the choices.

The *Homework Choices* section of this volume allows instructors to assign homework in a multiple choice format. Each problem has, in almost all cases, five choices that are labeled A, B, C, D, and E. The few problems that require a graph, drawing, or proof are not available for multiple choice solutions and are noted accordingly. The software is also available for preparing these multiple choice homework assignments.

My hope is that the new features and reorganization will substantially facilitate your test preparation and homework assignments.

*Acknowledgments: I would like to thank John D. Cutnell and Kenneth W. Johnson for their patience and unwavering attention to detail. I am also grateful to the following individuals for their support and assistance in the preparation of this manuscript: Cynthia Rhoads and Stuart Johnson at John Wiley & Sons and my wife, Cathie Marx. This book is dedicated to Frank R. Valentine (1910-1990).*

David T. Marx
Center for Advanced Friction Studies
Southern Illinois University at Carbondale
Carbondale, Illinois 62901-4343
*e-mail*: dtmarx@siu.edu

# CONTENTS

## Test Bank

**Chapter**

1. Introduction and Mathematical Concepts .................................... 1
2. Kinematics in One Dimension .............................................. 11
3. Kinematics in Two Dimensions ............................................. 25
4. Forces and Newton's Laws of Motion ....................................... 35
5. Dynamics of Uniform Circular Motion ...................................... 52
6. Work and Energy .......................................................... 60
7. Impulse and Momentum ..................................................... 71
8. Rotational Kinematics .................................................... 81
9. Rotational Dynamics ...................................................... 88
10. Elasticity and Simple Harmonic Motion ................................... 98
11. Fluids .................................................................. 107
12. Temperature and Heat .................................................... 117
13. The Transfer of Heat .................................................... 127
14. The Ideal Gas Law and Kinetic Theory .................................... 133
15. Thermodynamics .......................................................... 140
16. Waves and Sound ......................................................... 152
17. The Principle of Linear Superposition and Interference Phenomena ........ 161
18. Electric Forces and Electric Fields ..................................... 169
19. Electric Potential Energy and the Electric Potential ................... 181
20. Electric Circuits ....................................................... 192
21. Magnetic Forces and Magnetic Fields ..................................... 206
22. Electromagnetic Induction ............................................... 219
23. Alternating Current Circuits ............................................ 231
24. Electromagnetic Waves ................................................... 242
25. The Reflection of Light: Mirrors ........................................ 250
26. The Refraction of Light: Lenses and Optical Instruments ................. 256
27. Interference and the Wave Nature of Light ............................... 272

**28** *Special Relativity* .................................................. *281*

**29** *Particles and Waves* ............................................... *290*

**30** *The Nature of the Atom* .......................................... *297*

**31** *Nuclear Physics and Radioactivity* ............................ *306*

**32** *Ionizing Radiation, Nuclear Energy, and Elementary Particles* ............. *313*

**Test Bank Answers** ................................................ *320*

## Homework Choices

**Chapter**

**1** *Introduction and Mathematical Concepts* ................................. *330*

**2** *Kinematics in One Dimension* .............................................. *335*

**3** *Kinematics in Two Dimensions* ............................................. *341*

**4** *Forces and Newton's Laws of Motion* ..................................... *345*

**5** *Dynamics of Uniform Circular Motion* .................................... *353*

**6** *Work and Energy* ............................................................. *358*

**7** *Impulse and Momentum* .................................................... *364*

**8** *Rotational Kinematics* ....................................................... *369*

**9** *Rotational Dynamics* ........................................................ *375*

**10** *Elasticity and Simple Harmonic Motion* ................................. *381*

**11** *Fluids* ........................................................................... *387*

**12** *Temperature and Heat* ...................................................... *394*

**13** *The Transfer of Heat* ........................................................ *400*

**14** *The Ideal Gas Law and Kinetic Theory* .................................. *403*

**15** *Thermodynamics* ............................................................. *407*

**16** *Waves and Sound* ............................................................ *415*

**17** *The Principle of Linear Superposition and Interference Phenomena* ........ *421*

**18** *Electric Forces and Electric Fields* ....................................... *425*

**19** *Electric Potential Energy and the Electric Potential* .................. *430*

**20** *Electric Circuits* .............................................................. *434*

**21** *Magnetic Forces and Magnetic Fields* ................................... *443*

| | | |
|---|---|---|
| **22** | *Electromagnetic Induction* | *449* |
| **23** | *Alternating Current Circuits* | *455* |
| **24** | *Electromagnetic Waves* | *459* |
| **25** | *The Reflection of Light: Mirrors* | *463* |
| **26** | *The Refraction of Light: Lenses and Optical Instruments* | *467* |
| **27** | *Interference and the Wave Nature of Light* | *475* |
| **28** | *Special Relativity* | *479* |
| **29** | *Particles and Waves* | *482* |
| **30** | *The Nature of the Atom* | *486* |
| **31** | *Nuclear Physics and Radioactivity* | *491* |
| **32** | *Ionizing Radiation, Nuclear Energy, and Elementary Particles* | *496* |
| ***Homework Answers*** | | *500* |

# CHAPTER 1 — Introduction and Mathematical Concepts

*Section 1.2 Units*
*Section 1.3 The Role of Units in Problem Solving*

1. Which one of the following is an SI base unit?
   (a) gram
   (b) slug
   (c) newton
   (d) centimeter
   (e) kilogram

2. Complete the following statement: Today, the standard meter is defined in terms of
   (a) the distance from the earth's equator to the north pole.
   (b) the wavelength of light emitted from a krypton atom.
   (c) the wavelength of light emitted from a sodium atom.
   (d) a platinum-iridium bar kept in Sèvres, France.
   (e) the speed of light.

3. Complete the following statement: Today, the standard unit of mass is defined in terms of
   (a) a specified volume of water at 4 °C.
   (b) a standard platinum-iridium cylinder.
   (c) a specified number of cesium atoms.
   (d) a standard platinum bar.
   (e) the speed of light.

4. Complete the following statement: Today, the standard unit of time is defined in terms of
   (a) the electromagnetic waves emitted by cesium atoms.
   (b) the motion of the moon around the earth.
   (c) the motion of a precision pendulum.
   (d) the average solar day.
   (e) the speed of light.

5. A bead has a mass of one *milli*gram. Which one of the following statements indicates the correct mass of the bead in grams?
   (a) The bead has a mass of $1 \times 10^6$ grams.
   (b) The bead has a mass of $1 \times 10^3$ grams.
   (c) The bead has a mass of $1 \times 10^{-1}$ grams.
   (d) The bead has a mass of $1 \times 10^{-3}$ grams.
   (e) The bead has a mass of $1 \times 10^{-6}$ grams.

6. Which one of the following is the longest length?
   (a) $10^0$ meters
   (b) $10^2$ centimeters
   (c) $10^4$ millimeters
   (d) $10^5$ micrometers
   (e) $10^7$ nanometers

7. In the sport of platform diving, a platform is set at a height of 7.3 m above the surface of the water. What is the height, expressed in feet, of the platform?
   (a) 13 feet
   (b) 18 feet
   (c) 24 feet
   (d) 33 feet
   (e) 97 feet

8. A candy shop sells a pound of chocolate for $ 7.99. What is the price of 2.25 kg of chocolate at the shop?
   (a) $ 8.17
   (b) $ 12.51
   (c) $ 17.98
   (d) $ 29.66
   (e) $ 39.55

## INTRODUCTION AND MATHEMATICAL CONCEPTS

9. The ratio $\dfrac{1 \text{ kilogram}}{1 \text{ milligram}}$ is
    (a) $10^2$.
    (b) $10^3$.
    (c) $10^6$.
    (d) $10^{-3}$.
    (e) $10^{-6}$.

10. Which one of the following is equivalent to $2.0 \text{ m}^2$?
    (a) $2.0 \times 10^{-4} \text{ cm}^2$
    (b) $2.0 \times 10^4 \text{ cm}^2$
    (c) $2.0 \times 10^{-2} \text{ cm}^2$
    (d) $2.0 \times 10^2 \text{ cm}^2$
    (e) $2.0 \times 10^3 \text{ cm}^2$

11. Which one of the following pairs of units may *not* be added together, even after the appropriate unit conversions have been made?
    (a) grams and milligrams
    (b) slugs and kilograms
    (c) miles and kilometers
    (d) centimeters and yards
    (e) kilograms and kilometers

12. Which one of the following is equivalent to 24.8 m?
    (a) $2.48 \times 10^1 \text{ m}$
    (b) $2.48 \times 10^2 \text{ m}$
    (c) $24.8 \times 10^{-1} \text{ m}$
    (d) $24.8 \times 10^{-2} \text{ m}$
    (e) $2.48 \times 10^0 \text{ m}$

13. In the sport of horseshoe pitching, two stakes are separated by 40.0 feet. What is the distance in meters between the two stakes?
    (a) 24.4 m
    (b) 4.80 m
    (c) 18.3 m
    (d) 12.2 m
    (e) 15.7 m

14. The surface of a lake has an area of $15.5 \text{ km}^2$. What is the area of the lake in $\text{m}^2$?
    (a) $1.55 \times 10^4 \text{ m}^2$
    (b) $1.55 \times 10^5 \text{ m}^2$
    (c) $1.55 \times 10^6 \text{ m}^2$
    (d) $1.55 \times 10^7 \text{ m}^2$
    (e) $1.55 \times 10^8 \text{ m}^2$

15. The mathematical relationship between three physical quantities is given by $a = \dfrac{b^2}{c}$. If the dimension of $b$ is $\dfrac{[L]}{[T]}$; and the dimension of $c$ is $[L]$. Which one of the following is the dimension of $a$?
    (a) $[L]$
    (b) $[T]$
    (c) $\dfrac{[L]}{[T]}$
    (d) $\dfrac{[L]}{[T]^2}$
    (e) $\dfrac{[L]^2}{[T]^2}$

16. The distance $d$ that a certain particle moves may be calculated from the expression $d = at + bt^2$ where $a$ and $b$ are constants; and $t$ is the elapsed time. The dimensions of the quantities $a$ and $b$ are, respectively,
    (a) $\dfrac{[L]}{[T]}, \dfrac{[L]}{[T]^2}$
    (b) $[L], [L]^2$
    (c) $\dfrac{[L]}{[T]^2}, \dfrac{[L]}{[T]^3}$
    (d) $\dfrac{[L]}{[T]}, \dfrac{[L]^2}{[T]^2}$
    (e) $\dfrac{1}{[T]}, \dfrac{1}{[T]^2}$

17. Using the dimensions given for the variables in the table, determine which one of the following expressions is correct.

(a) $f = \dfrac{g}{2\pi l}$
(b) $f = 2\pi l/g$
(c) $2\pi f = \sqrt{\dfrac{g}{l}}$
(d) $2\pi f = \sqrt{\dfrac{l}{g}}$
(e) $f = 2\pi\sqrt{gl}$

| Variable | Dimension |
|---|---|
| $f$ | $\dfrac{1}{[T]}$ |
| $l$ | $[L]$ |
| $g$ | $\dfrac{[L]}{[T]^2}$ |

18. A certain physical quantity, $R$, is calculated using the formula: $R = 4a^2(b - c)$ where $a$, $b$, and $c$ are distances. What is the SI unit for $R$?
(a) cm
(b) $cm^2$
(c) m
(d) $m^2$
(e) $m^3$

## Section 1.4 Trigonometry

19. Which one of the following expressions may be used to correctly find the angle $\theta$ in the figure?

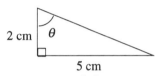

(a) $\theta = \cos^{-1}\left(\dfrac{5}{2}\right)$
(b) $\theta = \tan^{-1}\left(\dfrac{5}{2}\right)$
(c) $\theta = \sin^{-1}\left(\dfrac{2}{5}\right)$
(d) $\theta = \tan^{-1}\left(\dfrac{2}{5}\right)$
(e) $\theta = \sin^{-1}\left(\dfrac{5}{2}\right)$

20. The length of each side of a square is 4.0 m. What is the length of the diagonal of the square (shown as a dashed line in the figure)?
(a) 2.8 m
(b) 3.5 m
(c) 5.7 m
(d) 8.0 m
(e) 16 m

21. A 2.5-m ladder leans against a wall and makes an angle with the wall of 32° as shown in the figure. What is the height $h$ above the floor where the ladder makes contact with the wall?
(a) 2.1 m
(b) 1.3 m
(c) 2.4 m
(d) 1.6 m
(e) 1.9 m

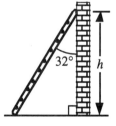

22. A pole is held vertically by attaching wires at a height of 13.4 m above the ground. The other end of each wire is anchored in the ground at a distance of 9.54 m from the base of the pole. The pole makes a right angle with the ground. What is the length of each wire?
(a) 14.1 m
(b) 19.7 m
(c) 11.5 m
(d) 16.4 m
(e) 22.8 m

4    INTRODUCTION AND MATHEMATICAL CONCEPTS

23. A surveyor wants to find the distance across a river. A stake is placed on each bank of the river as shown in the figure. She measures a distance of 30.0 m from one stake, thus finding a third vertex on a right triangle. She then measures the angle θ and finds it equal to 75.9°. What is the distance across the river?

(a) 89.2 m
(b) 119 m
(c) 268 m
(d) 15.3 m
(e) 29.0 m

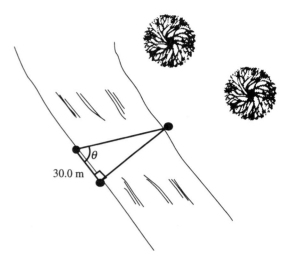

24. A certain mountain road is inclined 3.1° with respect to the horizon. What is the change in altitude of the car as a result of its traveling 2.90 km along the road?
(a) 157 m          (c) 116 m          (e) 289 m
(b) 181 m          (d) 203 m

## Section 1.5 The Nature of Physical Quantities: Scalars and Vectors
## Section 1.6 Vector Addition and Subtraction

25. Which one of the following is a vector quantity?
(a) mass                (c) time                (e) volume
(b) temperature         (d) displacement

26. Which one of the following is a vector quantity?
(a) the age of the earth
(b) the mass of a football
(c) the earth's pull on your body
(d) the temperature of an iron bar
(e) the number of people attending a baseball game

27. Which statement is true concerning scalar quantities?
(a) Scalar quantities must be represented by base units.
(b) Scalar quantities have both magnitude and direction.
(c) Scalar quantities can be added to vector quantities using rules of trigonometry.
(d) Scalar quantities can be added to other scalar quantities using rules of trigonometry.
(e) Scalar quantities can be added to other scalar quantities using rules of ordinary addition.

28. Two vectors **A** and **B** are added together to form a vector **C**. The relationship between the magnitudes of the vectors is given by: $A + B = C$. Which one of the following statements concerning these vectors is true?
(a) **A** and **B** must be displacements.
(b) **A** and **B** must have equal lengths.
(c) **A** and **B** must point in opposite directions.
(d) **A** and **B** must point in the same direction.
(e) **A** and **B** must be at right angles to each other.

29. Two vectors **A** and **B** are added together to form a vector **C**. The relationship between the magnitudes of the vectors is given by: $A^2 + B^2 = C^2$. Which statement concerning these vectors is true?
    (a) **A** and **B** must be at right angles to each other.
    (b) **A** and **B** could have any orientation relative to each other.
    (c) **A** and **B** must have equal lengths.
    (d) **A** and **B** must be parallel.
    (e) **A** and **B** could be antiparallel.

30. Three vectors **A**, **B**, and **C** add together to yield zero: **A** + **B** + **C** = 0. The vectors **A** and **C** point in *opposite* directions and their magnitudes are related by the expression : $A = 2C$. Which one of the following conclusions is correct?
    (a) **A** and **B** have equal magnitudes and point in opposite directions.
    (b) **B** and **C** have equal magnitudes and point in the same direction.
    (c) **B** and **C** have equal magnitudes and point in opposite directions.
    (d) **A** and **B** point in the same direction, but **A** has twice the magnitude of **B**.
    (e) **B** and **C** point in the same direction, but **C** has twice the magnitude of **B**.

31. What is the angle between the vectors **A** and −**A** when they are drawn from a common origin?
    (a) 0°   (c) 180°   (e) 360°
    (b) 90°  (d) 270°

32. What is the minimum number of vectors with *unequal* magnitudes whose vector sum can be zero?
    (a) two    (c) four   (e) six
    (b) three  (d) five

33. What is the minimum number of vectors with *equal* magnitudes whose vector sum can be zero?
    (a) two    (c) four   (e) six
    (b) three  (d) five

34. A physics student adds two displacement vectors with magnitudes of 8.0 km and 6.0 km. Which one of the following statements is true concerning the magnitude of the resultant displacement?
    (a) It must be 10.0 km.
    (b) It must be 14.0 km.
    (c) It could be zero depending on how the vectors are oriented.
    (d) No conclusion can be reached without knowing the directions of the vectors.
    (e) It could have any value between 2.0 km and 14.0 km depending on how the vectors are oriented.

35. A student adds two displacement vectors with magnitudes of 3.0 m and 4.0 m, respectively. Which one of the following could *not* be a possible choice for the resultant?
    (a) 1.3 m   (c) 5.0 m   (e) 7.8 m
    (b) 3.3 m   (d) 6.8 m

36. Two displacement vectors of magnitudes 21 cm and 79 cm are added. Which one of the following is the *only* possible choice for the magnitude of the resultant?
    (a) zero    (c) 37 cm   (e) 114 cm
    (b) 28 cm   (d) 82 cm

6   INTRODUCTION AND MATHEMATICAL CONCEPTS

37. Which expression is *false* concerning the vectors shown in the sketch?
    (a) **C** = **A** + **B**
    (b) **C** + **A** = −**B**
    (c) **A** + **B** + **C** = 0
    (d) $C < A + B$
    (e) $A^2 + B^2 = C^2$

38. City A lies 30 km directly south of city B. A bus, beginning at city A travels 50 km at 37° north of east to reach city C. How far, and in what direction must the bus go from city C to reach city B?
    (a) 20 km, west
    (b) 40 km, west
    (c) 80 km, west
    (d) 40 km, east
    (e) 80 km, east

39. Town A lies 20 km north of town B. Town C lies 13 km west of town A. A small plane flies directly from town B to town C. What is the displacement of the plane?
    (a) 33 km, 33° north of west
    (b) 19 km, 33° north of west
    (c) 24 km, 57° north of west
    (d) 31 km, 57° north of west
    (e) 6.6 km, 40° north of west

40. Four members of the Main Street Bicycle Club meet at a certain intersection on Main Street. The members then start from the same location, but travel in different directions. A short time later, displacement vectors for the four members are:
    **A** = 2.0 km, west;  **B** = 1.6 km, north;  **C** = 2.0 km, east;  **D** = 2.4 km, south
    What is the resultant displacement **R** of the members of the bicycle club: **R** = **A** + **B** + **C** + **D**?
    (a) 0.8 km, south
    (b) 0.4 km, 45° south of east
    (c) 3.6 km, 37° north of west
    (d) 4.0 km, east
    (e) 4.0 km, south

41. A force, **F**$_1$, of magnitude 2.0 N and directed due east is exerted on an object. A second force exerted on the object is **F**$_2$ = 2.0 N, due north. What is the magnitude and direction of a third force, **F**$_3$, which must be exerted on the object so that the resultant force is zero?
    (a) 1.4 N, 45° north of east
    (b) 1.4 N, 45° south of west
    (c) 2.8 N, 45° north of east
    (d) 2.8 N, 45° south of west
    (e) 4.0 N, 45° east of north

42. A sailboat leaves a harbor and sails 1.1 km in the direction 75° north of east. where the captain stops for lunch. A short time later, the boat sails 1.8 km in the direction 15° south of east. What is the magnitude of the resultant displacement?
    (a) 2.1 km
    (b) 1.5 km
    (c) 2.9 km
    (d) 1.2 km
    (e) 0.59 km

43. Three vectors **A**, **B**, and **C** have the following $x$ and $y$ components:
    $A_x = 1$ m,  $A_y = 0$
    $B_x = 1$ m,  $B_y = 1$ m
    $C_x = 0$,     $C_y = -1$ m
    According to the figure, how must **A**, **B**, and **C** be combined to form the vector **D**?
    (a) **D** = **A** − **B** − **C**
    (b) **D** = **A** − **B** + **C**
    (c) **D** = **A** + **B** − **C**
    (d) **D** = **A** + **B** + **C**
    (e) **D** = −**A** + **B** + **C**

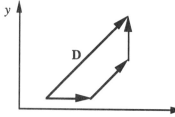

## Section 1.7 The Components of a Vector

44. Which one of the following statements concerning vectors and scalars is *false*?
    (a) In calculations, the vector components of a vector may be used in place of the vector itself.
    (b) It is possible to use vector components that are not perpendicular.
    (c) A scalar component may be either positive or negative.
    (d) A vector that is zero may have components other than zero.
    (e) Two vectors are equal only if they have the same magnitude and direction.

45. A displacement vector is 23 km in length and is directed 65° south of east. What are the components of this vector?

    |     | Eastward Component | Southward Component |
    |-----|-------------------|---------------------|
    | (a) | 21 km             | 9.7 km              |
    | (b) | 23 km             | 23 km               |
    | (c) | 23 km             | 0 km                |
    | (d) | 9.7 km            | 21 km               |
    | (e) | 0 km              | 23 km               |

46. The $x$ and $y$ components of a displacement vector are –3.00 m and +4.00 m, respectively. What angle does this vector make with the positive $x$ axis?
    (a) 233°
    (b) 127°
    (c) –53.0°
    (d) 53.0°
    (e) 37.0°

47. A race car makes one lap around a circular track of radius $R$. When the car has traveled *halfway* around the track, what is the magnitude of the car's displacement from the starting point?
    (a) $R$
    (b) $2R$
    (c) $\pi R$
    (d) $2\pi R$
    (e) zero

48. A bug crawls 4.25 m along the base of a wall. Upon reaching a corner, the bug's direction of travel changes from south to west. The bug then crawls 3.15 m before stopping. What is the magnitude of the bug's displacement?
    (a) 7.40 m
    (b) 2.72 m
    (c) 3.83 m
    (d) 4.91 m
    (e) 5.29 m

49. During the execution of a play, a football player carries the ball for a distance of 33 m in the direction 76° north of east. To determine the number of meters gained on the play, find the northward component of the ball's displacement.
    (a) 8.0 m
    (b) 16 m
    (c) 24 m
    (d) 28 m
    (e) 32 m

50. A bird flies 25.0 m in the direction 55° east of south to its nest. The bird then flies 75.0 m in the direction 55° west of north. What are the northward and westward components of the resultant displacement of the bird from its nest?

    |     | northward | westward |
    |-----|-----------|----------|
    | (a) | 29 m      | 41 m     |
    | (b) | 41 m      | 29 m     |
    | (c) | 35 m      | 35 m     |
    | (d) | 81 m      | 57 m     |
    | (e) | 57 m      | 81 m     |

# 8  INTRODUCTION AND MATHEMATICAL CONCEPTS

## Section 1.8 Addition of Vectors by Means of Components

51. Two vectors **A** and **B**, are added together to form the vector **C** = **A** + **B**.
    The relationship between the magnitudes of these vectors is given by:
    $$C_x = A \cos 30° + B$$
    $$C_y = -A \sin 30°$$
    Which statement best describes the orientation of these vectors?
    (a) **A** points in the negative $x$ direction while **B** points in the positive $y$ direction.
    (b) **A** points in the negative $y$ direction while **B** points in the positive $x$ direction.
    (c) **A** points 30° below the positive $x$ axis while **B** points in the positive $x$ direction.
    (d) **A** points 30° above the positive $x$ axis while **B** points in the positive $x$ direction.
    (e) **A** points 30° above the negative $x$ axis while **B** points in the positive $x$ direction.

52. Use the component method of vector addition to find the resultant of the following three vectors:
    $$A = 56 \text{ km, east}$$
    $$B = 11 \text{ km, } 22° \text{ south of east}$$
    $$C = 88 \text{ km, } 44° \text{ west of south}$$

    (a) 81 km, 14° west of south    (c) 52 km, 66° south of east    (e) 66 km, 7.1° west of south
    (b) 97 km, 62° south of east    (d) 68 km, 86° south of east

53. Use the component method of vector addition to find the components of the resultant of the four displacements shown in the figure. The magnitudes of the displacements are: $A = 2.25$ cm, $B = 6.35$ cm, $C = 5.47$ cm, and $D = 4.19$ cm.

    |   | x component | y component |
    |---|---|---|
    | (a) | 2.19 cm | −6.92 cm |
    | (b) | 3.71 cm | −1.09 cm |
    | (c) | 5.45 cm | −2.82 cm |
    | (d) | 1.09 cm | −3.71 cm |
    | (e) | 6.93 cm | −2.19 cm |

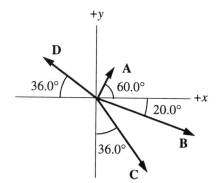

## Additional Problems

54. Which one of the following answers would give the correct number of significant figures when the following masses are added together: 3.6 kg, 113 kg, and 4.19 kg?
    (a) 121 kg            (c) 120.79 kg          (e) $120.8 \times 10^3$ kg
    (b) 120.8 kg          (d) $1.20 \times 10^2$ kg

55. A physics text has 1060 pages and is 33.5 millimeters thick between the inside front cover and the inside back cover. What is the thickness of a page? The answer should be expressed in scientific notation with the correct number of significant figures.
    (a) $3.55 \times 10^{-4}$ m       (c) $3.2 \times 10^{-3}$ m       (e) $3.16 \times 10^{-5}$ m
    (b) $3.160 \times 10^{-2}$ m      (d) $3.6 \times 10^{-6}$ m

56. Justine and her friends exit the physics classroom and walk 0.81 km to their math class. While walking, Justine's average step length is 58 cm. How many steps does she take in walking between these two classes?
   (a) 310
   (b) 720
   (c) 1400
   (d) 3100
   (e) 7200

*Questions 57 and 58 pertain to the situation described below:*

Two vectors, **A** and **B**, are added together to form the vector **C** = **A** + **B**. The relationship between the magnitudes of these vectors is given by:
$$C_x = 0$$
$$C_y = A \sin 60° + B \sin 30°$$
$A_x$ and $A_y$ point in the positive $x$ and $y$ directions, respectively.

57. Which one of the following statements best describes the orientation of vectors **A** and **B**?
   (a) **A** and **B** point in opposite directions.
   (b) **A** points 60° above the positive $x$ axis while **B** points 30° above the negative $x$ axis.
   (c) **A** points 60° above the negative $x$ axis while **B** points 30° above the positive $x$ axis.
   (d) **A** points 60° below the positive $x$ axis while **B** points 30° above the positive $y$ axis.
   (e) **A** points 60° below the positive $x$ axis while **B** points 30° below the positive $y$ axis.

58. How does the magnitude of **A** compare with that of **B**?
   (a) $A = B$
   (b) $A = 1.7B$
   (c) $A = 0.4B$
   (d) $A = 0.5B$
   (e) $A = 0.7B$

*Questions 59 through 61 pertain to the statement and table below:*

The table gives the $x$ and $y$ components of two vectors **A** and **B**:

| Vector | x component | y component |
|---|---|---|
| A | +15 units | +10 units |
| B | +15 units | −10 units |

59. Which one of the following statements concerning these vectors is true?
   (a) The vector **A** − **B** has no $x$ component.
   (b) The two vectors have different magnitudes.
   (c) **A** makes a 56° angle with the positive $x$ axis.
   (d) **B** makes a 34° angle with the positive $y$ axis.
   (e) The vector **A** + **B** makes a 34° angle with the positive $x$ axis.

60. Determine the magnitude of the vector sum **A** + **B**.
   (a) 5 units
   (b) 15 units
   (c) 20 units
   (d) 30 units
   (e) 50 units

61. Determine the magnitude of the vector difference **A** − **B**.
   (a) 5 units
   (b) 15 units
   (c) 20 units
   (d) 30 units
   (e) 50 units

# INTRODUCTION AND MATHEMATICAL CONCEPTS

*Questions 62 and 63 pertain to the situation described below.*

A boat radioed a distress call to a Coast Guard station. At the time of the call, a vector **A** from the station to the boat had a magnitude of 45.0 km and was directed 15.0° east of north. A vector from the station to the point where the boat was later found is **B** = 30.0 km, 15.0° north of east..

62. What are the components of the vector from the point where the distress call was made to the point where the boat was found? In other words, what are the components of vector **C** = **B** − **A**?

    | x component | y component |
    |---|---|
    | (a) 17.3 km, east | 35.7 km, south |
    | (b) 35.7 km, west | 17.4 km, north |
    | (c) 40.6 km, east | 51.2 km, south |
    | (d) 17.3 km, west | 51.2 km, south |
    | (e) 40.6 km, east | 35.7 km, north |

63. How far did the boat travel from the point where the distress call was made to the point where the boat was found? In other words, what is the magnitude of vector **C**?
    (a) 65.3 km
    (b) 39.7 km
    (c) 26.5 km
    (d) 54.0 km
    (e) 42.5 km

# CHAPTER 2
## Kinematics in One Dimension

*Section 2.1 Displacement*
*Section 2.2 Speed and Velocity*

1. A particle travels along a curved path between two points P and Q as shown. The *displacement* of the particle does *not* depend on
   (a) the location of P.
   (b) the location of Q.
   (c) the distance traveled from P to Q.
   (d) the shortest distance between P and Q.
   (e) the direction of Q from P.

2. For which one of the following situations will the path length equal the magnitude of the displacement?
   (a) A jogger is running around a circular path.
   (b) A ball is rolling down an inclined plane.
   (c) A train travels 5 miles east; and then, it stops and travels 2 miles west.
   (d) A ball rises and falls after being thrown straight up from the earth's surface.
   (e) A ball on the end of a string is moving in a vertical circle.

3. At time $t = 0$, an object is observed at $x = 0$; and its position along the $x$ axis follows this expression: $x = -3t + t^3$, where the units for distance and time are meters and seconds, respectively. What is the object's displacement $\Delta x$ between $t = 1.0$ s and $t = 3.0$ s?
   (a) +20 m
   (b) –20 m
   (c) +16 m
   (d) +2 m
   (e) –2 m

*Questions 4 and 5 pertain to the situation described below:*

Peter noticed a bug crawling along a meter stick and decided to record the bug's position in five second intervals. After the bug crawled off the meter stick, Peter created the table shown.

| time (s) | position (cm) |
|---|---|
| 0.00 | 49.6 |
| 5.00 | 39.2 |
| 10.0 | 42.5 |
| 15.0 | 41.0 |
| 20.0 | 65.7 |

4. What is the displacement of the bug between $t = 0.00$ and $t = 20.0$ s?
   (a) +39.9 cm
   (b) –39.9 cm
   (c) +65.7 cm
   (d) –16.1 cm
   (e) +16.1 cm

5. What is the total distance that the bug traveled between $t = 0.00$ and $t = 20.0$ s? Assume the bug only changed directions at the end of a five second interval.
   (a) 39.9 cm
   (b) 65.7 cm
   (c) 16.1 cm
   (d) 47.1 cm
   (e) 26.5 cm

6. Which one of the physical quantities listed below is *not* correctly paired with its SI unit and dimension?

   | Quantity | Unit | Dimension |
   |---|---|---|
   | (a) velocity | m/s | [L]/[T] |
   | (b) path length | m | [L] |
   | (c) speed | m/s | [L]/[T] |
   | (d) displacement | m/s² | [L]/[T]² |
   | (e) speed × time | m | [L] |

## KINEMATICS IN ONE DIMENSION

7. A car travels in a straight line covering a total distance of 90.0 miles in 60.0 minutes. Which one of the following statements concerning this situation is *necessarily* true?
   (a) The velocity of the car is constant.
   (b) The acceleration of the car must be non-zero.
   (c) The first 45 miles must have been covered in 30.0 minutes.
   (d) The speed of the car must be 90.0 miles per hour throughout the entire trip.
   (e) The average velocity of the car is 90.0 miles per hour in the direction of motion.

8. A Canadian goose flew 845 km from Southern California to Oregon with an average speed of 30.5 m/s. How long, in hours, did it take the goose to make this journey?
   (a) 27.7 h
   (b) 8.33 h
   (c) 66.1 h
   (d) 462 h
   (e) 7.70 h

9. Carol's hair grows with an average speed of $3.5 \times 10^{-9}$ m/s. How long does it take for her hair to grow 0.30 m? Note: 1 yr = $3.156 \times 10^7$ s.
   (a) 1.9 yr
   (b) 1.3 yr
   (c) 0.37 yr
   (d) 5.4 yr
   (e) 2.7 yr

10. Carl Lewis set a world record for the 100.0-m run with a time of 9.86 s. If, after reaching the finish line, Mr. Lewis walked directly back to his starting point in 90.9 s, what is the magnitude of his average velocity for the 200.0 m?
    (a) 1.10 m/s
    (b) 1.98 m/s
    (c) 5.60 m/s
    (d) 10.1 m/s
    (e) zero

11. During the first 18 minutes of a 1.0 hour trip, a car has an average speed of 11 m/s. What must the average speed of the car be during the last 42 minutes of the trip be if the car is to have an average speed of 21 m/s for the entire trip?
    (a) 21 m/s
    (b) 23 m/s
    (c) 25 m/s
    (d) 27 m/s
    (e) 29 m/s

12. A turtle takes 3.5 minutes to walk 18 m toward the south along a deserted highway. A truck driver stops and picks up the turtle. The driver takes the turtle to a town 1.1 km to the north with an average speed of 12 m/s. What is the magnitude of the average velocity of the turtle for its entire journey?
    (a) 3.6 m/s
    (b) 9.8 m/s
    (c) 6.0 m/s
    (d) 2.6 m/s
    (e) 11 m/s

*Questions 13 through 16 pertain to the situation described below:*

A race car, traveling at constant speed, makes one lap around a circular track of radius $r$ in a time $t$. Note: The circumference of a circle is given by $C = 2\pi r$.

13. When the car has traveled halfway around the track, what is the magnitude of its *displacement* from the starting point?
    (a) $r$
    (b) $2r$
    (c) $\pi r$
    (d) $2\pi r$
    (e) zero

14. What is the *average speed* of the car for one complete lap?
    (a) $\dfrac{r}{t}$
    (b) $\dfrac{2r}{t}$
    (c) $\dfrac{\pi r}{t}$
    (d) $\dfrac{2\pi r}{t}$
    (e) zero

15. Determine the *magnitude* of the *average velocity* of the car for one complete lap.
    (a) $\dfrac{r}{t}$
    (b) $\dfrac{2r}{t}$
    (c) $\dfrac{\pi r}{t}$
    (d) $\dfrac{2\pi r}{t}$
    (e) zero

16. Which one of the following statements concerning this car is true?
    (a) The displacement of the car does not change with time.
    (b) The instantaneous velocity of the car is constant.
    (c) The average speed of the car is the same over any time interval.
    (d) The average velocity of the car is the same over any time interval.
    (e) The average speed of the car over any time interval is equal to the magnitude of the average velocity over the same time interval.

## Section 2.3 Acceleration

17. In which one of the following situations does the car have a westward acceleration?
    (a) The car travels westward at constant speed.
    (b) The car travels eastward and speeds up.
    (c) The car travels westward and slows down.
    (d) The car travels eastward and slows down.
    (e) The car starts from rest and moves toward the east.

18. An elevator is moving upward with a speed of 11 m/s. Three seconds later, the elevator is still moving upward, but it's speed has been reduced to 5.0 m/s. What is the average acceleration of the elevator during the 3.0 s interval?
    (a) 2.0 m/s$^2$, downward
    (b) 2.0 m/s$^2$, upward
    (c) 5.3 m/s$^2$, downward
    (d) 5.3 m/s$^2$, upward
    (e) 2.7 m/s$^2$, downward

19. A landing airplane makes contact with the runway with a speed of 78.0 m/s and moves toward the south. After 18.5 seconds, the airplane comes to rest. What is the average acceleration of the airplane during the landing?
    (a) 2.11 m/s$^2$, north
    (b) 2.11 m/s$^2$, south
    (c) 4.22 m/s$^2$, north
    (d) 4.22 m/s$^2$, south
    (e) 14.3 m/s$^2$, north

20. A pitcher delivers a fast ball with a velocity of 43 m/s to the south. The batter hits the ball and gives it a velocity of 51 m/s to the north. What was the average acceleration of the ball during the 1.0 ms when it was in contact with the bat?
    (a) $4.3 \times 10^4$ m/s$^2$, south
    (b) $5.1 \times 10^4$ m/s$^2$, north
    (c) $9.4 \times 10^4$ m/s$^2$, north
    (d) $2.2 \times 10^3$ m/s$^2$, south
    (e) $7.0 \times 10^3$ m/s$^2$, north

21. A car is moving at a constant velocity when it is involved in a collision. The car comes to rest after 0.450 s with an average acceleration of 65.0 m/s$^2$ in the direction opposite that of the car's velocity. What was the speed, in km/h, of the car before the collision?
    (a) 29.2 km/h
    (b) 144 km/h
    (c) 44.8 km/h
    (d) 80.5 km/h
    (e) 105 km/h

## Section 2.4 Equations of Kinematics for Constant Acceleration
## Section 2.5 Applications of the Equations of Kinematics

22. Which one of the following is *not* a vector quantity?
    (a) acceleration
    (b) average speed
    (c) displacement
    (d) average velocity
    (e) instantaneous velocity

23. In which one of the following cases is the displacement of the object directly proportional to the time?
    (a) a ball rolls with constant velocity
    (b) a ball at rest is given a constant acceleration
    (c) a ball rolling with velocity $v_o$ is given a constant acceleration
    (d) a bead falling through oil experiences a decreasing acceleration
    (e) a rocket fired from the earth's surface experiences an increasing acceleration

24. Which one of the following situations is *not* possible?
    (a) A body has zero velocity and non-zero acceleration.
    (b) A body travels with a northward velocity and a northward acceleration.
    (c) A body travels with a northward velocity and a southward acceleration.
    (d) A body travels with a constant velocity and a time-varying acceleration.
    (e) A body travels with a constant acceleration and a time-varying velocity.

25. Starting from rest, a particle confined to move along a straight line is accelerated at a rate of 5.0 m/s$^2$. Which one of the following statements accurately describes the motion of this particle?
    (a) The particle travels 5.0 m during each second.
    (b) The particle travels 5.0 m *only* during the first second.
    (c) The speed of the particle increases by 5.0 m/s during each second.
    (d) The acceleration of the particle increases by 5.0 m/s$^2$ during each second.
    (e) The final speed of the particle will be proportional to the distance that the particle covers.

26. A car accelerates from rest at point A with constant acceleration of magnitude *a* and subsequently passes points **B** and **C** as shown in the figure.

    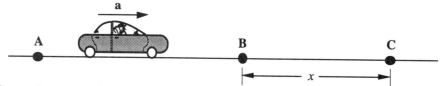

    The distance between points **B** and **C** is *x*, and the time required for the car to travel from **B** to **C** is *t*. Which expression determines the *average speed* of the car between the points **B** and **C**?
    (a) $v^2 = 2ax$
    (b) $v = \dfrac{x}{t}$
    (c) $v = xt$
    (d) $v = \dfrac{1}{2}at^2$
    (e) $v = at$

27. Two objects A and B accelerate from rest with the same constant acceleration. Object A accelerates for twice as much time as object B, however. Which one of the following statements is true concerning these objects at the end of their respective periods of acceleration?
    (a) Object A will travel twice as far as object B.
    (b) Object A will travel four times as far as object B.
    (c) Object A will travel eight times further than object B.
    (d) Object A will be moving four times faster than object B.
    (e) Object A will be moving eight times faster than object B.

28. Which one of the following statements must be true if the expression $x = v_0 t + \frac{1}{2} a t^2$ is to be used?
    (a) $x$ is constant.
    (b) $v$ is constant.
    (c) $t$ is constant.
    (d) $a$ is constant.
    (e) Both $v_0$ and $t$ are constant.

29. Two cars travel along a level highway. It is observed that the distance between the cars is *increasing*. Which one of the following statements concerning this situation is *necessarily* true?
    (a) The velocity of each car is increasing.
    (b) At least one of the cars has a *non-zero* acceleration.
    (c) The leading car has the greater acceleration.
    (d) The trailing car has the smaller acceleration.
    (e) Both cars could be accelerating at the same rate.

30. An object moving along a straight line is decelerating. Which one of the following statements concerning the object's acceleration is *necessarily* true?
    (a) The value of the acceleration is positive.
    (b) The direction of the acceleration is in the same direction as the displacement.
    (c) An object that is decelerating has a negative acceleration.
    (d) The direction of the acceleration is in the direction opposite to that of the velocity.
    (e) The acceleration changes as the object moves along the line.

31. A car, starting from rest, accelerates in a straight line path at a constant rate of 2.5 m/s². How far will the car travel in 12 seconds?
    (a) 180 m
    (b) 120 m
    (c) 30 m
    (d) 15 m
    (e) 4.8 m

32. A car starts from rest and accelerates at a constant rate in a straight line. In the *first* second the car covers a distance of 2.0 meters. How fast will the car be moving at the end of the *second* second?
    (a) 4.0 m/s
    (b) 16 m/s
    (c) 2.0 m/s
    (d) 32 m/s
    (e) 8.0 m/s

33. A car starts from rest and accelerates at a constant rate in a straight line. In the *first* second the car covers a distance of 2.0 meters. How much additional distance will the car cover during the *second* second of its motion?
    (a) 2.0 m
    (b) 4.0 m
    (c) 6.0 m
    (d) 8.0 m
    (e) 13 m

34. A car is initially traveling at 50.0 km/h. The brakes are applied and the car stops over a distance of 35 m. What was magnitude of the car's acceleration while it was braking?
    (a) 2.8 m/s²
    (b) 5.4 m/s²
    (c) 36 m/s²
    (d) 71 m/s²
    (e) 9.8 m/s²

35. The minimum takeoff speed for a certain airplane is 75 m/s. What minimum acceleration is required if the plane must leave a runway of length 950 m? Assume the plane starts from rest at one end of the runway.
    (a) 1.5 m/s²
    (b) 3.0 m/s²
    (c) 4.5 m/s²
    (d) 6.0 m/s²
    (e) 7.5 m/s²

36. A body initially at rest is accelerated at a constant rate for 5.0 seconds in the positive $x$ direction. If the final speed of the body is 20.0 m/s, what was the body's acceleration?
    (a) 0.25 m/s²
    (b) 2.0 m/s²
    (c) 4.0 m/s²
    (d) 9.8 m/s²
    (e) 1.6 m/s²

## 16  KINEMATICS IN ONE DIMENSION

37. A race car has a speed of 80 m/s when the driver releases a drag parachute. If the parachute causes a deceleration of –4 m/s$^2$, how far will the car travel before it stops?
    (a) 20 m
    (b) 200 m
    (c) 400 m
    (d) 800 m
    (e) 1000 m

38. A car traveling along a road begins accelerating with a constant acceleration of 1.5 m/s$^2$ in the direction of motion. After traveling 392 m at this acceleration, its speed is 35 m/s. Determine the speed of the car when it began accelerating.
    (a) 1.5 m/s
    (b) 7.0 m/s
    (c) 34 m/s
    (d) 49 m/s
    (e) 2.3 m/s

39. A train passes through a town with a constant speed of 16 m/s. After leaving the town, the train accelerates at 0.33 m/s$^2$ until it reaches a speed of 35 m/s. How far did the train travel while it was accelerating?
    (a) 0.029 km
    (b) 0.53 km
    (c) 1.5 km
    (d) 2.3 km
    (e) 3.0 km

40. A cheetah is walking at a speed of 1.10 m/s when it observes a gazelle 41.0 m directly ahead. If the cheetah accelerates at 9.55 m/s$^2$, how long does it take the cheetah to reach the gazelle if the gazelle doesn't move?
    (a) 4.29 s
    (b) 3.67 s
    (c) 3.05 s
    (d) 1.94 s
    (e) 2.82 s

*Questions 41 through 43 pertain to the situation described below:*

An object starts from rest and accelerates uniformly in a straight line in the positive *x* direction. After 11 seconds, its speed is 70.0 m/s.

41. Determine the acceleration of the object.
    (a) +3.5 m/s$^2$
    (b) +6.4 m/s$^2$
    (c) –3.5 m/s$^2$
    (d) –6.4 m/s$^2$
    (e) +7.7 m/s$^2$

42. How far does the object travel during the first 11 seconds?
    (a) 35 m
    (b) 77 m
    (c) 390 m
    (d) 590 m
    (e) 770 m

43. What is the *average velocity* of the object during the first 11 seconds?
    (a) +3.6 m/s
    (b) +6.4 m/s
    (c) +35 m/s
    (d) +72 m/s
    (e) –140 m/s

### Section 2.6 Freely Falling Bodies

44. Ball A is dropped from rest from a window. At the same instant, ball B is thrown downward; and ball C is thrown upward from the same window. Which statement concerning the balls is necessarily true if air resistance is neglected?
    (a) At some instant after it is thrown, the acceleration of ball C is zero.
    (b) All three balls strike the ground at the same time.
    (c) All three balls have the same velocity at any instant.
    (d) All three balls have the same acceleration at any instant.
    (e) All three balls reach the ground with the same velocity.

*TEST BANK* **Chapter 2** 17

45. A ball is thrown vertically upward from the surface of the earth. Consider the following quantities: (1) the speed of the ball; (2) the velocity of the ball; (3) the acceleration of the ball. Which of these is (are) zero when the ball has reached the maximum height?
    (a) 1 only
    (b) 2 only
    (c) 1 and 2
    (d) 1 and 3
    (e) 1, 2, and 3

46. A rock is thrown vertically upward from the surface of the earth. The rock rises to some maximum height and falls back toward the surface of the earth. Which one of the following statements concerning this situation is true if air resistance is neglected?
    (a) As the ball rises, its acceleration vector points upward.
    (b) The ball is a freely falling body for the duration of its flight.
    (c) The acceleration of the ball is zero when the ball is at its highest point.
    (d) The speed of the ball is negative while the ball falls back toward the earth.
    (e) The velocity and acceleration of the ball always point in the same direction.

47. A brick is dropped from rest from a height of 4.9 m. How long does it take for the brick to reach the ground?
    (a) 0.6 s
    (b) 1.0 s
    (c) 1.2 s
    (d) 1.4 s
    (e) 2.0 s

48. A ball is dropped from rest from a tower and strikes the ground 125 m below. Approximately how many seconds does it take for the ball to strike the ground after being dropped? Neglect air resistance.
    (a) 2.50 s
    (b) 3.50 s
    (c) 5.05 s
    (d) 12.5 s
    (e) 16.0 s

49. Water drips from rest from a leaf that is 20 meters above the ground. Neglecting air resistance, what is the speed of each water drop when it hits the ground?
    (a) 10 m/s
    (b) 15 m/s
    (c) 20 m/s
    (d) 30 m/s
    (e) 40 m/s

50. A rock is dropped from rest from a height $h$ above the ground in a vacuum. It falls and hits the ground with a speed of 11 m/s. From what height should it be dropped so that its speed on hitting the ground is 22 m/s? Neglect air resistance.
    (a) $1.4h$
    (b) $2.0h$
    (c) $3.0h$
    (d) $4.0h$
    (e) $0.71h$

51. A 5.0-kg rock is dropped from rest down a vertical mine shaft. How long does it take for the rock to reach a depth of 79 m? Neglect air resistance.
    (a) 2.8 s
    (b) 9.0 s
    (c) 4.9 s
    (e) 8.0 s
    (e) 4.0 s

52. What maximum height will be reached by a stone thrown straight up with an initial speed of 35 m/s?
    (a) 98 m
    (b) 160 m
    (c) 41 m
    (d) 62 m
    (e) 18 m

*18*    **KINEMATICS IN ONE DIMENSION**

*Questions 53 through 55 pertain to the situation described below:*

A ball is shot straight up from the surface of the earth with an initial speed of 19.6 m/s. Neglect any effects due to air resistance.

53. What is the magnitude of the ball's displacement from the starting point after 1.00 second has elapsed?
    (a) 9.80 m
    (b) 14.7 m
    (c) 19.6 m
    (d) 24.5 m
    (e) 58.8 m

54. What maximum height will the ball reach?
    (a) 9.80 m
    (b) 14.7 m
    (c) 19.6 m
    (d) 24.5 m
    (e) 58.8 m

55. How much time elapses between the throwing of the ball and its return to the original launch point?
    (a) 4.00 s
    (b) 2.00 s
    (c) 12.0 s
    (d) 8.00 s
    (e) 16.0 s

*Questions 56 through 59 pertain to the statement below:*

A tennis ball is shot vertically upward in an *evacuated chamber* with an initial speed of 20.0 m/s at time $t = 0$.

56. How high does the ball rise?
    (a) 10.2 m
    (b) 20.4 m
    (c) 40.8 m
    (d) 72.4 m
    (e) 98.0 m

57. Approximately how long does it take the tennis ball to reach its maximum height?
    (a) 0.50 s
    (b) 2.04 s
    (c) 4.08 s
    (d) 6.08 s
    (e) 9.80 s

58. Determine the velocity of the ball at $t = 3.00$ seconds.
    (a) 9.40 m/s, downward
    (b) 9.40 m/s, upward
    (c) 29.4 m/s, downward
    (d) 38.8 m/s, upward
    (e) 38.8 m/s, downward

59. What is the magnitude of the acceleration of the ball when it is at its highest point?
    (a) zero
    (b) 9.80 m/s$^2$
    (c) 19.6 m/s$^2$
    (d) 4.90 m/s$^2$
    (e) 3.13 m/s$^2$

## Section 2.7 *Graphical Analysis of Velocity and Acceleration*

60. Starting from rest, a particle which is confined to move along a straight line is accelerated at a rate of 5.0 m/s$^2$. Which statement concerning the *slope* of the *position versus time* graph for this particle is true?
    (a) The slope has a constant value of 5.0 m/s.
    (b) The slope has a constant value of 5.0 m/s$^2$.
    (c) The slope is both constant and negative.
    (d) The slope is not constant and *increases* with increasing time.
    (e) The slope is not constant and *decreases* with increasing time.

*Questions 61 through 63 pertain to the graph below:*

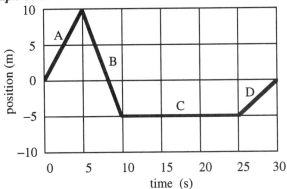

An object is moving along the x axis. The graph shows its position from the starting point as a function of time. Various segments of the graph are identified by the letters A, B, C, and D.

61. During which interval(s) is the object *moving* in the negative x direction?
    (a) during interval B only
    (b) during intervals B and C
    (c) during intervals C and D
    (d) during intervals B and D
    (e) during intervals B, C, and D

62. What is the *velocity* of the object at $t = 7.0$ s?
    (a) +3.0 m/s
    (b) −1.0 m/s
    (c) −2.0 m/s
    (d) −3.0 m/s
    (e) zero

63. What is the *acceleration* of the object at $t = 7.0$ s?
    (a) zero
    (b) −2.0 m/s$^2$
    (c) −3.0 m/s$^2$
    (d) +9.8 m/s$^2$
    (e) +4.0 m/s$^2$

*Questions 64 through 67 pertain to the statement and graph below:*

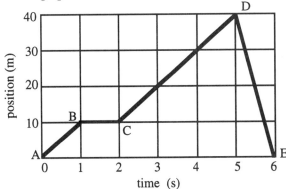

An object is moving along a straight line. The graph shows the object's position from the starting point as a function of time.

64. In which segment(s) of the graph does the object's *average velocity* (measured from $t = 0$) *decrease* with time?
    (a) AB only
    (b) BC only
    (c) DE only
    (d) AB and CD
    (e) BC and DE

65. What was the *instantaneous velocity* of the object at $t = 4$ seconds?
    (a) +6 m/s
    (b) +8 m/s
    (c) +10 m/s
    (d) +20 m/s
    (e) +40 m/s

66. In which segments(s) of the graph does the object have the highest speed?
    (a) AB
    (b) BC
    (c) CD
    (d) DE
    (e) AB and CD

## 20 KINEMATICS IN ONE DIMENSION

67. At which time(s) does the object reverse its direction of motion?
    (a) 1 s and 2 s
    (b) 2 s and 5 s
    (c) 1 s
    (d) 2 s
    (e) 5 s

*Questions 68 through 71 pertain to the statement and graph below:*

An object is moving along a straight line. The graph shows the object's velocity as a function of time.

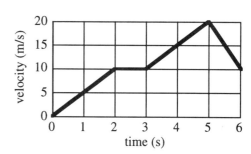

68. During which interval(s) of the graph does the object travel *equal distances* in *equal times*?
    (a) 0 to 2 s
    (b) 2 s to 3 s
    (c) 3 s to 5 s
    (d) 0 to 2 s and 3 s to 5 s
    (e) 0 to 2 s, 3 to 5 s, and 5 to 6 s

69. During which interval(s) of the graph does the speed of the object *increase by equal amounts* in *equal times*?
    (a) 0 to 2 s
    (b) 2 s to 3 s
    (c) 3 s to 5 s
    (d) 0 to 2 s and 3 s to 5 s
    (e) 0 to 2 s, 3 to 5 s, and 5 to 6 s

70. How far does the object move in the interval from $t = 0$ to $t = 2$ s?
    (a) 7.5 m
    (b) 10 m
    (c) 15 m
    (d) 20 m
    (e) 25 m

71. What is the acceleration of the object in the interval from $t = 5$ s to $t = 6$ s?
    (a) $-40$ m/s$^2$
    (b) $+40$ m/s$^2$
    (c) $-20$ m/s$^2$
    (d) $+20$ m/s$^2$
    (e) $-10$ m/s$^2$

*Questions 72 through 74 pertain to the situation described below:*

An object is moving along a straight line in the positive $x$ direction. The graph shows its position from the starting point as a function of time. Various segments of the graph are identified by the letters A, B, C, and D.

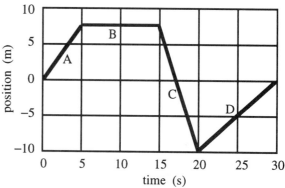

72. Which segment(s) of the graph represent(s) a *constant velocity* of $+1.0$ m/s?
    (a) A
    (b) B
    (c) C
    (d) D
    (e) A and C

73. What was the *instantaneous velocity* of the object at the end of the eighth second?
    (a) +7.5 m/s
    (b) +0.94 m/s
    (c) −0.94 m/s
    (d) +1.1 m/s
    (e) zero

74. During which interval(s) did the object move in the negative *x* direction?
    (a) only during interval B
    (b) only during interval C
    (c) only during interval D
    (d) during both intervals C and D
    (e) The object never moved in the negative *x* direction.

## Additional Problems

75. The rate at which the acceleration of an object changes with time is called the *jerk*. What is the dimension of the jerk?
    (a) $\dfrac{[L]}{[T]}$
    (b) $\dfrac{[L]}{[T]^2}$
    (c) $\dfrac{[L]^2}{[T]^2}$
    (d) $\dfrac{[L]}{[T]^3}$
    (e) $\dfrac{[L]^2}{[T]^3}$

76. In a race, José runs 1.00 mile in 4.02 min, mounts a bicycle, and rides back to his starting point, which is also the finish line, in 3.02 min. What is the magnitude of José's average velocity for the race?
    (a) zero
    (b) 12.1 mi/h
    (c) 14.9 mi/h
    (d) 17.0 mi/h
    (e) 19.9 mi/h

*Questions 77 and 78 pertain to the situation described below:*

A motorist travels due north at 30 mi/h for 2.0 hours. She then reverses her direction and travels due south at 60 mi/h for 1.0 hour.

77. What is the average speed of the motorist?
    (a) zero
    (b) 30 mi/h
    (c) 40 mi/h
    (d) 50 mi/h
    (e) 60 mi/h

78. What is the average velocity of the motorist?
    (a) zero
    (b) 40 mi/h, north
    (c) 40 mi/h, south
    (d) 45 mi/h, north
    (e) 45 mi/h, south

*Questions 79 through 81 pertain to the statement below:*

Starting from rest, a particle confined to move along a straight line is accelerated at a rate of 4 m/s².

79. Which statement accurately describes the motion of the particle?
    (a) The particle travels 4 meters during each second.
    (b) The particle travels 4 meters during the first second only.
    (c) The speed of the particle increases by 4 m/s during each second.
    (d) The acceleration of the particle increases by 4 m/s² during each second.
    (e) The final velocity of the particle will be proportional to the distance that the particle covers.

## KINEMATICS IN ONE DIMENSION

80. After 10 seconds, how far will the particle have traveled?
    (a) 20 m
    (b) 40 m
    (c) 100 m
    (d) 200 m
    (e) 400 m

81. What is the speed of the particle after it has traveled 8 m?
    (a) 4 m/s
    (b) 8 m/s
    (c) 30 m/s
    (d) 60 m/s
    (e) 100 m/s

*Questions 82 through 85 pertain to the situation described below:*

A rock, dropped from rest near the surface of an atmosphere-free planet, attains a speed of 20.0 m/s after falling 8.0 meters.

82. What is the magnitude of the acceleration due to gravity on the surface of this planet?
    (a) 0.40 m/s$^2$
    (b) 1.3 m/s$^2$
    (c) 2.5 m/s$^2$
    (d) 25 m/s$^2$
    (e) 160 m/s$^2$

83. How long did it take the object to fall the 8.0 meters mentioned?
    (a) 0.40 s
    (b) 0.80 s
    (c) 1.3 s
    (d) 2.5 s
    (e) 16 s

84. How long would it take the object, falling from rest, to fall 16 m on this planet?
    (a) 0.8 s
    (b) 1.1 s
    (c) 2.5 s
    (d) 3.5 s
    (e) 22 s

85. Determine the speed of the object after falling from rest through 16 m on this planet.
    (a) 28 m/s
    (b) 32 m/s
    (c) 56 m/s
    (d) 64 m/s
    (e) 320 m/s

*Questions 86 through 90 pertain to the situation described below:*

A tennis ball is shot vertically upward from the surface of an atmosphere-free planet with an initial speed of 20.0 m/s. One second later, the ball has an instantaneous velocity in the upward direction of 15.0 m/s.

86. What is the magnitude of the acceleration due to gravity on the surface of this planet?
    (a) 5.0 m/s$^2$
    (b) 9.8 m/s$^2$
    (c) 12 m/s$^2$
    (d) 15 m/s$^2$
    (e) 24 m/s$^2$

87. How long does it take the ball to reach its maximum height?
    (a) 2.0 s
    (b) 2.3 s
    (c) 4.0 s
    (d) 4.6 s
    (e) 8.0 s

88. How high does the ball rise?
    (a) 70.0 m
    (b) 10.0 m
    (c) 50.0 m
    (d) 20.0 m
    (e) 40.0 m

89. Determine the velocity of the ball when it returns to its original position.
    Note: assume the upward direction is positive.
    (a) +20 m/s
    (b) −20 m/s
    (c) +40 m/s
    (c) −40 m/s
    (e) zero

90. How long is the ball in the air when it returns to its original position?
    (a) 4.0 s
    (b) 4.6 s
    (c) 8.0 s
    (d) 9.2 s
    (e) 16 s

*Questions 91 and 92 pertain to the situation described below:*

A small object is released from rest and falls 100 feet near the surface of the earth. Neglect air resistance.

91. How long will it take to fall through the 100 feet mentioned?
    (a) 2.49 s
    (b) 3.12 s
    (c) 4.50 s
    (d) 6.25 s
    (e) 10.0 s

92. How fast will the object be moving after falling through the 100 feet mentioned?
    (a) 9.8 ft/s
    (b) 40 ft/s
    (c) 80 ft/s
    (d) 160 ft/s
    (e) 320 ft/s

*Questions 93 through 96 pertain to the situation described below:*

The figure shows the speed as a function of time for an object in free fall near the surface of the earth.

The object was dropped from rest in a long evacuated cylinder.

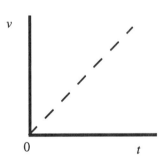

93. Which one of the following statements best explains why the graph goes through the origin?
    (a) The object was in a vacuum.
    (b) The object was dropped from rest.
    (c) The velocity of the object was constant.
    (d) All $v$ vs. $t$ curves pass through the origin.
    (e) The acceleration of the object was constant.

94. What is the numerical value of the slope of the line?
    (a) 1.0 m/s$^2$
    (b) 2.0 m/s$^2$
    (c) 7.7 m/s$^2$
    (d) 9.8 m/s$^2$
    (e) This cannot be determined from the information given since the speed and time values are unknown.

95. What is the speed of the object 3.0 seconds after it is dropped?
    (a) 3.0 m/s
    (b) 7.7 m/s
    (c) 9.8 m/s
    (d) 29 m/s
    (e) This cannot be determined since there is no specified value of height.

**24** KINEMATICS IN ONE DIMENSION

96. If the same object were released in air, the magnitude of its acceleration would begin at the free-fall value, but it would decrease continuously to zero as the object continued to fall.

In which choice below does the solid line best represent the speed of the object as a function of time when it is dropped from rest in air?

*Note:* The dashed line shows the free-fall under vacuum graph for comparison.

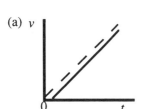 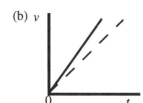 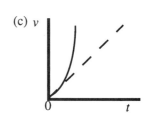

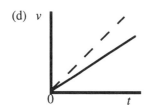 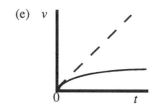

# CHAPTER 3
# Kinematics in Two Dimensions

## Section 3.1 Displacement, Velocity, and Acceleration

1. A park ranger wanted to measure the height of a tall tree. The ranger stood 6.10 m from the base of the tree; and he observed that his line of sight made an angle of 73.5° above the horizontal as he looked at the top of the tree. What is the height of the tree?
   (a) 5.84 m
   (b) 8.77 m
   (c) 11.7 m
   (d) 17.3 m
   (e) 20.6 m

2. A toolbox is carried from the base of a ladder at point A as shown in the figure. The toolbox comes to a rest on a scaffold 5.0 m above the ground. What is the magnitude of the displacement of the toolbox in its movement from point A to point B?
   (a) 15 m
   (b) 19 m
   (c) 8.1 m
   (d) 11 m
   (e) 13 m

   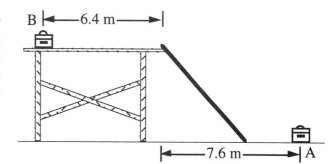

3. A delivery truck leaves a warehouse and travels 2.60 km north. The truck makes a right turn and travels 1.33 km east before making another right turn and traveling 1.45 km south to arrive at its destination. What is the magnitude and direction of the truck's displacement from the warehouse?
   (a) 1.76 km, 40.8° north of east
   (b) 1.15 km, 59.8° north of east
   (c) 1.33 km, 30.2° north of east
   (d) 2.40 km, 45.0° north of east
   (e) 5.37 km, 49.2° north of east

4. A car travels due east at 20 m/s. It makes a turn due south and continues to travel at 20 m/s. What is the change in velocity of the car?
   (a) 20 m/s, due east
   (b) 20 m/s, due south
   (c) 28 m/s, 45° south of west
   (d) 28 m/s, 45° south of east
   (e) zero

5. A car travels along a highway with a velocity of 24 m/s, west. The car exits the highway and 4.0 s later its instantaneous velocity is 16 m/s, 45° north of west. What is the magnitude of the average acceleration of the car during the five second interval?
   (a) 2.4 m/s$^2$
   (b) 4.2 m/s$^2$
   (c) 1.2 m/s$^2$
   (d) 11 m/s$^2$
   (e) 17 m/s$^2$

6. A train travels due south at 60 m/s. It reverses its direction and travels due north at 60 m/s. What is the change in velocity of the train?
   (a) 120 m/s, due north
   (b) 120 m/s, due south
   (c) 60 m/s, due north
   (d) 60 m/s, due south
   (e) zero

# KINEMATICS IN TWO DIMENSIONS

*Questions 7 through 10 pertain to the situation described below:*

A small boat travels 80.0 km north and then travels 60.0 km east in 1.0 hour.

7. What is the boat's displacement for the one hour trip?
   (a) 20 km
   (b) 100 km
   (c) 140 km
   (d) 280 km
   (e) 10 000 km

8. What is the boat's *average speed* during the one hour trip?
   (a) 20 km/h
   (b) 100 km/h
   (c) 140 km/h
   (d) 280 km/h
   (e) 10 000 km/h

9. What is the *magnitude* of the boat's *average velocity* for the one hour trip?
   (a) 20 km/h
   (b) 100 km/h
   (c) 140 km/h
   (d) 10 000 km/h
   (e) 20 000 km/h

10. What is the *direction* of the boat's *average velocity* for the one hour trip?
    (a) due east
    (b) 36.9° north of east
    (c) 41.7° north of east
    (d) 49.2° north of east
    (e) 53.1° north of east

## Section 3.2 Equations of Kinematics in Two Dimensions
## Section 3.3 Projectile Motion

11. An airplane traveling north at 400 m/s is accelerated due east at a rate of 50 m/s$^2$ for 6 s. If the effects of air resistance *and* gravity are ignored, what is the final speed of the plane?
    (a) 300 m/s
    (b) 400 m/s
    (c) 500 m/s
    (d) 700 m/s
    (e) 800 m/s

*Questions 12 and 13 pertain to the following situation.*

A spaceship is observed traveling in the positive $x$ direction with a speed of 150 m/s when it begins accelerating at a constant rate. The spaceship is observed 25 s later traveling with an instantaneous velocity of 1500 m/s at an angle of 55° above the $+x$ axis.

12. What was the magnitude of the acceleration of the spaceship during the 25 seconds?
    (a) 1.5 m/s$^2$
    (b) 7.3 m/s$^2$
    (c) 28 m/s$^2$
    (d) 48 m/s$^2$
    (e) 57 m/s$^2$

13. What was the magnitude of the displacement of the spaceship during the 25 seconds?
    (a) $2.0 \times 10^4$ m
    (b) $1.4 \times 10^4$ m
    (c) $2.8 \times 10^4$ m
    (d) $1.0 \times 10^4$ m
    (e) $1.7 \times 10^2$ m

14. A football is kicked at an angle θ with respect to the horizontal. Which one of the following statements best describes the *acceleration* of the football during this event if air resistance is neglected?
    (a) The acceleration is zero at all times.
    (b) The acceleration is 9.8 m/s$^2$ at all times.
    (c) The acceleration is zero when the football has reached the highest point in its trajectory.
    (d) The acceleration is positive as the football rises, and its is negative as the football falls.
    (e) The acceleration starts at 9.8 m/s$^2$ and drops to some constant lower value as the ball approaches the ground.

15. A baseball is hit upward and travels along a parabolic arc before it strikes the ground. Which one of the following statements is necessarily true?
    (a) The acceleration of the ball decreases as the ball moves upward.
    (b) The velocity of the ball is zero when the ball is at the highest point in the arc.
    (c) The acceleration of the ball is zero when the ball is at the highest point in the arc.
    (d) The x-component of the velocity of the ball is the same throughout the ball's flight.
    (e) The velocity of the ball is a maximum when the ball is at the highest point in the arc.

16. A physics student standing on the edge of a cliff throws a stone vertically *downward* with an initial speed of 10 m/s. The instant before the stone hits the ground below, it is traveling at a speed of 30 m/s. If the physics student were to throw the rock *horizontally outward* from the cliff instead, with the same initial speed of 10 m/s, how fast would the stone be traveling just before it hits the ground?
    (a) 10 m/s
    (b) 20 m/s
    (c) 30 m/s
    (d) 40 m/s
    (e) The height of the cliff must be specified to answer this question.

17. A tennis ball is thrown from ground level with velocity $v_0$ directed 30° above the horizontal. If it takes the ball 1.0 s to reach the top of its trajectory, what is the magnitude of the initial velocity?
    (a) 4.9 m/s
    (b) 9.8 m/s
    (c) 11.3 m/s
    (d) 19.6 m/s
    (e) 34.4 m/s

*Questions 18 and 19 pertain to the situation described below:*

A projectile is fired at an angle of 60.0° above the horizontal with an initial speed of 30.0 m/s.

18. What is the magnitude of the *horizontal* component of the projectile's displacement at the end of 2 s?
    (a) 30 m
    (b) 40 m
    (c) 10 m
    (d) 20 m
    (e) 50 m

19. How long does it take the projectile to reach the highest point in its trajectory?
    (a) 1.5 s
    (b) 2.7 s
    (c) 4.0 s
    (d) 6.2 s
    (e) 9.8 s

*Questions 20 through 25 pertain to the situation described below:*

A projectile is fired from a gun and has initial horizontal and vertical components of velocity equal to 30 m/s and 40 m/s, respectively.

20. Determine the initial speed of the projectile.
    (a) 40 m/s
    (b) 50 m/s
    (c) 60 m/s
    (d) 70 m/s
    (e) 80 m/s

21. At what angle is the projectile fired (measured with respect to the horizontal)?
    (a) 37°
    (b) 40°
    (c) 45°
    (d) 53°
    (e) 60°

22. *Approximately* how long does it take the projectile to reach the highest point in its trajectory?
    (a) 1 s
    (b) 2 s
    (c) 4 s
    (d) 8 s
    (e) 16 s

28  KINEMATICS IN TWO DIMENSIONS

23. What is the speed of the projectile when it is at the highest point in its trajectory?
    (a) zero
    (b) 20 m/s
    (c) 30 m/s
    (d) 40 m/s
    (e) 50 m/s

24. What is the acceleration of the projectile when it reaches its maximum height?
    (a) zero
    (b) 9.8 m/s$^2$, downward
    (c) 4.9 m/s$^2$, downward
    (d) smaller than 9.8 m/s$^2$ and non-zero.
    (e) Its magnitude is 9.8 m/s$^2$ and; its direction is changing.

25. What is the magnitude of the projectile's velocity just before it strikes the ground?
    (a) zero
    (b) 9.8 m/s
    (c) 30 m/s
    (d) 40 m/s
    (e) 50 m/s

26. A spring-loaded gun is aimed horizontally and is used to launch identical balls with *different initial speeds*. The gun is at a fixed position above the floor. The balls are fired one at a time.

    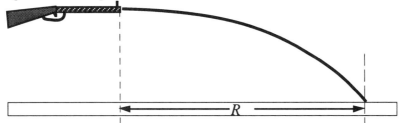

    If the speed of the second projectile fired is *twice the speed* of the first projectile fired, how is the horizontal range (denoted $R$ in the figure) affected?
    (a) The range for both projectiles will be the same.
    (b) The range of the second projectile will be half as much as that of the first projectile.
    (c) The range of the second projectile will be twice as large as that of the first projectile.
    (d) The range of the second projectile is about 1.4 times larger than that of the first projectile.
    (e) The range of the second projectile will be smaller than that of the first projectile by a factor of 1.4.

27. A projectile is fired horizontally with an initial speed of 57 m/s. What are the horizontal and vertical components of its displacement 3.0 s after it is fired?

    | | horizontal | vertical |
    |---|---|---|
    | (a) | 44 m | 29 m |
    | (b) | 170 m | 29 m |
    | (c) | 170 m | 44 m |
    | (d) | 210 m | 44 m |
    | (e) | 210 m | 0 m |

28. A quarterback throws a pass at an angle of 35° above the horizontal with an initial speed of 25 m/s. The ball is caught by the receiver 2.55 seconds later. Determine the distance the ball was thrown.
    (a) 13 m
    (b) 18 m
    (c) 36 m
    (d) 52 m
    (e) 72 m

*Questions 29 through 31 pertain to the situation described below:*

A shell is fired with a horizontal velocity in the positive $x$ direction from the top of an 80 m high cliff. The shell strikes the ground 1330 m from the base of the cliff.

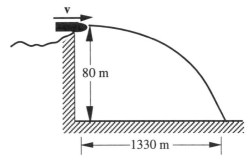

29. Determine the initial speed of the shell.
    (a) 4.0 m/s
    (b) 9.8 m/s
    (c) 82 m/s
    (d) 170 m/s
    (e) 330 m/s

30. What is the speed of the shell as it hits the ground?
    (a) 4.0 m/s
    (b) 9.8 m/s
    (c) 82 m/s
    (d) 170 m/s
    (e) 330 m/s

31. What is the magnitude of the acceleration of the shell just before it strikes the ground?
    (a) 4.0 m/s$^2$
    (b) 9.8 m/s$^2$
    (c) 82 m/s$^2$
    (d) 170 m/s$^2$
    (e) 330 m/s$^2$

32. A cannonball is aimed 30.0° above the horizontal and is fired with an initial speed of 125 m/s at ground level. How far away from the cannon will the cannonball hit the ground?
    (a) 125 m
    (b) 138 m
    (c) 695 m
    (d) 1040 m
    (e) 1380 m

33. A puck slides across a smooth, level tabletop at height $H$ at a constant speed $v_0$. It slides off the edge of the table and hits the floor a distance $x$ away as shown in the figure.

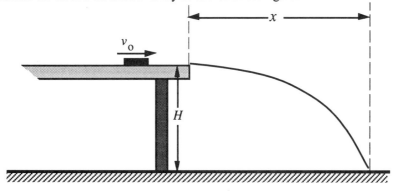

What is the relationship between the distances $x$ and $H$?

(a) $x = v_0 \sqrt{\dfrac{2H}{g}}$

(b) $x = \dfrac{v_0^2}{2gH}$

(c) $x = \dfrac{v_0^2}{gH}$

(d) $H = v_0 \sqrt{\dfrac{2x}{g}}$

(e) $x = v_0 \dfrac{H}{g}$

## 30  KINEMATICS IN TWO DIMENSIONS

34. An arrow is shot horizontally from a height of 4.9 m above the ground. The initial speed of the arrow is 45 m/s. Neglecting friction, how long will it take for the arrow to hit the ground?
    (a) 9.2 s
    (b) 6.0 s
    (c) 1.0 s
    (d) 1.4 s
    (e) 4.6 s

35. A bullet is aimed at the "mark" on the wall a distance L away from the firing position. Because of gravity, the bullet strikes the wall a distance $\Delta y$ below the mark as suggested in the figure. **Note**: *The drawing is not to scale.*

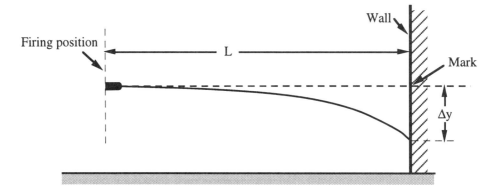

If the distance L were half as large, and the bullet had the same initial velocity, how would $\Delta y$ be affected?
(a) It will double.
(b) It will be half as large.
(c) It will be four times larger.
(d) It will be one fourth as large.
(e) It is not possible to determine unless numerical values are given for the distances.

*Questions 36 through 39 pertain to the statement and diagram below:*

A tennis ball is thrown upward at an angle from point $A$. It follows a parabolic trajectory and hits the ground at point $D$. At the instant shown, the ball is at point $B$. Point $C$ represents the highest position of the ball above the ground.

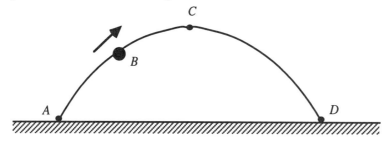

36. How do the $x$ and $y$ components of the velocity vector of the ball compare at the points $B$ and $C$?
    (a) The velocity components are non-zero at $B$ and zero at $C$.
    (b) The $x$ components are the same; the $y$ component at $C$ is zero.
    (c) The $x$ components are the same; the $y$ component has a larger magnitude at $C$ than at $B$.
    (d) The $x$ component is larger at $C$ than at $B$; the $y$ component at $B$ points up while at $C$, it points downward.
    (e) The $x$ component is larger at $B$ than at $C$; the $y$ component at $B$ points down while at $C$, it points upward.

37. How do the *x* and *y* components of the velocity vector of the ball compare at the points *A* and *D*?
    (a) The velocity components are non-zero at *A* and are zero at *D*.
    (b) The velocity components are the same in magnitude and direction at both points.
    (c) The velocity components have the same magnitudes at both points, but their directions are reversed.
    (d) The velocity components have the same magnitudes at both points, but the directions of the *x* components are reversed.
    (e) The velocity components have the same magnitudes at both points, but the directions of the *y* components are reversed.

38. Which statement is true concerning the ball when it is at *C*, the highest point in its trajectory?
    (a) The ball's velocity and acceleration are both zero.
    (b) The ball's velocity is perpendicular to its acceleration.
    (c) The ball's velocity is not zero, but its acceleration is zero.
    (d) The ball's velocity is zero, but its acceleration is not zero.
    (e) The horizontal and vertical components of the ball's velocity are equal.

39. At which point is the velocity vector changing most rapidly with time?
    (a) *A*
    (b) *B*
    (c) *C*
    (d) *D*
    (e) It is changing at the same rate at all four points.

*Questions 40 through 42 refer to the statement below:*

A football is kicked with a speed of 18 m/s at an angle of 65° to the horizontal.

40. What are the respective *horizontal* and *vertical* components of the initial velocity of the football?
    (a) 7.6 m/s, 16 m/s
    (b) 16 m/s, 7.6 m/s
    (c) 8.4 m/s, 13 m/s
    (d) 13 m/s, 8.4 m/s
    (e) 9 m/s, 9 m/s

41. How long is the football in the air? Neglect air resistance.
    (a) 1.1 s
    (b) 1.6 s
    (c) 2.0 s
    (d) 3.3 s
    (e) 4.0 s

42. How far does the football travel *horizontally* before it hits the ground?
    (a) 18 m
    (b) 25 m
    (c) 36 m
    (d) 48 m
    (e) 72 m

*Questions 43 and 44 pertain to the situation described below:*

A rock is kicked *horizontally* at a speed of 10 m/s from the edge of a cliff. The rock strikes the ground 55 m from the foot of the cliff of height *H* as suggested in the figure. Neglect air resistance.

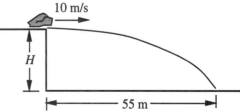

43. How long is the rock in the air?
    (a) 1.0 s
    (b) 1.2 s
    (c) 3.4 s
    (d) 5.5 s
    (e) 11.2 s

## 32  KINEMATICS IN TWO DIMENSIONS

44. What is the approximate value of H, the height of the cliff?
    (a) 27 m
    (b) 54 m
    (c) 150 m
    (d) 300 m
    (e) 730 m

*Questions 45 through 47 refer to the statement below:*

A projectile is fired horizontally with an initial speed of 50.0 m/s. Neglect air resistance.

45. What is the magnitude of the displacement of the projectile 3.00 s after it is fired?
    (a) 29.4 m
    (b) 44.1 m
    (c) 150 m
    (d) 156 m
    (e) 194 m

46. What is the speed of the projectile 3.00 s after it is fired?
    (a) 29.4 m/s
    (b) 50.0 m/s
    (c) 58.0 m/s
    (d) 79.4 m/s
    (e) 98.6 m/s

47. What is the magnitude of the acceleration of the projectile 3.00 s after it is fired?
    (a) 9.8 m/s$^2$
    (b) 16.6 m/s$^2$
    (c) 29.4 m/s$^2$
    (d) 5.42 m/s$^2$
    (e) 4.07 m/s$^2$

## Section 3.4  Relative Velocity

48. A tennis ball is thrown vertically upward in an evacuated chamber with an initial speed of 20.0 m/s at time $t = 0$. Approximately, what is the *initial* speed of the ball relative to an observer in a car that moves *horizontally* past the evacuated chamber with a constant speed of 30 m/s?
    (a) 10 m/s
    (c) 20 m/s
    (c) 30 m/s
    (d) 36 m/s
    (e) 50 m/s

*Questions 49 and 50 pertain to the statement below:*

A football is kicked with a speed of 22 m/s at an angle of 60.0° to the positive x direction. At that instant, an observer rides past the football in a car that moves with a constant speed of 11 m/s in the positive x direction.

49. Determine the initial velocity of the ball relative to the observer in the car.
    (a) 19 m/s in the +y direction
    (b) 11 m/s in the +y direction
    (c) 19 m/s at 60° to the +x direction
    (d) 33 m/s at 60° to the +x direction
    (e) 17 m/s at 60° to the +x direction

50. Complete the following statement: According to the observer in the car, the ball will
    (a) follow a parabolic path.
    (b) follow a hyperbolic path.
    (c) follow a path that is straight up and down in the y direction.
    (d) follow a path that is straight across in the +x direction.
    (e) follow a straight line that is angled (less than 90°) with respect to the x direction.

*Questions 51 and 52 pertain to the situation described below:*

Two cars approach each other on a straight and level road. Car A is traveling at 75 km/h north and car B is traveling at 45 km/h south. Both velocities are measured relative to the ground.

51. What is the velocity of car A relative to an observer in car B?
    (a) 75 km/h, north
    (b) 30 km/h, north
    (c) 30 km/h, south
    (d) 120 km/h, south
    (e) 120 km/h, north

52. At a certain instant, the distance between the cars is 15 km. How long will it take from that instant for the two cars to meet?
    (a) 450 s
    (b) 900 s
    (c) 720 s
    (d) 1200 s
    (e) 1900 s

53. A ferry can travel at an optimal speed of 8 km/h in still water measured relative to the shore. What is the optimal speed of the ferry, relative to the shore, if it moves perpendicular to a 6 km/h current?
    (a) 4 km/h
    (b) 8 km/h
    (c) 10 km/h
    (d) 14 km/h
    (e) 28 km/h

54. A boat that can travel at 4.0 km/h in still water crosses a river with a current of 2.0 km/h. At what angle must the boat be pointed upstream (that is, relative to its actual path) to go straight across the river?
    (a) 27°
    (b) 30°
    (c) 60°
    (d) 63°
    (e) 90°

*Questions 55 and 56 pertain to the situation described below:*

A man points his rowboat north from A to B, straight across a river of width 100 m. The river flows due east. The man starts across, rowing steadily at 0.75 m/s and reaches the other side of the river at point C, 150 m downstream from his starting point.

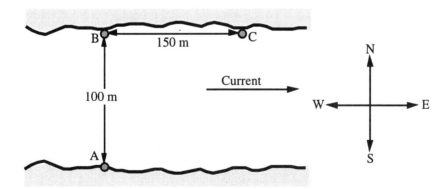

55. What is the speed of the river?
    (a) 0.38 m/s
    (b) 0.67 m/s
    (c) 1.1 m/s
    (d) 6.7 m/s
    (e) 7.5 m/s

56. While the man is crossing the river, what is his velocity relative to the shore?
    (a) 1.35 m/s, 34° north of east
    (b) 2.00 m/s, 56° north of east
    (c) 1.74 m/s, 34° north of east
    (d) 2.11 m/s, 34° north of east
    (e) 2.50 m/s, 42° north of east

## KINEMATICS IN TWO DIMENSIONS

57. Two cars A and B approach each other at an intersection. Car A is traveling south at 20 m/s, while car B is traveling east at 17 m/s. What is the velocity of car A as described by the passengers in car B?
    (a) 37 m/s, eastward
    (b) 26 m/s, 40° south of east
    (c) 26 m/s, 50° south of east
    (d) 26 m/s, 50° south of west
    (e) 26 m/s, 40° south of west

58. A passenger at rest on a moving subway train fires a bullet straight up from his seat. The event is viewed by observers at rest on the station platform as the train moves by the platform with constant velocity. What is the trajectory of the bullet as described by the observers on the platform?
    (a) a straight horizontal path in the direction of the train's velocity
    (b) a straight vertical path up and down
    (c) a circular path centered on the gun
    (d) a straight diagonal path
    (e) a parabolic path

59. Airplane One flies due east at 250 km/h relative to the ground. At the same time, Airplane Two flies 325 km/h, 35° north of east relative to the ground. What is the velocity of Airplane One relative to Airplane Two?
    (a) 16 km/h, due west
    (b) 270 km/h, due north
    (c) 210 km/h, 55° south of east
    (d) 170 km/h, 15° north of east
    (e) 190 km/h, 85° south of west

60. A police car is in pursuit of a stolen pickup truck. At one instant, the car has a speed of 34 m/s and is 164 m behind the truck. At the same time, the truck has a speed of 32 m/s. If neither vehicle accelerates, how long will it take for the police to catch up to the truck?
    (a) 4.8 s
    (b) 24 s
    (c) 51 s
    (d) 82 s
    (e) 96 s

## Additional Problems

61. A bicyclist is riding at a constant speed along a straight line path. The rider throws a ball straight up to a height a few meters above her head. Ignoring air resistance, where will the ball land?
    (a) behind the rider
    (b) in front of the rider
    (c) in the same hand that threw the ball
    (d) in the opposite hand to the one that threw it
    (e) This cannot be determined without knowing the speed of the rider and the maximum height of the ball.

62. A basketball player is running at a constant speed of 2.5 m/s when he tosses a basketball upward with a speed of 6.0 m/s. How far does the player run before he catches the ball? Ignore air resistance.
    (a) 3.1 m
    (b) 4.5 m
    (c) The ball cannot be caught because it will fall behind the player.
    (d) 6.0 m
    (e) 7.5 m

# CHAPTER 4
# Forces and Newton's Laws of Motion

*Section 4.1 The Concepts of Force and Mass*
*Section 4.2 Newton's First Law of Motion*
*Section 4.3 Newton's Second Law of Motion*

1. With one exception, each of the following units can be used to express mass. What is the exception?
   (a) newton
   (b) slug
   (c) gram
   (d) N·s$^2$/m
   (e) kilogram

2. Complete the following statement: The term *net force* most accurately describes
   (a) the mass of an object
   (b) the inertia of an object.
   (c) the quantity that causes displacement.
   (d) the quantity that keeps an object moving.
   (e) the quantity that changes the velocity of an object.

3. Which one of the following terms is used to indicate the natural tendency of an object to remain at rest or in motion at a constant speed along a straight line?
   (a) velocity
   (b) force
   (c) acceleration
   (d) equilibrium
   (e) inertia

4. Complete the following statement: An inertial reference frame is one in which
   (a) Newton's first law of motion is valid.
   (b) the inertias of objects within the frame are zero.
   (c) the frame is accelerating.
   (d) the acceleration due to gravity is greater than zero.
   (e) Newton's third law of motion is not valid.

5. A net force **F** is required to give an object with mass $m$ an acceleration **a**. If a net force 6 **F** is applied to an object with mass $2m$, what is the acceleration on this object?
   (a) **a**
   (b) 2**a**
   (c) 3**a**
   (d) 4**a**
   (e) 6**a**

6. When the net force that acts on a hockey puck is 10 N, the puck accelerates at a rate of 50 m/s$^2$. Determine the mass of the puck.
   (a) 0.2 kg
   (b) 1.0 kg
   (c) 5 kg
   (d) 10 kg
   (e) 50 kg

7. The figure shows the velocity versus time curve for a car traveling along a straight line. Which of the following statements is false?
   (a) No net force acts on the car during interval **B**.
   (b) Net forces act on the car during intervals **A** and **C**.
   (c) Opposing forces may be acting on the car during interval **B**.
   (d) Opposing forces may be acting on the car during interval **C**.
   (e) The magnitude of the net force acting during interval **A** is less than that during **C**.

8. The graph shows the velocities of two objects of equal mass as a function of time. Forces $F_A$, $F_B$, and $F_C$ acted on the objects during intervals A, B, and C, respectively.

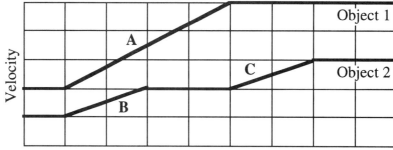

Which one of the following choices is the correct relationship between the magnitudes of the forces?
(a) $F_B = F_C > F_A$
(b) $F_C > F_B > F_A$
(c) $F_A > F_B = F_C$
(d) $F_A = F_B = F_C$
(e) $F_A > F_B > F_C$

9. A 15-N net force is applied for 6.0 s to a 12-kg box initially at rest. What is the speed of the box at the end of the 6.0 s interval?
(a) 1.8 m/s
(b) 15 m/s
(c) 3.0 m/s
(d) 30 m/s
(e) 7.5 m/s

10. A 2150-kg truck is traveling along a straight, level road at a constant speed of 55.0 km/h when the driver removes his foot from the accelerator. After 21.0 s, the truck's speed is 33.0 km/h. What is the magnitude of the average net force acting on the truck during the 21.0 s interval?
(a) 2250 N
(b) 626 N
(c) 1890 N
(d) 972 N
(e) 229 N

*Questions 11 through 14 pertain to the situation described below:*

A 2.0-kg object moves in a straight line on a horizontal frictionless surface.

The graph shows the velocity of the object as a function of time. The various equal time intervals are labeled using Roman numerals: I, II, III, IV, and V.

The net force on the object always acts along the line of motion of the object.

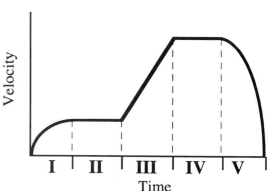

11. Which section(s) of the graph correspond to a condition of *zero net force*?
(a) V only
(b) III only
(c) II and IV
(d) II, III, and IV
(e) I, III, and V

12. Which section of the graph corresponds to the application of the *largest constant* net force?
(a) I
(b) II
(c) III
(d) IV
(e) V

13. In which section of the graph is the magnitude of the net force decreasing?
(a) I
(b) II
(c) III
(d) IV
(e) V

14. In which section(s) of the graph is the net force changing?
    (a) I and III
    (b) II and IV
    (c) III
    (d) IV
    (e) I and V

## Section 4.4 The Vector Nature of Newton's Second Law of Motion
## Section 4.5 Newton's Third Law of Motion

15. An object moves in the eastward direction at constant speed. A net force directed northward acts on the object for 5.0 s. At the end of the 5.0 second period, the net force drops to zero. Which one of the following statements is necessarily true?
    (a) The object will be moving eastward when the force drops to zero.
    (b) The final velocity of the object will be directed north of east.
    (c) The change in the velocity of the object will be directed north of west.
    (d) The direction of the object's acceleration depends on how fast the object was initially moving.
    (e) The magnitude of the object's acceleration depends on how fast the object was initially moving.

16. Two forces act on a 16-kg object. The first force has a magnitude of 68 N and is directed 24° north of east. The second force is 32 N, 48° north of west. What is the acceleration of the object resulting from the action of these two forces?
    (a) 1.6 m/s², 5.5° north of east
    (b) 1.9 m/s², 18° north of west
    (c) 2.4 m/s², 34° north of east
    (d) 3.6 m/s², 5.5° north of west
    (e) 4.1 m/s², 52° north of east

17. An apple crate with a weight of 225 N accelerates along a *frictionless* surface as the crate is pulled with a force of 14.5 N as shown in the drawing. What is the horizontal acceleration of the crate?
    (a) 1.40 m/s²
    (b) 0.427 m/s²
    (c) 1.29 m/s²
    (d) 0.597 m/s²
    (e) 0.644 m/s²

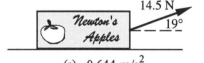

18. Two forces act on a hockey puck. For which orientation of the forces will the puck acquire an acceleration with the *largest* magnitude?

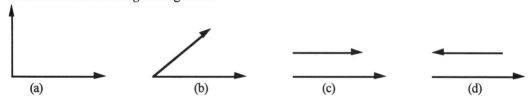

    (e) The magnitude of the acceleration will be the same in all four cases shown above.

*Questions 19 and 20 pertain to the situation described below:*

A horse pulls a cart along a flat road. Consider the following four forces that arise in this situation.
  (*1*) the force of the horse pulling on the cart
  (*2*) the force of the cart pulling on the horse
  (*3*) the force of the horse pushing on the road
  (*4*) the force of the road pushing on the horse

19. Which two forces form an "action-reaction" pair which obeys Newton's third law?
    (a) *1* and *4*
    (b) *1* and *3*
    (c) *2* and *4*
    (d) *3* and *4*
    (e) *2* and *3*

## FORCES AND NEWTON'S LAWS OF MOTION

20. Suppose that the horse and cart have started from rest; and as time goes on, their speed increases in the same direction. Which one of the following conclusions is correct concerning the magnitudes of the forces mentioned above?
    (a) force *1* exceeds force *2*
    (b) force *2* is less than force *3*
    (c) force *2* exceeds force *4*
    (d) force *3* exceeds force *4*
    (e) force *1* and *2* cannot have equal magnitudes

*Questions 21 and 22 pertain to the situation described below:*

A physics student in a hot air balloon ascends vertically at constant speed. Consider the following four forces that arise in this situation:
$\mathbf{F}_1$ = the weight of the hot air balloon
$\mathbf{F}_2$ = the weight of the student
$\mathbf{F}_3$ = the force of the student pulling on the earth
$\mathbf{F}_4$ = the force of the hot air balloon pulling on the student

21. Which two forces form an "action-reaction" pair that obeys Newton's third law?
    (a) $\mathbf{F}_1$ and $\mathbf{F}_2$
    (b) $\mathbf{F}_2$ and $\mathbf{F}_3$
    (c) $\mathbf{F}_1$ and $\mathbf{F}_3$
    (d) $\mathbf{F}_2$ and $\mathbf{F}_4$
    (e) $\mathbf{F}_3$ and $\mathbf{F}_4$

22. Which one of the following relationships concerning the forces or their magnitudes is true?
    (a) $F_4 > F_2$
    (b) $F_1 < F_2$
    (c) $F_4 > F_1$
    (d) $F_2 = -\mathbf{F}_4$
    (e) $\mathbf{F}_3 = -\mathbf{F}_4$

*Questions 23 and 24 pertain to the situation described below:*

A book is resting on the surface of a table. Consider the following four forces that arise in this situation:
(*1*) the force of the earth pulling on the book
(*2*) the force of the table pushing on the book
(*3*) the force of the book pushing on the table
(*4*) the force of the book pulling on the earth

23. Which two forces form an "action-reaction" pair which obeys Newton's third law?
    (a) *1* and *2*
    (b) *1* and *3*
    (c) *1* and *4*
    (d) *2* and *4*
    (e) *3* and *4*

24. The book has zero acceleration. Which pair of forces, excluding "action-reaction" pairs, must be equal in magnitude and opposite in direction?
    (a) *1* and *2*
    (b) *1* and *3*
    (c) *1* and *4*
    (d) *2* and *3*
    (e) *2* and *4*

*Questions 25 and 26 pertain to the statement and figure below:*

A 70.0-kg astronaut pushes to the left on a spacecraft with a force **F** in "gravity-free" space. The spacecraft has a total mass of $1.0 \times 10^4$ kg.

During the push, the astronaut accelerates to the right with an acceleration of 0.36 m/s².

25. Which one of the following statements is true concerning this situation?
    (a) The spacecraft does not move, but the astronaut moves to the right with a constant speed.
    (b) The astronaut stops moving after he stops pushing on the spacecraft.
    (c) The force exerted on the astronaut is larger than the force exerted on the spacecraft.
    (d) The force exerted on the spacecraft is larger than the force exerted on the astronaut.
    (e) The velocity of the astronaut increases while he is pushing on the spacecraft.

26. Determine the magnitude of the acceleration of the spacecraft.
    (a) $51.4 \text{ m/s}^2$
    (b) $0.36 \text{ m/s}^2$
    (c) $2.5 \times 10^{-3} \text{ m/s}^2$
    (d) $7.0 \times 10^{-3} \text{ m/s}^2$
    (e) $3.97 \times 10^{-4} \text{ m/s}^2$

*Section 4.6   Types of Forces: An Overview*
*Section 4.7   The Gravitational Force*

27. Consider the following forces.
    (1) frictional
    (2) gravitational
    (3) tension
    (4) strong nuclear
    (5) normal
    (6) electroweak
    Which of the forces listed are considered fundamental forces?
    (a) *1, 2*, and *4*
    (b) *1, 2, 3*, and *5*
    (c) *1, 3*, and *5*
    (d) *2, 4*, and *6*
    (e) *2, 3, 4*, and *6*

28. A rock is thrown straight up from the earth's surface. Which one of the following statements concerning the *net force* acting on the rock at the top of its path is true?
    (a) It is equal to the weight of the rock.
    (b) It is instantaneously equal to zero.
    (c) Its direction changes from up to down.
    (d) It is greater than the weight of the rock.
    (e) It is less than the weight of the rock, but greater than zero.

29. Two point masses $m$ and $M$ are separated by a distance $d$. If the distance between the masses is increased to $3d$, how does the gravitational force between them change?
    (a) The force will be one-third as great.
    (b) The force will be one-ninth as great.
    (c) The force will be three times as great.
    (d) The force will be nine times as great.
    (e) It is impossible to determine without knowing the numerical values of $m$, $M$, and $d$.

30. Two point masses $m$ and $M$ are separated by a distance $d$. If the separation $d$ remains fixed and the masses are increased to the values $3m$ and $3M$ respectively, how does the gravitational force between them change?
    (a) The force will be one-third as great.
    (b) The force will be one-ninth as great.
    (c) The force will be three times as great.
    (d) The force will be nine times as great.
    (e) It is impossible to determine without knowing the numerical values of $m$, $M$, and $d$.

# FORCES AND NEWTON'S LAWS OF MOTION

31. Which one of the following statements concerning the two "gravitational constants" $G$, the universal gravitational constant, and $g$ the magnitude of the acceleration due to gravity is true?
    (a) The values of $g$ and $G$ do not depend on location.
    (b) The values of $g$ and $G$ depend on location.
    (c) The value of $G$ is the same everywhere in the universe, but the value of $g$ is not.
    (d) The value of $g$ is the same everywhere in the universe, but the value of $G$ is not.
    (e) The values of $g$ and $G$ are equal on the surface of any planet, but in general, vary with location in the universe.

32. Two satellites of different masses are in the same circular orbit about the earth. Which one of the following statements is true concerning the magnitude of the gravitational force that acts on each of them?
    (a) The magnitude of the gravitational force is zero for both satellites.
    (b) The magnitude of the gravitational force is the same for both satellites, but not zero.
    (c) The magnitude of the gravitational force is zero for one, but not for the other.
    (d) The magnitude of the gravitational force depends on their masses.
    (e) The magnitude of the gravitational force varies from point to point in their orbits.

33. An astronaut orbits the earth in a space capsule whose height above the earth is equal to the earth's radius. How does the weight of the astronaut in the capsule compare to her weight on the earth?
    (a) It is equal to her weight on earth.
    (b) It is equal to one-half of her weight on earth.
    (c) It is equal to one-third of her weight on earth.
    (d) It is one-fourth her weight on earth.
    (e) It is equal to one-sixteenth her weight on earth.

34. An astronaut orbits the earth in a space capsule whose height above the earth is equal to the earth's radius. How does the mass of the astronaut in the capsule compare to her mass on the earth?
    (a) It is equal to her mass on earth.
    (b) It is equal to one-half of her mass on earth.
    (c) It is equal to one-third of her mass on earth.
    (d) It is one-fourth her mass on earth.
    (e) It is equal to one-sixteenth her mass on earth.

35. Which statement best explains why the weight of an object of mass $m$ is different on Mars than it is on the Earth?
    (a) The mass of Mars is different from that of Earth.
    (b) The masses and radii of Mars and Earth are not the same.
    (c) The mass $m$ is further from the Earth's center when it is on Mars.
    (d) The constant $G$ is different on Mars.
    (e) The mass $m$ will be different on Mars.

36. A marble is dropped straight down from a distance $h$ above the floor.
    Let $F_m$ = the magnitude of the gravitational force *on the marble* due to the earth;
    $F_e$ = the magnitude of the gravitational force *on the earth* due to the marble;
    $a_m$ = the magnitude of the acceleration *of the marble* toward the earth;
    $a_e$ = the magnitude of the acceleration *of the earth* toward the marble.
    Which set of conditions is true as the marble falls toward the earth?
    (a) $F_m = F_e$ and $a_m < a_e$
    (b) $F_m = F_e$ and $a_m > a_e$
    (c) $F_m < F_e$ and $a_m = a_e$
    (d) $F_m > F_e$ and $a_m = a_e$
    (e) $F_m < F_e$ and $a_m > a_e$

37. What is the weight of a 2.50-kg bag of sand on the surface of the earth?
    (a) 2.50 N
    (b) 9.80 N
    (c) 24.5 N
    (d) 49.0 N
    (e) 98.0 N

38. A 2.00-kg projectile is fired at an angle of 20.0°. What is the magnitude of the force exerted on the projectile when it is at the highest position in its trajectory?
    (a) 19.6 N
    (b) 14.7 N
    (c) 9.80 N
    (d) 4.90 N
    (e) zero

39. What is the magnitude of the gravitational force acting on a 79.5-kg student due to a 60.0-kg student sitting 2.25 m away in the lecture hall?
    (a) $3.14 \times 10^{-9}$ N
    (b) $2.82 \times 10^{-8}$ N
    (c) $7.91 \times 10^{-10}$ N
    (d) $1.41 \times 10^{-7}$ N
    (e) $6.29 \times 10^{-8}$ N

## Section 4.8  The Normal Force
## Section 4.9  Static and Kinetic Frictional Forces

*Questions 40 and 41 pertain to the situation described below:*

A force **P** pulls on a crate of mass $m$ on a rough surface. The figure shows the magnitudes and directions of the forces that act on the crate in this situation. **W** represents the weight of the crate. $F_N$ represents the normal force on the crate, and **f** represents the frictional force.

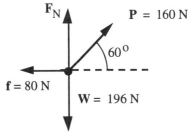

40. Which statement best describes the motion of the crate?
    (a) The crate must be at rest.
    (b) The crate must be moving with constant velocity.
    (c) The crate must be moving with constant acceleration.
    (d) The crate may be either at rest or moving with constant velocity.
    (e) The crate may be either at rest or moving with constant acceleration.

41. What is the magnitude of $F_N$, the normal force on the crate?
    (a) 57 N
    (b) 80 N
    (c) 160 N
    (d) 196 N
    (e) 230 N

42. A 10-kg block is set moving with an initial speed of 6 m/s on a rough horizontal surface. If the force of friction is 20 N, approximately how far does the block travel before it stops?
    (a) 1.5 m
    (b) 3 m
    (c) 6 m
    (d) 9 m
    (e) 18 m

43. An automobile's wheels are locked as it slides to a stop from an initial speed of 30.0 m/s. If the coefficient of kinetic friction is 0.200 and the road is horizontal, approximately how long does it take the car to stop?
    (a) 6.00 s
    (b) 7.57 s
    (c) 15.3 s
    (d) 22.5 s
    (e) 30.0 s

# FORCES AND NEWTON'S LAWS OF MOTION

*Questions 44 through 46 pertain to the situation described below:*

Two blocks rest on a horizontal *frictionless* surface as shown. The surface between the top and bottom blocks is roughened so that there is no slipping between the two blocks. A 30-N force is applied to the bottom block as suggested in the figure.

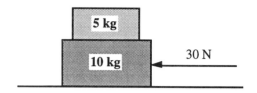

44. What is the acceleration of the "two block" system?
    (a) 1 m/s²
    (b) 2 m/s²
    (c) 3 m/s²
    (d) 6 m/s²
    (e) 15 m/s²

45. What is the force of static friction between the top and bottom blocks?
    (a) zero
    (b) 10 N
    (c) 20 N
    (d) 25 N
    (e) 30 N

46. What is the minimum coefficient of static friction necessary to keep the top block from slipping on the bottom block?
    (a) 0.05
    (b) 0.10
    (c) 0.20
    (d) 0.30
    (e) 0.40

47. Note the following situations:

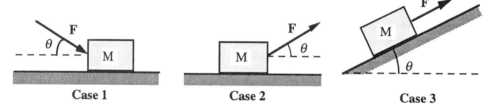

In which case will the magnitude of the normal force on the block be equal to $(Mg + F \sin \theta)$?
    (a) case *1* only
    (b) case *2* only
    (c) both cases *1* and *2*
    (d) both cases *2* and *3*
    (e) cases *1*, *2*, and *3*

*Questions 48 and 49 pertain to the situation described below:*

A block is pulled along a rough level surface at constant speed by the force **P**. The figure shows the free-body diagram for the block.

$F_N$ represents the normal force on the block; and **f** represents the force of kinetic friction.

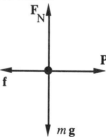

48. What is the magnitude of $F_N$?
    (a) *mg*
    (b) *P*
    (c) *f*
    (d) 2*mg*
    (e) This cannot be determined from the information given.

49. If the coefficient of kinetic friction, $\mu_k$, between the block and the surface is 0.30 and the magnitude of the frictional force is 80.0 N, what is the weight of the block?
   (a) 1.6 N
   (b) 4.0 N
   (c) 160 N
   (d) 270 N
   (e) 410 N

50. Two identical blocks are pulled along a rough surface as suggested in the figure.

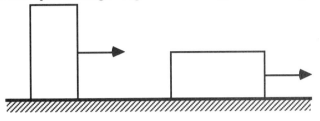

   Which one of the following statements is *false*?
   (a) The coefficient of kinetic friction is the same in each case.
   (b) A force of the same magnitude is needed to keep each block moving.
   (c) A force of the same magnitude was required to start each block moving.
   (d) The magnitude of the force of kinetic friction is greater for the block on the right.
   (e) The normal force exerted on the blocks by the surface is the same for both blocks.

51. A crate rests on the flatbed of a truck which is initially traveling at 15 m/s on a level road. The driver applies the brakes and the truck is brought to a halt in a distance of 38 m. If the deceleration of the truck is constant, what is the minimum coefficient of friction between the crate and the truck that is required to keep the crate from sliding?
   (a) 0.20
   (b) 0.30
   (c) 0.39
   (d) 0.59
   (e) This cannot be determined without knowing the mass of the crate.

*Questions 52 and 53 pertain to the statement below:*

A 2.0-N force acts horizontally on a 10-N block that is initially at rest on a horizontal surface. The coefficient of static friction between the block and the surface is 0.50.

52. What is the magnitude of the frictional force that acts on the block?
   (a) zero
   (b) 2 N
   (c) 5 N
   (d) 8 N
   (e) 10 N

53. Suppose that the block now moves across the surface with constant speed under the action of a horizontal 3.0 N force. Which statement concerning this situation is *not* true?
   (a) The block is not accelerated.
   (b) The net force on the block is zero.
   (c) The frictional force on the block has magnitude 3.0 N.
   (d) The coefficient of kinetic friction between the block and the surface is 0.30.
   (e) The direction of the total force that the surface exerts on the block is vertically upward.

44   FORCES AND NEWTON'S LAWS OF MOTION

54. A 2.0-N rock slides on a frictionless inclined plane. Which one of the following statements is true concerning the normal force that the plane exerts on the rock?

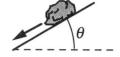

(a) The normal force is zero.
(b) The normal force is 2.0 N.
(c) The normal force is greater than 2.0 N.
(d) It *increases* as the angle of inclination, $\theta$, is *increased*.
(e) The normal force is less than 2.0 N, but greater than zero.

55. A boy pulls a sled of mass 5.0 kg with a rope that makes an 60.0° angle with respect to the horizontal surface of a frozen pond. The boy pulls on the rope with a force of 10.0 N; and the sled moves with constant velocity. What is the coefficient of friction between the sled and the ice?
(a) 0.10      (c) 0.18      (e) 1.0
(b) 0.12      (d) 0.20

56. In an experiment with a block of wood on an inclined plane, with dimensions shown in the figure, the following observations are made:
  (1) If the block is placed on the inclined plane, it remains there at rest.
  (2) If the block is given a small push, it will accelerate toward the bottom of the incline without any further pushing.

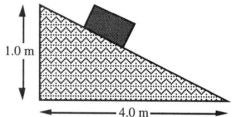

Which is the *best* conclusion that can be drawn from these observations?
(a) The coefficient of kinetic friction must be negative.
(b) Both coefficients of friction must be less than 0.25.
(c) Both coefficients of friction must be greater than 0.25.
(d) The coefficient of static friction must be less than the coefficient of kinetic friction.
(e) The coefficient of static friction is greater than 0.25 while the coefficient of kinetic friction is less than 0.25.

*Section 4.10   The Tension Force*
*Section 4.11   Equilibrium Applications of Newton's Laws of Motion*

57. A rock is suspended from a string and moves downward at constant speed. Which statement is true concerning the tension in the string if air resistance is ignored?
(a) The tension is zero.
(b) The tension points downward.
(c) The tension is equal to the weight of the rock.
(d) The tension is less than the weight of the rock.
(e) The tension is greater than the weight of the rock.

58. A rock is suspended from a string and moves downward at constant speed. Which one of the following statements is true concerning the tension in the string *if air resistance is not ignored?*
(a) The tension is zero.
(b) The tension points downward.
(c) The tension is equal to the weight of the rock.
(d) The tension is less than the weight of the rock.
(e) The tension is greater than the weight of the rock.

59. A rock is suspended from string and accelerates downward. Which one of the following statements concerning the tension in the string is true?
   (a) The tension points downward.
   (b) The tension is less than the weight of the rock.
   (c) The tension is equal to the weight of the rock.
   (d) The tension is greater than the weight of the rock.
   (e) The tension is independent of the magnitude of the rock's acceleration.

60. A rock is suspended from string and accelerates upward. Which statement is true concerning the tension in the string?
   (a) The tension points downward.
   (b) The tension is less than the weight of the rock.
   (c) The tension is equal to the weight of the rock.
   (d) The tension is greater than the weight of the rock.
   (e) The tension is independent of the magnitude of the rock's acceleration.

61. In a tug-of-war, 5 men on each team pull with an average force of 500 N each. What is the tension in the center of the rope?
   (a) zero
   (b) 100 N
   (c) 500 N
   (d) 2500 N
   (e) 5000 N

62. Under what condition(s) will an object be in *equilibrium*?
   (a) If it is either at rest or moving with constant velocity.
   (b) If it is either moving with constant velocity or with constant acceleration.
   (c) Only if it is at rest.
   (d) Only if it is moving with constant velocity.
   (e) Only if it is moving with constant acceleration.

63. A 4-kg block is connected by means of a *massless* rope to a 2-kg block as shown in the figure. Complete the following statement: If the 4-kg block is to begin sliding, the coefficient of static friction between the 4-kg block and the surface must be
   (a) less than zero
   (b) greater than 2
   (c) greater than 1, but less than 2
   (d) greater than 0.5, but less than 1
   (e) less than 0.5, but greater than zero

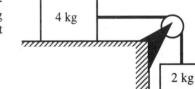

64. A block of mass $M$ is hung by ropes as shown. The system is in equilibrium. The point O represents the knot, the junction of the three ropes. Which of the following statements is true concerning the magnitudes of the three forces in equilibrium?
   (a) $F_1 = F_2 = F_3$
   (b) $F_2 = 2F_3$
   (c) $F_2 < F_3$
   (d) $F_1 = F_2 = \dfrac{F_3}{2}$
   (e) $F_1 > F_3$

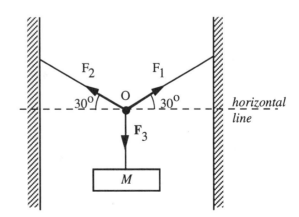

## FORCES AND NEWTON'S LAWS OF MOTION

65. A small plane climbs with a constant velocity of 250 m/s at an angle of 28° with respect to the horizontal. Which statement is true concerning the magnitude of the *net force* on the plane?
    (a) It is equal to zero.
    (b) It is equal to the weight of the plane.
    (c) It is equal to the magnitude of the force of air resistance.
    (d) It is less than the weight of the plane but greater than zero.
    (e) It is equal to the component of the weight of the plane in the direction of motion.

66. A 20-kg crate is suspended from a fixed beam by two vertical ropes. What is the tension in each rope?
    (a) 10 N
    (b) 40 N
    (c) 100 N
    (d) 200 N
    (e) 390 N

67. A muscle builder holds the ends of a massless rope. At the center of the rope, a 15-kg ball is hung as shown. What is the tension in the rope if the angle θ in the drawing is 4.5°?
    (a) 1900 N
    (b) 940 N
    (c) 470 N
    (d) 230 N
    (e) 150 N

*Questions 68 through 70 pertain to the situation described below:*

A block of mass $M$ is held motionless on a frictionless inclined plane by means of a string attached to a vertical wall as shown in the figure.

68. What is the magnitude of the tension **T** in the string?
    (a) zero
    (b) $Mg$
    (c) $Mg \cos \theta$
    (d) $Mg \sin \theta$
    (e) $Mg \tan \theta$

69. If the string breaks, what is the magnitude of the acceleration of the block as it slides down the inclined plane?
    (a) zero
    (b) $g$
    (c) $g \cos \theta$
    (d) $g \sin \theta$
    (e) $g \tan \theta$

70. Assuming the plane were not frictionless and the string were to break, what minimum value of the coefficient of static friction, $\mu_s$, would prevent the block from sliding down the inclined plane?
    (a) zero
    (b) 1
    (c) $\cos \theta$
    (d) $\sin \theta$
    (e) $\tan \theta$

71. Two sleds are hooked together in tandem as shown in the figure. The front sled is twice as massive as the rear sled.

The sleds are pulled along a frictionless surface by an applied force **F**. The tension in the rope between the sleds is **T**. Determine the ratio of the magnitudes of the two forces, { EMBED Equation.2 }.
(a) 0.25
(b) 0.33
(c) 0.50
(d) 0.67
(e) 2.0

72. A block of weight $W$ is suspended by a string of fixed length. The ends of the string are held at various positions as shown in the figures below. In which case, if any, is magnitude of the tension the along the string the largest?

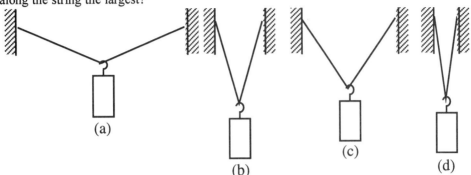

(e) It will be the same in all four cases, since the string must support the entire weight of the block.

*Questions 73 and 74 pertain to the system described below:*

A system of two cables supports a 150-N ball as shown.

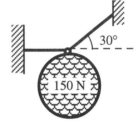

73. What is the tension in the right-hand cable?
(a) 87 N
(b) 150 N
(c) 170 N
(d) 260 N
(e) 300 N

74. What is the tension in the horizontal cable?
(a) 87 N
(b) 150 N
(c) 170 N
(d) 260 N
(e) 300 N

## Section 4.12  Nonequilibrium Applications of Newton's Laws of Motion

75. A woman stands on a bathroom scale in an elevator which is not moving. The scale reads 500 N. The elevator then moves downward at a constant velocity of 5 m/s. What does the scale read while the elevator descends with constant velocity?
(a) 100 N
(b) 250 N
(c) 450 N
(d) 500 N
(e) 750 N

76. A 20.0-kg package is dropped from a high tower in still air and is "tracked" by a radar system. When the package is 25 m above the ground, the radar tracking indicates that its acceleration is 7.0 m/s$^2$. Determine the force of air resistance on the package.
(a) 28 N
(b) 56 N
(c) 140 N
(d) 196 N
(e) 340 N

77. A 10-kg block is pushed against a vertical wall by a horizontal force of 100 N as shown in the figure. The coefficient of static friction, $\mu_s$, between the block and the wall is 0.60; and the coefficient of kinetic friction, $\mu_k$, is 0.40. Which one of the following statements is true if the block is initially at rest?

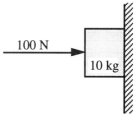

(a) The total force exerted on the block by the wall is horizontally directed.
(b) The block slides down the wall with an acceleration of magnitude 3.8 m/s$^2$.
(c) The block will slide down the wall because the force of static friction can be no larger than 60 N.
(d) The block will remain at rest because the coefficient of static friction is greater than the coefficient of kinetic friction.
(e) The block will slide down the wall because the coefficient of kinetic friction is less than the coefficient of static friction.

78. A certain crane can provide a maximum lifting force of 25 000 N. It hoists a 2000-kg load starting at ground level by applying the maximum force for a 2 second interval; then, it applies just sufficient force to keep the load moving upward at constant speed. Approximately how long does it take to raise the load from ground level to a height of 30 m?
(a) 2 s
(b) 5 s
(c) 7 s
(d) 9 s
(e) 10 s

*Questions 79 and 80 pertain to the situation described below:*

A 10-kg block is connected to a 40-kg block as shown in the figure. The surface on which the blocks slide is frictionless. A force of 50 N pulls the blocks to the right.

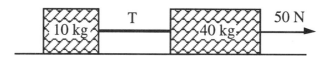

79. What is the magnitude of the acceleration of the 40-kg block?
(a) 0.5 m/s$^2$
(b) 1 m/s$^2$
(c) 2 m/s$^2$
(d) 4 m/s$^2$
(e) 5 m/s$^2$

80. What is the magnitude of the tension **T** in the rope that connects the two blocks?
(a) zero
(b) 10 N
(c) 20 N
(d) 40 N
(e) 50 N

81. A 71-kg man stands on a bathroom scale in an elevator. What does the scale read if the elevator is ascending with an acceleration of 3.0 m/s$^2$?
(a) 140 N
(b) 480 N
(c) 690 N
(d) 830 N
(e) 910 N

82. A 4-kg block and a 2-kg block can move on the horizontal frictionless surface. The blocks are accelerated by a +12-N force that pushes the larger block against the smaller one. Determine the force that the 2-kg block exerts on the 4-kg block.

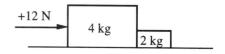

(a) −12 N
(b) −4 N
(c) zero
(d) +4 N
(e) +8 N

83. A man stands on a spring scale in a moving elevator and notices that the scale reading is 20% larger than when he weighs himself in his bathroom. Which statement can *not* be true?
    (a) The tension in the supporting cable must exceed the weight of the elevator and its contents.
    (b) The speed of the elevator changes by equal amounts in equal times.
    (c) The elevator could be moving upward with increasing speed.
    (d) The elevator could be moving downward with decreasing speed.
    (e) The elevator could be moving upward at constant speed.

## Additional Problems

*Questions 84 through 86 pertain to the situation described below:*

A block is pulled at constant speed along a rough level surface by a rope that makes an angle of 30° with respect to the horizontal. The applied force along the rope is **P**. The force of kinetic friction between the block and the surface is 10 N.

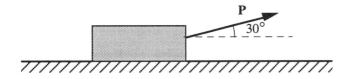

84. Which one of the following expressions gives the magnitude of the force **P** in SI units?
    (a) 10/(cos 30°)
    (b) 10/(sin 30°)
    (c) 10(cos 30°)
    (d) 10(sin 30°)
    (e) tan 30°

85. Complete the following statement: The magnitude of the normal force could be reduced by
    (a) increasing the speed of the block.
    (b) decreasing the coefficient of kinetic friction.
    (c) decreasing the velocity of the block.
    (d) decreasing the angle made by the rope.
    (e) increasing the angle made by the rope.

86. Which one of the following actions will increase the frictional force on the block?
    (a) increasing the contact surface area
    (b) decreasing the contact surface area
    (c) increasing the weight of the block
    (d) decreasing the speed of the block
    (e) increasing the angle made by the rope

*Questions 87 through 89 pertain to the situation described below:*

A rope holds a 10-kg rock at rest on a *frictionless* inclined plane as shown.

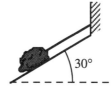

87. Determine the tension in the rope.
    (a) 9.8 N
    (b) 20 N
    (c) 49 N
    (d) 85 N
    (e) 98 N

88. Which one of the following statements concerning the force exerted *on* the plane *by* the rock is true?
    (a) It is zero.
    (b) It is 98 N.
    (c) It is greater than 98 N.
    (d) It is less than 98 N, but greater than zero.
    (e) It *increases* as the angle of inclination is *increased*.

50   FORCES AND NEWTON'S LAWS OF MOTION

89. Determine the magnitude of the acceleration of the rock down the inclined plane if the rope breaks?
    (a) zero
    (b) 4.9 m/s$^2$
    (c) 5.7 m/s$^2$
    (d) 8.5 m/s$^2$
    (e) 9.8 m/s

90. Three spring scales are attached along a straight line as shown. The scale on the left is attached to a wall. A force of 15 N is applied to the scale at the right.

    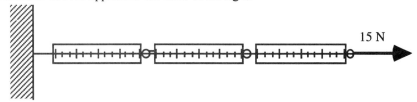

    What is the reading on the middle scale?
    (a) zero
    (b) 45 N
    (c) 10 N
    (d) 5 N
    (e) 15 N

91. A *massless horizontal strut* is attached to the wall at the hinge O. Which one of the following phrases best describes the force that the hinge pin supplies *to the strut* if the weight of the cables is also neglected?
    (a) 100 lb, straight up
    (b) 50 lb, to the right
    (c) 200 lb, to the right
    (d) 244 lb, 27° above the strut
    (e) 56 lb, to the left

    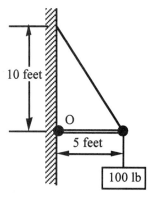

*Questions 92 and 93 pertain to the situation described below:*

Two 5-N boxes are attached to opposite ends of a spring scale and suspended from pulleys as shown.

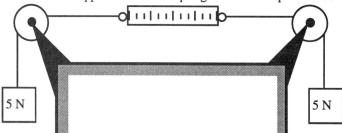

92. What is the reading on the scale?
    (a) zero
    (b) 2.5 N
    (c) 5 N
    (d) 10 N
    (e) 25 N

93. Suppose that the system were placed in an elevator that accelerated downward at 2 m/s$^2$. What would the scale read?
    (a) zero
    (b) 4 N
    (c) 6 N
    (d) 8 N
    (e) 10 N

94. A spring scale is loosely fastened to the ceiling of a railway car. When a 1.0-kg block is hung from the scale, it reads 12 N and is oriented as shown in the figure.

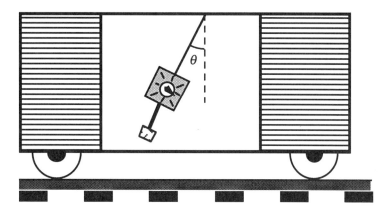

What is the approximate acceleration of the car as measured by an observer at rest on the ground outside of the car?
(a) 7 m/s² to the right
(b) 7 m/s² to the left
(c) 12 m/s² to the right
(d) 12 m/s² to the left
(e) It is impossible to calculate since the angle $\theta$ has not been given.

*Questions 95 through 97 pertain to the following situation:*

A block is at rest on a rough inclined plane and is connected to an object with the same mass as shown. The rope may be considered massless; and the pulley may be considered frictionless. The coefficient of static friction between the block and the plane is $\mu_s$; and the coefficient of kinetic friction is $\mu_k$.

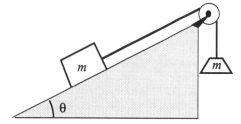

95. What is the magnitude of the static frictional force acting on the block?
(a) $mg \sin \theta$
(b) $mg \cos \theta$
(c) $mg (1 - \sin \theta)$
(d) $mg (1 - \cos \theta)$
(e) $mg$

96. If the rope were cut between the block and the pulley, what would be the magnitude of the acceleration of the block down the plane?
(a) $g$
(b) $g - \mu_k \sin \theta$
(c) $g - \mu_k \cos \theta$
(d) $g(\tan \theta - \mu_k \sin \theta)$
(e) $g(\sin \theta - \mu_k \cos \theta)$

97. If the mass of the suspended object is doubled, what will be the acceleration of the block up the plane?
(a) $g(2 - \mu_k \sin \theta)$
(b) $2g(\mu_k \sin \theta - \cos \theta)$
(c) $g(2\tan \theta - \mu_k \sin \theta)$
(d) $g(2 - \sin \theta - \mu_k \cos \theta)$
(e) $g(2\cos \theta - \mu_k \sin \theta)$

# CHAPTER 5
# Dynamics of Uniform Circular Motion

## Section 5.1 Uniform Circular Motion
## Section 5.2 Centripetal Acceleration

1. A ball moves with a constant speed of 4 m/s around a circle of radius 0.25 m. What is the period of the motion?
   (a) 0.1 s
   (b) 0.4 s
   (c) 0.7 s
   (d) 1 s
   (e) 2 s

2. The second hand on a watch has a length of 4.50 mm and makes one revolution in 60.00 s. What is the speed of the end of the second hand as it moves in uniform circular motion?
   (a) $9.42 \times 10^{-4}$ m/s
   (b) $2.67 \times 10^{-3}$ m/s
   (c) $5.34 \times 10^{-3}$ m/s
   (d) $4.71 \times 10^{-4}$ m/s
   (e) $2.36 \times 10^{-5}$ m/s

3. A race car is traveling at constant speed around a circular track. What happens to the centripetal acceleration of the car if the speed is doubled?
   (a) It remains the same.
   (b) It increases by a factor of 2.
   (c) It increases by a factor of 4.
   (d) It is decreased by a factor of one-half.
   (e) It is decreased by a factor of one-fourth.

4. A ball is whirled on the end of a string in a horizontal circle of radius R at constant speed v. The centripetal acceleration of the ball can be increased by a factor of 4 by
   (a) keeping the speed fixed and increasing the radius by a factor of 4.
   (b) keeping the radius fixed and increasing the speed by a factor of 4.
   (c) keeping the radius fixed and increasing the period by a factor of 4.
   (d) keeping the radius fixed and decreasing the period by a factor of 4.
   (e) keeping the speed fixed and decreasing the radius by a factor of 4.

5. A rock is whirled on the end of a string in a horizontal circle of radius $R$ with a constant period $T$. If the radius of the circle is reduced to $R/2$, while the period remains $T$, what happens to the centripetal acceleration of the rock?
   (a) It remains the same.
   (b) It increases by a factor of 2.
   (c) It increases by a factor of 4.
   (d) It decreases by a factor of 2.
   (e) It decreases by a factor of 4.

6. A car traveling at 20 m/s rounds a curve so that its centripetal acceleration is 5 m/s$^2$. What is the radius of the curve?
   (a) 4 m
   (b) 8 m
   (c) 80 m
   (d) 160 m
   (e) 640 m

*Questions 7 and 8 pertain to the following situation:*

The world's largest Ferris wheel with a radius of 50.0 m is located in Yokohama City, Japan. Each of the sixty gondolas on the wheel takes 1.00 minute to complete one revolution when it is running at full speed. **Note**: Ignore gravitational effects.

7. What is the uniform speed of a gondola when the Ferris wheel is running at full speed?
   (a) 314 m/s
   (b) 1.67 m/s
   (c) 10.5 m/s
   (d) 18.6 m/s
   (e) 5.24 m/s

8. What is the centripetal acceleration of the gondola when the Ferris wheel is running at full speed?
   (a) 0.548 m/s$^2$
   (b) 6.91 m/s$^2$
   (c) 2.21 m/s$^2$
   (d) 0.732 m/s$^2$
   (e) 6.28 m/s$^2$

## Section 5.3  Centripetal Force
## Section 5.4  Banked Curves

9. A boy is whirling a stone around his head by means of a string. The string makes one complete revolution every second, and the tension in the string is $F_T$. The boy then speeds up the stone, keeping the radius of the circle unchanged, so that the string makes two complete revolutions every second.  What happens to the tension in the sting?
   (a) The tension is unchanged.
   (b) The tension reduces to half of its original value.
   (c) The tension increases to twice its original value.
   (d) The tension increases to four times its original value.
   (e) The tension reduces to one-fourth of its original value.

10. A 0.25-kg ball attached to a string is rotating in a horizontal circle of radius 0.5 m.  If the ball revolves twice every second, what is the tension in the sting?
    (a) 2 N
    (b) 5 N
    (c) 7 N
    (d) 10 N
    (e) 20 N

11. A certain string just breaks when it is under 400 N of tension.  A boy uses this string to whirl a 10-kg stone in a horizontal circle of radius 10 m.  The boy continuously increases the speed of the stone.  At approximately what speed will the string break?
    (a) 10 m/s
    (b) 20 m/s
    (c) 80 m/s
    (d) 100 m/s
    (e) 400 m/s

12. In an amusement park ride, a small child stands against the wall of a cylindrical room which is then made to rotate.  The floor drops downward and the child remains pinned against the wall.  If the radius of the device is 2.15 m and the relevant coefficient of friction between the child and the wall is 0.400, with what minimum speed is the child moving if he is to remain pinned against the wall?
    (a) 7.26 m/s
    (b) 3.93 m/s
    (c) 12.1 m/s
    (d) 5.18 m/s
    (e) 9.80 m/s

13. Which force is responsible for holding a car in a *frictionless* banked curve?
    (a) the reaction force to the car's weight
    (b) the vertical component of the car's weight
    (c) the vertical component of the normal force
    (d) the horizontal component of the car's weight
    (e) the horizontal component of the normal force

14. Which force is responsible for holding a car in an *unbanked* curve?
    (a) the car's weight
    (b) the force of friction
    (c) the reaction force to the car's weight
    (d) the vertical component of the normal force
    (e) the horizontal component of the normal force

15. The maximum speed at which a car can safely negotiate an unbanked curve depends on all of the following factors except
    (a) the diameter of the curve.
    (b) the acceleration due to gravity.
    (c) the coefficient of static friction between the road and the tires.
    (d) the coefficient of kinetic friction between the road and the tires.
    (e) the ratio of the static frictional force between the road and the tires and the normal force exerted on the car.

16. The maximum speed at which a car can safely negotiate a *frictionless* banked curve depends on all of the following *except*
    (a) the mass of the car.
    (b) the angle of banking.
    (c) the diameter of the curve.
    (d) the radius of the curve.
    (e) the acceleration due to gravity.

17. Determine the minimum angle at which a roadbed should be banked so that a car traveling at 20.0 m/s can safely negotiate the curve if the radius of the curve is $2.00 \times 10^2$ m.
    (a) 0.200°  (c) 11.5°  (e) 78.2°
    (b) 0.581°  (d) 19.6°

18. A car enters a horizontal, curved roadbed of radius 50 m. The coefficient of static friction between the tires and the roadbed is 0.20. What is the maximum speed with which the car can safely negotiate the unbanked curve?
    (a) 5 m/s  (c) 20 m/s  (e) 100 m/s
    (b) 10 m/s  (d) 40 m/s

*Questions 19 through 21 pertain to the statement below:*

A 1000-kg car travels along a straight 500-m portion of highway (from **A** to **B**) at a constant speed of 10 m/s. At **B**, the car encounters an unbanked curve of radius 50 m. The car follows the road from **B** to **C** traveling at a constant speed of 10 m/s while the direction of the car changes from east to south.

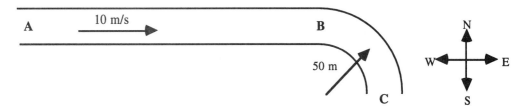

19. What is the magnitude of the acceleration of the car as it travels from A to B?
    (a) 2 m/s²  (c) 10 m/s²  (e) zero
    (b) 5 m/s²  (d) 20 m/s²

20. What is the magnitude of the acceleration of the car as it travels from B to C?
    (a) 2 m/s²  (c) 10 m/s²  (e) zero
    (b) 5 m/s²  (d) 20 m/s²

21. What is the magnitude of the frictional force between the tires and the road as the car negotiates the curve from B to C?
   (a) 20 000 N
   (b) 10 000 N
   (c) 5000 N
   (d) 2000 N
   (e) 1000 N

## Section 5.5 Satellites in Circular Orbit
## Section 5.6 Apparent Weightlessness and Artificial Gravity

22. The earth exerts the necessary centripetal force on an orbiting satellite to keep it moving in a circle at constant speed. Which statement best explains why the speed of the satellite does not change even though there is a net force exerted on it?
   (a) The satellite is in equilibrium.
   (b) The acceleration of the satellite is zero.
   (c) The centripetal force has magnitude $mv^2/r$.
   (d) The centripetal force is canceled by the reaction force.
   (e) The centripetal force is always perpendicular to the velocity.

23. Callisto and Io are two of Jupiter's moons. The distance from Callisto to the center of Jupiter is approximately 4.5 times as far as that of Io. How does Callisto's orbital period, $T_C$, compare to that of Io, $T_I$?
   (a) $T_C = 4.5 T_I$
   (b) $T_C = 21 T_I$
   (c) $T_C = 9.5 T_I$
   (d) $T_C = 0.2 T_I$
   (e) $T_C = 2.7 T_I$

24. Consider a hypothetical planet in our solar system whose average distance from the Sun is about four times that of our earth. Determine the orbital period for this hypothetical planet.
   (a) 0.25 year
   (b) 2.5 years
   (c) 4 years
   (d) 8 years
   (e) 16 years

25. Consider a satellite in a circular orbit around the earth. If it were at an altitude equal to twice the radius of the earth, $2R_E$, how would its speed $v$ be related to the earth's radius $R_E$, and the magnitude $g$ of the acceleration due to gravity on the earth's surface?
   (a) $v^2 = \dfrac{gR_E}{9}$
   (b) $v^2 = 2gR_E$
   (c) $v^2 = \dfrac{gR_E}{3}$
   (d) $v^2 = \dfrac{gR_E}{4}$
   (e) $v^2 = \dfrac{gR_E}{2}$

26. A satellite in orbit around the earth has a period of one hour. An identical satellite is placed in an orbit having a radius which is nine times larger than the first satellite. What is the period of the second satellite?
   (a) 0.04 h
   (b) 3 h
   (c) 4 h
   (d) 9 h
   (e) 27 h

27. The orbital radius about the Sun of Saturn is about 10 times that of Earth. Complete the following statement: The period of Saturn is about
   (a) 6 yr.
   (b) 30 yr.
   (c) 40 yr.
   (d) 90 yr.
   (e) 160 yr.

## DYNAMICS OF UNIFORM CIRCULAR MOTION

28. An artificial satellite in a circular orbit around the Sun has a period of 8 earth years. Determine the ratio of the satellite's orbital radius about the Sun to the earth's orbital radius about the Sun. Assume that the earth's orbit about the Sun is circular.
    (a) 1
    (b) 2
    (c) 4
    (d) 8
    (e) 23

29. The mass and radius of the moon are $7.4 \times 10^{22}$ kg and $1.7 \times 10^6$ m, respectively. What is the weight of a 1.0-kg object on the surface of the moon?
    (a) 1.0 N
    (b) 1.7 N
    (c) 3.7 N
    (d) 8.8 N
    (e) 9.8 N

30. An object weighs 10 N on the earth's surface. What is the weight of the object on a planet which has one tenth the earth's mass and one half the earth's radius?
    (a) 4 N
    (b) 2 N
    (c) 1 N
    (d) 10 N
    (e) 20 N

*Questions 31 through 33 pertain to the situation described below:*

A 2400-kg satellite is in a circular orbit around a planet. The satellite travels with a constant speed of $6.67 \times 10^3$ m/s.

The radius of the circular orbit is $8.92 \times 10^6$ m.

31. At the instant shown in the figure, which arrow indicates the direction of the net force on the satellite?

    (a) →   (b) ←   (c) ↑   (d) ↓   (e) ↘

32. What is the acceleration of the satellite?
    (a) 2.5 m/s²
    (b) 21 m/s²
    (c) 9.8 m/s²
    (d) 5.0 m/s²
    (e) zero

33. Determine the magnitude of the gravitational force exerted on the satellite by the planet.
    (a) $1.2 \times 10^4$ N
    (b) $2.4 \times 10^4$ N
    (c) $5.0 \times 10^{-3}$ N
    (d) $7.5 \times 10^{-4}$ N
    (e) This cannot be determined since the mass and radius of the planet are not specified.

34. What is the acceleration due to gravity at an altitude of $1.00 \times 10^6$ m above the earth's surface?
    **Note**: the radius of the earth is $6.38 \times 10^6$ m.
    (a) 3.99 m/s²
    (b) 9.80 m/s²
    (c) 5.00 m/s²
    (d) 6.77 m/s²
    (e) 7.32 m/s²

35. The radius of the earth is $6.38 \times 10^6$ m and its mass is $5.98 \times 10^{24}$ kg. What is the acceleration due to gravity at a height of $1.28 \times 10^7$ m above the earth's surface?
   (a) 1.08 m/s²
   (b) 2.15 m/s²
   (c) 9.80 m/s²
   (d) 0.659 m/s²
   (e) 0.114 m/s²

36. A spaceship is in orbit around the earth at an altitude of 12 000 miles. Which one of the following statements best explains why the astronauts experience "weightlessness?"
   (a) The centripetal force of the earth on the astronaut in orbit is zero.
   (b) The pull of the earth on the spaceship is canceled by the pull of the other planets.
   (c) The spaceship is in free fall and its floor cannot press upwards on the astronauts.
   (d) The force of gravity decreases as the inverse square of the distance from the earth's center.
   (e) The force of the earth on the spaceship and the force of the spaceship on the earth cancel because they are equal in magnitude but opposite in direction.

37. The radius of the earth is 6400 km. An incoming meteorite approaches the earth along the trajectory shown. The point C in the figure is 6400 km above the earth's surface. The point A is located at the earth's center. At point C, what acceleration would the meteorite experience due to the earth's gravity?

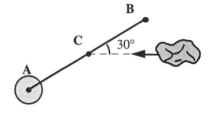

   (a) 9.8 m/s² toward A
   (b) 2.5 m/s² toward A
   (c) 2.5 m/s² toward B
   (d) 5.0 m/s² toward B
   (e) 5.0 m/s² toward A

38. A space station is designed in the shape of a large, uniformly rotating hollow donut. The outer radius of the station is 350 m. With what period must the station rotate so that a person sitting on the outer wall experiences "artificial gravity," i.e. an acceleration of 9.8 m/s²?
   (a) 230 s
   (b) 170 s
   (c) 110 s
   (d) 76 s
   (e) 38 s

### Section *5.7 Vertical Circular Motion

39. A plane is traveling at 200 m/s following the arc of a vertical circle of radius R. At the top of its path, the passengers experience "weightlessness." To one significant figure, what is the value of R?
   (a) 200 m
   (b) 1000 m
   (c) 2000 m
   (d) 4000 m
   (e) 40 000 m

40. A 0.75-kg ball is attached to a 1.0-m rope and whirled in a vertical circle. The rope will break when the tension exceeds 450 N. What is the maximum speed the ball can have at the bottom of the circle without breaking the rope?
   (a) 24 m/s
   (b) 12 m/s
   (c) 32 m/s
   (d) 16 m/s
   (e) 38 m/s

## DYNAMICS OF UNIFORM CIRCULAR MOTION

*Questions 41 and 42 pertain to the situation described below:*

A small car of mass $M$ travels along a straight, horizontal track.

As suggested in the figure, the track then bends into a vertical circle of radius $R$.

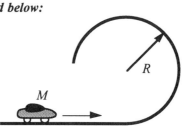

41. What is the minimum acceleration that the car must have at the top of the track if it is to remain in contact with the track?
    (a) 4.9 m/s², downward
    (b) 4.9 m/s², upward
    (c) 9.8 m/s², upward
    (d) 9.8 m/s², downward
    (e) 19.6 m/s², upward

42. Which expression determines the minimum speed that the car must have at the top of the track if it is to remain in contact with the track?
    (a) $v = MgR$
    (b) $v = 2gR$
    (c) $v^2 = 2gR$
    (d) $v^2 = gR$
    (e) $v = gR$

## Additional Problems

*Questions 43 through 46 pertain to the situation described below:*

A 1500-kg car travels at a constant speed of 22 m/s around a circular track which has a radius of 85 m.

43. Which statement is true concerning this car?
    (a) The velocity of the car is changing.
    (b) The car is characterized by constant velocity.
    (c) The car is characterized by constant acceleration.
    (d) The car has a velocity vector that points along the radius of the circle.
    (e) The car has an acceleration vector that is tangent to the circle at all times.

44. What is the magnitude of the acceleration of the car?
    (a) 5.7 m/s²
    (b) 0.26 m/s²
    (c) 9.8 m/s²
    (d) 1.2 m/s²
    (e) zero

45. What is the *average velocity* of the car during one revolution?
    (a) 8.0 m/s
    (b) 12 m/s
    (c) 26 m/s
    (d) 44 m/s
    (e) zero

46. Determine the magnitude of the net force that acts on the car.
    (a) 390 N
    (b) 1800 N
    (c) 8.5 × 10³ N
    (d) 1.5 × 10⁴ N
    (e) zero

47. Jupiter has a mass that is roughly 320 times that of the Earth and a radius equal to 11 times that of the Earth. What is the acceleration due to gravity on the surface of Jupiter?
    (a) 2.7 m/s²
    (b) 9.8 m/s²
    (c) 26 m/s²
    (d) 87 m/s²
    (e) 260 m/s²

*Questions 48 through 50 pertain to the situation described below:*

A rocket orbits a planet in a circular orbit at a constant speed as shown in the drawing.
**Note** the arrows shown:

[1]　　　　[2]　　　　[3]　　　　[4]　　　　[5]

48. At the instant shown in the drawing, which arrow indicates the direction of the acceleration of the rocket?
    (a) 1
    (b) 2
    (c) 3
    (d) 4
    (e) 5

49. At the instant shown in the drawing, which arrow shows the reaction force exerted on the planet by the rocket?
    (a) 1
    (b) 2
    (c) 3
    (d) 4
    (e) 5

50. Suppose that the radius of the circular path is $R$ when the speed of the rocket is $v$ and the acceleration of the rocket has magnitude $a$. If the radius and speed are increased to $2R$ and $2v$ respectively, what is the magnitude of the rocket's subsequent acceleration?
    (a) $\frac{a}{2}$
    (b) $2a$
    (c) $a$
    (d) $4a$
    (e) $8a$

51. The record for the highest speed achieved in a laboratory for a uniformly rotating object was $2.01 \times 10^3$ m/s for a 0.15-m long carbon rod. What was the period of rotation of the rod?
    (a) $7.4 \times 10^{-5}$ s
    (b) $3.1 \times 10^{-4}$ s
    (c) $4.7 \times 10^{-4}$ s
    (d) $5.2 \times 10^{-3}$ s
    (e) $1.3 \times 10^{-3}$ s

*Questions 52 and 53 pertain to the following situation.*

An airplane flying at 115 m/s flying due east makes a gradual turn following a circular path to fly south. The turn takes 15 seconds to complete.

52. What is the radius of the curve that the plane follows in making the turn?
    (a) 280 m
    (b) 350 m
    (c) 830 m
    (d) 1100 m
    (e) 1600 m

53. What is the magnitude of the centripetal acceleration during the turn?
    (a) zero
    (b) 6.9 m/s$^2$
    (c) 8.1 m/s$^2$
    (d) 9.8 m/s$^2$
    (e) 12 m/s$^2$

# CHAPTER 6  Work and Energy

*Section 6.1 Work Done by a Constant Force*
*Section 6.2 The Work-Energy Theorem and Kinetic Energy*

1. In which situation is zero net work done?
   (a) A ball rolls down an inclined plane.
   (b) A physics student stretches a spring.
   (c) A projectile falls toward the earth's surface.
   (d) A box is pulled across a rough floor at constant velocity.
   (e) A child pulls a wagon across a rough surface causing it to accelerate.

2. Work may be expressed using all of the following units *except*:
   (a) N•m          (c) erg          (e) watt
   (b) joule        (d) ft•lb

3. A concrete block is pulled 7.0 m across a frictionless surface by means of a rope. The tension in the rope is 40 N; and the net work done on the block is 247 J.
   What angle does the rope make with the horizontal?
   (a) 28°          (c) 47°          (e) 88°
   (b) 41°          (d) 62°

   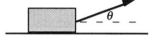

4. A 5.00-kg block of ice is sliding across a frozen pond at 2.00 m/s. A 7.60 N force is applied in the direction of motion. The ice block slides 15.0 m; and then the force is removed. The work done by the applied force is
   (a) −114 J       (c) −735 J       (e) +19.7 J
   (b) +114 J       (d) +735 J

5. A force of magnitude 25 N directed at an angle of 37° above the horizontal moves a 10-kg crate along a horizontal surface at constant velocity. How much work is done by this force in moving the crate a distance of 15 m?
   (a) zero         (c) 40 J         (e) 300 J
   (b) 1.7 J        (d) 98 J

6. A constant force of 25 N is applied as shown to a block which undergoes a displacement of 7.5 m to the right along a frictionless surface while the force acts. What is the work done by the force?

   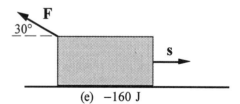

   (a) zero         (c) −94 J        (e) −160 J
   (b) +94 J        (d) +160 J

7. A 1.0-kg ball on the end of a string is whirled at a constant speed of 2.0 m/s in a horizontal circle of radius 1.5 m. What is the work done by the centripetal force during one revolution?
   (a) zero         (c) 6.0 J        (e) 33 J
   (b) 2.7 J        (d) 25 J

8. Brenda carries an 8.0-kg suitcase as she walks 25 m along a horizontal walkway to her room at a constant speed of 1.5 m/s. How much work does Brenda do in carrying her suitcase?
   (a) zero
   (b) 40 J
   (c) 200 J
   (d) 300 J
   (e) 2000 J

9. Which one of the following is an example of an object with a non-zero kinetic energy?
   (a) a drum of diesel fuel on a parked truck
   (b) a stationary pendulum
   (c) a satellite in geosynchronous orbit
   (d) a car parked at the top of a hill
   (e) a boulder resting at the bottom of a cliff

10. Which one of the following statements concerning kinetic energy is true?
    (a) It can be measured in watts.
    (b) It is always equal to the potential energy.
    (c) It is always positive.
    (d) It is a quantitative measure of inertia.
    (e) It is directly proportional to velocity.

11. Which one of the following has the largest kinetic energy?
    (a) a raindrop falling
    (b) a woman swimming
    (c) a jet airplane flying at maximum speed
    (d) the earth moving in its orbit around the sun
    (e) the space shuttle orbiting the Earth

12. In which one of the following situations will there be an *increase* in kinetic energy?
    (a) A projectile approaches its maximum height.
    (b) A box is pulled across a rough floor at constant speed.
    (c) A child pushes a merry go round causing it to rotate faster.
    (d) A satellite travels in a circular orbit around the earth at fixed altitude.
    (e) A stone at the end of a string is whirled in a horizontal circle at constant speed.

13. A 1500-kg car travels at a constant speed of 22 m/s around a circular track which is 80 m across. What is the kinetic energy of the car?
    (a) zero
    (b) $3.6 \times 10^5$ J
    (c) $3.3 \times 10^4$ J
    (d) $1.6 \times 10^4$ J
    (e) $7.2 \times 10^5$ J

14. A car with kinetic energy $8 \times 10^6$ J travels along a horizontal road. How much work is required to stop the car in 10 s?
    (a) zero
    (b) $8 \times 10^4$ J
    (c) $8 \times 10^5$ J
    (d) $8 \times 10^6$ J
    (e) $8 \times 10^7$ J

15. How much energy is dissipated in braking a 1000-kg car to a stop from an initial speed of 20 m/s?
    (a) 20 000 J
    (b) 200 000 J
    (c) 400 000 J
    (d) 800 000 J
    (e) 10 000 J

16. A 10.0-g bullet traveling horizontally at 755 m/s strikes a stationary target and stops after penetrating 14.5 cm into the target. What is the average force of the target on the bullet?
    (a) $1.97 \times 10^4$ N
    (b) $2.07 \times 10^5$ N
    (c) $6.26 \times 10^3$ N
    (d) $3.13 \times 10^4$ N
    (e) $3.93 \times 10^4$ N

62  WORK AND ENERGY

17. A car is traveling at 7.0 m/s when the driver applies the brakes. The car moves 1.5 m before it comes to a complete stop. If the car had been moving at 14 m/s, how far would it have continued to move after the brakes were applied? Assume the braking force is constant.
    (a) 1.5 m
    (b) 3.0 m
    (c) 4.5 m
    (d) 6.0 m
    (e) 7.5 m

18. Use the work-energy theorem to find the force required to accelerate an electron ($m = 9.11 \times 10^{-31}$ kg) from rest to a speed of $1.50 \times 10^7$ m/s in a distance of 0.0125 m.
    (a) $8.20 \times 10^{-15}$ N
    (b) $5.47 \times 10^{-22}$ N
    (c) $8.20 \times 10^{-17}$ N
    (d) $1.64 \times 10^{-14}$ N
    (e) $3.56 \times 10^{-19}$ N

*Section 6.3 Gravitational Potential Energy*
*Section 6.4 Conservative Forces and Nonconservative Forces*

19. In which one of the following systems is there a *decrease* in gravitational potential energy?
    (a) a boy stretches a horizontal spring
    (b) a girl jumps down from a bed
    (c) a crate rests at the bottom of an inclined plane
    (d) a car ascends a steep hill
    (e) water is forced upward through a pipe

20. An elevator supported by a single cable descends a shaft at a constant speed. The only forces acting on the elevator are the tension in the cable and the gravitational force. Which one of the following statements is true?
    (a) The magnitude of the work done by the tension force is larger than that done by the gravitational force.
    (b) The magnitude of the work done by the gravitational force is larger than that done by the tension force.
    (c) The work done by the tension force is zero.
    (d) The work done by the gravitational force is zero.
    (e) The net work done by the two forces is zero.

21. A 40-kg block is lifted vertically 20 meters from the surface of the earth. To one significant figure, what is the change in the gravitational potential energy of the block?
    (a) +800 J
    (b) –800 J
    (c) +8000 J
    (d) –8000 J
    (e) zero

22. Two balls of equal size are dropped from the same height from the roof of a building. One ball has twice the mass of the other. When the balls reach the ground, how do the kinetic energies of the two balls compare?
    (a) The lighter one has one fourth as much kinetic energy as the other.
    (b) The lighter one has one half as much kinetic energy as the other.
    (c) The lighter one has the same kinetic energy as the other.
    (d) The lighter one has twice as much kinetic energy as the other.
    (e) The lighter one has four times as much kinetic energy as the other.

23. A 1500-kg elevator moves upward with constant speed through a vertical distance of 25 m. How much work was done by the tension in the cable?
    (a) 990 J
    (b) 8100 J
    (c) 140 000 J
    (d) 370 000 J
    (e) 430 000 J

24. A 12-kg crate is pushed up an incline from point A to point B as shown in the figure. What is the change in the gravitational potential energy of the crate?
    (a) +590 J
    (b) −590 J
    (c) +1200 J
    (d) −1200 J
    (e) +60 J

25. A woman stands on the edge of a cliff and throws a stone *vertically downward* with an initial speed of 10 m/s. The instant before the stone hits the ground below, it has 450 J of kinetic energy. If she were to throw the stone *horizontally outward* from the cliff with the same initial speed of 10 m/s, how much kinetic energy would it have just before it hits the ground?
    (a) 50 J
    (b) 100 J
    (c) 450 J
    (d) 800 J
    (e) 950 J

26. A woman pulls a crate up a rough inclined plane at constant speed. Which one of the following statements concerning this situation is false?
    (a) The gravitational potential energy of the crate is increasing.
    (b) The net work done by all the forces acting on the crate is zero.
    (c) The work done on the crate by the normal force of the plane is zero.
    (d) The woman does "positive" work in pulling the crate up the incline.
    (e) The work done on the object by gravity is zero.

27. A block is dropped from a high tower and is falling freely under the influence of gravity. Which one of the following statements is true concerning this situation? Neglect air resistance.
    (a) As the block falls, the net work done by all of the forces acting on the block is zero.
    (b) The kinetic energy increases by equal amounts over equal distances.
    (c) The kinetic energy of the block increases by equal amounts in equal times.
    (d) The potential energy of the block decreases by equal amounts in equal times.
    (e) The total energy of the block increases by equal amounts over equal distances.

28. A rock is thrown straight up from the earth's surface. Which one of the following statements describes the energy transformation of the rock as it rises? Neglect air resistance.
    (a) The total energy of the rock increases.
    (b) The kinetic energy increases and the potential energy decreases.
    (c) Both the potential energy and the total energy of the rock increase.
    (d) The kinetic energy decreases and the potential energy increases.
    (e) Both the kinetic energy and the potential energy of the rock remain the same.

29. Complete the following statement: A force that acts on an object is said to be *conservative* if
    (a) it obeys Newton's laws of motion.
    (b) it results in a change in the object's kinetic energy.
    (c) it always acts in the direction of motion of the object.
    (d) the work it does on the object is independent of the path of the motion.
    (e) the work it does on the object is equal to the increase in the object's kinetic energy.

30. Which one of the following is an example of a conservative force?
    (a) tension
    (b) normal force
    (c) static frictional force
    (d) motor propulsion force
    (e) elastic spring force

## Section 6.5 Conservation of Mechanical Energy

31. A helicopter ($m = 1250$ kg) is cruising at a speed of 25.0 m/s at an altitude of 185 m. What is the total mechanical energy of the helicopter?
    (a) $3.91 \times 10^5$ J
    (b) $2.66 \times 10^6$ J
    (c) $2.27 \times 10^6$ J
    (d) $6.18 \times 10^5$ J
    (e) $1.88 \times 10^6$ J

*Questions 32 and 33 pertain to the situation described below:*

A 2.0-kg projectile is fired with initial velocity components $v_{ox} = 30$ m/s and $v_{oy} = 40$ m/s from a point on the earth's surface. Neglect any effects due to air resistance.

32. What is the kinetic energy of the projectile when it reaches the highest point in its trajectory?
    (a) zero
    (b) 900 J
    (c) 1600 J
    (d) 2500 J
    (e) 4900 J

33. How much work was done in firing the projectile?
    (a) 900 J
    (b) 1600 J
    (c) 2500 J
    (d) 4900 J
    (e) 9800 J

34. A roller-coaster car is moving at 20 m/s along a straight horizontal track. What will its speed be after climbing the 15-m hill shown in the figure if friction is ignored?

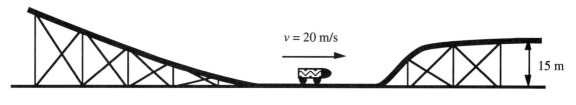

   (a) 17 m/s
   (b) 7 m/s
   (c) 5 m/s
   (d) 10 m/s
   (e) 14 m/s

35. Two boxes are connected to each other as shown. The system is released from rest and the 1.00-kg box falls through a distance of 1.00 m. The surface of the table is frictionless.
What is the kinetic energy of box B just before it reaches the floor?
    (a) 2.45 J
    (b) 4.90 J
    (c) 9.80 J
    (d) 29.4 J
    (e) 39.2 J

36. A 3.0-kg block falls from rest through a distance of 6.0 m in an evacuated tube near the surface of the earth. What is its speed after it has fallen the 6.0 m distance?
    (a) 8.0 m/s
    (b) 11 m/s
    (c) 13 m/s
    (d) 26 m/s
    (e) 120 m/s

37. A bicyclist is traveling at a speed of 20.0 m/s as he approaches the bottom of a hill. He decides to coast up the hill and stops upon reaching the top. Determine the vertical height of the hill.
    (a) 28.5 m
    (b) 3.70 m
    (c) 11.2 m
    (d) 40.8 m
    (e) 20.4 m

38. A skier leaves the top of a slope with an initial speed of 5.0 m/s. Her speed at the bottom of the slope is 13 m/s. What is the height of the slope?
    (a) 1.1 m
    (b) 4.6 m
    (c) 6.4 m
    (d) 7.3 m
    (e) 11 m

39. A roller coaster starts from rest at the top of a hill 18 m high as shown. The car travels to the bottom of the hill and continues up the next hill which is 10.0 m high.

    How fast is the car moving at the top of the 10.0 m hill if friction is ignored?
    (a) 6.4 m/s
    (b) 8.1 m/s
    (c) 13 m/s
    (d) 18 m/s
    (e) 27 m/s

40. An engineer is asked to design a playground slide such that the speed a child reaches at the bottom does not exceed 6.0 m/s. Determine the maximum height that the slide can be.
    (a) 1.8 m
    (b) 2.9 m
    (c) 3.2 m
    (d) 4.5 m
    (e) 14 m

*Questions 41 and 42 pertain to the statement and figure below:*

A block of mass $m$ is released from rest at a height $R$ above a horizontal surface. The acceleration due to gravity is $g$.

The block slides along the inside of a frictionless circular hoop of radius $R$.

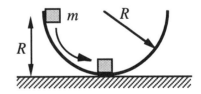

41. Which one of the following expressions determines the speed of the mass at the bottom of the hoop?
    (a) zero
    (b) $v = mgR$
    (c) $v = mg/(2R)$
    (d) $v^2 = g^2/R$
    (e) $v^2 = 2gR$

42. What is the magnitude of the normal force exerted on the block by the hoop when the block has reached the bottom?
    (a) zero
    (b) $mg^2/R$
    (c) $mg$
    (d) $2\,mg$
    (e) $3mg$

43. A care package is dropped from rest from a helicopter hovering 25 m above the ground. What is the speed of the package just before it reaches the ground? Neglect air resistance.
    (a) 22 m/s
    (b) 16 m/s
    (c) 12 m/s
    (d) 8.0 m/s
    (e) 5.0 m/s

## Section 6.6 Nonconservative Forces and the Work-Energy Theorem

● 44. A physics student shoves a 0.50-kg block from the bottom of a frictionless 30.0° inclined plane. The student performs 4.0 J of work and the block slides a distance s along the incline before it stops. Determine the value of s.

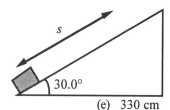

(a) 8.0 cm
(b) 16 cm
(c) 82 cm
(d) 160 cm
(e) 330 cm

● 45. An automobile approaches a barrier at a speed of 20 m/s along a level road. The driver locks the brakes at a distance of 50m from the barrier. What minimum coefficient of kinetic friction is required to stop the automobile before it hits the barrier?
(a) 0.4
(b) 0.5
(c) 0.6
(d) 0.7
(e) 0.8

*Questions 46 and 47 pertain to the situation described below.*

A 9.0-kg box of oranges slides from rest down a frictionless incline from a height of 5.0 m. A constant frictional force, introduced at point **A**, brings the block to rest at point **B**, 19 m to the right of point **A**.

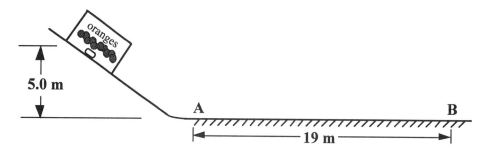

◐ 46. What is the speed of the block just before it reaches point A?
(a) 98 m/s
(b) 21 m/s
(c) 9.9 m/s
(d) 5.7 m/s
(e) 4.4 m/s

● 47. What is the coefficient of kinetic friction, $\mu_k$, of the surface from A to B?
(a) 0.11
(b) 0.26
(c) 0.33
(d) 0.47
(e) 0.52

## Section 6.7 Power

○ 48. A car with kinetic energy $8 \times 10^6$ J travels along a horizontal road. How much power is required to stop the car in 10 s?
(a) zero
(b) $8 \times 10^4$ W
(c) $8 \times 10^5$ W
(d) $8 \times 10^6$ W
(e) $8 \times 10^7$ W

◐ 49. A 51-kg woman runs up a flight of stairs in 5.0 s. Her net upward displacement is 5.0 m. Approximately, what average power did the woman exert while she was running?
(a) 5.0 kW
(b) 1.0 kW
(c) 0.75 kW
(d) 0.50 kW
(e) 0.25 kW

50. What power is needed to lift a 49-kg person a vertical distance of 5.0 m in 20.0 s?
    (a) 12.5 W
    (b) 25 W
    (c) 60 W
    (d) 120 W
    (e) 210 W

51. An escalator is 30.0 meters long and slants at 30.0° relative to the horizontal. If it moves at 1.00 m/s, at what rate does it do work in lifting a 50.0 kg man from the bottom to the top of the escalator?
    (a) 49.3 W
    (b) 98.0 W
    (c) 245 W
    (d) 292 W
    (e) 495 W

52. How much power is needed to lift a 75-kg student vertically upward at a constant speed of 0.33 m/s?
    (a) 12.5 W
    (b) 25 W
    (c) 115 W
    (d) 230 W
    (e) 243 W

53. A warehouse worker uses a forklift to lift a crate of pickles on a platform to a height 2.75 m above the floor. The combined mass of the platform and the crate is 207 kg. If the power expended by the forklift is 1440 W, how long does it take to lift the crate?
    (a) 37.2 s
    (b) 5.81 s
    (c) 3.87 s
    (d) 18.6 s
    (e) 1.86 s

54. A dam is used to block the passage of a river and to generate electricity. Approximately $5.73 \times 10^4$ kg of water fall each second through a height of 19.6 m. If one half of the gravitational potential energy of the water were converted to electrical energy, how much power would be generated?
    (a) $5.50 \times 10^6$ W
    (b) $2.70 \times 10^9$ W
    (c) $1.10 \times 10^7$ W
    (d) $1.35 \times 10^9$ W
    (e) $1.25 \times 10^5$ W

55. The amount on energy needed to power a 0.10 kW bulb for one minute would be just sufficient to lift a 1.0-kg object through a vertical distance of
    (a) 12 m
    (b) 75 m
    (c) 100 m
    (d) 120 m
    (e) 610 m

## Section 6.9 Work Done by a Variable Force

56. The graph shows the force component along the displacement as a function of the magnitude of the displacement. Determine the work done by the force during the interval from 2 to 10 m.
    (a) 140 J
    (b) 190 J
    (c) 270 J
    (d) 450 J
    (e) 560 J

## Additional Problems

57. Which one of the following is *not* a unit of energy?
    (a) foot • pound
    (b) kilowatt • hour
    (c) newton • meter
    (d) watt
    (e) joule

68  WORK AND ENERGY

*Questions 58 and 59 pertain to the situation described below:*

A 0.50-kg ball on the end of a rope is moving in a vertical circle of radius 3.0 m near the surface of the earth where the acceleration due to gravity, g, is 9.8 m/s².

Point **A** is at the top of the circle; **C** is at the bottom. Points **B** and **D** are exactly halfway between **A** and **C**.

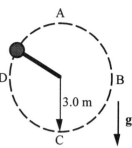

58. Which one of the following statements concerning the tension in the rope is true?
    (a) It is smallest at **A**.
    (b) It is smallest at **C**.
    (c) It is smallest at both **B** and **D**.
    (d) It is the same at **A** and **C**.
    (e) It is the same at **A**, **B**, **C**, and **D**.

59. The ball moves on the circle from A to C under the influence of gravity alone. If the kinetic energy of the ball is 35 J at A, what is its kinetic energy at C?
    (a) zero
    (b) 29 J
    (c) 35 J
    (d) 44 J
    (e) 64 J

60. A motorist driving a 1000-kg car wishes to increase her speed from 20 m/s to 30 m/s in 5 s. Determine the horsepower required to do this. Neglect friction.
    (a) 20 hp
    (b) 30 hp
    (c) 70 hp
    (d) 80 hp
    (e) 90 hp

61. A top fuel dragster with a mass of 500.0 kg starts from rest and completes a quarter mile (402 m) race in a time of 5.0 s. The dragster's final speed is 130 m/s. Neglecting friction, what average power was needed to produce this final speed?
    (a) 140 hp
    (b) 750 hp
    (c) 1100 hp
    (d) $2.7 \times 10^4$ hp
    (e) $8.5 \times 10^5$ hp

*Questions 62 through 66 pertain to the situation described below:*

A 10.0-kg crate slides along a horizontal frictionless surface at a constant speed of 4.0 m/s. The crate then slides down a frictionless incline and across a second horizontal surface as shown in the figure.

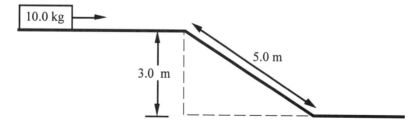

62. What is the kinetic energy of the crate as it slides on the upper surface?
    (a) 30 J
    (b) 80 J
    (c) 140 J
    (d) 290 J
    (e) 490 J

TEST BANK　　　　　　　　　　　　Chapter 6　　69

63. While the crate slides along the upper surface, how much gravitational potential energy does it have compared to what it would have on the lower surface?
    (a) 30 J
    (b) 80 J
    (c) 140 J
    (d) 290 J
    (e) 490 J

64. What is the speed of the crate when it arrives at the lower surface?
    (a) 7.7 m/s
    (b) 8.6 m/s
    (c) 59 m/s
    (d) 75 m/s
    (e) 98 m/s

65. What is the kinetic energy of the crate as it slides on the lower surface?
    (a) 290 J
    (b) 320 J
    (c) 370 J
    (d) 490 J
    (e) 570 J

66. What minimum coefficient of kinetic friction is required to bring the crate to a stop over a distance of 5.0 m along the lower surface?
    (a) 0.30
    (b) 0.32
    (c) 0.60
    (d) 0.66
    (e) 0.76

*Questions 67 and 68 pertain to the situation described below:*

A rope exerts a force **F** on a 20.0-kg crate. The crate starts from rest and accelerates upward at 5.00 m/s$^2$ near the surface of the earth.

67. What is the kinetic energy of the crate when it is 4.0 m above the floor?
    (a) 400 J
    (b) 250 J
    (c) 116 J
    (d) 704 J
    (e) 1180 J

68. How much work was done by the force F in raising the crate 4.0 m above the floor?
    (a) 399 J
    (b) 250 J
    (c) 116 J
    (d) 704 J
    (e) 1180 J

*Questions 69 through 71 pertain to the statement and diagram below:*

A 300-N force accelerates a 50.0-kg crate from rest along a horizontal frictionless surface for a distance of 20.0 m as shown in the figure.

69. What is the final speed of the crate?
    (a) 5.0 m/s
    (b) 15 m/s
    (c) 60 m/s
    (d) 120 m/s
    (e) 240 m/s

70. How much work is done on the crate?
    (a) 160 J
    (b) 750 J
    (c) 1000 J
    (d) 6000 J
    (e) 15 000 J

71. What coefficient of friction would be required to keep the crate moving at constant speed under the action of the 300-N force?
    (a) 0.30
    (b) 0.32
    (c) 0.52
    (d) 0.61
    (e) 0.98

# CHAPTER 7     Impulse and Momentum

## Section 7.1 The Impulse Momentum Theorem

1. Which one of the following is true concerning momentum?
   (a) Momentum is a force.
   (b) Momentum is a scalar quantity.
   (c) The SI unit of momentum is kg • m$^2$/s.
   (d) The momentum of an object is always positive.
   (e) Momentum and impulse are measured in the same units.

2. A rock is dropped from a high tower and falls freely under the influence of gravity. Which one of the following statements is true concerning the rock as it falls?
   (a) It will gain an equal amount of momentum during each second.
   (b) It will gain an equal amount of kinetic energy during each second.
   (c) It will gain an equal amount of speed for each meter through which it falls.
   (d) It will gain an equal amount of momentum for each meter through which it falls.
   (e) The amount of momentum it gains will be proportional to the amount of potential energy that it loses.

3. A stunt person jumps from the roof of a tall building, but no injury occurs because the person lands on a large, air-filled bag. Which one of the following best describes why no injury occurs?
   (a) The bag provides the necessary force to stop the person.
   (b) The bag reduces the impulse to the person.
   (c) The bag increases the amount of time the force acts on the person and reduces the change in momentum.
   (d) The bag decreases the amount of time during which the momentum is changing and reduces the average force on the person.
   (e) The bag increases the amount of time during which the momentum is changing and reduces the average force on the person.

4. A 1.0-kg ball has a velocity of 12 m/s downward just before it strikes the ground and bounces up with a velocity of 12 m/s upward. What is the change in momentum of the ball?
   (a) zero
   (b) 12 kg • m/s, downward
   (c) 12 kg • m/s, upward
   (d) 24 kg • m/s, downward
   (e) 24 kg • m/s, upward

5. A projectile is launched with 200 kg • m/s of momentum and 1000 J of kinetic energy. What is the mass of the projectile?
   (a) 5 kg
   (b) 10 kg
   (c) 20 kg
   (d) 40 kg
   (e) 50 kg

6. A 0.1-kg steel ball is dropped straight down onto a hard horizontal floor and bounces straight up. Its speed just before *and just after* impact with the floor is 10 m/s. Determine the magnitude of the impulse delivered to the floor by the steel ball.
   (a) zero
   (b) 1 N • s
   (c) 2 N • s
   (d) 10 N • s
   (e) 100 N • s

7. A machine gun fires 50-g bullets at the rate of 4 bullets per second. The bullets leave the gun at a speed of 1000 m/s. What is the average recoil force experienced by the machine gun?
   (a) 10 N
   (b) 20 N
   (c) 100 N
   (d) 200 N
   (e) 1000 N

# 72  IMPULSE AND MOMENTUM

8. A 0.065-kg tennis ball moving to the right with a speed of 15 m/s is struck by a tennis racket, causing it to move to the left with a speed of 15 m/s. If the ball remains in contact with the racquet for 0.020 s, what is the magnitude of the average force experienced by the ball?
   (a) zero
   (b) 98 N
   (c) 160 N
   (d) $1.6 \times 10^5$ N
   (e) $9.8 \times 10^4$ N

9. A baseball of mass $m$ is struck by a bat so that it acquires a speed $v$. If $t$ represents the duration of the collision between the bat and the ball, which expression determines the magnitude of the average force exerted on the ball?
   (a) $(1/2)mv^2$
   (b) $mvt$
   (c) $(1/2)mv^2 t$
   (d) $mt^2/(2v)$
   (e) $mv/t$

10. An airplane is traveling at 225 m/s when it strikes a weather balloon ($m = 1.82$ kg), which can be considered to be at rest relative to the ground below. After the collision, the balloon is caught on the fuselage and is traveling with the airplane. The collision takes place over a time interval of $4.44 \times 10^{-3}$ s. What is the average force that the balloon exerts on the airplane?
    (a) 415 N
    (b) $2.78 \times 10^4$ N
    (c) $9.22 \times 10^4$ N
    (d) $4.61 \times 10^5$ N
    (e) $5.07 \times 10^6$ N

11. While a car is stopped at a traffic light in a storm, raindrops strike the roof of the car. The area of the roof is 5.0 m$^2$. Each raindrop has a mass of $3.7 \times 10^{-4}$ kg and speed of 2.5 m/s before impact and has zero speed after the impact. If, on average at a given time, 150 raindrops strike each square meter, what is the impulse of the rain striking the car?
    (a) 0.69 N·s
    (b) 0.046 N·s
    (c) 0.14 N·s
    (d) 11 N·s
    (e) 21 N·s

12. A bat strikes a 0.050-kg baseball so that its velocity changes by +30 m/s in 0.10 s. With what average force was the ball struck?
    (a) +15 N
    (b) −15 N
    (c) +300 N
    (d) −300 N
    (e) +150 N

*Questions 13 and 14 pertain to the situation described below:*

A 4.0-kg block slides along a frictionless surface with a constant speed of 5.0 m/s. Two seconds *after* it begins sliding, a horizontal, time-dependent force is applied to the mass. The force is removed eight seconds later. The graph shows how the force on the block varies with time.

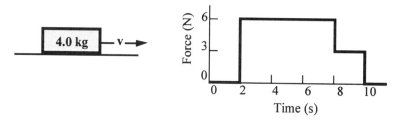

13. What is the magnitude of the total impulse of the force acting on the block?
    (a) 20 N·s
    (b) 42 N·s
    (c) 48 N·s
    (d) 54 N·s
    (e) 60 N·s

14. What, approximately, is the speed of the block at $t = 11$ seconds?
    (a) 5.0 m/s
    (b) 16 m/s
    (c) 25 m/s
    (d) 65 m/s
    (e) 75 m/s

15. The head of a hammer ($m = 1.5$ kg) moving at 4.5 m/s strikes a nail and bounces back with the same speed after an elastic collision lasting 0.075 s. What is the magnitude of the average force the hammer exerts on the nail?
    (a) 6.8 N
    (b) 60 N
    (c) 90 N
    (d) 180 N
    (e) 240 N

16. A football player kicks a 0.41-kg football initially at rest; and the ball flies through the air. If the kicker's foot was in contact with the ball for 0.051 s and the ball's initial speed after the collision is 21 m/s, what was the magnitude of the average force on the football?
    (a) 9.7 N
    (b) 46 N
    (c) 81 N
    (d) 170 N
    (e) 210 N

## Section 7.2 The Principle of Conservation of Linear Momentum

17. In which one of the following situations is linear momentum *not conserved*?
    (a) A bomb suspended by a string explodes into one hundred fragments.
    (b) A bowling ball collides with ten pins.
    (c) A golf ball is struck by a club.
    (d) An astronaut floating in space throws a hammer away and subsequently moves in the opposite direction.
    (e) A tree limb is struck by lightning and falls to the ground.

18. A stationary bomb explodes in gravity-free space breaking into a number of small fragments. Which one of the following statements concerning this event is true?
    (a) Kinetic energy is conserved in this process.
    (b) The fragments must have equal kinetic energies.
    (c) The sum of the kinetic energies of the fragments must be zero.
    (d) The vector sum of the linear momenta of the fragments must be zero.
    (e) The velocity of any one fragment must be equal to the velocity of any other fragment.

19. An object of mass $3m$, initially at rest, explodes breaking into two fragments of mass $m$ and $2m$ respectively. Which one of the following statements concerning the fragments *after the explosion* is true?
    (a) They may fly off at right angles.
    (b) They may fly off in the same direction.
    (c) The smaller fragment will have twice the speed of the larger fragment.
    (d) The larger fragment will have twice the speed of the smaller fragment.
    (e) The smaller fragment will have four times the speed of the larger fragment.

20. A sled of mass $m$ is coasting on the icy surface of a frozen river. While it is passing under a bridge, a package of equal mass $m$ is dropped straight down and lands on the sled (without causing any damage). The sled plus the added load then continue along the original line of motion. How does the kinetic energy of the sled + load compare with the original kinetic energy of the sled?
    (a) It is 1/4 the original kinetic energy of the sled.
    (b) It is 1/2 the original kinetic energy of the sled.
    (c) It is 3/4 the original kinetic energy of the sled.
    (d) It is the same as the original kinetic energy of the sled.
    (e) It is twice the original kinetic energy of the sled.

74  IMPULSE AND MOMENTUM

21. A 50-kg toboggan is coasting on level snow. As it passes beneath a bridge, a 20-kg parcel is dropped straight down and lands in the toboggan. If $(KE)_1$ is the original kinetic energy of the toboggan and $(KE)_2$ is the kinetic energy after the parcel has been added, what is the ratio $(KE)_2/(KE)_1$.
    (a) 0.36
    (b) 0.60
    (c) 0.71
    (d) 1.00
    (e) 1.67

22. A bullet of mass $m$ is fired at speed $v_0$ into a wooden block of mass $M$. The bullet instantaneously comes to rest in the block. The block with the embedded bullet slides along a horizontal surface with a coefficient of kinetic friction $\mu$.

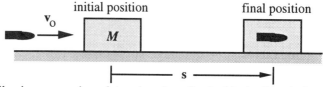

Which one of the following expressions determines how far the block slides before it comes to rest (the magnitude of displacement **s** in the figure)?

(a) $\dfrac{mv_0^2}{M\mu g}$

(b) $\left(\dfrac{m}{m+M}\right)\dfrac{v_0^2}{\mu g}$

(c) $\left(\dfrac{m}{m+M}\right)^2 \dfrac{v_0^2}{2\mu g}$

(d) $\left(\dfrac{m}{m+M}\right)^2 \sqrt{\dfrac{v_0^2}{2\mu g}}$

(e) $\dfrac{v_0^2}{\mu g}$

23. A 100-kg cannon at rest contains a 10-kg cannon ball. When fired, the cannon ball leaves the cannon with a speed of 90 m/s. What is the recoil speed of the cannon?
    (a) 4.5 m/s
    (b) 9 m/s
    (c) 45 m/s
    (d) 90 m/s
    (e) zero

24. An 80-kg astronaut carrying a 20-kg tool kit is initially drifting toward a stationary space shuttle at a speed of 2 m/s. If she throws the tool kit toward the shuttle with a speed of 6 m/s as seen from the shuttle, her final speed is
    (a) 1 m/s toward the shuttle.
    (b) 1 m/s away from the shuttle.
    (c) 2 m/s toward the shuttle.
    (d) 4 m/s toward the shuttle.
    (e) 6 m/s away from the shuttle.

*Section 7.3 Collisions in One Dimension*
*Section 7.4 Collisions in Two Dimensions*

25. Different types of collisions between interacting bodies are *categorized* on the basis of
    (a) kinetic energy conservation.
    (b) mechanical energy conservation.
    (c) linear momentum conservation.
    (d) the magnitude of the forces involved.
    (e) the temporal duration of the collision.

26. Two objects of equal mass traveling toward each other with equal speeds undergo a head on collision. Which one of the following statements concerning their velocities *after the collision* is *necessarily* true?
    (a) They will exchange velocities.
    (b) Their velocities will be reduced.
    (c) Their velocities will be unchanged.
    (d) Their velocities will be zero.
    (e) Their velocities may be zero.

27. Complete the following statement: A collision is *elastic* if
    (a) the final velocities are zero.
    (b) the objects stick together.
    (c) the final kinetic energy is zero.
    (d) the final momentum is zero.
    (e) the total kinetic energy is conserved.

28. Which one of the following is characteristic of an *inelastic collision*?
    (a) Total mass is not conserved.
    (b) Total energy is not conserved.
    (c) Linear momentum is not conserved.
    (d) Kinetic energy is not conserved.
    (e) The change in momentum is less than the total impulse.

29. Complete the following statement: Momentum will be conserved in a two-body collision *only if*
    (a) both bodies come to rest.
    (b) the collision is perfectly elastic.
    (c) the kinetic energy of the system is conserved.
    (d) the net external force acting on the two-body system is zero.
    (e) the internal forces of the two body system cancel in action-reaction pairs.

30. Two cars of equal mass collide on a horizontal frictionless surface. Before the collision, car **A** is at rest while car **B** has a constant velocity of 12 m/s. After the collision the two bodies are stuck together. What is the speed of the composite body (**A** + **B**) after the collision?
    (a) 3 m/s
    (b) 6 m/s
    (c) 12 m/s
    (d) 24 m/s
    (e) 30 m/s

31. A stone of mass 2 kg falls 100 meters near the surface of the earth. It strikes the ground *without any rebound* thereby making a perfectly *inelastic* collision with the earth. Approximately how much kinetic energy is transferred to the earth in this process?
    (a) zero
    (b) 200 J
    (c) 2000 J
    (d) 10 000 J
    (e) 20 000 J

32. A tennis ball has a velocity of 12 m/s downward just before it strikes the ground and bounces up with an velocity of 12 m/s upward. Which statement is true concerning this situation?
    (a) The momentum of the ball and the momentum of the earth both change.
    (b) Neither the momentum of the ball nor the momentum of the earth changes.
    (c) The momentum of the ball is changed; the momentum of the earth is not changed.
    (d) The momentum of the ball is unchanged; the momentum of the earth is changed.
    (e) Both the momentum and the kinetic energy of the ball change because of the collision.

33. A 3.0-kg cart moving to the right with a speed of 1.0 m/s has a head-on collision with a 5.0-kg cart that is initially moving to the left with a speed of 2 m/s. After the collision, the 3.0-kg cart is moving to the left with a speed of 1 m/s. What is the final velocity of the 5.0-kg cart?
    (a) zero
    (b) 0.8 m/s to the right
    (c) 0.8 m/s to the left
    (d) 2.0 m/s to the right
    (e) 2.0 m/s to the left

34. A 1000-kg car traveling east at 20 m/s collides with a 1500-kg car traveling west at 10 m/s. The cars stick together after the collision. What is their common velocity after the collision?
    (a) 16 m/s, east
    (b) 6 m/s, west
    (c) 4 m/s, east
    (d) 2 m/s, east
    (e) 1 m/s, west

76  IMPULSE AND MOMENTUM

35. A 0.050-kg lump of clay moving horizontally at 12 m/s strikes *and sticks to* a stationary 0.10-kg cart which can move on a frictionless air track. Determine the speed of the cart and clay after the collision.
    (a) 2 m/s
    (b) 4 m/s
    (c) 6 m/s
    (d) 8 m/s
    (e) 12 m/s

36. A 0.10-kg cart traveling in the positive $x$ direction at 10.0 m/s collides with a 0.30-kg cart at rest. The collision is elastic. What is the velocity of the 0.10-kg cart after the collision?
    (a) +2.5 m/s
    (b) −2.5 m/s
    (c) +5 m/s
    (d) −5 m/s
    (e) +3.3 m/s

37. A 50.0-kg boy runs at a speed of 10.0 m/s and jumps onto a cart as shown in the figure. The cart is initially at rest.

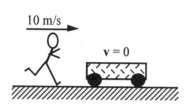
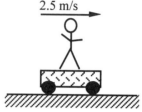

If the speed of the cart with the boy on it is 2.50 m/s, what is the mass of the cart?
    (a) 150 kg
    (b) 175 kg
    (c) 210 kg
    (d) 260 kg
    (e) 300 kg

38. A 0.015-kg marble moving to the right at 0.40 m/s has a head-on, elastic collision with a 0.045-kg marble sitting at rest on a smooth, level surface. Which of the following are the correct magnitudes and directions of the velocities of the two marbles after the collision?

| | 0.015 kg marble | 0.045 kg marble |
|---|---|---|
| (a) | 0.15 m/s, left | 0.25 m/s, right |
| (b) | 0.10 m/s, right | 0.30 m/s, right |
| (c) | zero | 0.25 m/s, right |
| (d) | 0.20 m/s, left | 0.20 m/s, right |
| (e) | 0.40 m/sec, left | zero |

39. Two objects constitute an isolated system. In an elastic collision between the two objects, which one of the following statements is a false statement?
    (a) The total kinetic energy is conserved.
    (b) The kinetic energy of each object is the same before and after the collision.
    (c) The total momentum is conserved.
    (d) The magnitude of the force exerted by each object on the other object is equal.
    (e) The total kinetic energy before the collision is equal to the total kinetic energy after the collision.

40. A 35-kg girl is standing near and to the left of a 43-kg boy on the frictionless surface of a frozen pond. The boy tosses a 0.75-kg ice ball to the girl with a horizontal speed of 6.2 m/s. What are the velocities of the boy and the girl immediately after the girl catches the ice ball?

| | girl | boy |
|---|---|---|
| (a) | 0.81 m/s, left | 0.67 m/s, right |
| (b) | 0.17 m/s, left | 0.14 m/s, left |
| (c) | 0.18 m/s, right | 0.13 m/s, left |
| (d) | 0.42 m/s, left | 0.49 m/s, right |
| (e) | 0.13 m/s, left | 0.11 m/s, right |

41. A 2.5-kg ball and a 5.0-kg ball have an elastic collision. Before the collision, the 2.5-kg ball was at rest and the other ball had a speed of 3.5 m/s. What is the kinetic energy of the 2.5-kg ball after the collision?
    (a) 1.7 J
    (b) 3.4 J
    (c) 8.1 J
    (d) 14 J
    (e) 27 J

*Questions 42 and 43 pertain to the following situation.*

A comet fragment of mass $1.96 \times 10^{13}$ kg is moving at $6.50 \times 10^4$ m/s when it crashes into Callisto, a moon of Jupiter. The mass of Callisto is $1.08 \times 10^{23}$ kg. The collision is completely inelastic.

42. Assuming for this calculation that Callisto's initial momentum is zero, what is the recoil speed of Callisto immediately after the collision?
    (a) $3.34 \times 10^{-18}$ m/s
    (b) $1.27 \times 10^{-14}$ m/s
    (c) $3.58 \times 10^{-12}$ m/s
    (d) $6.13 \times 10^{-7}$ m/s
    (e) $1.18 \times 10^{-5}$ m/s

43. How much kinetic energy was released in the collision?
    (a) $8.28 \times 10^{22}$ J
    (b) $3.51 \times 10^{27}$ J
    (c) $7.02 \times 10^{27}$ J
    (d) $4.14 \times 10^{22}$ J
    (e) $1.50 \times 10^{13}$ J

44. Car One is traveling due north and Car Two is traveling due east. After the collision shown, Car One rebounds due south. Which of the lettered arrows is the only one which can represent the final direction of Car Two?

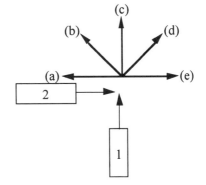

45. A 0.50-kg bomb is sliding along an icy pond (frictionless surface) with a velocity of 2.0 m/s to the west. The bomb explodes into two pieces. After the explosion, a 0.20-kg piece moves south at 4.0 m/s. What are the components of the velocity of the 0.30-kg piece?
    (a) 4.0 m/s north, 0 m/s
    (b) 2.7 m/s north, 3.3 m/s west
    (c) 4.0 m/s north, 2.7 m/s west
    (d) 0 m/s, 2.0 m/s east
    (e) 4.0 m/s north, 2.0 m/s east

46. In the game of billiards, all the balls have approximately the same mass, about 0.17 kg. In the figure, the cue ball strikes another ball such that it follows the path shown. The other ball has a speed of 1.5 m/s immediately after the collision. What is the speed of the cue ball after the collision?

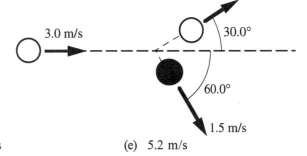

    (a) 1.5 m/s
    (b) 1.8 m/s
    (c) 2.6 m/s
    (d) 4.3 m/s
    (e) 5.2 m/s

## Section 7.5 Center of Mass

47. Which one of the following statements concerning center of mass is true?
    (a) All of an object's mass is located at its center of mass.
    (b) The center of mass of an object must be located within the object.
    (c) The center of mass of a system of objects cannot change even if there are forces acting on the objects.
    (d) The velocity of the center of mass of a system of objects is greatly affected by a collision of objects within the system.
    (e) The velocity of the center of mass of a system of objects is constant when the sum of the external forces acting on the system is zero.

48. The drawing shows two 4.5-kg balls located on the $y$ axis at 1.0 and 9.0 m, respectively; a third ball with a mass 2.3 kg is located at 6.0 m. What is the location of the center of mass of this system?
    (a) 4.8 m
    (b) 5.2 m
    (c) 5.6 m
    (d) 6.0 m
    (e) 6.4 m

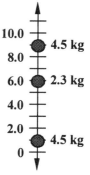

49. A juggler demonstrates his abilities by keeping a 2.3-kg pipe wrench, a 1.5-kg hatchet, and a 1.0-kg hammer flying through the air above his head. The white circles on the graph represent the positions of the center of mass of each of the flying objects at one instant. What are the $x$ and $y$ coordinates of the center of mass for the system of these three objects?

|     | $x$     | $y$     |
|-----|---------|---------|
| (a) | 0.46 m  | 0.47 m  |
| (b) | 0.30 m  | 0.54 m  |
| (c) | 0.42 m  | 0.60 m  |
| (d) | 0.47 m  | 0.26 m  |
| (e) | 0.60 m  | 0.42 m  |

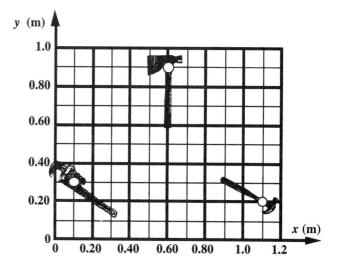

50. During hockey practice, two pucks are sliding across the ice. At one instant, a 0.18-kg puck is moving at 16 m/s while the other puck has a mass of 0.14 kg and a speed of 3.8 m/s. What is the velocity of the center of mass of the two pucks?
    (a) 5.0 m/s
    (b) 7.0 m/s
    (c) 9.0 m/s
    (d) 11 m/s
    (e) 13 m/s

## Additional Problems

*Questions 51 through 53 pertain to the following situation:*

A space vehicle of mass $m$ has a speed $v$. At some instant, it separates into two pieces, each of mass $0.5m$. One of the pieces is at rest just after the separation.

51. Which one of the following statements concerning this situation is true?
    (a) The moving piece has speed 2v.
    (b) This process conserves kinetic energy.
    (c) The piece at rest possesses kinetic energy.
    (d) The process does not conserve total energy.
    (e) This process does not conserve momentum.

52. What is the kinetic energy of the moving piece just after the separation?
    (a) zero
    (b) $(1/4)mv^2$
    (c) $(1/2)mv^2$
    (d) $mv^2$
    (e) $2mv^2$

53. How much work was done by the internal forces that caused the separation?
    (a) zero
    (b) $(1/4)mv^2$
    (c) $(1/2)mv^2$
    (d) $mv^2$
    (e) $2mv^2$

54. A mother is holding her 4.5-kg baby in her arms while riding in a car moving at 22 m/s. The car is involved in a head-on collision and stops within 1.5 seconds. What is the magnitude of the force exerted by the baby on his mother's arms?
    (a) 45 N
    (b) 66 N
    (c) 90 N
    (d) 99 N
    (e) 150 N

*Questions 55 through 57 pertain to the situation described below:*

A 2.0-kg pistol fires a 1.0-g bullet with a muzzle speed of 1000 m/s. The bullet then strikes a 10-kg wooden block resting on a horizontal frictionless surface. The block and the embedded bullet then slide across the surface.

55. What is the kinetic energy of the bullet as it travels toward the block?
    (a) 100 J
    (b) 500 J
    (c) 1000 J
    (d) 5000 J
    (e) 10 000 J

56. The explosive charge in the pistol acts for 0.001 s. What is the average force exerted on the bullet while it is being fired?
    (a) 0.001 N
    (b) 1.0 N
    (c) 100 N
    (d) 500 N
    (e) 1000 N

57. What is the speed of the "bullet + block" system immediately after the bullet is embedded in the block?
    (a) 0.1 m/s
    (b) 10 m/s
    (c) 1000 m/s
    (d) 10 000 m/s
    (e) zero

58. A rocket is launched vertically from rest and burns fuel at a constant rate of 136 kg/s. Exhaust gases are expelled with a speed of $5.25 \times 10^3$ m/s relative to the rocket. What is the magnitude of the thrust?
    (a) $7.14 \times 10^5$ N
    (b) $3.64 \times 10^6$ N
    (c) $2.59 \times 10^{-2}$ N
    (d) 808 N
    (e) 38.6 N

## IMPULSE AND MOMENTUM

59. A 100-kg fisherman and a 500-kg supply crate are on a frozen pond that is essentially frictionless. The man and the crate are initially separated by a distance of 600 meters. The fisherman uses a very light rope to pull the crate closer to him. How far has the man moved when the crate reaches the fisherman?
    (a) zero
    (b) 10 m
    (c) 50 m
    (d) 100 m
    (e) 500 m

*Questions 60 and 61 pertain to the following situation:*

A stationary 4-kg shell explodes into three pieces. Two of the fragments have a mass of 1 kg each and move along the paths shown with a speed of 10 m/s. The third fragment moves upward as shown.

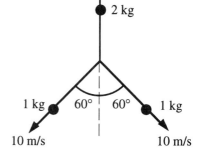

60. What is the speed of the third fragment?
    (a) zero
    (b) 1 m/s
    (c) 5 m/s
    (d) 10 m/s
    (e) 20 m/s

61. What is the speed of the center of mass of this system after the explosion?
    (a) zero
    (b) 1 m/s
    (c) 3 m/s
    (d) 5 m/s
    (e) 7 m/s

# CHAPTER 8  Rotational Kinematics

*Section 8.1 Rotational Motion and Angular Displacement*
*Section 8.2 Angular Velocity and Angular Acceleration*

1. The angular measure *1.0 radian* is equal to
   (a) 0.0175°
   (b) 57.3°
   (c) 1.57°
   (d) 3.14°
   (e) 6.28°

2. A bicycle travels 141 m along a circular track of radius 15 m. What is the angular displacement (in radians) of the bicycle from its starting position?
   (a) 1.0 rad
   (b) 1.5 rad
   (c) 3.0 rad
   (d) 4.7 rad
   (e) 9.4 rad

3. What is the angular speed in *rad/s* of the second hand of a watch?
   (a) $1.7 \times 10^{-3}$ rad/s
   (b) 0.10 rad/s
   (c) 0.02 rad/s
   (d) 6.28 rad/s
   (e) 60 rad/s

4. The earth takes slightly less than one day to complete one rotation about the axis passing through its poles. The actual time is $8.616 \times 10^4$ s. Given this information, what is the angular speed of the earth about its axis?
   (a) $7.292 \times 10^{-5}$ rad/s
   (b) $2.321 \times 10^{-6}$ rad/s
   (c) $9.951 \times 10^{-5}$ rad/s
   (d) $6.334 \times 10^{-4}$ rad/s
   (e) $1.990 \times 10^{-7}$ rad/s

5. A wheel with a radius of 0.10 m is rotating at 35 rev/s and slows down uniformly to 15 rev/s over a 3.0 s interval. What is the angular acceleration of a point on the wheel?
   (a) $-2.0$ rev/s$^2$
   (b) 0.67 rev/s$^2$
   (c) $-6.7$ rev/s$^2$
   (d) 42 rev/s$^2$
   (e) $-17$ rev/s$^2$

*Section 8.3 The Equations of Rotational Kinematics*

6. Which equation is valid *only* when the angular measure is expressed in *radians*?
   (a) $\alpha = \dfrac{\Delta \omega}{\Delta t}$
   (b) $\omega = \dfrac{\Delta \theta}{\Delta t}$
   (c) $\omega^2 = \omega_0^2 + 2\alpha\theta$
   (d) $\omega = \dfrac{v_T}{r}$
   (e) $\theta = \dfrac{1}{2}\alpha t^2 + \omega_0 t$

7. During the spin-dry cycle of a washing machine, the motor slows from 95 rad/s to 30 rad/s while the turning the drum through an angle of 402 radians. What is the magnitude of the angular acceleration of the motor?
   (a) 64 rad/s$^2$
   (b) 32 rad/s$^2$
   (c) 10 rad/s$^2$
   (d) 20 rad/s$^2$
   (e) 1.0 rad/s$^2$

8. A fan rotating with an initial angular velocity of 1000 rev/min is switched off. In 2 seconds, the angular velocity decreases to 200 rev/min. Assuming that the angular acceleration is constant, how many revolutions does the blade undergo during this time?
   (a) 10
   (b) 20
   (c) 100
   (d) 125
   (e) 1200

## ROTATIONAL KINEMATICS

9. An airplane engine starts from rest; and 2 seconds later, it is rotating with an angular speed of 300 rev/min. If the angular acceleration is constant, how many revolutions does the propeller undergo during this time?
   (a) 5
   (b) 10
   (c) 50
   (d) 300
   (e) 600

*Questions 10 through 12 pertain to the statement below:*

A grindstone of radius 4.0 m is initially spinning with a constant angular speed of 8.0 rad/s. The angular speed is then increased to 10 rad/s over the next 4.0 seconds. Assume that the angular acceleration is constant.

10. What is the average angular speed of the grindstone?
    (a) 0.5 rad/s
    (b) 2.0 rad/s
    (c) 4.5 rad/s
    (d) 9.0 rad/s
    (e) 18 rad/s

11. What is the magnitude of the angular acceleration of the grindstone?
    (a) 0.50 rad/s$^2$
    (b) 2.0 rad/s$^2$
    (c) 4.5 rad/s$^2$
    (d) 9.0 rad/s$^2$
    (e) 18 rad/s$^2$

12. Through how many revolutions does the grindstone turn during the 4.0 second interval?
    (a) 0.64
    (b) 3.82
    (c) 4.00
    (d) 5.73
    (e) 36.0

13. A grindstone, initially at rest, is given a constant angular acceleration so that it makes 20.0 revolutions in the first 8.00 s. What is its angular acceleration?
    (a) 0.313 rad/s$^2$
    (b) 0.625 rad/s$^2$
    (c) 2.50 rad/s$^2$
    (d) 1.97 rad/s$^2$
    (e) 3.94 rad/s$^2$

14. A wheel, originally rotating at 126 rad/s undergoes a constant angular deceleration of 5.00 rad/s$^2$. What is its angular speed after it has turned through an angle of 628 radians?
    (a) 15 rad/s
    (b) 19 rad/s
    (c) 98 rad/s
    (d) 121 rad/s
    (e) 150 rad/s

15. A wheel turns through an angle of 188 radians in 8.0 s; and its angular speed at the end of the period is 40 rad/s. If the angular acceleration is constant, what was the angular speed of the wheel at the beginning of the 8.0 s interval?
    (a) 4.8 rad/s
    (b) 7.0 rad/s
    (c) 9.1 rad/s
    (d) 23.5 rad/s
    (e) 32.5 rad/s

16. A stereo turntable rotating at 33.3 rev/min slows down and stops 31 s after the motor is turned off. How many revolutions did the turntable make during this time?
    (a) 8.6
    (b) 17
    (c) 54
    (d) 160
    (e) 1000

17. A roulette wheel with a 1.0 m radius reaches a maximum angular speed of 18 rad/s before it slows to a complete stop 35 revolutions (220 rad) after attaining the maximum speed. How long did it take for the wheel to stop?
    (a) 12 s
    (b) 48 s
    (b) 3.7 s
    (d) 8.8 s
    (e) 24 s

## Section 8.4 Angular Variables and Tangential Variables

18. A 0.254-m diameter circular saw blade rotates at a constant angular speed of 117 rad/s. What is the tangential speed of the tip of a saw tooth at the edge of the blade?
    (a) 29.7 m/s
    (b) 14.9 m/s
    (c) 9.46 m/s
    (d) 7.45 m/s
    (e) 2.17 m/s

19. What is the tangential speed of Nairobi, Kenya, a city near the equator? The earth makes one revolution every 23.93 h and has an equatorial radius of 6380 km.
    (a) 74.0 m/s
    (b) 116 m/s
    (c) 148 m/s
    (d) 232 m/s
    (e) 465 m/s

*Questions 20 and 21 pertain to the following situation:*

On an amusement park ride, passengers are seated in a horizontal circle of radius 7.5 m. The seats begin from rest and are uniformly accelerated for 21 seconds to a maximum rotational speed of 1.4 rad/s.

20. What is the tangential acceleration of the passengers during the first 21 s of the ride?
    (a) 0.067 m/s$^2$
    (b) 0.50 m/s$^2$
    (c) 1.4 m/s$^2$
    (d) 7.5 m/s$^2$
    (e) 11 m/s$^2$

21. What is the instantaneous tangential speed of the passengers 15 s after the acceleration begins?
    (a) 0.067 m/s
    (b) 0.50 m/s
    (c) 1.4 m/s
    (d) 7.5 m/s
    (e) 11 m/s

## Section 8.5 Centripetal Acceleration and Tangential Acceleration

22. A rigid body rotates about a fixed axis with a constant angular acceleration. Which one of the following statements is true concerning the tangential acceleration of any point on the body?
    (a) It is zero.
    (b) It depends on the angular velocity.
    (c) It is equal to the centripetal acceleration.
    (d) It is constant in both magnitude and direction.
    (e) It depends on the change in the angular velocity.

23. Two points are located on a rigid wheel that is rotating with a *decreasing* angular velocity about a fixed axis. Point A is located on the rim of the wheel and point B is halfway between the rim and the axis. Which one of the following statements is true concerning this situation?
    (a) Both points have the same centripetal acceleration.
    (b) Both points have the same tangential acceleration.
    (c) The angular velocity of A is greater than that of B.
    (d) Both points have the same instantaneous angular velocity.
    (e) Each second, A turns through a greater angle than B.

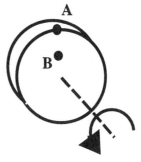

24. A circular disk of radius 0.010 m rotates with a constant angular speed of 5.0 rev/s. What is the acceleration of a point on the edge of the disk?
    (a) 0.31 m/s$^2$
    (b) 1.6 m/s$^2$
    (c) 9.9 m/s$^2$
    (d) 2500 m/s$^2$
    (e) zero

## 84  ROTATIONAL KINEMATICS

25. A circular disk of radius 2.0 m rotates with a constant angular acceleration of 20.0 rad/s². What is the acceleration of a point on the edge of the disk at the instant that its angular speed is 1.0 rev/s?
    (a) 40 m/s²
    (b) 79 m/s²
    (c) 110 m/s²
    (d) 120 m/s²
    (e) zero

*Questions 26 through 28 pertain to the statement below:*

A wheel of radius 0.5 m rotates with a constant angular speed about an axis perpendicular to its center. A point on the wheel that is 0.2 m from the center has a tangential speed of 2 m/s.

26. Determine the angular speed of the wheel.
    (a) 0.4 rad/s
    (b) 2.0 rad/s
    (c) 4.0 rad/s
    (d) 10 rad/s
    (e) 20 rad/s

27. Determine the tangential speed of a point 0.4 m from the center of the wheel.
    (a) 0.4 m/s
    (b) 2.0 m/s
    (c) 4.0 m/s
    (d) 10.0 m/s
    (e) 20 m/s

28. Determine the tangential acceleration of the point that is 0.2 m from the center.
    (a) 0.4 m/s²
    (b) 2.0 m/s²
    (c) 4.0 m/s²
    (d) 10 m/s²
    (e) zero

29. The original Ferris wheel had a radius of 38 m and completed a full revolution ($2\pi$ radians) every two minutes when operating at its maximum speed. If the wheel were uniformly slowed from its maximum speed to a stop in 35 seconds, what would be the magnitude of the tangential acceleration at the outer rim of the wheel during its deceleration?
    (a) 0.0015 m/s²
    (b) 0.056 m/s²
    (c) 0.54 m/s²
    (d) 1.6 m/s²
    (e) 6.8 m/s²

30. A ball attached to a string starts at rest and undergoes a constant angular acceleration as it travels in a horizontal circle of radius 0.30 m. After 0.65 sec, the angular speed of the ball is 9.7 rad/s. What is the tangential acceleration of the ball?
    (a) 4.5 m/s²
    (b) 0.32 m/s²
    (c) 15 m/s²
    (d) 7.6 m/s²
    (e) 28 m/s²

## Section 8.6  Rolling Motion

31. A bicycle has tires of radius 0.35 meters. If the bicycle is traveling at a constant speed of 12 m/s, at approximately what angular speed are the tires rotating?
    (a) 85 rev/min
    (b) 197 rev/min
    (c) 214 rev/min
    (d) 327 rev/min
    (e) 423 rev/min

32. A bicycle with wheels of radius 0.4 m travels on a level road at a speed of 8 m/s. What is the angular speed of the wheels?
    (a) 10 rad/s
    (b) 20 rad/s
    (c) ($\pi / 10$) rad/s
    (d) ($10\pi$) rad/s
    (e) ($20 / \pi$) rad/s

*Questions 33 through 36 pertain to the statement below:*

A 2.0-kg solid disk rolls without slipping on a horizontal surface so that its center proceeds to the right with speed 5.0 m/s.

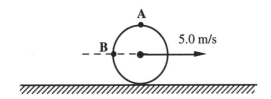

The point **A** is the uppermost point on the disk and the point **B** is along the horizontal line that connects the center of the disk to the rim.

33. Which one of the following statements concerning the direction of the disk's *angular velocity* is true?
    (a) It points to the left.
    (b) It points to the right.
    (c) It points into the paper.
    (d) It points out of the paper.
    (e) It varies from point to point on this disk.

34. What is the *instantaneous* speed of point **A** *with respect to the ground*?
    (a) zero
    (b) 2.5 m/s
    (c) 5.0 m/s
    (d) 7.1 m/s
    (e) 10.0 m/s

35. What is the *instantaneous* speed of point **B** *with respect to the ground*?
    (a) zero
    (b) 5.0 m/s
    (c) 7.1 m/s
    (d) 7.5 m/s
    (e) 10.0 m/s

36. What is the *instantaneous* speed of the point of the disk that makes contact with the surface?
    (a) zero
    (b) 5.0 m/s
    (c) 7.1 m/s
    (d) 7.5 m/s
    (e) 10.0 m/s

37. A car travels in a circular path with constant speed. Which one of the following quantities is constant and *non-zero* for this car?
    (a) linear velocity
    (b) angular velocity
    (c) centripetal acceleration
    (d) angular acceleration
    (e) total acceleration

38. Which statement concerning a wheel undergoing rolling motion is true?
    (a) The angular acceleration of the wheel must be zero.
    (b) The tangential velocity is the same for all points on the wheel.
    (c) The linear velocity for all points on the rim of the wheel is non-zero.
    (d) The tangential velocity is the same for all points on the rim of the wheel.
    (e) There is no slipping at the point where the wheel touches the surface on which it is rolling.

39. A circular hoop rolls without slipping on a flat horizontal surface. Which one of the following is necessarily true?
    (a) All points on the rim of the hoop have the same speed.
    (b) All points on the rim of the hoop have the same velocity.
    (c) Every point on the rim of the wheel has a different velocity.
    (d) All points on the rim of the hoop have acceleration vectors that are tangent to the hoop.
    (e) All points on the rim of the hoop have acceleration vectors that point toward the center of the hoop.

## 86 ROTATIONAL KINEMATICS

40. A bicycle wheel of radius 0.70 m is turning at an angular speed of 6.3 rad/s as it rolls on a horizontal surface without slipping. What is the *linear speed* of the wheel?
   (a)  1.4 m/s
   (b)  28 m/s
   (b)  0.11 m/s
   (d)  4.4 m/s
   (e)  9.1 m/s

### Section 8.7 The Vector Nature of Angular Variables

41. A top is spinning counterclockwise as shown in the figure. It is also moving to the right with a speed $v$. With reference to the side view, what is the direction of the angular velocity?
   (a) downward
   (b) upward
   (c) left
   (d) right
   (e) into the paper

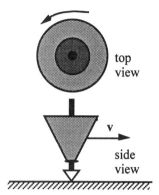

### Additional Problems

*Questions 42 through 44 pertain to the statement below:*

A grindstone of radius 1.0 m is spinning with a constant angular speed of 2.0 rad/s.

42. What is the *tangential speed* of a point on the rim of the grindstone?
   (a) zero
   (b) 0.5 m/s
   (c) 1.0 m/s
   (d) 2.0 m/s
   (e) 4.0 m/s

43. What is the magnitude of the *centripetal acceleration* of a point on the rim of the grindstone?
   (a) zero
   (b) 0.5 m/s$^2$
   (c) 1.0 m/s$^2$
   (d) 2.0 m/s$^2$
   (e) 4.0 m/s$^2$

44. What is the magnitude of the *tangential acceleration* of a point on the rim of the wheel?
   (a) zero
   (b) 0.5 m/s$^2$
   (c) 1.0 m/s$^2$
   (d) 2.0 m/s$^2$
   (e) 4.0 m/s$^2$

*Questions 45 through 48 pertain to the statement below:*

A long thin rod of length $2L$ rotates with a constant angular acceleration of 10 rad/s$^2$ about an axis that is perpendicular to the rod and passes through its center.

45. What is the ratio of the tangential acceleration of a point on the end of the rod to that of a point a distance $L/2$ from the end of the rod?
   (a) 1:1
   (b) 1:2
   (c) 2:1
   (d) 4:1
   (e) 1:4

46. What is the ratio of the centripetal acceleration of a point on the end of the rod to that of a point a distance $L/2$ from the end of the rod?
    - (a) 1:1
    - (b) 1:2
    - (c) 2:1
    - (d) 4:1
    - (e) 1:4

47. What is the ratio of the angular speed (at any instant) of a point on the end of the rod to that of a point a distance $L/2$ from the end of the rod?
    - (a) 1:1
    - (b) 1:2
    - (c) 2:1
    - (d) 4:1
    - (e) 1:4

48. What is the ratio of the tangential speed (at any instant) of a point on the end of the rod to that of a point a distance $L/2$ from the end of the rod?
    - (a) 1:1
    - (b) 1:2
    - (c) 2:1
    - (d) 4:1
    - (e) 1:4

*Questions 49 and 50 pertain to the following situation:*

A bicycle wheel of radius 0.70 m is rolling without slipping on a horizontal surface with an angular speed of 2.0 rev/s when the cyclist begins to uniformly apply the brakes. The bicycle stops in 5.0 s.

49. Through how many revolutions did the wheel turn during the 5.0 seconds of braking?
    - (a) 10 rev
    - (b) 2.0 rev
    - (c) 9.6 rev
    - (d) 5.0 rev
    - (e) 0.4 rev

50. How far did the bicycle travel during the 5.0 seconds of braking?
    - (a) 1.8 m
    - (b) 8.8 m
    - (c) 22 m
    - (d) 42 m
    - (e) 44 m

# CHAPTER 9 — Rotational Dynamics

## Section 9.1 The Effects of Forces and Torques on the Motion of Rigid Objects

1. Complete the following statement: When a net torque is applied to a rigid object, it always produces a
   (a) constant acceleration.
   (b) rotational equilibrium.
   (c) constant angular velocity.
   (d) constant angular momentum.
   (e) change in angular velocity.

2. A wrench is used to tighten a nut as shown in the figure. A 12-N force is applied 7.0 cm from the axis of rotation. What is the torque due to the applied force?
   (a) 0.58 N • m
   (b) 0.84 N • m
   (c) 1.71 N • m
   (d) 14 N • m
   (e) 58 N • m

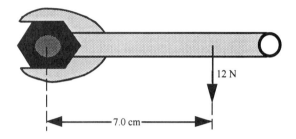

3. A string is tied to a door knob 0.79 m from the hinge as shown in the figure. At the instant shown, the force applied to the string is 5.0 N. What is the torque on the door?
   (a) 3.3 N • m
   (b) 2.2 N • m
   (c) 1.1 N • m
   (d) 0.84 N • m
   (e) 0.40 N • m

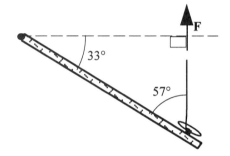

## Section 9.2 Rigid Objects in Equilibrium
## Section 9.3 Center of Gravity

4. Complete the following statement: A body is in *translational* equilibrium
   (a) only if it is at rest.
   (b) only if it is moving with constant velocity.
   (c) only if it is moving with constant acceleration.
   (d) if it is either at rest or moving with constant velocity.
   (e) if it is moving with either constant velocity or constant acceleration.

5. A horizontal, 10-m plank weighs 100 N. It rests on two supports that are placed 1 m from each end as shown in the figure.

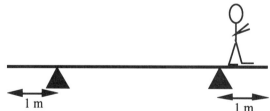

How close to one end can an 800-N person stand without causing the plank to tip?
(a) 0 m
(b) 0.2 m
(c) 0.5 m
(d) 0.6 m
(e) 0.8 m

6. A 3.0-kg ball and a 1.0-kg ball are placed at opposite ends of a *massless* beam so that the system is in equilibrium as shown. **Note**: *The sketch is not drawn to scale.*

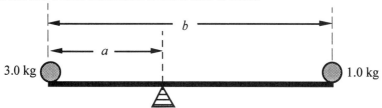

What is the ratio of the lengths $b/a$ ?
(a) 2.0
(b) 2.5
(c) 3.0
(d) 4.0
(e) 5.0

7. A meter stick is pivoted at the 0.50 m line. A 3.0-kg object is hung from the 0.10 m line. Where should a 5.0-kg object be hung to achieve equilibrium?
(a) 0.06 m line
(b) 0.24 m line
(c) 0.56 m line
(d) 0.74 m line
(e) A 5.0-kg object cannot be used to sustain equilibrium in this system.

8. In the drawing at the right, the large wheel has a radius of 8.5 m. A rope is wrapped around the edge of the wheel and a 7.6 kg-box hangs from the rope. A smaller disk of radius 1.9 m is attached to the wheel. A rope is wrapped around the edge of the disk as shown. An axis of rotation passes through the center of the wheel-disk system.

What is the value of the mass M which will prevent the wheel from rotating?

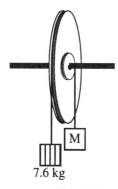

(a) 1.7 kg
(b) 3.8 kg
(c) 12 kg
(d) 34 kg
(e) 46 kg

*Questions 9 and 10 pertain to the situation described below:*

An 80-kg man balances the boy on a teeter-totter as shown.
**Note**: *Ignore the weight of the board.*

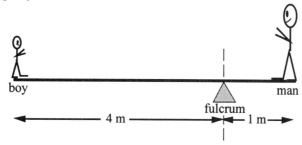

9. What is the approximate mass of the boy?
(a) 10 kg
(b) 20 kg
(c) 40 kg
(d) 45 kg
(e) 50 kg

# 90 ROTATIONAL DYNAMICS

10. What, approximately, is the magnitude of the downward force exerted on the fulcrum?
    (a) zero
    (b) 100 N
    (c) 600 N
    (d) 800 N
    (e) 1000 N

11. Consider four point masses located as shown in the sketch.

    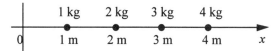

    What is the x coordinate of the center of gravity for this system?
    (a) 2.0 m
    (b) 2.7 m
    (c) 3.0 m
    (d) 3.3 m
    (e) 3.8 m

12. A 14-kg beam is hinged at one end. A 6.0-kg triangular object and a 7.5-kg I-shaped object are positioned as shown. The dots indicate the individual centers of gravity of the beam and the two objects.

    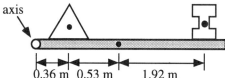

    What is the distance from the axis of rotation to the center of gravity for this system?
    (a) 1.3 m
    (b) 1.1 m
    (c) 0.96 m
    (d) 0.89 m
    (e) 0.71 m

13. Which one of the following statements most accurately describes the *center of gravity* of an object?
    (a) It is the point where gravity acts on the object.
    (b) It is the point where all the mass is concentrated.
    (c) It must be experimentally determined for all objects.
    (d) It is the point on the object where all the weight is concentrated.
    (e) It is the point from which the torque produced by the weight of the object can be calculated.

## Section 9.4 Newton's Second Law for Rotational Motion about a Fixed Axis

14. Consider the following four objects:
    a hoop            a solid sphere
    a flat disk       a hollow sphere

    Each of the objects has mass M and radius R. The axis of rotation passes through the center of each object, and is perpendicular to the plane of the hoop and the plane of the flat disk. Which object requires the largest torque to give it the same angular acceleration?
    (a) the hoop
    (b) the flat disk
    (c) the solid sphere
    (d) the hollow sphere
    (e) both the solid and the hollow spheres

15. A 50 N·m torque acts on a wheel of moment of inertia 150 kg·m². If the wheel starts from rest, how long will it take the wheel to make one revolution?
    (a) 0.33 s
    (b) 0.66 s
    (c) 2.4 s
    (d) 6.1 s
    (e) 10 s

16. A string is wrapped around a pulley of radius 0.05 m and moment of inertia 0.2 kg·m². If the string is pulled with a force **F**, the resulting angular acceleration α of the pulley is 2 rad/s².

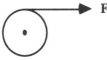

Determine the magnitude of the force **F**.
(a) 0.4 N
(b) 2 N
(c) 8 N
(d) 16 N
(e) 40 N

17. A *massless* frame in the shape of a square with 2 meter sides has a 1-kg ball at each corner. What is the moment of inertia of the four balls about an axis through the corner marked $O$ and perpendicular to the plane of the paper?
(a) 4 kg·m²
(b) 8 kg·m²
(c) 10 kg·m²
(d) 12 kg·m²
(e) 16 kg·m²

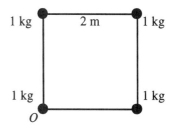

18. A certain merry-go-round is accelerated uniformly from rest and attains an angular speed of 0.4 rad/s in the first 10 seconds. If the net applied torque is 2000 N·m, what is the moment of inertia of the merry-go-round?
(a) 400 kg·m²
(b) 800 kg·m²
(c) 5000 kg·m²
(d) 50 000 kg·m²
(e) This cannot be determined since the radius is not specified.

19. The drawing shows the top view of a door that is 2 m wide. Two forces are applied to the door as indicated.

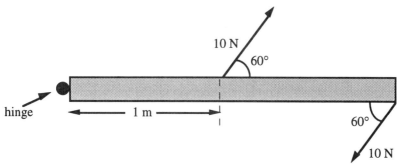

What is the magnitude of the net torque on the door with respect to the hinge?
(a) zero
(b) 5.0 N·m
(c) 8.7 N·m
(d) 10.0 N·m
(e) 26.0 N·m

20. Which one of the following statements concerning the moment of inertia $I$ is false?
(a) $I$ may be expressed in units of kg·m².
(b) $I$ depends on the angular acceleration of the object as it rotates.
(c) $I$ depends on the location of the rotation axis relative to the particles that make up the object.
(d) $I$ depends on the orientation of the rotation axis relative to the particles that make up the object.
(e) Of the particles that make up an object, the particle with the smallest mass may contribute the greatest amount to $I$.

## 92  ROTATIONAL DYNAMICS

21. Two uniform solid spheres, **A** and **B** have the same mass. The radius of sphere **B** is twice that of sphere **A**. The axis of rotation passes through each sphere. Which one of the following statements concerning the moments of inertia of these spheres is true?
    (a) The moment of inertia of **A** is one-fourth that of **B**.
    (b) The moment of inertia of **A** is one-half that of **B**.
    (c) The moment of inertia of **A** is 5/4 that of **B**.
    (d) The moment of inertia of **A** is 5/8 that of **B**.
    (e) The two spheres have equal moments of inertia.

22. Three objects are attached to a *massless* rigid rod which has an axis of rotation as shown. Assuming all of the mass of each object is located at the point shown for each, calculate the moment of inertia of this system.

    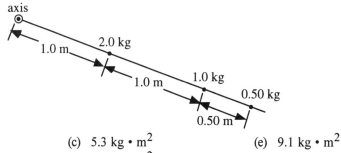

    (a) 1.3 kg • m²        (c) 5.3 kg • m²        (e) 9.1 kg • m²
    (b) 3.1 kg • m²        (d) 7.2 kg • m²

23. Three children are pulling on a rotatable platform on a playground. The platform has a radius of 3.65 m.
    In the picture, two children are pulling with equal forces of 40.0 N in an attempt to make the platform rotate clockwise. The third child applies a force of 60 N as shown. What is the net torque on the platform?

    Note: "ccw" is counterclockwise and "cw" is clockwise.

    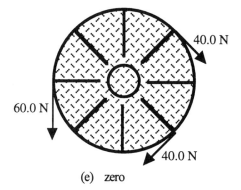

    (a) 73 N • m, ccw      (c) 511 N • m, ccw     (e) zero
    (b) 73 N • m, cw       (d) 511 N • m, cw

24. A string is wrapped around a pulley of radius 0.10 m and moment of inertia 0.15 kg • m². The string is pulled with a force of 12 N. What is the magnitude of the resulting angular acceleration of the pulley?
    (a)  18 rad/s²         (c) 80 rad/s²          (e) 8.0 rad/s²
    (b) 0.13 rad/s²        (d) 0.055 rad/s²

25. A 45-N brick is suspended by a light string from a 2.0-kg pulley. The brick is released from rest and falls to the floor below as the pulley rotates through 5.0 rad. The pulley may be considered a solid disk of radius 1.5 m. What is the angular speed of the pulley?
    (a) 7.3 rad/s          (d) 15 rad/s
    (b) 8.1 rad/s          (e) 19 rad/s
    (c) 9.4 rad/s

    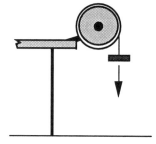

## Section 9.5 Rotational Work and Energy

26. A hollow cylinder of mass $M$ and radius $R$ rolls down an inclined plane. A block of mass $M$ slides down an identical inclined plane. If both objects are released at the same time,
    (a) the cylinder will reach the bottom first.
    (b) the block will reach the bottom first.
    (c) the block will reach the bottom with the greater kinetic energy.
    (d) the cylinder will reach the bottom with the greater kinetic energy.
    (e) both the block and the cylinder will reach the bottom at the same time.

27. A solid sphere and a hollow sphere each of mass $M$ and radius $R$ are released at the same time from the top of an inclined plane. Which one of the following statements is necessarily true?
    (a) The solid sphere will reach the bottom first.
    (b) The hollow sphere will reach the bottom first.
    (c) Both spheres will reach the bottom at the same time.
    (d) The solid sphere will reach the bottom with the greater kinetic energy.
    (e) The hollow sphere will reach the bottom with the greater kinetic energy.

28. Consider the following three objects, each of the same mass and radius:

    *(1)* a solid sphere      *(2)* a solid disk      *(3)* a hoop

    All three are released from rest at the top of an inclined plane. The three objects proceed down the incline undergoing rolling motion without slipping. In which order do the objects reach the bottom of the incline?
    (a) 1, 2, 3
    (b) 2, 3, 1
    (c) 3, 1, 2
    (d) 3, 2, 1
    (e) All three reach the bottom at the same time.

29. A 50-kg rider on a moped of mass 75 kg is traveling with a speed of 20 m/s. The two wheels of the moped have a radius of 0.2 m and a moment of inertia of 0.2 kg m$^2$. What is the total *rotational* kinetic energy of the wheels?
    (a) 80 J
    (b) 100 J
    (c) 500 J
    (d) 2000 J
    (e) 4000 J

30. A 1.0 kg wheel in the form of a solid disk rolls along a horizontal surface with a speed of 6.0 m/s. What is the total kinetic energy of the wheel?
    (a) 9.0 J
    (b) 18 J
    (c) 27 J
    (d) 36 J
    (e) 54 J

31. A 2.0-kg solid cylinder of radius 0.5 m rotates at a rate of 40 rad/s about its cylindrical axis. What *power* is required to bring the cylinder to rest in 10 s?
    (a) 20 W
    (b) 40 W
    (c) 160 W
    (d) 200 W
    (e) 400 W

32. A solid cylinder of radius 0.35 m is released from rest from a height of 1.8 m and rolls down the incline as shown. What is the angular speed of the cylinder when it reaches the horizontal surface?
    (a) 8.2 rad/s
    (b) 14 rad/s
    (c) 34 rad/s
    (d) 67 rad/s
    (e) This cannot be determined because the mass is unknown.

*94*  **ROTATIONAL DYNAMICS**

33. A solid sphere rolls without slipping along a horizontal surface. What percentage of its total kinetic energy is rotational kinetic energy?
    (a) 33 %
    (b) 50 %
    (c) 12 %
    (d) 75 %
    (e) 29 %

34. A hollow sphere of radius 0.25 m is rotating at 13 rad/s about an axis that passes through its center. The mass of the sphere is 3.8 kg. Assuming a constant net torque is applied to the sphere, how much work is required to bring the sphere to a stop?
    (a) 1.0 J
    (b) 3.8 J
    (c) 13 J
    (d) 25 J
    (e) 38 J

*Section 9.6 Angular Momentum*

35. A child standing on the edge of a freely spinning merry-go-round moves quickly to the center. Which one of the following statements is necessarily true concerning this event and why?
    (a) The angular speed of the system decreases because the moment of inertia of the system has increased.
    (b) The angular speed of the system increases because the moment of inertia of the system has increased.
    (c) The angular speed of the system decreases because the moment of inertia of the system has decreased.
    (d) The angular speed of the system increases because the moment of inertia of the system has decreased.
    (e) The angular speed of the system remains the same because the net torque on the merry-go-round is zero.

36. What happens when a spinning ice skater draws in her outstretched arms?
    (a) Her angular momentum decreases.
    (b) Her angular momentum increases.
    (c) Her moment of inertia decreases causing her to speed up.
    (d) Her moment of inertia decreases causing her to slow down.
    (e) The torque that she exerts increases her moment of inertia.

37. A spinning star begins to collapse under its own gravitational pull. Which one of the following occurs as the star becomes smaller?
    (a) Its angular velocity decreases.
    (b) Its angular momentum increases.
    (c) Its angular velocity remains constant.
    (d) Its angular momentum remains constant.
    (e) Both its angular momentum and its angular velocity remain constant.

38. A spinning skater draws in her outstretched arms thereby *reducing* her moment of inertia by a factor of 2. Determine the ratio of her final kinetic energy to her initial kinetic energy.
    (a) 0.5
    (b) 1
    (c) 2
    (d) 4
    (e) 16

39. A 1500-kg satellite orbits a planet in a circular orbit of radius $6.2 \times 10^6$ m. What is the angular momentum of the satellite in its orbit around the planet if the satellite completes one orbit every $1.5 \times 10^4$ seconds.?
    (a) $3.9 \times 10^6$ kg·m²/s
    (b) $1.4 \times 10^{14}$ kg·m²/s
    (c) $6.2 \times 10^8$ kg·m²/s
    (d) $8.1 \times 10^{11}$ kg·m²/s
    (e) $2.4 \times 10^{13}$ kg·m²/s

40. A 60.0-kg skater begins a spin with an angular speed of 6.0 rad/s. By changing the position of her arms, the skater decreases her moment of inertia by 50%. What is the skater's final angular speed?
   (a) 3.0 rad/s
   (b) 4.5 rad/s
   (c) 9.0 rad/s
   (d) 12 rad/s
   (e) 18 rad/s

41. Two equal spheres, labeled **A** and **B** in the figure, are attached to a massless rod with a frictionless pivot at the point P. The system is made to rotate clockwise with angular speed $\omega$ on a horizontal, frictionless tabletop.

   Sphere **A** collides with and sticks to another sphere of which is at rest on the tabletop.

   **Note**: *the masses of all three spheres are equal.*

   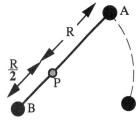

   What is the angular speed of the system immediately after the collision?
   (a) $\omega$
   (b) $0.82\omega$
   (c) $0.60\omega$
   (d) $0.56\omega$
   (e) $0.29\omega$

42. Planets **A** and **B** are uniform solid spheres that rotate at a constant speed about axes through their centers. Although **B** has twice the mass and three times the radius of **A**, each planet has the same rotational kinetic energy. What is the ratio $\omega_B/\omega_A$ of their angular speeds?
   (a) 0.055
   (b) 0.093
   (c) 0.165
   (d) 0.191
   (e) 0.236

43. A solid sphere of radius $R$ rotates about a diameter with an angular speed $\omega$. The sphere then collapses under the action of internal forces to a final radius $R/2$. What is the final angular speed of the sphere?
   (a) $\omega/4$
   (b) $\omega/2$
   (c) $\omega$
   (d) $2\omega$
   (e) $4\omega$

44. A ball of mass $M$ moves in a circular path on a horizontal, frictionless surface. It is attached to a light string that passes through a hole in the center of the table. If the string is pulled down, thereby reducing the radius of the ball's path, the speed of the ball is observed to increase. Complete the following sentence: This occurs because
   (a) the linear momentum of the ball is conserved.
   (b) it is required by Newton's first law of motion.
   (c) the angular momentum of the ball is conserved.
   (d) the angular momentum of the ball must increase.
   (e) the total mechanical energy of the ball must remain constant.

   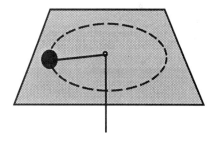

45. A 3.0-kg ball moves in a straight line at 10 m/s as shown in the figure to the right. At the instant shown, what is its angular momentum about the point **P**?
   (a) 30 kg·m²/s
   (b) 90 kg·m²/s
   (c) 120 kg·m²/s
   (d) 150 kg·m²/s
   (e) zero

   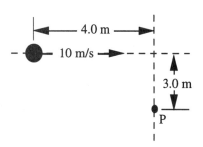

## ROTATIONAL DYNAMICS

*Questions 46 and 47 pertain to the situation described below:*

Two skaters, each of mass 40 kg, approach each other along parallel paths that are separated by a distance of 2 m. Both skaters have a speed of 10 m/s.

The first skater carries a 2-m pole that may be considered massless. As he passes the pole, the second skater catches hold of the end. The two skaters then go around in a circle about the center of the pole.

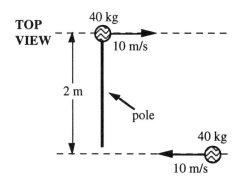

46. What is the angular speed of the skaters after they have linked together?
    (a) 4 rad/s
    (b) 5 rad/s
    (c) 10 rad/s
    (d) 20 rad/s
    (e) 40 rad/s

47. What is their combined angular momentum about the center of the pole?
    (a) 2 kg·m²/s
    (b) 40 kg·m²/s
    (c) 80 kg·m²/s
    (d) 400 kg·m²/s
    (e) 800 kg·m²/s


*Questions 48 and 49 pertain to the situation described below:*

A 2.0-kg hoop rolls without slipping on a horizontal surface so that its center proceeds to the right with a constant linear speed of 6.0 m/s.

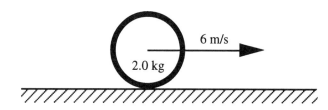

48. Which one of the following statements is true concerning the *angular momentum* of this hoop?
    (a) It points into the paper.
    (b) It points out of the paper.
    (c) It points to the left.
    (d) It points to the right.
    (e) It varies from point to point on the hoop.

49. What is the total kinetic energy of the hoop?
    (a) 36 J
    (b) 54 J
    (c) 72 J
    (d) 96 J
    (e) 140 J


### Additional Problems

50. A compact disc rotates about its center at constant angular speed. Which one of the following quantities is *constant* and *non-zero* for a dust particle near the edge of the disc?
    (a) linear velocity
    (b) torque about the center of the disc
    (c) centripetal acceleration
    (d) angular acceleration
    (e) angular momentum

51. A steady horizontal force **F** of magnitude 21 N is applied at the axle of a solid disk as shown. The disk has mass 2.0 kg and diameter 0.10 m.

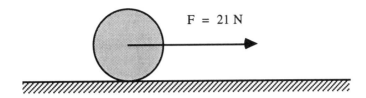

What is the linear speed of the center of the disk after it has moved 12 m?
(a) 9.0 m/s
(b) 13 m/s
(c) 18 m/s
(d) 40 m/s
(e) 80 m/s

52. A uniform disk of radius 1.2 m and mass 0.60 kg is rotating at 25 rad/s around an axis that passes through its center and is perpendicular to the disk.

A rod makes contact with the rotating disk with a force of 4.5 N at a point 0.75 m from the axis of rotation as shown and brings the disk to a stop in 5.0 s. What is the coefficient of kinetic friction for the two materials in contact?
(a) 0.22
(b) 0.15
(c) 0.64
(d) 0.37
(e) 0.55

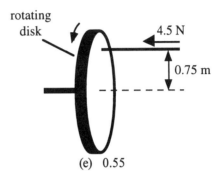

# CHAPTER 10

# Elasticity and Simple Harmonic Motion

*Section 10.1 Elastic Deformation*
*Section 10.2 Stress, Strain, and Hooke's Law*

1. What are the SI units of the shear modulus?
   (a) $N/m^2$
   (b) $N/m$
   (c) $N \cdot m$
   (d) $N/m^3$
   (e) $N \cdot m^2$

2. A cable stretches by an amount $d$ when it supports a crate of mass $M$. The cable is then cut in half. If the same crate is supported by either half of the cable, by how much will the cable stretch?
   (a) $d$
   (b) $d/2$
   (c) $d/4$
   (d) $2d$
   (e) $4d$

3. A cable stretches by an amount d when it supports a crate of mass $M$. The cable is cut in half. What is the mass of the load that can be supported by either half of the cable, if the cable stretches by an amount $d$?
   (a) $M/4$
   (b) $M/2$
   (c) $M$
   (d) $2M$
   (e) $4M$

4. A cable stretches by an amount $d$ when it supports a crate of mass $M$. The cable is replaced by another cable of the same material having the same length but twice the diameter. If the same crate is supported by the thicker cable, by how much will the cable stretch?
   (a) $d/4$
   (b) $d/2$
   (c) $d$
   (d) $2d$
   (e) $4d$

5. A cable stretches by an amount $d$ when it supports a crate of mass $M$. The cable is replaced by another cable of the same material having the same length but twice the diameter. What is the mass of the load that can be supported by the thicker cable if it stretches by an amount $d$?
   (a) $M/4$
   (b) $M/2$
   (c) $M$
   (d) $2M$
   (e) $4M$

6. Young's modulus cannot be applied to
   (a) a stretched wire
   (b) a compressed rod
   (c) a bending beam
   (d) a compressed liquid
   (e) a stretched rubber band

7. The maximum compressional stress that a bone can withstand is $1.6 \times 10^8$ $N/m^2$ before it breaks. A thigh bone (femur), which is the largest and longest bone in the human body, has a cross sectional area of $7.7 \times 10^{-4}$ $m^2$. What is the maximum compressional force that can be applied to the thigh bone?
   (a) $2.1 \times 10^{11}$ N
   (b) $1.2 \times 10^5$ N
   (c) $4.8 \times 10^{12}$ N
   (d) $3.0 \times 10^3$ N
   (e) This cannot be determined since Young's modulus is not given.

8. Young's modulus of nylon is $5 \times 10^9$ $N/m^2$. A force of $5 \times 10^5$ N is applied to a 2-m length of nylon of cross sectional area 0.1 $m^2$. By how much does the nylon stretch?
   (a) $2 \times 10^{-1}$ m
   (b) $2 \times 10^{-2}$ m
   (c) $2 \times 10^{-3}$ m
   (d) $2 \times 10^{-4}$ m
   (e) $2 \times 10^{-5}$ m

TEST BANK  Chapter 10  99

9. The radius of a sphere of lead ($B = 4.2 \times 10^{10}$ N/m$^2$) is 1.000 m on the surface of the earth where the pressure is $1.01 \times 10^5$ N/m$^2$. The sphere is taken by submarine to the deepest part of the ocean to a depth of $1.10 \times 10^4$ m where it is exposed to a pressure is $1.25 \times 10^8$ N/m$^2$. What is the volume of the sphere at the bottom of the ocean?
   (a)  4.176 m$^3$
   (b)  4.189 m$^3$
   (c)  $1.25 \times 10^{-2}$ m$^3$
   (d)  0.134 m$^3$
   (e)  0.988 m$^3$

10. The shear modulus of aluminum is $2.4 \times 10^{10}$ N/m$^2$. An aluminum nail of radius $7.5 \times 10^{-4}$ m projects 0.035 m horizontally outward from a wall. A man hangs a wet raincoat of weight 25.5 N from the end of the nail. Assuming the wall holds its end of the nail, what is the vertical deflection of the other end of the nail?
   (a)  $1.8 \times 10^{-3}$ m
   (b)  $3.3 \times 10^{-2}$ m
   (c)  $7.9 \times 10^{-6}$ m
   (d)  $4.2 \times 10^{-4}$ m
   (e)  $2.1 \times 10^{-5}$ m

*Questions 11 through 13 pertain to the situation described below:*

A 500-N weight is hung from the end of a wire of cross-sectional area 0.010 cm$^2$. The wire stretches from its original length of 200.00 cm to 200.50 cm.

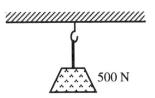

11. What is the stress on the wire?
    (a)  $5.0 \times 10^2$ N/m$^2$
    (b)  $1.0 \times 10^6$ N/m$^2$
    (c)  $1.0 \times 10^8$ N/m$^2$
    (d)  $5.0 \times 10^6$ N/m$^2$
    (e)  $5.0 \times 10^8$ N/m$^2$

12. What is the elongation strain on the wire?
    (a)  $1.0 \times 10^2$
    (b)  $1.0 \times 10^{-2}$
    (c)  $5.0 \times 10^2$
    (d)  $2.5 \times 10^{-3}$
    (e)  $4.0 \times 10^2$

13. Determine the Young's modulus of the wire.
    (a)  $1.0 \times 10^{11}$ N/m$^2$
    (b)  $1.0 \times 10^9$ N/m$^2$
    (c)  $2.0 \times 10^{11}$ N/m$^2$
    (d)  $2.0 \times 10^9$ N/m$^2$
    (e)  $5.0 \times 10^{-12}$ N/m$^2$

*Questions 14 and 15 pertain to the situation described below:*

A plastic box has an initial volume of 2.00 m$^3$. It is then submerged below the surface of a liquid and its volume decreases to 1.96 m$^3$.

14. What is the volume strain on the box?
    (a)  0.02
    (b)  0.04
    (c)  0.2
    (d)  0.4
    (e)  0.98

# 100   ELASTICITY AND SIMPLE HARMONIC MOTION

15. Complete the following statement: In order to calculate the "stress" on the box, in addition to the information given, one must also know
    (a) the mass of the box.
    (b) the bulk modulus of the material from which the box is made.
    (c) the shear modulus of the material from which the box is made.
    (d) the Young's modulus of the material from which the box is made.
    (e) the bulk modulus of the liquid.

16. Complete the following statement: In general, the term *stress* refers to
    (a) a change in volume.
    (b) a change in length.
    (c) a force per unit area.
    (d) a fractional change in length.
    (e) a force per unit length.

17. Which one of the following statements concerning Hooke's law is false?
    (a) It relates stress and strain.
    (b) It is valid only for springs.
    (c) It can be verified experimentally.
    (d) It can be applied to a wide range of materials.
    (e) It is valid only within the elastic limit of a given material.

18. A cylinder with a radius of 0.120 m is placed between the plates of an hydraulic press as illustrated in the drawing. A $4.45 \times 10^5$-N force is applied to the cylinder. What is the pressure on the end of the cylinder due to the applied force?
    (a) $9.84 \times 10^6$ Pa
    (b) $3.13 \times 10^5$ Pa
    (c) $6.18 \times 10^6$ Pa
    (d) $5.34 \times 10^4$ Pa
    (e) $3.71 \times 10^6$ Pa

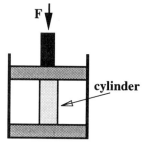

## Section 10.3  The Ideal Spring and Simple Harmonic Motion

19. A block is suspended from the ceiling by a long, thin strip of tungsten metal. The strip behaves as a spring. To produce a 0.25 m horizontal deflection of the block, a force of 6.5 N is required. Calculate the spring constant for the tungsten strip.
    (a) 0.038 N/m
    (b) 1.2 N/m
    (c) 1.6 N/m
    (d) 13 N/m
    (e) 26 N/m

20. A 25-coil spring with a spring constant of 350 N/m is cut into five equal springs with five coils each. What is the spring constant of each of the 5-coil springs?
    (a) 14 N/m
    (b) 70 N/m
    (c) 350 N/m
    (d) 700 N/m
    (e) 1750 N/m

21. In the produce section of a supermarket, five pears are placed on a spring scale. The placement of the pears stretches the spring and causes the dial to move from zero to a reading of 2.0 kg. If the spring constant is 450 N/m, what is the displacement of the spring due to the weight of the pears?
   (a) 0.0044 m
   (b) 0.0088 m
   (c) 0.018 m
   (d) 0.044 m
   (e) 0.088 m

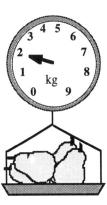

22. A vertical block-spring system on earth has a period of 6.0 s. What is the period of this same system on the moon where the acceleration due to gravity is roughly 1/6 that of earth?
   (a) 1.0 s
   (b) 2.4 s
   (c) 6.0 s
   (d) 15 s
   (e) 36 s

## Section 10.4 Simple Harmonic Motion and the Reference Circle

23. Which one of the following statements is true concerning an object executing simple harmonic motion?
   (a) Its velocity is never zero.
   (b) Its acceleration is never zero.
   (c) Its velocity and acceleration are simultaneously zero.
   (d) Its velocity is zero when its acceleration is a maximum.
   (e) Its maximum acceleration is equal to its maximum velocity.

24. When a force of 20.0 N is applied to a spring, it elongates 0.20 m. Determine the period of oscillation of a 4.0-kg object suspended from this spring.
   (a) 0.6 s
   (b) 1.3 s
   (c) 3.1 s
   (d) 4.1 s
   (e) 6.3 s

25. The position of a simple harmonic oscillator is given by $x(t) = (0.5 \text{ m}) \cos\left(\dfrac{\pi}{3}t\right)$ where $t$ is in seconds. What is the maximum velocity of this oscillator?
   (a) 0.17 m/s
   (b) 0.50 m/s
   (c) 0.67 m/s
   (d) 1.0 m/s
   (e) 2.0 m/s

26. The position of a simple harmonic oscillator is given by $x(t) = (0.5 \text{ m}) \cos\left(\dfrac{\pi}{3}t\right)$ where $t$ is in seconds. What is the period of the oscillator?
   (a) 0.17 s
   (b) 0.67 s
   (c) 1.5 s
   (d) 3.0 s
   (e) 6.0 s

27. A ball hung from a vertical spring oscillates in simple harmonic motion with an angular frequency of 2.6 rad/s and an amplitude of 0.075 m. What is the maximum acceleration of the ball?
   (a) 0.13 m/s$^2$
   (b) 0.20 m/s$^2$
   (c) 0.51 m/s$^2$
   (d) 2.6 m/s$^2$
   (e) 35 m/s$^2$

*102* **ELASTICITY AND SIMPLE HARMONIC MOTION**

28. The acceleration of a certain simple harmonic oscillator is given by $a = -(15.8 \text{ m/s}^2) \cos(2.51t)$. What is the amplitude of the simple harmonic motion?
    (a) 2.51 m
    (b) 4.41 m
    (c) 6.30 m
    (d) 11.1 m
    (e) 15.8 m

## Section 10.5 Energy and Simple Harmonic Motion

29. A 1.0-kg object is suspended from a spring with $k = 16$ N/m. The mass is pulled 0.25 m downward from its equilibrium position and allowed to oscillate. What is the maximum kinetic energy of the object?
    (a) 0.25 J
    (b) 0.50 J
    (c) 1.0 J
    (d) 2.0 J
    (e) 4.0 J

30. A spring required a force of 1.0 N to compress it 0.1 m. How much work is required to stretch the 0.4 m?
    (a) 0.4 J
    (b) 0.6 J
    (c) 0.8 J
    (d) 2 J
    (e) 4 J

31. A 10-kg box is at rest at the end of an unstretched spring with constant $k = 4000$ N/m. The mass is struck with a hammer giving it a velocity of 6.0 m/s to the right across a frictionless surface. What is the amplitude of the resulting oscillations of this system?

    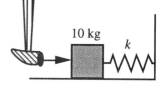

    (a) 0.3 m
    (b) 0.4 m
    (c) 0.5 m
    (d) 0.6 m
    (e) 2 m

32. A certain spring compressed 0.20 m has 10 J of elastic potential energy. The spring is then cut into two halves and one of the halves is compressed by 0.20 m. How much potential energy is stored in the compressed half of the spring?
    (a) 5 J
    (b) 10 J
    (c) 14 J
    (d) 20 J
    (e) 40 J

33. A 1.0-kg block oscillates with a frequency of 10 Hz at the end of a certain spring. The spring is then cut into two halves. The 1.0-kg block is then made to oscillate at the end of one of the halves. What is the frequency of oscillation of the mass?
    (a) 5 Hz
    (b) 10 Hz
    (c) 14 Hz
    (d) 20 Hz
    (e) 40 Hz

34. A 0.2-kg block is held in place by a force **F** that results in a 0.10-m compression of a spring beneath the block. The spring constant is 100 N/m. Assuming the mass of the spring is negligible compared to that of the block, to what maximum height would the block rise if the force **F** were removed?
    (a) 0.26 m
    (b) 0.52 m
    (c) 2.5 m
    (d) 5 m
    (e) 10 m

    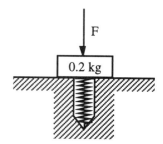

# TEST BANK     Chapter 10     103

● 35. A spring with constant $k = 40.0$ N/m is at the base of a frictionless, 30.0°-inclined plane. A 0.50-kg block is pressed against the spring, compressing it 0.20 m from its equilibrium position. The block is then released. If the block is not attached to the spring, how far along the incline will it travel before it stops?

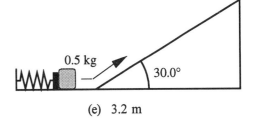

(a) 0.080 m
(b) 0.16 m
(c) 0.32 m
(d) 1.6 m
(e) 3.2 m

◒ 36. A ping-pong ball weighs $2.5 \times 10^{-2}$ N. The ball is placed inside a cup which sits on top of a vertical spring. If the spring is compressed 0.055 m and released, the maximum height above the compressed position that the ball reaches is 2.84 m. Neglect air resistance and determine the spring constant.

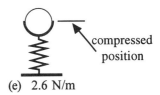

(a) 47 N/m
(b) 24 N/m
(c) 11 N/m
(d) 5.2 N/m
(e) 2.6 N/m

**Questions 37 and 38 pertain to the situation described below:**

A relaxed spring protrudes from an opening 0.050 meters as shown in Figure A.
A 1.00-kg block is found to just force the spring completely into the opening as shown in Figure B.

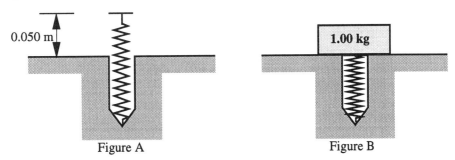

Figure A         Figure B

◒ 37. Determine the spring constant $k$.
(a) 20.0 N/m
(b) 196 N/m
(c) 392 N/m
(d) 3920 N/m
(e) 7840 N/m

◒ 38. How much potential energy is stored in the spring in figure B?
(a) 0.245 J
(b) 0.490 J
(c) 4.90 J
(d) 9.80 J
(e) 19.6 J

## Section 10.6 The Pendulum

● 39. An iron ball hangs from a 24-m steel cable and is used in the demolition of a building at a location where the acceleration due to gravity is 9.9 m/s². The ball is swung outward from its equilibrium position for a distance of 4.5 m. Assuming the system behaves as a simple pendulum, find the maximum speed of the ball during its swing.
(a) 1.9 m/s
(b) 7.0 m/s
(c) 11 m/s
(d) 2.9 m/s
(e) 9.8 m/s

## 104 ELASTICITY AND SIMPLE HARMONIC MOTION

40. A simple pendulum on earth has a period of 6.0 s. What is the approximate period of this pendulum on the moon where the acceleration due to gravity is roughly 1/6 that of earth?
    (a) 1.0 s
    (b) 2.4 s
    (c) 6.0 s
    (d) 15 s
    (e) 36 s

41. A pendulum is transported from sea-level, where the acceleration due to gravity $g = 9.80$ m/s$^2$, to the bottom of Death Valley. At this location, the period of the pendulum is decreased by 3.00%. What is the value of $g$ in Death Valley?
    (a) 9.22 m/s$^2$
    (b) 9.51 m/s$^2$
    (c) 9.80 m/s$^2$
    (d) 10.1 m/s$^2$
    (e) 10.4 m/s$^2$

42. In a certain clock, a pendulum of length $L_1$ has a period $T_1 = 0.95$ s. The length of the pendulum is adjusted to a new value $L_2$ such that $T_2 = 1.0$ s. What is the ratio $L_2/L_1$?
    (a) 0.90
    (b) 0.95
    (c) 1.0
    (d) 1.1
    (e) 1.3

43. What is the period of pendulum consisting of a 6-kg object oscillating on a 4-m string?
    (a) 0.25 s
    (b) 0.50 s
    (c) 1.0 s
    (d) 2.0 s
    (e) 4.0 s

44. A simple pendulum consists of a ball of mass $m$ suspended from the ceiling using a string of length $L$. The ball is displaced from its equilibrium position by an angle $\theta$. What is the magnitude of the restoring force that moves the ball toward its equilibrium position and produces simple harmonic motion?
    (a) $kx$
    (b) $mg$
    (c) $mg (\cos \theta)$
    (d) $mg (\sin \theta)$
    (e) $mgL(\sin \theta)$

## Section 10.7 Damped Harmonic Motion
## Section 10.8 Driven Harmonic Motion and Resonance

45. Which one of the following terms is used to describe a system in which the degree of damping is just enough to stop the system from oscillating?
    (a) slightly damped
    (b) underdamped
    (c) critically damped
    (d) overdamped
    (e) resonance

46. A simple harmonic oscillator with a period of 2.0 s is subject to damping so that it loses one percent of its amplitude per cycle. About how much energy does this oscillator lose per cycle?
    (a) 0.5%
    (b) 1.0%
    (c) 1.4%
    (d) 2.0%
    (e) 4.0%

47. Complete the following sentence: Resonance occurs in harmonic motion when
    (a) the system is overdamped.
    (b) the system is critically damped.
    (c) the energy in the system is a minimum.
    (d) the driving frequency is the same as the natural frequency of the system.
    (e) the energy in the system is proportional to the square of the motion's amplitude.

## Additional Problems

48. A 0.750-kg object hanging from a vertical spring is observed to oscillate with a period of 2.00 s. When the 0.750-kg object is removed and replaced by a 1.25-kg object, what will be the period of oscillation?
    (a) 1.55 s
    (b) 2.58 s
    (c) 3.32 s
    (d) 4.38 s
    (e) 7.45 s

*Questions 49 through 55 pertain to the situation described below:*

When a 0.20-kg block is suspended from a vertically hanging spring, it stretches the spring from its original length of 0.050 m to 0.060 m.

The same block is attached to the same spring and placed on a horizontal, frictionless surface. The block is then pulled so that the spring stretches to a total length of 0.10 m.  The block is released at time $t = 0$ and undergoes simple harmonic motion.

49. What is the frequency of the motion?
    (a) 0.50 Hz
    (b) 1.00 Hz
    (c) 5.0 Hz
    (d) 10.0 Hz
    (e) 31 Hz

50. Complete the following statement: In order to increase the frequency of the motion, one would have to
    (a) reduce the spring constant.
    (b) decrease the mass of the block on the end of the spring.
    (c) increase the length of the spring.
    (d) reduce the distance that the spring is initially stretched.
    (e) increase the distance that the spring is initially stretched

51. Which one of the following statements is true concerning the motion of the block?
    (a) Its acceleration is constant.
    (b) The period of its motion depends on its amplitude.
    (c) Its acceleration is greatest when the spring returns to the 5.0 cm position.
    (d) Its velocity is greatest when the mass has reached its maximum displacement.
    (e) Its acceleration is greatest when the mass has reached its maximum displacement.

52. Which one of the following expressions gives the displacement of the block as a function of time? Let $x$ be in cm and $t$ in seconds.
    (a) $x = 5 \cos (10\pi\, t)$
    (b) $x = 5 \sin (10\pi\, t)$
    (c) $x = 10 \cos (5\pi\, t)$
    (d) $x = 10 \sin (5\pi\, t)$
    (e) $x = 10 \cos (10\pi\, t)$

53. What is the speed of the block each time the spring is 5.0 cm long?
    (a) zero
    (b) 5 cm/s
    (c) 16 cm/s
    (d) 100 cm/s
    (e) 157 cm/s

54. What is the maximum acceleration of the block?
    (a) 1.57 m/s$^2$
    (b) 3.14 m/s$^2$
    (c) 49 m/s$^2$
    (d) 98 m/s$^2$
    (e) 157 m/s$^2$

55. What is the total mechanical energy of the system at any instant?
    (a) 0.25 J
    (b) 0.49 J
    (c) 0.98 J
    (d) 4.9 J
    (e) 9.8 J

## 106 ELASTICITY AND SIMPLE HARMONIC MOTION

56. A 2.0-kg object is attached to a spring ($k = 55.6$ N/m) which hangs vertically from the ceiling. The object is displaced 0.045 m vertically. When the object is released, the system undergoes simple harmonic motion. What is the magnitude of the object's maximum acceleration?
    (a) 1.3 m/s$^2$
    (b) 2.3 m/s$^2$
    (c) 4.4 m/s$^2$
    (d) 9.8 m/s$^2$
    (e) 11 m/s$^2$

57. The brick shown in the drawing is glued to the floor. A 150-N force is applied to the top surface of the brick as shown. If the brick has a shear modulus of $5.4 \times 10^9$ N/m$^2$, how far to the right does the top face move relative to the stationary bottom face?

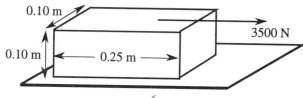

    (a) $5.8 \times 10^{-6}$ m
    (b) $2.6 \times 10^{-6}$ m
    (c) $1.1 \times 10^{-6}$ m
    (d) $6.5 \times 10^{-7}$ m
    (e) $3.4 \times 10^{-7}$ m

# CHAPTER 11    Fluids

## Section 11.1  Mass Density

1. The density of mercury is $1.36 \times 10^4$ kg/m$^3$. What is the mass of a $4.00 \times 10^{-4}$ m$^3$ sample of mercury?
   (a) 0.0343 kg
   (b) 0.002 94 kg
   (c) 2.94 kg
   (d) 5.44 kg
   (e) 6.29 kg

2. At standard temperature and pressure, carbon dioxide has a density of 1.98 kg/m$^3$. What volume does 0.85 kg of carbon dioxide occupy at standard temperature and pressure?
   (a) 0.43 m$^3$
   (b) 0.86 m$^3$
   (c) 1.7 m$^3$
   (d) 2.3 m$^3$
   (e) 4.8 m$^3$

3. What mass of water (at 4.0 °C) can be contained in a rectangular box whose dimensions are 10.0 cm by 5.00 cm by 1.00 cm?  The density of water at 4.0 °C is $1.000 \times 10^3$ kg/m$^3$.
   (a) 5.0 g
   (b) 10.0 g
   (c) 25.0 g
   (d) 50.0 g
   (e) 0.25 kg

4. The density of iron is 7860 kg/m$^3$. What is the mass of an iron sphere whose diameter is 0.50 m?
   (a) 123 kg
   (b) 164 kg
   (c) 514 kg
   (d) 983 kg
   (e) 4110 kg

5. Within a certain type of star called a neutron star, the material at the center has a mass density of $1.0 \times 10^{18}$ kg/m$^3$.  If a small sphere of this material of radius $1.0 \times 10^{-5}$ m were somehow transported to the surface of the earth, what would be the weight of this sphere?
   (a) 1000 N
   (b) 4200 N
   (c) $4.1 \times 10^4$ N
   (d) $7.0 \times 10^4$ N
   (e) $3.8 \times 10^9$ N

6. A solid cylinder has a radius of 0.051 m and a height of 0.0030 m. The cylinder is composed of two different materials with mass densities of 1950 kg/m$^3$ and 1470 kg/m$^3$. If each of the two materials occupies an *equal volume*, what is the mass of the cylinder?
   (a) $8.4 \times 10^{-2}$ kg
   (b) $7.1 \times 10^{-2}$ kg
   (c) $6.5 \times 10^{-2}$ kg
   (d) $5.3 \times 10^{-2}$ kg
   (e) $4.2 \times 10^{-2}$ kg

## Section 11.2  Pressure

7. A child wants to pump up a bicycle tire so that its pressure is $2.5 \times 10^5$ Pa above that of atmospheric pressure. If the child uses a pump with a circular piston 0.035 m in diameter, what force must the child exert?
   (a) 120 N
   (b) 240 N
   (c) 340 N
   (d) 930 N
   (e) 1300 N

8. How much force does the atmosphere exert on one side of a vertical wall 4.00-m high and 10.0-m long?
   (a) $2.53 \times 10^3$ N
   (b) $1.01 \times 10^5$ N
   (c) $4.05 \times 10^5$ N
   (d) $4.05 \times 10^6$ N
   (e) $1.01 \times 10^6$ N

## 108 FLUIDS

9. A gas sample is confined to a chamber with a movable piston. A small load is placed on the piston and the system is allowed to reach equilibrium.

   If the weight of the piston and the load is 70.0 N and the piston has an area of $5.0 \times 10^{-4}$ m$^2$, what is the pressure exerted on the piston by the gas?

   **Note**: Atmospheric pressure is $1.013 \times 10^5$ Pa.

   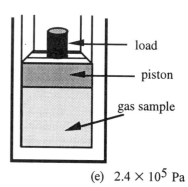

   (a) $2.8 \times 10^4$ Pa
   (b) $5.6 \times 10^4$ Pa
   (c) $7.3 \times 10^4$ Pa
   (d) $1.4 \times 10^5$ Pa
   (e) $2.4 \times 10^5$ Pa

10. Using the value of atmospheric pressure at sea level, $1 \times 10^5$ Pa, estimate the total mass of the earth's atmosphere above a 5-m$^2$ area.
    (a) $5 \times 10^4$ kg
    (b) $9 \times 10^2$ kg
    (c) $2 \times 10^{-4}$ kg
    (d) $4 \times 10^{-2}$ kg
    (e) $3 \times 10^5$ kg

*Questions 11 and 12 pertain to the following situation:*

In a classroom demonstration, a 73.5-kg physics professor lies on a "bed of nails." The bed consists of a large number of evenly spaced, relatively sharp nails mounted in a board so that the points extend vertically outward from the board. While the professor is lying down, approximately 1900 nails make contact with his body.

11. What is the average force exerted by each nail on the professor's body?
    (a) 0.0201 N
    (b) 0.379 N
    (c) 1.42 N
    (d) 723 N
    (e) $1.42 \times 10^6$ N

12. If the area of contact at the head of each nail is $1.26 \times 10^{-6}$ m$^2$, what is the average pressure at each contact?
    (a) $1.59 \times 10^4$ Pa
    (b) $5.71 \times 10^8$ Pa
    (c) $1.11 \times 10^{12}$ Pa
    (d) $1.11 \times 10^6$ Pa
    (e) $3.01 \times 10^5$ Pa

*Questions 13 and 14 pertain to the situation described below:*

A swimming pool has the dimensions shown in the drawing. It is filled with water to a uniform depth of 8.00 m.

The density of water = $1.00 \times 10^3$ kg/m$^3$.

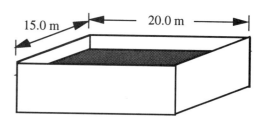

13. What is the total pressure exerted on the bottom of the swimming pool?
    (a) $0.79 \times 10^5$ Pa
    (b) $1.48 \times 10^5$ Pa
    (c) $1.80 \times 10^5$ Pa
    (d) $1.97 \times 10^5$ Pa
    (e) $2.49 \times 10^5$ Pa

14. What is the total force exerted on the bottom of the swimming pool?
    (a) $2.40 \times 10^7$ N
    (b) $5.40 \times 10^7$ N
    (c) $5.90 \times 10^7$ N
    (d) $7.50 \times 10^7$ N
    (e) $8.40 \times 10^7$ N

*Section 11.3 Pressure and Depth in a Static Fluid*
*Section 11.4 Pressure Gauges*

15. A submarine is operating at a depth of 100.0 m below the surface of the ocean. If the air inside the submarine is maintained at a pressure of 1.0 atmosphere, what is the magnitude of the force that acts on the rectangular hatch 2.0 m × 1.0 m on the deck of the submarine?
    (a) 980 N
    (b) $2.0 \times 10^3$ N
    (c) $5.0 \times 10^4$ N
    (d) $9.8 \times 10^5$ N
    (e) $2.0 \times 10^6$ N

16. The two dams are identical with the exception that the water reservoir behind dam **A** extends twice the horizontal distance behind it as that of dam **B**. Which one of the following statements regarding these dams is correct?
    (a) The force exerted by the water on dam **A** is greater than that on dam **B**.
    (b) The force exerted by the water on dam **B** is greater than that on dam **A**.
    (c) Dam **A** is more likely to collapse than dam **B** if the water level rises.
    (d) Dam **B** is more likely to collapse than dam **A** if the water level rises.
    (e) The horizontal distance of the water behind the two dams does not determine the force on them.

17. At a location where the acceleration due to gravity is 9.807 m/s$^2$, the atmospheric pressure is $9.891 \times 10^4$ Pa. A barometer at the same location is filled with an unknown liquid. What is the density of the unknown liquid if its height in the barometer is 1.163 m?
    (a) 210 kg/m$^3$
    (b) 4336 kg/m$^3$
    (c) 5317 kg/m$^3$
    (d) 8672 kg/m$^3$
    (e) 9688 kg/m$^3$

18. A column of water 70.0 cm high supports a column of an unknown liquid as suggested in the figure (not drawn to scale). Assume that both liquids are at rest and that the density of water is $1.0 \times 10^3$ kg/m$^3$.

    Determine the density of the unknown liquid.

    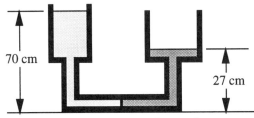

    (a) $3.9 \times 10^2$ kg/m$^3$
    (b) $1.2 \times 10^3$ kg/m$^3$
    (c) $2.6 \times 10^3$ kg/m$^3$
    (d) $3.3 \times 10^3$ kg/m$^3$
    (e) $3.9 \times 10^3$ kg/m$^3$

19. A woman is enjoying a tropical drink while lying on a beach. The acceleration due to gravity at her location is 9.85 m/s$^2$. What gauge pressure must exist in the woman's mouth if she is drinking through a straw extending 0.085 m above the surface of the drink?
    **Note**: Assume the drink has a density of 1015 kg/m$^3$.
    (a) 850 Pa
    (b) 970 Pa
    (c) 1100 Pa
    (d) 2100 Pa
    (e) $1.0 \times 10^5$ Pa

*110* **FLUIDS**

20. The largest barometer ever built was an oil-filled barometer constructed in Leicester, England in 1991. The oil had a height of 12.2 m. Assuming a pressure of $1.013 \times 10^5$ Pa, what was the density of the oil used in the barometer?
    (a) 798 kg/m³
    (b) 847 kg/m³
    (c) 981 kg/m³
    (d) 1150 kg/m³
    (e) 1210 kg/m³

## Section 11.5 Pascal's Principle

21. Complete the following sentence: The operation of a hydraulic jack is an application of
    (a) Pascal's principle.
    (b) Bernoulli's principle.
    (c) Archimedes' principle.
    (d) irrotational flow.
    (e) the continuity equation.

22. A force of 250 N is applied to a hydraulic jack piston which is 0.01 m in diameter. If the piston which supports the load has a diameter of 0.10 m, approximately how much mass can be lifted by the jack? Ignore any difference in height between the pistons.
    (a) 255 kg
    (b) 500 kg
    (c) 800 kg
    (d) 2550 kg
    (e) 6300 kg

23. In a car lift, compressed air with a gauge pressure of $4.0 \times 10^5$ Pa is used to raise a piston with a circular cross-sectional area. If the radius of the piston is 0.17 m, what is the maximum mass that can be raised using this piston?
    (a) 530 kg
    (b) 3700 kg
    (c) 9800 kg
    (d) 22 000 kg
    (e) 41 000 kg

24. Which one of the following statements concerning a completely enclosed fluid is true?
    (a) Any change in the applied pressure of the fluid produces a change in pressure that depends on direction.
    (b) The pressure at all points within the fluid is independent of any pressure applied to it.
    (c) Any change in applied pressure produces an equal change in pressure at all points within the fluid.
    (d) An increase in pressure in one part of the fluid results in an equal decrease in pressure in another part.
    (e) The pressure in the fluid is the same at all points within the fluid.

## Section 11.6 Archimedes' Principle

25. Which one of the following statements concerning the buoyant force on an object submerged in a liquid is true?
    (a) The buoyant force depends on the mass of the object.
    (b) The buoyant force depends on the weight of the object.
    (c) The buoyant force is independent of the density of the liquid.
    (d) The buoyant force depends on the volume of the liquid displaced.
    (e) The buoyant force will increase with depth if the liquid is incompressible.

26. A 2-kg block displaces 10 kg of water when it is held fully immersed. The object is then tied down as shown in the figure; and it displaces 5 kg of water. What is the tension in the string?
    (a) 10 N
    (b) 20 N
    (c) 30 N
    (d) 70 N
    (e) 100 N

27. An object weighs 15 N in air and 13 N when submerged in water. Determine the density of the object.
    (a) 330 kg/m$^3$
    (b) 500 kg/m$^3$
    (c) $1.2 \times 10^3$ kg/m$^3$
    (d) $6.0 \times 10^3$ kg/m$^3$
    (e) $7.5 \times 10^3$ kg/m$^3$

28. The density of ice is 0.92 g/cm$^3$; and the density of seawater is 1.03 g/cm$^3$. A large iceberg floats in Arctic waters. What fraction of the volume of the iceberg is exposed?
    (a) 0.08 %
    (b) 11 %
    (c) 89 %
    (d) 92 %
    (e) 99 %

29. A small sculpture made of brass ($\rho_{brass}$= 8470 kg/m$^3$) is believed to have a secret central cavity. The weight of the sculpture in air is 15.76 N. When it is submerged in water, the weight is 13.86 N. What is the volume of the secret cavity?
    (a) $1 \times 10^{-4}$ m$^3$
    (b) $2 \times 10^{-4}$ m$^3$
    (c) $3 \times 10^{-5}$ m$^3$
    (d) $2 \times 10^{-6}$ m$^3$
    (e) $4 \times 10^{-6}$ m$^3$

30. A balloon inflated with helium gas (density = 0.2 kg/m$^3$) has a volume of $6 \times 10^{-3}$ m$^3$. If the density of air is 1.3 kg/m$^3$, what is the buoyant force exerted on the balloon?
    (a) 0.01 N
    (b) 0.08 N
    (c) 0.8 N
    (d) 1.3 N
    (e) 7.8 N

31. Three blocks labeled A, B, and C are floating in water as shown in the drawing. Blocks A and B have the same mass and volume. Block C has the same volume, but is submerged to a greater depth than the other two blocks. Which one of the following statements concerning this situation is false?

    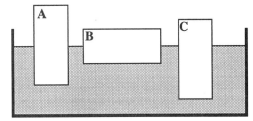

    (a) The density of block A is less than that of block C.
    (b) The buoyant force acting on block A is equal to that acting on block B.
    (c) The volume of water displaced by block C is greater than that displaced by block B.
    (d) The buoyant force acting on block C is greater than that acting on block B.
    (e) The volume of water displaced by block A is greater than that displaced by block B.

*Questions 32 and 33 pertain to the following situation:*

When a block of volume $1.00 \times 10^{-3}$ m$^3$ is hung from a spring scale as shown in Figure A, the scale reads 10.0 N. When the same block is then placed in an unknown liquid, it floats with 2/3 of its volume submerged as suggested in Figure B. The density of water is $1.00 \times 10^3$ kg / m$^3$.

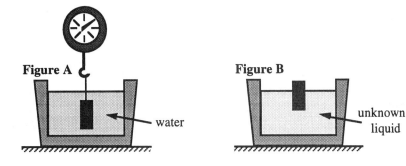

*112* **FLUIDS**

32. Determine the mass of the block.
    (a) 1.02 kg
    (b) 2.02 kg
    (c) 3.02 kg
    (d) 4.04 kg
    (e) 9.80 kg

33. Determine the density of the unknown liquid.
    (a) $3.03 \times 10^3$ kg/m$^3$
    (b) $4.62 \times 10^3$ kg/m$^3$
    (c) $6.16 \times 10^3$ kg/m$^3$
    (d) $8.01 \times 10^3$ kg/m$^3$
    (e) $1.57 \times 10^4$ kg/m$^3$

*Questions 34 through 36 pertain to the situation described below:*

A balloon is released from a tall building. Its total mass of the balloon including the enclosed gas is 2.0 kg. Its volume is 5.0 m$^3$. The density of air is 1.3 kg/m$^3$.

34. What is the average density of the balloon?
    (a) 0.2 kg/m$^3$
    (b) 0.4 kg/m$^3$
    (c) 0.8 kg/m$^3$
    (d) 1.0 kg/m$^3$
    (e) 1.2 kg/m$^3$

35. Will the balloon *rise, fall*, or *remain stationary*, and why?
    (a) The balloon will fall because its density is greater than that of air.
    (b) The balloon will remain stationary because its density is less than that of air.
    (c) The balloon will rise because the upward buoyant force is greater than its weight.
    (d) The balloon will fall because the upward buoyant force is less than its weight.
    (e) The balloon will fall because the downward buoyant force is greater than the upward buoyant force.

36. What is the magnitude of the initial acceleration of the balloon?
    (a) zero
    (b) 9.8 m/s$^2$
    (c) 10.9 m/s$^2$
    (d) 22.1 m/s$^2$
    (e) 43.6 m/s$^2$

## Section 11.7 Fluids in Motion
## Section 11.8 The Equation of Continuity

37. Which one of the following statements concerning *streamline flow* is true?
    (a) At any given point in the fluid, the velocity is constant in time.
    (b) Streamline flow occurs when there are sharp obstacles in the path of a fast-moving fluid.
    (c) Streamline flow is described by Pascal's principle.
    (d) Streamline flow is described by Archimedes' principle.
    (e) The velocity vectors are the same for all particles in the fluid.

38. Which one of the following statements concerning fluid streamlines is true?
    (a) In steady flow, streamlines must remain parallel and equally spaced.
    (b) In turbulent flow, two or more streamlines may cross.
    (c) In turbulent flow, streamlines can begin or end at any point.
    (d) Streamlines are perpendicular to the velocity of the fluid at every point.
    (e) In steady flow, the pattern of streamlines does not change with time.

39. Ann uses a hose to water her garden. The water enters the hose through a faucet with a 6.0-cm diameter. The speed of the water at the faucet is 5 m/s. If the faucet and the nozzle are at the same height, and the water leaves the nozzle with a speed of 20 m/s, what is the diameter of the nozzle?
    (a) 1.5 cm
    (b) 2.0 cm
    (c) 3.0 cm
    (d) 4.0 cm
    (e) 6.0 cm

40. Water flows through a pipe of diameter 8.0 cm with a speed of 10.0 m/s. It then enters a smaller pipe of diameter 3.0 cm. What is the speed of the water as it flows through the smaller pipe?
    (a) 1.4 m/s
    (b) 2.8 m/s
    (c) 27 m/s
    (d) 54 m/s
    (e) 71 m/s

41. Water enters a pipe of diameter 3.0 cm with a velocity of 3.0 m/s. The water encounters a constriction where its velocity is 12 m/s. What is the diameter of the constricted portion of the pipe?
    (a) 0.33 cm
    (b) 0.75 cm
    (c) 1.0 cm
    (d) 1.5 cm
    (e) 12 cm

42. Water is flowing through a channel which is 12 m wide with a speed of 0.75 m/s. The water then flows into four identical channels which have a width of 4.0 m wide. The depth of the water does not change as it flows into the four channels. What is the speed of the water in one of the smaller channels?
    (a) 0.56 m/s
    (b) 2.3 m/s
    (c) 0.25 m/s
    (d) 0.75 m/s
    (e) 0.12 m/s

## Section 11.9 Bernoulli's Equation
## Section 11.10 Applications of Bernoulli's Equation

43. Complete the following statement: Bernoulli's principle is a statement of
    (a) hydrostatic equilibrium.
    (b) thermal equilibrium in fluids.
    (c) mechanical equilibrium in fluids.
    (d) energy conservation in dynamic fluids.
    (e) momentum conservation in dynamic fluids.

44. A curtain hangs straight down in front of an open window. A sudden gust of wind blows past the window; and the curtain is pulled out of the window. Which law, principle, or equation can be used to explain this movement of the curtain?
    (a) Poiseuille's law
    (b) Bernoulli's equation
    (c) the equation of continuity
    (d) Archimedes' principle
    (e) Pascal's principle

45. Which one of the following statements is false concerning the derivation or usage of Bernoulli's equation?
    (a) The fluid must be non-viscous.
    (b) Streamline flow is assumed.
    (c) The fluid must be incompressible.
    (d) The work-energy theorem is used to derive Bernoulli's equation.
    (e) Vertical distances are always measured relative to the lowest point within the fluid.

*114* **FLUIDS**

○ 46. A horizontal piping system that delivers a constant flow of water is constructed from pipes with different diameters as shown in the figure.

At which of the labeled points is the water in the pipe under the greatest pressure?
(a) A
(b) B
(c) C
(d) D
(e) E

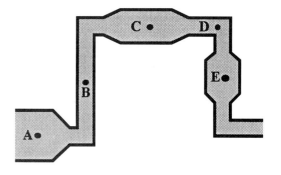

◐ 47. A large tank is filled with water to a depth of 15 m. A spout located 10.0 m above the bottom of the tank is then opened as shown in the drawing. With what speed will water emerge from the spout?
(a) 3.1 m/s
(b) 9.9 m/s
(c) 14 m/s
(d) 17 m/s
(e) 31 m/s

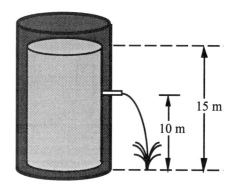

◐ 48. Oil ($\rho = 925$ kg/m$^3$) is flowing through a pipeline at a constant speed when it encounters a vertical bend in the pipe raising it 4.0 m. The cross sectional area of the pipe does not change. What is the difference in pressure ($P_B - P_A$) in the portions of the pipe before and after the rise?

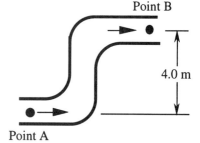

(a) $+2.4 \times 10^4$ Pa
(b) $-3.6 \times 10^4$ Pa
(c) $+5.1 \times 10^5$ Pa
(d) $-7.2 \times 10^5$ Pa
(e) $-1.8 \times 10^3$ Pa

◐ 49. The density of the liquid flowing through the horizontal pipe in the drawing is 1500 kg/m$^3$. The speed of the fluid at point A is 5.5 m/s while at point B it is 8.0 m/s. What is the difference in pressure, $P_B - P_A$, between points B and A?
(a) $-1.9 \times 10^3$ Pa
(b) $+3.8 \times 10^3$ Pa
(c) $-2.5 \times 10^4$ Pa
(d) $+5.0 \times 10^4$ Pa
(e) $-7.6 \times 10^5$ Pa

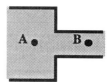

## Section 11.11 Viscous Flow

50. Which law, principle, or equation can be used to determine the volume flow rate of a liquid that is forced through a hypodermic needle?
    (a) Poiseuille's law
    (b) Pascal's principle
    (c) Bernoulli's equation
    (d) Archimedes' principle
    (e) the equation of continuity

51. A hypodermic needle consists of a plunger of circular cross-section that slides inside a hollow cylindrical syringe. When the plunger is pushed, the contents of the syringe are forced through a hollow needle (also of circular cross-section). If a 4.0 N force is applied to the plunger and the diameters of the plunger and the needle are 1.2 cm and 2.5 mm, respectively, what force is needed *to prevent* fluid flow at the needle?
    (a) 0.17 N
    (b) 0.27 N
    (c) 0.43 N
    (d) 0.83 N
    (e) 2.7 N

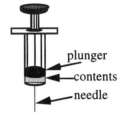

## Additional Problems

52. The figure shows two fish tanks, each having sides of width 1 foot. Tank **A** is 3 feet long while tank **B** is 6 feet long. Both tanks are filled with 1 foot of water.

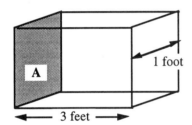 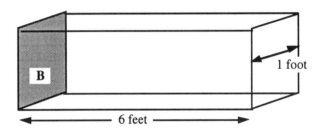

$S_A$ = the magnitude of the force of the water on the side of tank **A**
$S_B$ = the magnitude of the force of the water on the side of tank **B**
$B_A$ = the magnitude of the force of the water on the bottom of tank **A**
$B_B$ = the magnitude of the force of the water on the bottom of tank **B**

Using the notation given above, which one of the following sets of equations below is correct for this situation?
(a) $S_A = S_B$ and $B_A = B_B$
(b) $S_A = 2S_B$ and $B_A = B_B$
(c) $2S_A = S_B$ and $2B_A = B_B$
(d) $S_A = S_B$ and $2B_A = B_B$
(e) $S_A = 2S_B$ and $B_A = 2B_B$

***Questions 53 through 58 pertain to the situation described below:***

A glass tube has several different cross-sectional areas with the values indicated in the figure. A piston at the left end of the tube exerts pressure so that the mercury sample within the tube flows from the right end with a speed of 8.0 m/s. Three points within the tube are labeled **A**, **B**, and **C**.

# FLUIDS

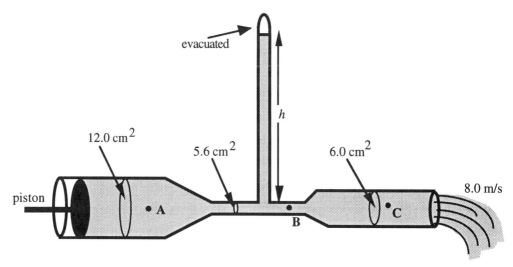

**Notes:** The drawing is not drawn to scale.
Atmospheric pressure is $1.01 \times 10^5$ N/m$^2$; and the density of mercury is 13 600 kg/m$^3$.

53. At what speed is mercury flowing past the point labeled **A**?
    (a) 2.0 m/s
    (b) 4.0 m/s
    (c) 8.0 m/s
    (d) 12 m/s
    (e) 16 m/s

54. What is the total pressure at point **A**?
    (a) $1.01 \times 10^5$ Pa
    (b) $2.02 \times 10^5$ Pa
    (c) $2.25 \times 10^5$ Pa
    (d) $3.26 \times 10^5$ Pa
    (e) $4.27 \times 10^5$ Pa

55. At what speed is mercury flowing past the point labeled **B**?
    (a) 2.27 m/s
    (b) 4.27 m/s
    (c) 7.47 m/s
    (d) 8.57 m/s
    (e) 9.28 m/s

56. What is the total pressure at point **B**?
    (a) $1.01 \times 10^5$ Pa
    (b) $3.71 \times 10^5$ Pa
    (c) $4.27 \times 10^5$ Pa
    (d) $6.44 \times 10^5$ Pa
    (e) $7.45 \times 10^5$ Pa

57. What is the total pressure at point **C**?
    (a) $1.01 \times 10^5$ Pa
    (b) $3.26 \times 10^5$ Pa
    (c) $3.66 \times 10^5$ Pa
    (d) $6.44 \times 10^5$ Pa
    (e) $7.45 \times 10^5$ Pa

58. Determine the height $h$ of mercury in the manometer with the evacuated upper end.
    (a) 136 mm
    (b) 269 mm
    (c) 278 mm
    (d) 366 mm
    (e) 371 mm

# CHAPTER 12  Temperature and Heat

*Section 12.1 Common Temperature Scales*
*Section 12.2 The Kelvin Temperature Scale*
*Section 12.3 Thermometers*

1. Which one of the following temperatures is approximately equal to "room temperature?"
   (a) 0 K
   (b) 0 °C
   (c) 100 °C
   (d) 100 K
   (e) 293 K

2. Complete the following statement: A temperature decrease of 30 C° is equal to a temperature decrease of
   (a) 30 F°.
   (b) 30 K.
   (c) 17 F°.
   (d) 26 F°.
   (e) 303 K.

3. Absolute zero on the Celsius temperature scale is –273.15 °C. What is absolute zero on the Fahrenheit temperature scale?
   (a) –331.67 °F
   (b) –363.67 °F
   (c) –395.67 °F
   (d) –427.67 °F
   (e) –459.67 °F

4. The digital sign outside a local bank reports that the temperature is 44 °C. What is the temperature in degrees Fahrenheit?
   (a) 56 °F
   (b) 79 °F
   (c) 99 °F
   (d) 111 °F
   (e) 120 °F

5. Three thermometers are in the same water bath. After thermal equilibrium is established, it is found that the Celsius thermometer reads 100 °C, the Fahrenheit thermometer reads 212 °F, and the Kelvin thermometer reads 273 K. Which one of the following statements is the most reasonable conclusion?
   (a) The Kelvin thermometer is incorrect.
   (b) The Celsius thermometer is incorrect.
   (c) The Fahrenheit thermometer is incorrect.
   (d) All three thermometers are incorrect.
   (e) The three thermometers are at different temperatures.

6. Three thermometers are placed in a closed insulated box and are allowed to reach thermal equilibrium. One is calibrated in Fahrenheit degrees, one in Celsius degrees, and one in Kelvins. The Celsius thermometer reads –40 °C and the Kelvin thermometer reads 233 K. Which one of the following statements is necessarily true?
   (a) The Kelvin thermometer should read –233 K.
   (b) The Kelvin thermometer should read –313 K.
   (c) The Fahrenheit thermometer must read –40 °F.
   (d) If water were found within the box, it must be in the liquid state.
   (e) If the temperature of the contents is increased by 10 C°, the reading on the Kelvin thermometer should increase by 273 K.

7. Which one of the following properties could *not* be used as a temperature sensitive property in the construction of a thermometer?
   (a) the change in mass of a solid
   (b) the change in volume of a liquid
   (c) the change in length of a metal rod
   (d) the change in electrical resistance of a wire
   (e) the change in pressure of a gas at constant volume

*118* TEMPERATURE AND HEAT

## Section 12.4 Linear Thermal Expansion

8. Which one of the following statements is the best explanation for the fact that metal pipes that carry water often burst during cold winter months?
   (a) Water contracts upon freezing while the metal expands at lower temperatures.
   (b) The metal contracts to a greater extent than the water.
   (c) The interior of the pipe contracts less than the outside of the pipe.
   (d) Water expands upon freezing while the metal contracts at lower temperatures.
   (e) Both the metal and the water expand, but the water expands to a greater extent.

9. The coefficient of linear expansion of steel is $12 \times 10^{-6}/C°$. A railroad track is made of single rails of steel 1.0 km in length. By what length would these single rails change between a cold day when the temperature is $-10$ °C and a hot day at 30 °C?
   (a) 0.62 cm
   (b) 24 cm
   (c) 48 cm
   (d) 480 cm
   (e) 620 cm

10. The coefficient of linear expansion of aluminum is $23 \times 10^{-6}/C°$. A circular hole in an aluminum plate is 2.725 cm in diameter at 0 °C. What is the diameter of the hole if the temperature of the plate is raised to 100 °C ?
    (a) 0.0063 cm
    (b) 2.728 cm
    (c) 2.731 cm
    (d) 2.757 cm
    (e) 2.788 cm

11. A metal rod 40.0000 cm long at 40 °C is heated to 60 °C. The length of the rod is then measured to be 40.0105 cm. What is the coefficient of linear expansion of the metal?
    (a) $13 \times 10^{-6}/C°$
    (b) $22 \times 10^{-6}/C°$
    (c) $44 \times 10^{-6}/C°$
    (d) $53 \times 10^{-6}/C°$
    (e) $71 \times 10^{-6}/C°$

12. A copper plate has a length of 0.12 m and a width of 0.10 m at 25 °C. The plate is uniformly heated to 175 °C. If the linear expansion coefficient for copper is $1.7 \times 10^{-5}/C°$, what is the *change* in the area of the plate as a result of the increase in temperature?
    (a) $2.6 \times 10^{-5}$ m$^2$
    (b) $6.1 \times 10^{-5}$ m$^2$
    (c) $3.2 \times 10^{-6}$ m$^2$
    (d) $4.9 \times 10^{-7}$ m$^2$
    (e) $7.8 \times 10^{-9}$ m$^2$

13. At a certain temperature, a simple pendulum has a period of 1.500 seconds. The support wire is made of silver and has a coefficient of linear thermal expansion of $1.90 \times 10^{-5}/C°$. How much must the temperature be increased to increase the period to 1.506 seconds?
    (a) 118 C°
    (b) 221 C°
    (c) 316 C°
    (d) 434 C°
    (e) 528 C°

14. Complete the following statement: *Bimetallic strips* used as adjustable switches in electric appliances consist of metallic strips that must have different
    (a) mass.
    (b) length.
    (c) volume.
    (d) expansion coefficients.
    (e) specific heat capacities.

15. A steel string guitar is strung so that there is negligible tension in the strings at a temperature of 24.9 °C. The guitar is taken to an outdoor winter concert where the temperature of the strings decreases to –15.1 °C. The cross-sectional area of a particular string is $5.5 \times 10^{-6}$ m$^2$. The distance between the points where the string is attached does not change. For steel, Young's modulus is $2.0 \times 10^{11}$ N/m$^2$; and the coefficient of linear expansion is $1.2 \times 10^{-5}$/C°. Use your knowledge of linear thermal expansion and stress to calculate the tension in the string at the concert.
    (a) 530 N
    (b) 240 N
    (c) 120 N
    (d) 60 N
    (e) 30 N

## Section 12.5 Volume Thermal Expansion

16. The coefficient of linear expansion of a certain solid is $9 \times 10^{-6}$/C°. Assuming that this solid behaves like most solids, what is its coefficient of volume expansion?
    (a) $9.1 \times 10^{-6}$/C°
    (b) $27 \times 10^{-36}$/C°
    (c) $27 \times 10^{-6}$/C°
    (d) $729 \times 10^{-6}$/C°
    (e) $729 \times 10^{-36}$/C°

17. Which one of the following statements explains why it is difficult to measure the coefficient of volume expansion for a liquid?
    (a) Liquids are more compact than solids.
    (b) Liquids are more compact than gases.
    (c) Liquids tend to expand more slowly than solids.
    (d) The liquid will lose heat to the containing vessel.
    (e) The volume of the containing vessel will also increase.

18. A steel gas tank of volume 0.070 m$^3$ is filled to the top with gasoline at 20 °C. The tank is placed inside a chamber with an interior temperature of 50 °C. The coefficient of volume expansion for gasoline is $9.50 \times 10^{-4}$/C°; and the coefficient of linear expansion of steel is $12 \times 10^{-6}$/C°. After the tank and its contents reach thermal equilibrium with the interior of the chamber, how much gasoline has spilled?
    (a) $2.52 \times 10^{-5}$ m$^3$
    (b) $7.56 \times 10^{-5}$ m$^3$
    (c) $1.69 \times 10^{-3}$ m$^3$
    (d) $1.92 \times 10^{-3}$ m$^3$
    (e) $2.00 \times 10^{-3}$ m$^3$

19. The coefficient of volumetric expansion for gold is $4.2 \times 10^{-5}$/C°. The density of gold is 19 300 kg/m$^3$ at 0.0 °C. What is the density of gold at 1050 °C?
    (a) 20 200 kg/m$^3$
    (b) 19 300 kg/m$^3$
    (c) 19 000 kg/m$^3$
    (d) 18 800 kg/m$^3$
    (e) 18 500 kg/m$^3$

20. A tanker ship is filled with $2.25 \times 10^5$ m$^3$ of gasoline at a refinery in southern Texas when the temperature is 17.2 °C. When the ship arrives in New York City, the temperature is 1.3 °C. If the coefficient of volumetric expansion for gasoline is $9.50 \times 10^{-4}$/C°, how much has the volume of the gasoline decreased when it is unloaded in New York?
    (a) $1.50 \times 10^{-2}$ m$^3$
    (b) 66.2 m$^3$
    (c) 1290 m$^3$
    (d) 3400 m$^3$
    (e) $1.05 \times 10^4$ m$^3$

## Section 12.6 Heat and Internal Energy
## Section 12.7 Heat and Temperature Change: Specific Heat Capacity

21. Complete the following statement: The term *heat* most accurately describes
    (a) the internal energy of an object.
    (b) a measure of how hot an object is.
    (c) the absolute temperature of an object.
    (d) the molecular motion inside of an object.
    (e) the flow of energy due to a temperature difference.

22. A soft drink manufacturer claims that a new diet drink is "low Joule." The label indicates the available energy is 6300 J. What is the equivalent of this energy in Calories (1 Calorie = 1000 cal)?
    (a) 0.015 Cal      (c) 1.0 Cal      (e) 4.8 Cal
    (b) 0.48 Cal       (d) 1.5 Cal

23. The units of heat are equivalent to those of which one of the following quantities?
    (a) force/time          (c) temperature              (e) power
    (b) work                (d) specific heat capacity • time

24. The specific heat capacity of iron is approximately half that of aluminum. Two balls of equal mass, one made of iron and the other of aluminum, both at 80 °C, are dropped into a thermally insulated jar that contains an equal mass of water at 20 °C. Thermal equilibrium is eventually reached. Which one of the following statements concerning the final temperatures is true?
    (a) Both balls will reach the same final temperature.
    (b) The iron ball will reach a higher final temperature than the aluminum ball.
    (c) The aluminum ball will reach a higher final temperature than the iron ball.
    (d) The difference in the final temperatures of the balls depends on the initial mass of the water.
    (e) The difference in the final temperatures of the balls depends on the initial temperature of the water.

25. Two cubes, one silver and one iron, have the same mass and temperature. A quantity $Q$ of heat is removed from each cube. Which one of the following properties causes the final temperatures of the cubes to be different?
    (a) density                      (c) specific heat capacity      (e) volume
    (b) latent heat of vaporization  (d) coefficient of volume expansion

26. A 2.00-kg metal object requires $5.02 \times 10^3$ J of heat to raise its temperature from 20.0 °C to 40.0 °C. What is the specific heat capacity of the metal?
    (a) 63.0 J/(kg • C°)        (c) 251 J/(kg • C°)        (e) $1.00 \times 10^3$ J/(kg • C°)
    (b) 126 J/(kg • C°)         (d) 502 J/(kg • C°)

27. A 200.0-kg object is attached via an ideal pulley system to paddle wheels that are submerged in 0.480 kg of water at 20.0 °C in an insulated container as shown in the figure at the right. Then, the object falls through a distance of 5.00 m causing the paddle wheel to turn. Assuming that all of the mechanical energy lost by the falling object goes into the water, determine the final temperature of the water.

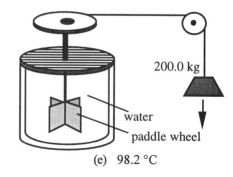

    (a) 4.90 °C       (c) 24.9 °C       (e) 98.2 °C
    (b) 20.5 °C       (d) 40.4 °C

28. A 0.20-kg lead shot is heated to 90.0 °C and dropped into an ideal calorimeter containing 0.50 kg of water initially at 20.0 °C. What is the final equilibrium temperature of the lead shot? The specific heat capacity of lead is 128 J/(kg • C°); and the specific heat of water is 4186 J/(kg • C°).
   (a) 4.8 °C
   (b) 20.8 °C
   (c) 22.4 °C
   (d) 27.8 °C
   (e) 42.1 °C

29. A gold sphere has a radius of 1.000 cm at 25.0 °C. If 7650 J of heat are added to the sphere, what will be the final volume of the sphere? Gold has a density of 19 300 kg/m$^3$ at 25.0 °C, a specific heat capacity of 129 J/(kg • C°), and a coefficient of volume expansion of $42.0 \times 10^{-6}$/C°.
   (a) $2.88 \times 10^{-6}$ m$^3$
   (b) $3.01 \times 10^{-6}$ m$^3$
   (c) $3.33 \times 10^{-6}$ m$^3$
   (d) $3.91 \times 10^{-6}$ m$^3$
   (e) $4.32 \times 10^{-6}$ m$^3$

*Questions 30 and 31 pertain to the situation described below:*

A 2.00-kg metal block slides on a rough, horizontal surface inside an insulated box. After sliding a distance of 500.0 m, its temperature is increased by 2.00 °C. **Note:** Assume that all of the heat generated by frictional heating goes into the metal block. For this metal, the specific heat capacity is 0.150 cal/(g • C°).

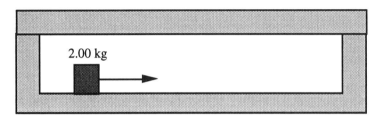

30. How much work does the force of friction do *on* the block?
    (a) zero
    (b) 300 J
    (c) −300 J
    (d) 2510
    (e) −2510 J

31. What is the coefficient of sliding friction between the block and the surface?
    (a) zero
    (b) 0.061
    (c) 0.100
    (d) 0.256
    (e) 0.299

## Section 12.8 Heat and Phase Change: Latent Heat

32. Which one of the following phrases is an example of *sublimation*?
    (a) the fumes produced by moth balls
    (b) the mist produced by liquid nitrogen
    (c) the formation of dew on blades of grass
    (d) the formation of raindrops in the atmosphere
    (e) the condensation of steam on a kitchen window

33. Complete the following statement: When a substance undergoes *fusion* it
    (a) freezes.
    (b) sublimes.
    (c) condenses.
    (d) vaporizes.
    (e) evaporates.

34. Complete the following statement: When solid NH$_3$ passes directly to the gaseous state it is said to
    (a) melt.
    (b) sublime.
    (c) condense.
    (d) evaporate.
    (e) fuse.

*122* **TEMPERATURE AND HEAT**

35. A 1.0-g sample of steam at 100 °C loses 560 calories of heat. What is the resulting temperature of the sample?
    (a) 20 °C
    (b) 80 °C
    (c) 88 °C
    (d) 96 °C
    (e) 99 °C

36. Heat is added to a substance, but its temperature does not rise. Which one of the following statements provides the best explanation for this observation?
    (a) The substance must be a gas.
    (b) The substance must be a non-perfect solid.
    (c) The substance undergoes a change of phase.
    (d) The substance has unusual thermal properties.
    (e) The substance must be cooler than its environment.

37. Which would cause a more serious burn: 30 g of steam or 30 g of liquid water, both at 100 °C; and why is this so?
    (a) Water, because it is more dense than steam.
    (b) Steam, because of its specific heat capacity.
    (c) Steam, because of its latent heat of vaporization.
    (d) Water, because its specific heat is greater than that of steam.
    (e) Either one would cause a burn of the same severity since they are both at the same temperature.

38. Heat is added to a sample of water in an insulated container. Which one of the following statements is *necessarily* true?
    (a) The temperature of the water will rise.
    (b) The volume of the water must decrease.
    (c) The mass of the system must decrease.
    (d) Under certain conditions, the temperature of the water can decrease.
    (e) The type of change that will occur depends on the original temperature of the water.

39. A 5.0-g sample of ice at 0.0 °C falls through a distance of 20.0 meters and undergoes a completely inelastic collision with the earth. If all of the lost mechanical energy is absorbed by the ice, how much of it melts?
    (a) $2.9 \times 10^{-3}$ g
    (b) $4.3 \times 10^{-3}$ g
    (c) $7.6 \times 10^{-3}$ g
    (d) $1.8 \times 10^{-2}$ g
    (e) $2.1 \times 10^{-2}$ g

40. Using the data in the table, determine how many calories are needed to change 100 g of solid $X$ at 10 °C to a vapor at 210 °C.

    | Thermodynamic Constants for Substance $X$ | |
    |---|---|
    | heat of fusion | 40.0 cal/g |
    | heat of vaporization | 150.0 cal/g |
    | melting point | 10.0 °C |
    | boiling point | 210.0 °C |
    | specific heat capacity (liquid $X$) | 0.500 cal/(g • C°) |

    (a) 4000 cal
    (b) 10 000 cal
    (c) 15 000 cal
    (d) 29 000 cal
    (e) 39 000 cal

41. A 0.030-kg ice cube at 0 °C is placed in an insulated box that contains a fixed quantity of steam at 100 °C. When thermal equilibrium of this closed system is established, its temperature is found to be 23 °C.
    Determine the original mass of the steam at 100 °C.
    (a) 0.17 g
    (b) 1.7 g
    (c) 2.5 g
    (d) 4.8 g
    (e) 5.0 g

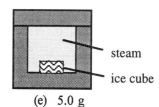

42. A 0.040-kg ice cube at 0 °C is placed in an insulated box that contains 0.0075 kg of steam at 100 °C. What is the equilibrium temperature reached by this closed system?
Note: Assume that all of the ice melts.
(a) 22.7 °C
(b) 33.6 °C
(c) 44.9 °C
(d) 50.7 °C
(e) 66.4 °C

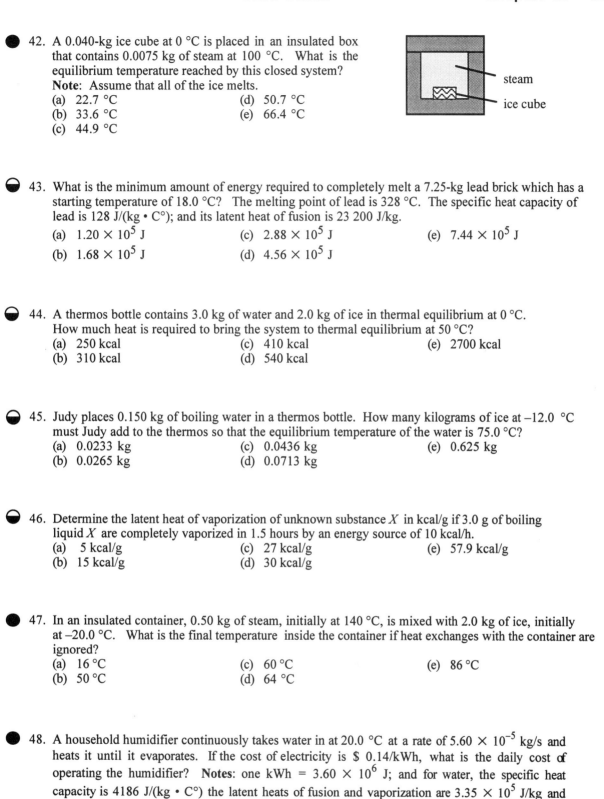

43. What is the minimum amount of energy required to completely melt a 7.25-kg lead brick which has a starting temperature of 18.0 °C? The melting point of lead is 328 °C. The specific heat capacity of lead is 128 J/(kg · C°); and its latent heat of fusion is 23 200 J/kg.
(a) $1.20 \times 10^5$ J
(b) $1.68 \times 10^5$ J
(c) $2.88 \times 10^5$ J
(d) $4.56 \times 10^5$ J
(e) $7.44 \times 10^5$ J

44. A thermos bottle contains 3.0 kg of water and 2.0 kg of ice in thermal equilibrium at 0 °C. How much heat is required to bring the system to thermal equilibrium at 50 °C?
(a) 250 kcal
(b) 310 kcal
(c) 410 kcal
(d) 540 kcal
(e) 2700 kcal

45. Judy places 0.150 kg of boiling water in a thermos bottle. How many kilograms of ice at –12.0 °C must Judy add to the thermos so that the equilibrium temperature of the water is 75.0 °C?
(a) 0.0233 kg
(b) 0.0265 kg
(c) 0.0436 kg
(d) 0.0713 kg
(e) 0.625 kg

46. Determine the latent heat of vaporization of unknown substance $X$ in kcal/g if 3.0 g of boiling liquid $X$ are completely vaporized in 1.5 hours by an energy source of 10 kcal/h.
(a) 5 kcal/g
(b) 15 kcal/g
(c) 27 kcal/g
(d) 30 kcal/g
(e) 57.9 kcal/g

47. In an insulated container, 0.50 kg of steam, initially at 140 °C, is mixed with 2.0 kg of ice, initially at –20.0 °C. What is the final temperature inside the container if heat exchanges with the container are ignored?
(a) 16 °C
(b) 50 °C
(c) 60 °C
(d) 64 °C
(e) 86 °C

48. A household humidifier continuously takes water in at 20.0 °C at a rate of $5.60 \times 10^{-5}$ kg/s and heats it until it evaporates. If the cost of electricity is $ 0.14/kWh, what is the daily cost of operating the humidifier? Notes: one kWh = $3.60 \times 10^6$ J; and for water, the specific heat capacity is 4186 J/(kg · C°) the latent heats of fusion and vaporization are $3.35 \times 10^5$ J/kg and $2.26 \times 10^6$ J/kg, respectively.
(a) $ 0.31
(b) $ 0.49
(c) $ 0.58
(d) $ 0.65
(e) $ 0.70

*Questions 49 through 51 pertain to the situation described below:*

A 0.0500-kg lead bullet of volume $5.00 \times 10^{-6}$ m$^3$ at 20.0 °C hits a block that is made of an ideal thermal insulator and comes to rest at its center. Then, the temperature of the bullet is found to be 327 °C. Use the following information for lead:

coefficient of linear expansion: $\alpha = 2.0 \times 10^{-5}$/C°
specific heat capacity: $c = 128$ J/(kg · C°)
latent heat of fusion: $L_f = 23\,200$ J/kg
melting point: $T_{melt} = 327$ °C

49. How much heat was needed to raise the bullet to its final temperature?
    (a) 963 J
    (b) 1930 J
    (c) 3640 J
    (d) 3880 J
    (e) 4440 J

50. What is the volume of the bullet after it comes to rest.
    (a) $5.00 \times 10^{-6}$ m$^3$
    (b) $5.01 \times 10^{-6}$ m$^3$
    (c) $5.04 \times 10^{-6}$ m$^3$
    (d) $5.07 \times 10^{-6}$ m$^3$
    (e) $5.09 \times 10^{-6}$ m$^3$

51. How much additional heat would be needed to melt the bullet?
    (a) 420 J
    (b) 628 J
    (c) 837 J
    (d) 1170 J
    (e) 2010 J

## Section 12.9 Equilibrium Between Phases of Matter

52. A liquid is in equilibrium with its vapor in a closed vessel. Which one of the following statements is *necessarily* true?
    (a) The rate of condensation is greater than the evaporation rate.
    (b) The rate of evaporation is greater than the condensation rate.
    (c) The temperature of the vapor is greater than that of the liquid.
    (d) Molecules of the liquid do not have enough energy to vaporize.
    (e) The temperature of the vapor is the same as that of the liquid.

53. The graph shows the equilibrium vapor pressure versus temperature for a certain liquid and its vapor within an open container. If the container is at sea level, at approximately what temperature will the liquid boil?
    (a) 50 °C
    (b) 65 °C
    (c) 75 °C
    (d) 85 °C
    (e) 100 °C

## Section 12.10 Humidity

54. On a warm summer day, the relative humidity is 20 % when the temperature is 32 °C. Which one of the following statements is true if the temperature suddenly drops to 26 °C and all other conditions remain the same?
    (a) The relative humidity will decrease.
    (b) The relative humidity will increase.
    (c) The dew point will change.
    (d) The partial pressure of water vapor will decrease.
    (e) The vaporization curve of water will change.

## Additional Problems

55. An ordinary mercury thermometer at room temperature is quickly placed in a beaker of hot water. The mercury column is observed to drop slightly before it rises to the final equilibrium temperature. Which one of the following statements is the best explanation for this behavior?
    (a) The glass envelope expands before the heat reaches the mercury.
    (b) The expansion coefficient of glass is larger than that of mercury.
    (c) Both the mercury and the glass initially expand, but at different rates.
    (d) Initially, the mercury contracts.
    (e) Initially the glass envelope contracts.

*Questions 56 through 60 pertain to the situation described below:*

Heat is added to a 1.0-kg solid sample of a material at –200 °C. The figure shows the temperature of the material as a function of the heat added.

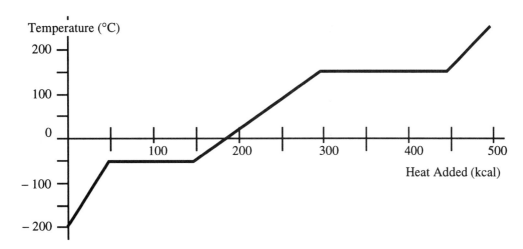

56. Which one of the following statements concerning this substance is true?
    (a) It boils at 300 °C.
    (b) It melts at –200 °C.
    (c) It is a liquid at 200 °C.
    (d) It can coexist as a solid and a liquid at –50 °C.
    (e) It can exist as a solid, liquid, and gas at 150 °C.

57. What is the latent heat of fusion of this material?
    (a) 50 cal/g
    (b) 100 cal/g
    (c) 150 cal/g
    (d) 300 cal/g
    (e) 450 cal/g

58. What is the latent heat of vaporization of this material?
    (a) 50 cal/g
    (b) 100 cal/g
    (c) 150 cal/g
    (d) 300 cal/g
    (e) 450 cal/g

**TEMPERATURE AND HEAT**

59. What is the specific heat capacity of this substance in its solid state?
    - (a) 0.33 cal/(g · C°)
    - (b) 0.75 cal/(g · C°)
    - (c) 1.00 cal/(g · C°)
    - (d) 1.33 cal/(g · C°)
    - (e) 3.00 cal/(g · C°)

60. What is the specific heat capacity of this substance in its liquid state?
    - (a) 0.33 cal/(g · C°)
    - (b) 0.75 cal/(g · C°)
    - (c) 1.00 cal/(g · C°)
    - (d) 1.33 cal/(g · C°)
    - (e) 3.00 cal/(g · C°)

# CHAPTER 13     The Transfer of Heat

## Section 13.1 Convection

1. Complete the following statement: The transfer of heat by *convection* will occur
   - (a) only in metals.
   - (b) only in a vacuum.
   - (c) only in non-metallic solids.
   - (d) with or without the presence of matter.
   - (e) only in the presence of a liquid or a gas.

2. Which one of the following statements best explains why convection does not occur in solids?
   - (a) Molecules in a solid are more closely spaced than in a gas.
   - (b) The molecules in a solid are not free to move throughout its volume.
   - (c) Molecules in a solid vibrate at a lower frequency than those of a liquid.
   - (d) Solids are more compressible than liquids.
   - (e) Solids are less compressible than gases.

3. Which one of the following is *not* an example of convection?
   - (a) Smoke rises above a fire.
   - (b) An eagle soars on an updraft of wind.
   - (c) A person gets a suntan on a beach.
   - (d) Water cooks spaghetti.
   - (e) An electric heater warms a room.

4. Which one of the following processes does not occur during the convective transfer of heat within a container of air?
   - (a) The volume of a warmed part of the air is reduced and its density increases.
   - (b) A continuous flow of warmer and cooler parts of air is established.
   - (c) The flow of air molecules results in a flow of heat.
   - (d) The cooler portion of the air surrounding a warmer part exerts a buoyant force on it.
   - (e) As the warmer part of the air moves, it is replaced by cooler air which is subsequently warmed.

## Section 13.2 Conduction

5. Suppose you are sitting next to a fireplace in which there is a fire burning. One end of a metal poker has been left in the fire. Which one of the following statements concerning this situation is true?
   - (a) You can feel the heat of the fire primarily because of convection.
   - (b) The other end of the poker is warmed through conduction.
   - (c) Heat escapes through the chimney primarily through conduction.
   - (d) You can feel the heat of the fire primarily because of conduction.
   - (e) The other end of the poker is warmed through convection.

6. The two ends of an iron rod are maintained at different temperatures. The amount of heat that flows through the rod by conduction during a given time interval does *not* depend upon
   - (a) the length of the iron rod.
   - (b) the thermal conductivity of iron.
   - (c) the temperature difference between the ends of the rod.
   - (d) the mass of the iron rod.
   - (e) the duration of the time interval.

7. The ends of a cylindrical steel rod are maintained at two different temperatures. The rod conducts heat from one end to the other at a rate of 10 cal/s. At what rate would a steel rod twice as long and twice the diameter conduct heat between the same two temperatures?
   - (a) 5 cal/s
   - (b) 10 cal/s
   - (c) 20 cal/s
   - (d) 40 cal/s
   - (e) 80 cal/s

*128*  **THE TRANSFER OF HEAT**

○ 8. Which one of the following materials is the *best* thermal conductor?
(a) diamond
(b) Styrofoam
(c) nitrogen gas
(d) concrete
(e) goose down

○ 9. Which of the following materials is the *poorest* thermal conductor?
(a) gold
(b) ice
(c) copper
(d) wood
(e) air

○ 10. The space between the inner walls of a thermos bottle is evacuated to minimize heat transfer by
(a) radiation.
(b) conduction.
(c) conduction and convection.
(d) conduction and radiation.
(e) conduction, convection, and radiation.

◐ 11. At what rate is heat lost through a 1.0 m by 1.5 m rectangular glass window pane that is 0.5 cm thick when the inside temperature is 20 °C and the outside temperature is 5 °C? The thermal conductivity for glass is 0.80 W/(m · C°).
(a) 18 W
(b) 36 W
(c) 720 W
(d) 3600 W
(e) 7200 W

● 12. Two cylindrical steel rods **A** and **B** have *radii* of 0.02 m and 0.04 m, respectively. It is found that the two steel rods conduct the same amount of heat per unit time for the same temperature differences between their two ends. What is the ratio of the lengths of the rods, $L_A/L_B$?
(a) 0.25
(b) 0.50
(c) 1.00
(d) 2.00
(e) 4.00

◐ 13. A cabin has a 0.159 m thick wooden floor [$k$ = 0.141 W/(m · C°)] with an area of 13.4 m². A roaring fire keeps the interior of the cabin at a comfortable 18.0 °C while the air temperature in the crawl space below the cabin is –20.6 °C. What is the rate of heat conduction through the wooden floor?
(a) 31 J/s
(b) 138 J/s
(c) 214 J/s
(d) 245 J/s
(e) 459 J/s

◐ 14. A granite wall has a thickness of 0.61 m and a thermal conductivity of 2.1 W/(m · C°). The temperature on one face of the wall is 3.2 °C and 20.0 °C on the opposite face. How much heat is transferred in one hour through each square meter of the granite wall?
(a) 210 000 J/m²
(b) 106 000 J/m²
(c) 77 000 J/m²
(d) 1800 J/m²
(e) 58 J/m²

● 15. On a cold winter day the outside temperature is –5.0 °C while the interior of a well-insulated garage is maintained at 20.0 °C by an electric heater. Assume the walls have a total area of 75 m², a thickness of 0.15 m, and a thermal conductivity of 0.042 W/(m · C°). What is the cost to heat the garage for six hours at these temperatures if the cost of electricity is $ 0.14/kWh? **Note:** 1 kWh = 3.6 × 10⁶ J.
(a) $ 0.18
(b) $ 0.44
(c) $ 0.61
(d) $ 0.74
(e) $ 1.09

◐ 16. In an experiment to determine the thermal conductivity of a bar of a new alloy, one end of the bar is maintained at 0.0 °C and the other end at 100.0 °C. The bar has a cross-sectional area of 1.0 cm² and a length of 15 cm. If the rate of heat conduction through the bar is 24 W, what is the thermal conductivity of the bar?
(a) 24 W/(m · C°)
(b) 360 W/(m · C°)
(c) 160 W/(m · C°)
(d) 63 W/(m · C°)
(e) 0.029 W/(m · C°)

17. A slab of insulation is made of three layers, as Drawing I indicates. Each of the layers **A**, **B**, and **C** has the same thickness, but a different thermal conductivity. Heat flows through the slab, and the temperatures are as shown. What are the temperatures $T_1$ and $T_2$ In Drawing II where the layers are arranged in a different order?

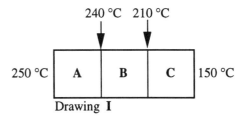

Drawing I

(a) $T_1 = 230$ °C and $T_2 = 170$ °C
(b) $T_1 = 200$ °C and $T_2 = 180$ °C
(c) $T_1 = 220$ °C and $T_2 = 160$ °C
(d) $T_1 = 180$ °C and $T_2 = 160$ °C
(e) $T_1 = 210$ °C and $T_2 = 190$ °C

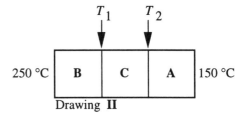

Drawing II

*Questions 18 through 21 pertain to the situation described below*:

Two bars, **A** and **B**, each of length 2.0 m and cross sectional area 1.0 m², are placed end to end as shown in the figure.

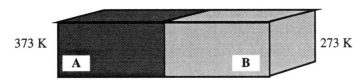

Their thermal conductivities are $k_A = 339$ J/(s • m • K) and $k_B = 837$ J/(s • m • K). The left end of bar **A** is maintained at 373 K while the right end of **B** is maintained at 273 K.

18. Which one of the following statements is true concerning the above situation?
    (a) For a given time *t*, the amount of heat transferred through bar **A** is the same as that through **B**.
    (b) For a given time *t*, the amount of heat transferred through bar **A** is smaller than that through **B**.
    (c) For a given time *t*, the amount of heat transferred through bar **A** is less than that through **B**.
    (d) The time for a quantity of heat $Q$ to pass through bar **A** is less than that for **B**.
    (e) Time is not a factor in determining the quantity of heat conducted through these bars.

19. What is the temperature at the interface between **A** and **B**?
    (a) 332 K
    (b) 323 K
    (c) 296 K
    (d) 313 K
    (e) 302 K

20. What is the rate of heat transfer through this combination?
    (a) $3.0 \times 10^3$ J/s
    (b) $6.0 \times 10^3$ J/s
    (c) $9.0 \times 10^3$ J/s
    (d) $1.2 \times 10^4$ J/s
    (e) $2.4 \times 10^4$ J/s

21. How much heat passes the interface between **A** and **B** in 5.0 s?
    (a) $6.0 \times 10^4$ J
    (b) $1.5 \times 10^3$ J
    (c) $1.2 \times 10^5$ J
    (d) $4.5 \times 10^4$ J
    (e) $3.0 \times 10^4$ J

## Section 13.3 Radiation

22. Which one of the following objects, initially all at the same temperature, will be most efficient in losing heat?
    (a) a dull black box in vacuum
    (b) a dull black box in air
    (c) a box with an emissivity of 0.1
    (d) a polished silver box in air
    (e) a polished silver box in vacuum

23. Complete the following statement: Most of the heat that is lost to space from the earth occurs by
    (a) conduction.
    (b) convection.
    (c) radiation.
    (d) both conduction and radiation.
    (e) both conduction and convection.

24. Which object will emit more electromagnetic radiation than it absorbs from its surroundings?
    (a) a 600 °C lead sphere in a 700 °C oven
    (b) a girl scout sitting close to a campfire
    (c) an ice cube in beaker of water at 50 °C
    (d) a 200 °C copper coin in a beaker of water at 98 °C
    (e) an ice cube in thermal equilibrium with the interior of a freezer

25. Complete the following statement: The interior of a thermos bottle is silvered to minimize heat transfer due to
    (a) radiation.
    (b) conduction.
    (c) conduction and convection.
    (d) conduction and radiation.
    (e) conduction, convection, and radiation.

26. The sun continuously radiates energy into space, some of which is intercepted by the earth. The average temperature of the surface of the earth remains about 300 K. Why doesn't the earth's temperature rise as it intercepts the sun's energy?
    (a) The earth reflects the sun's light.
    (b) The earth radiates an amount of energy into space equal to the amount it receives.
    (c) The energy only raises the temperature of the upper atmosphere and never reaches the surface.
    (d) The thermal conductivity of the earth is low.
    (e) The heat is carried away from the earth by convection currents.

27. Two identical solid spheres have the same temperature. One of the spheres is cut into two identical pieces. These two hemispheres are then separated. The intact sphere radiates an energy $Q$ during a given time interval. During the same interval, the two hemispheres radiate a total energy $Q'$. What is the ratio $Q'/Q$?
    (a) 2.0
    (b) 0.50
    (c) 4.0
    (d) 1.5
    (e) 0.25

28. Which one of the following statements concerning the Stefan-Boltzmann equation is true?
    (a) This equation applies only to perfect radiators.
    (b) This equation applies only to perfect absorbers.
    (c) This equation is valid with any temperature units.
    (d) This equation describes the transport of thermal energy by conduction.
    (e) The equation can be used to calculate the power absorbed by any surface.

29. Which one of the following statements concerning *emissivity* is false?
    (a) The emissivity is 1.0 for a perfect radiator.
    (b) The emissivity is 1.0 for a perfect absorber.
    (c) Emissivity depends on the condition of the surface.
    (d) Emissivity is a dimensionless quantity.
    (e) Emissivity depends on the surface area of the object.

30. A hot metal ball is hung in an oven that is maintained at 700 K; and it cools. When the temperature of the ball is 900 K, it is losing heat at a rate of 0.10 J/min. At what rate will the ball loose heat when the ball reaches 800 K? Assume that the emissivity of the ball does not change appreciably with temperature.
    (a) 0.1 J/min
    (b) 0.2 J/min
    (c) 0.4 J/min
    (d) 0.6 J/min
    (e) 0.01 J/min

31. Assume that the sun is a sphere of radius $6.96 \times 10^8$ m and that its surface temperature is $5.8 \times 10^3$ K. If the sun radiates at a rate of $3.90 \times 10^{26}$ W and is a perfect emitter, at what rate is energy emitted *per square meter* at the sun's surface?
    (a) $5.6 \times 10^7$ W/m$^2$
    (b) $6.4 \times 10^7$ W/m$^2$
    (c) $5.6 \times 10^{17}$ W/m$^2$
    (d) $12.8 \times 10^7$ W/m$^2$
    (e) $25.6 \times 10^7$ W/m$^2$

32. Which one of the following graphs shows the rate at which heat is emitted from a hot body as a function of its Kelvin temperature $T$?

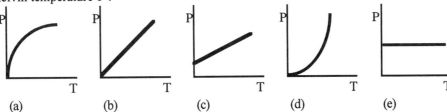

33. A blue supergiant star has a radius of $7.0 \times 10^{10}$ m. The spherical surface behaves as a blackbody radiator. If the surface temperature is $2.2 \times 10^4$ K, what is the rate at which energy is radiated from the star?
    (a) $1.2 \times 10^{-3}$ J/s
    (b) $7.7 \times 10^{19}$ J/s
    (c) $2.0 \times 10^{23}$ J/s
    (d) $8.2 \times 10^{32}$ J/s
    (e) $1.9 \times 10^{43}$ J/s

34. Assuming a filament in a 100 W light bulb acts like a perfect blackbody, what is the temperature of the hottest portion of the filament if it has a surface area of $6.3 \times 10^{-5}$ m$^2$? The Stefan-Boltzmann constant is $5.67 \times 10^{-8}$ W/(m$^2 \cdot$ K$^4$).
    (a) 130 K
    (b) 1100 K
    (c) 2300 K
    (d) 5800 K
    (e) 30 000 K

35. A person steps out of the shower and dries off. The person's skin with an emissivity of 0.70 has a total area of 1.2 m$^2$ and a temperature of 33 °C. What is the net rate at which energy is lost to the room through radiation by the naked person if the room temperature is 24 °C?
    (a) 0.041 W
    (b) 25 W
    (c) 81 W
    (d) 67 W
    (e) 47 W

## 132 THE TRANSFER OF HEAT

36. Object **A** has an emissivity of 0.95; and its temperature is 25 °C. At what temperature (in degrees Celsius) does object **B**, whose emissivity is 0.60, emit radiation at the same rate as object **A** if both objects have the same surface area?
    (a) 28 °C
    (b) 40 °C
    (c) 61 °C
    (d) 73 °C
    (e) 97 °C

## Additional Problems

37. A beaker of water is placed on a Bunsen burner. As the lower layers of water are heated, they become less dense and rise. This permits cooler layers to move downward and be heated. Eventually, the water boils. Which method(s) of heat transfer is (are) primarily responsible for boiling the sample?
    (a) conduction
    (b) convection
    (c) radiation
    (d) both conduction and radiation
    (e) both conduction and convection

38. If a beaker of water is placed *under* a broiler so that the heating coil is above the beaker. It is observed that only the surface layer boils. The water at the bottom of the beaker remains close to the initial temperature of the water. Which of the following statements is the most reasonable conclusion to be drawn from these observations?
    (a) Water is a poor conductor of heat.
    (b) The sample must contain impurities.
    (c) Water is not easily heated by radiation.
    (d) Water exhibits anomalous thermal behavior.
    (e) The molecular motion in the sample is not random.

*Questions 39 and 40 pertain to the situation described below:*

2.0 kg of ice at 0.0 °C are placed in a closed glass beaker [$k = 0.80$ W/(m · C°)]. The closed beaker is then placed in a 30 °C room. The surface area of the beaker is $4.5 \times 10^{-6}$ m$^2$; and its walls are 0.005 m thick.

39. How much heat is required to melt the ice?
    (a) 16 J
    (b) 1080 J
    (c) $1.6 \times 10^5$ J
    (d) $6.7 \times 10^5$ J
    (e) $4.5 \times 10^6$ J

40. How long does it take for the ice to melt?
    (a) 4.7 min
    (b) 7.8 min
    (c) 24 min
    (d) 2.7 h
    (e) 7.8 h

# CHAPTER 14
# The Ideal Gas Law and Kinetic Theory

## Section 14.1 Molecular Mass, the Mole, and Avogadro's Number

1. Complete the following statement: The *atomic mass unit* (u) is defined so that 1 u is exactly equal to the mass of
   (a) a single hydrogen atom.
   (b) 1/4 of a helium molecule.
   (c) 1/16 of an oxygen-16 atom.
   (d) 1/32 of an oxygen molecule.
   (e) 1/12 of a carbon-12 atom.

2. Which one of the following statements concerning the *mole* is false?
   (a) It is related to Avogadro's number.
   (b) It is defined in terms of the carbon-12 isotope.
   (c) It is the SI base unit for expressing the "amount" of a substance.
   (d) One mole of a substance has the same mass as one mole of any other substance.
   (e) One mole of a substance contains the same number of particles as one mole of any other substance.

3. How many molecules are in 0.064 kg of sulfur dioxide, $SO_2$? (atomic masses: S = 32 u; O = 16 u)
   (a) 3
   (b) 64
   (c) $2.00 \times 10^{24}$
   (d) $6.02 \times 10^{23}$
   (e) $3.85 \times 10^{25}$

4. What is the mass of 1.5 moles of hydrogen molecules, $H_2$? (atomic mass of H = 1 u)
   (a) 0.0063 g
   (b) 0.0030 kg
   (c) 0.0082 kg
   (d) 0.015 kg
   (e) 0.022 kg

5. How many moles are in a 0.23-kg sample of carbon dioxide, $CO_2$? (C = 12 u; O = 16 u)
   (a) 5.2
   (b) 52
   (c) 230
   (d) $1.1 \times 10^4$
   (e) $1.4 \times 10^{26}$

6. Any given samples of oxygen and carbon that have the same number of atoms must have masses that occur in what ratio, carbon/oxygen? Note the following atomic masses: C = 12 u; O = 16 u.
   (a) 1:1
   (b) 1:2
   (c) 2:1
   (d) 8:3
   (e) 4:3

7. A gold coin has a mass of 24.75 g. If the atomic mass of gold is 192.967 u, how many gold atoms are in the coin?
   (a) $7.724 \times 10^{22}$
   (b) $9.158 \times 10^{22}$
   (c) $3.812 \times 10^{23}$
   (d) $5.526 \times 10^{24}$
   (e) $1.390 \times 10^{25}$

## Section 14.2 The Ideal Gas Law

8. Which one of the following statements best explains why gases are not commercially sold by volume?
   (a) Gas volume is negligible.
   (b) Gas volume is difficult to measure.
   (c) Gas volume depends on the type of gas.
   (d) Gases have comparatively low densities.
   (e) Gas volume depends on temperature and pressure.

## 134  THE IDEAL GAS LAW AND KINETIC THEORY

9. Note the following six properties:
   (1) number of molecules    (3) temperature    (5) latent heat
   (2) moment of inertia      (4) pressure       (6) volume
   Which four of the listed properties are needed to describe an ideal gas?
   (a) 1, 2, 4, 6
   (b) 1, 3, 5, 6
   (c) 1, 3, 4, 6
   (d) 1, 4, 5, 6
   (e) 2, 4, 5, 6

10. A sample of a monatomic ideal gas is originally at 20 °C. What is the final temperature of the gas if both the pressure and volume are doubled?
    (a) 5 °C
    (b) 20 °C
    (c) 80 °C
    (d) 900 °C
    (e) 1200 °C

11. An ideal gas at 0 °C is contained within a rigid vessel. The temperature of the gas is increased by 1 C°. What is $P_f/P_i$, the ratio of the final to initial pressure?
    (a) 273/274
    (b) 274/273
    (c) 1/2
    (d) 1/10
    (e) 1/273

12. Neon gas at 20 °C is confined within a rigid vessel. It is then heated until its pressure is doubled. What is the final temperature of the gas?
    (a) 10 °C
    (b) 20 °C
    (c) 40 °C
    (d) 313 °C
    (e) 586 °C

13. Argon gas at 305 K is confined within a constant volume at a pressure $P_1$. If the gas has a pressure $P_2$ when it is cooled to 195 K, what is the ratio of $P_2$ to $P_1$?
    (a) 0.410
    (b) 0.639
    (c) 0.717
    (d) 1.28
    (e) 1.56

14. A sample of neon gas at 20 °C is confined to a cylinder with a movable piston. It is then heated until its pressure is doubled. What is the final temperature of the gas?
    (a) 10 °C
    (b) 40 °C
    (c) 313 °C
    (d) 586 °C
    (e) This cannot be found since the final and initial volumes are unknown.

15. Heat is supplied to a sample of a monatomic ideal gas at 40 °C. It is observed that the gas expands until its volume is doubled and the pressure drops to half of its original value. What is the final temperature of the gas?
    (a) 10 °C
    (b) 20 °C
    (c) 40 °C
    (d) 80 °C
    (e) 1600 °C

16. In the space between two stars, the temperature of a gas cloud is 12 K; and the density of the gas is $1.2 \times 10^{-8}$ atom/m$^3$. What is the absolute pressure of the gas?
    (a) $2.0 \times 10^{-30}$ Pa
    (b) $1.2 \times 10^{-28}$ Pa
    (c) $2.0 \times 10^{-17}$ Pa
    (d) $1.2 \times 10^{-6}$ Pa
    (e) $1.4 \times 10^{-4}$ Pa

17. An ideal gas with a fixed number of molecules is maintained at a constant pressure. At 30.0 °C, the volume of the gas is 1.50 m³. What is the volume of the gas when the temperature is increased to 75.0 °C?
    (a) 0.60 m³
    (b) 1.30 m³
    (c) 1.72 m³
    (d) 2.45 m³
    (e) 3.75 m³

18. The volume of a carbon dioxide bubble rising in a glass of beer is observed to nearly double as the bubble rises from the bottom to the top of the glass. Why, according to our textbook, does the volume nearly double?
    (a) The temperature at the bottom is cooler than it is at the top.
    (b) The amount of carbon dioxide in the bubble increases.
    (c) The fluid pressure of the beer is greater at the bottom of the glass than at the top.
    (d) The pressure inside the bubble decreases as it rises.
    (e) The shape of the glass determines the net force exerted on the bubble.

19. A sealed container has a volume of 0.020 m³ and contains 15.0 g of molecular nitrogen ($N_2$) which has a molecular mass of 28.0 u. The gas is at a temperature of 525 K. What is the absolute pressure of the nitrogen gas?
    (a) $3.9 \times 10^{-19}$ Pa
    (b) $4.3 \times 10^{-5}$ Pa
    (c) $1.2 \times 10^5$ Pa
    (d) $1.9 \times 10^5$ Pa
    (e) $4.3 \times 10^6$ Pa

20. An ideal gas is contained in a vessel with a moveable piston. Initially, the gas has a volume of 0.018 m³, an absolute pressure of 1.5 atm, and a temperature of 30 °C. The pressure is 0.75 atm when the volume of the container is decreased to 0.009 m³. What is the final temperature of the gas?
    (a) 76 K
    (b) 98 K
    (c) 170 K
    (d) 240 K
    (e) 300 K

## Section 14.3 Kinetic Theory of Gases

21. Complete the following statement: The absolute temperature of an ideal gas is directly proportional to
    (a) the number of molecules in the sample.
    (b) the average momentum of a molecule of the gas.
    (c) the average translational kinetic energy of the gas.
    (d) the amount of heat required to raise the temperature of the gas by 1 C°.
    (e) the relative increase in volume of the gas for a temperature increase of 1 C°.

22. At what temperature would one mole of molecular oxygen ($O_2$) have $5.0 \times 10^3$ J of *translational* kinetic energy? Note: the atomic mass of O is 16 u.
    (a) 130 °C
    (b) 390 °C
    (c) 400 °C
    (d) 670 °C
    (e) 1000 °C

23. A five liter tank contains 2.00 moles of oxygen gas, $O_2$, at 40 °C. What pressure is exerted on the sides of the tank by the oxygen molecules?
    (a) 83.3 Pa
    (b) $4.01 \times 10^3$ Pa
    (c) $1.33 \times 10^5$ Pa
    (d) $4.01 \times 10^5$ Pa
    (e) $1.04 \times 10^6$ Pa

## 136 THE IDEAL GAS LAW AND KINETIC THEORY

24. A $1.00 \times 10^{-2}$ m$^3$ flask contains 0.0160 kg of oxygen gas, $O_2$, at 77.0 °C. What is the pressure exerted on the inner walls of the flask by the oxygen gas? **Note:** the atomic mass of O is 15.9994 u.
    (a) $3.19 \times 10^4$ Pa
    (b) $1.45 \times 10^5$ Pa
    (c) $2.90 \times 10^5$ Pa
    (d) $5.79 \times 10^5$ Pa
    (e) $4.87 \times 10^6$ Pa

25. Which one of the following properties of a gas is not consistent with kinetic theory?
    (a) The average speed of the gas molecules is smaller at high temperatures.
    (b) Gas molecules are widely separated.
    (c) Gases fill whatever space is available to them.
    (d) Gas molecules move rapidly in a random fashion.
    (e) Gas molecules make elastic collisions with the walls of the containing vessel.

26. Which one of the following statements concerning a collection of gas molecules at a certain temperature is true?
    (a) All molecules move with the same velocity.
    (b) Most of the molecules have the same kinetic energy.
    (c) The lower the temperature, the greater are the molecular speeds.
    (d) All molecules possess the same momentum.
    (e) The molecules have a range of kinetic energies.

27. Consider two ideal gases, **A** and **B**, at the same temperature. The *rms* speed of the molecules of gas **A** is twice that of gas **B**. How does the molecular mass of **A** compare to that of **B**?
    (a) It is twice that of **B**.
    (b) It is one half that of **B**.
    (c) It is four times that of **B**.
    (d) It is one fourth that of **B**.
    (e) It is 1.4 times that of **B**.

28. Under which of the following conditions would you expect real gases to approach ideal behavior?
    (a) low temperature and low pressure
    (b) high temperature and low pressure
    (c) low temperature and high pressure
    (d) high temperature and high pressure
    (e) high temperature and high density

29. Which one of the following factors is directly responsible for the *pressure* exerted by a confined gas?
    (a) the atomic mass of the gas
    (b) the density of the sample of molecules
    (c) the temperature of the sample of molecules
    (d) the collision of gas molecules with the sides of the containing vessel
    (e) the average translational kinetic energy of the molecules

30. Complete the following statement: A bicycle tire explodes after lying in the hot afternoon sun. This is an illustration of
    (a) Charles' law.
    (b) Boyle's law.
    (c) Fick's law.
    (d) the ideal gas law.
    (e) the Maxwell speed distribution.

31. Complete the following statement: The internal energy of an ideal monatomic gas is
    (a) proportional to the pressure and inversely proportional to the volume of the gas.
    (b) independent of the number of moles of the gas.
    (c) proportional to the Kelvin temperature of the gas.
    (d) dependent on both the pressure and the temperature of the gas.
    (e) a constant that is independent of pressure, volume or temperature.

32. A mixture of two ideal gases **A** and **B** is in thermal equilibrium at 600 K. A molecule of **A** has one-fourth the mass of a molecule of **B** and the *rms* speed of molecules of **A** is 400 m/s. Determine the *rms* speed of molecules of **B**.
   (a) 100 m/s
   (b) 200 m/s
   (c) 400 m/s
   (d) 800 m/s
   (e) 1600 m/s

33. Complete the following statement: The pressure exerted by a monatomic, ideal gas on the walls of its containing vessel is a measure of
   (a) the molecular kinetic energy per unit volume.
   (b) the average random kinetic energy per molecule.
   (c) the temperature of the gas, regardless of the volume of the vessel.
   (d) the total internal energy of the gas, regardless of the volume of the vessel.
   (e) the momentum per unit volume.

34. A canister containing 150 kg of an ideal gas has a volume of 8.0 m$^3$. If the gas exerts a pressure of $5.0 \times 10^5$ Pa, what is the *rms* speed of the molecules?
   (a) 160 m/s
   (b) 280 m/s
   (c) 350 m/s
   (d) 390 m/s
   (e) 420 m/s

35. A flask contains 1.00 mole of oxygen gas, O$_2$, at 0.00 °C and $1.013 \times 10^5$ Pa. What is the *rms* speed of the molecules? Note: the atomic mass of O is 16 u.
   (a) 230 m/s
   (b) 460 m/s
   (c) 651 m/s
   (d) 920 m/s
   (e) 1302 m/s

36. A tank contains 135 moles of the monatomic gas argon at a temperature of 15.3 °C. How much energy must be added to the gas to increase its temperature to 45.0 °C?
   (a) $2.50 \times 10^3$ J
   (b) $3.33 \times 10^4$ J
   (c) $5.00 \times 10^4$ J
   (d) $5.70 \times 10^5$ J
   (e) $7.50 \times 10^6$ J

37. A gas molecule with a molecular mass of 32.0 u has a speed of 325 m/s. What is the temperature of the gas molecule?
   (a) 72.0 K
   (b) 136 K
   (c) 305 K
   (d) 459 K
   (e) A temperature cannot be assigned to a single molecule.

38. A physics student looks into a microscope and observes that small particles suspended in water are moving about in an irregular motion. Which of the following statements is the best explanation for this observation?
   (a) Water molecules strike the particles giving them the same average kinetic energy as the water.
   (b) The particles are carried by convection currents in the water.
   (c) The small particles may be considered a fluid; and thus, move about randomly.
   (d) The actual motion is regular, but the speeds of particles are too large to observe the regular motion.
   (e) The particles are moving to the fill the volume of the water.

39. Calculate the rms speed of the oxygen molecules in the air if the temperature is 5.00 °C.
   **Note:** The mass of the oxygen molecule is 31.9988 u.
   (a) 62.0 m/s
   (b) 86.3 m/s
   (c) 328 m/s
   (d) 465 m/s
   (e) 487 m/s

138  THE IDEAL GAS LAW AND KINETIC THEORY

40. What is the internal energy of 1.75 kg of helium (atomic mass = 4.00260 u) with a temperature of 100 °C?
    (a) $4.65 \times 10^3$ J
    (b) $5.44 \times 10^5$ J
    (c) $2.03 \times 10^6$ J
    (d) $8.16 \times 10^6$ J
    (e) $1.22 \times 10^7$ J

## Section 14.4 Diffusion

41. Which one of the following statements concerning *diffusion* is true?
    (a) Diffusion occurs only in gases.
    (b) Diffusion occurs in solids and fluids.
    (c) Diffusion is described by Boyle's law.
    (d) The SI unit for the diffusion constant $D$ is m/s$^2$.
    (e) Diffusion occurs when molecules move from regions of lower concentration to regions of higher concentration.

42. A concentration difference of a certain solute of $1.0 \times 10^{-2}$ kg/m$^3$ is maintained between the ends of a tube with a length of 3.5 m and a cross-sectional area of $2.5 \times 10^{-1}$ m$^2$. When 0.0040 g of the solute is introduced to the tube, it takes 350 minutes for this solute to diffuse through the solvent to the opposite end of the tube. What is the diffusion constant for the solute?
    (a) $2.7 \times 10^{-7}$ m$^2$/s
    (b) $4.5 \times 10^{-9}$ m$^2$/s
    (c) $7.5 \times 10^{-10}$ m$^2$/s
    (d) $6.3 \times 10^{-8}$ m$^2$/s
    (e) $1.1 \times 10^{-10}$ m$^2$/s

43. A lower tank contains water while an upper tank contains air with a negligible concentration of water vapor. A tube with a circular cross section and radius 0.050 m is inserted between the two tanks. The length of the tube is 0.25 m. The diffusion coefficient of water vapor through air is $2.4 \times 10^{-5}$ m$^2$/s. The concentration of water vapor just above the surface of the water is $1.7 \times 10^{-2}$ kg/m$^3$. If the water vapor is removed from the upper tank so that the concentration there remains nearly zero, what mass of water vapor diffuses through the tube each hour?
    (a) $4.6 \times 10^{-5}$ kg
    (b) $2.0 \times 10^{-6}$ kg
    (c) $7.1 \times 10^{-7}$ kg
    (d) $1.3 \times 10^{-8}$ kg
    (e) $8.2 \times 10^{-8}$ kg

## Additional Problems

44. How many air molecules are in a room at temperature 23.8 °C and standard pressure if the dimensions of the room are 3.66 m × 3.66 m × 2.43 m?
    (a) 1330
    (b) 16 600
    (c) $3.03 \times 10^{24}$
    (d) $8.02 \times 10^{26}$
    (e) $1.00 \times 10^{28}$

45. An automobile tire is inflated to a gauge pressure of 32 lb/in$^2$ at a temperature of −10.0 °C. Under strenuous driving, the tire heats up to 40.0 °C. What is the new gauge pressure if the volume of the tire remains essentially the same? (atmospheric pressure = 14.7 lb/in$^2$)
    (a) 17.5 lb/in$^2$
    (b) 20.6 lb/in$^2$
    (c) 38.0 lb/in$^2$
    (d) 40.9 lb/in$^2$
    (e) 55.6 lb/in$^2$

46. An air bubble 20 m below the surface of a lake has a volume of 0.02 m³. The bubble then rises to the surface. If the temperature of the lake is uniform, what is the volume of the bubble just before it breaks through the surface?
    (a) 0.02 m³
    (b) 0.04 m³
    (c) 0.06 m³
    (d) 0.08 m³
    (e) 0.10 m³

47. A bubble with a volume of 1.0 cm³ forms at the bottom of a lake that is 20.0 m deep. The temperature at the bottom of the lake is 10.0 °C. The bubble rises to the surface where the temperature is 25 °C. Assume that the bubble is small enough that its temperature always matches that of its surroundings. What is the volume of the bubble just before it breaks the surface of the water?
    (a) 2.1 cm³
    (b) 2.8 cm³
    (c) 3.1 cm³
    (d) 6.0 cm³
    (e) 7.7 cm³

48. On a cold day (–3 °C) the gauge pressure on a tire is 2.0 atm. If the tire is heated to 27 °C, without changing its volume, what will the gauge pressure be?
    (a) 2.1 atm
    (b) 2.3 atm
    (c) 2.9 atm
    (d) 3.3 atm
    (e) 27 atm

49. On a cold day (–3 °C) the gauge pressure on a tire reads 2 atm. If the tire is heated to 27 °C, what will be the absolute pressure of the air inside the tire?
    (a) 2.0 atm
    (b) 2.2 atm
    (c) 2.4 atm
    (d) 2.9 atm
    (e) 3.3 atm

50. What is the density of methane, $CH_4$, (molecular mass = 16 u) at STP?
    (a) 0.357 kg/m³
    (b) 0.386 kg/m³
    (c) 0.431 kg/m³
    (d) 0.712 kg/m³
    (e) 0.951 kg/m³

# CHAPTER 15 Thermodynamics

*Section 15.1 Thermodynamic Systems and Their Surroundings*
*Section 15.2 The Zeroth Law of Thermodynamics*
*Section 15.3 The First Law of Thermodynamics*

1. Complete the following statement: Walls that separate a system from its surroundings and permit heat to flow through them are called
   (a) diathermal walls.
   (b) adiabatic walls.
   (c) entropic walls.
   (d) isobaric walls.
   (e) isochoric walls.

2. Which one of the following situations is a direct application of the Zeroth Law of Thermodynamics?
   (a) Block A has twice the temperature of block B before they are brought into contact. Upon contact, heat flows from block A to block B.
   (b) A sample of gas within a cylinder with a piston is held at constant temperature and pressure while it is allowed to expand. During this process, the gas absorbs heat from its surroundings.
   (c) The motor of a refrigerator uses electric energy to remove heat from inside the refrigerator and transfer it to the room.
   (d) A physicist removes energy from a system in her laboratory until she reaches a temperature of $3 \times 10^{-10}$ K, a temperature very close to (but still greater than) absolute zero.
   (e) A thermometer is calibrated by placing it in an ice water bath within an adiabatic container until the thermometer is in thermal equilibrium with the ice water.

3. Complete the following statement: The first law of thermodynamics states that
   (a) heat is a form of energy.
   (b) entropy is a function of state.
   (c) the entropy of the universe is increasing.
   (d) the change in the internal energy of a system is given by $Q - W$.
   (e) no engine can be more efficient than a Carnot engine operating between the same two temperatures.

4. When the gas enclosed beneath the piston shown in the figure receives 1930 J of heat, $Q$, from its surroundings, it performs 2250 J of work in raising the piston. What is the change in the internal energy of the gas?
   (a) −320 J
   (b) +320 J
   (c) −4180 J
   (d) +4180 J
   (e) zero

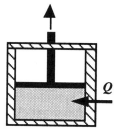

5. Enclosed beneath the moveable piston in the figure is 4.8 moles of a monatomic ideal gas. The gas performs work on the piston as 2300 J of heat are added from the surroundings. During the process, the temperature of the gas decreases by 45 K. How much work does the gas perform?
   (a) 5000 J
   (b) 3200 J
   (c) 1400 J
   (d) 600 J
   (e) 400 J

# TEST BANK

## Chapter 15

6. An ideal gas absorbs 750 J of heat as it performs 625 J of work. What is the resulting change in temperature if there is 1.3 moles of the gas?
   - (a) −8.6 K
   - (b) −4.3 K
   - (c) 7.7 K
   - (d) 9.6 K
   - (e) 23 K

## Section 15.4 Thermal Processes

7. What are the SI units of the product of pressure and volume, $PV$?
   - (a) newton
   - (b) kilogram • meter/second
   - (c) joule
   - (d) meter$^2$
   - (e) newton • second

8. An *isobaric* process is represented on a *pressure-volume* graph by which one of the following curves?
   - (a) a parabola
   - (b) a hyperbola
   - (c) a vertical line
   - (d) a horizontal line
   - (e) a circle

9. A match is placed in an oxygen-filled cylinder that has a movable piston. The piston is moved so quickly that no heat escapes as the match ignites. What kind of change is demonstrated in this process?
   - (a) an isobaric change
   - (b) an adiabatic change
   - (c) an isothermal change
   - (d) an isochoric change
   - (e) a change of heat capacity

10. A container is divided into two chambers that are separated by a valve. The left chamber contains one mole of a monatomic ideal gas. The right chamber is evacuated. At some instant, the valve is opened and the gas rushes freely into the right chamber. Which one of the following statements concerning this process is true?
    - (a) Work is done by the gas.
    - (b) The temperature of the gas decreases.
    - (c) The change in the entropy of the gas is zero.
    - (d) The walls of the containing vessel must get colder.
    - (e) The change in the internal energy of the gas is zero.

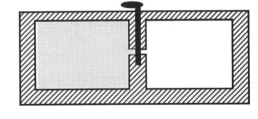

11. A thermally isolated sample of an ideal gas at a fixed temperature is confined to one half of a container by an impermeable membrane. The other half of the container is evacuated. The membrane is then pierced and the gas is allowed to expand freely and to double its volume as shown. Which one of the following statements is true concerning this situation?
    - (a) The process is reversible.
    - (b) This is an isothermal process.
    - (c) The entropy of the gas decreases.
    - (d) The internal energy of the gas must decrease.
    - (e) The temperature of the gas decreases to one-half of its original value.

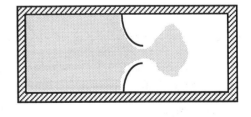

*142* **THERMODYNAMICS**

*Questions 12 through 14 pertain to the following situation:*

5.00 kg of liquid water is heated to 100.0 °C in a closed system. At this temperature, the density of liquid water is 958 kg/m$^3$. The pressure is maintained at atmospheric pressure of $1.01 \times 10^5$ Pa. A moveable piston of negligible weight rests on the surface of the water. The water is then converted to steam by adding an additional amount of heat to the system. When all of the water is converted, the final volume of the steam is 8.50 m$^3$. The latent heat of vaporization of water is $2.26 \times 10^6$ J/kg.

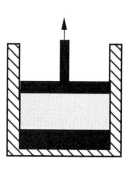

12. How much work is done by this closed system during this isothermal process?
    (a) 8.37 J
    (b) $4.20 \times 10^3$ J
    (c) $1.21 \times 10^4$ J
    (d) $8.58 \times 10^5$ J
    (e) $1.94 \times 10^6$ J

13. How much heat added to the system in the isothermal process of converting all of the water into steam?
    (a) $2.17 \times 10^3$ J
    (b) $1.70 \times 10^4$ J
    (c) $4.52 \times 10^5$ J
    (d) $3.78 \times 10^6$ J
    (e) $1.13 \times 10^7$ J

14. What is the change in the internal energy during this isothermal process?
    (a) zero
    (b) $1.28 \times 10^4$ J
    (c) $4.40 \times 10^5$ J
    (d) $2.93 \times 10^6$ J
    (e) $1.04 \times 10^7$ J

15. Which one of the following pressure-volume graphs represents an *isochoric* process?

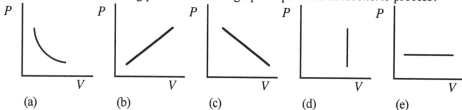

### Section 15.5 Thermal Processes that Utilize an Ideal Gas

16. Heat is added to a sample of an ideal monatomic gas. Which one of the following statements is necessarily true?
    (a) The gas must expand.
    (b) The gas must do work.
    (c) The type of change that will occur depends on the conditions of the gas when the heat was added.
    (d) The gas must change phase.
    (e) The temperature of the gas must increase.

17. If one complete cycle of a reversible process is carried out on a sample of an ideal gas so that its final state is the same as its initial state, which of the following quantities is the *only* one which can be *non-zero*?
    (a) the change in volume of the sample
    (b) the net heat absorbed by the sample
    (c) the change in the entropy of the sample
    (d) the change in temperature of the sample
    (e) the change in the internal energy of the sample

18. In one stage of a reversible process, the temperature of an ideal gas remains constant as its volume is decreased. Which one of the following statements concerning this situation is true?
    (a) The process is adiabatic.
    (b) The pressure of the gas decreases in this process.
    (c) Heat flows out of the gas and into the surroundings.
    (d) The gas does "positive" work on its surroundings.
    (e) The average kinetic energy of the gas molecules increases.

19. A fixed amount of ideal gas is compressed adiabatically. Which entry in the table below correctly depicts the *sign* of the *work done*, the *change in the internal energy*, and the *heat exchanged* with the environment?

    |     | work done | change in internal energy | heat exchanged |
    | --- | --- | --- | --- |
    | (a) | positive | negative | zero |
    | (b) | negative | zero | positive |
    | (c) | negative | negative | zero |
    | (d) | positive | positive | zero |
    | (e) | negative | positive | zero |

20. A fixed amount of ideal gas is compressed isothermally. Which entry in the table below correctly depicts the sign of the work done, the change in the internal energy, and the heat exchanged with the environment?

    |     | work done | change in internal energy | heat exchanged |
    | --- | --- | --- | --- |
    | (a) | negative | zero | negative |
    | (b) | positive | negative | zero |
    | (c) | negative | zero | positive |
    | (d) | negative | negative | zero |
    | (e) | positive | zero | positive |

21. An ideal monatomic gas undergoes an adiabatic process. Its internal energy *increases* by 50 J. Which pair of choices below is the correct for this process?

    |     | work done | heat exchanged |
    | --- | --- | --- |
    | (a) | 50 J by the system | zero |
    | (b) | 50 J on the system | zero |
    | (c) | 50 J by the system | 100 J supplied |
    | (d) | zero | 50 J removed |
    | (e) | zero | 50 J added |

22. Enclosed beneath the moveable piston in the figure is 1.5 moles of a monatomic ideal gas at 314 K. The initial volume of the gas is 3.0 m$^3$. The gas is compressed isothermally to a final volume of 1.0 m$^3$. How much heat is removed from the gas?
    (a) −6450 J
    (b) −4300 J
    (c) −2900 J
    (d) −1450 J
    (e) zero

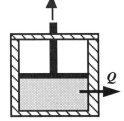

23. Neon is a monatomic gas with a molar heat capacity at constant volume of 12.66 J/(mol · K). Two moles of neon gas enclosed in a constant volume system receive 5250 J of heat. If the gas was initially at 293 K, what is the final temperature of the neon?
    (a) 200 K
    (b) 300 K
    (c) 400 K
    (d) 500 K
    (e) 600 K

## 144 THERMODYNAMICS

24. A paddle wheel frictionally adds thermal energy to 5.0 moles of an ideal monatomic gas in a sealed insulated container. The paddle wheel is driven by a cord connected to a falling object as shown in the figure. How far has the 2.0-kg object fallen when the temperature of the gas increases by 10 K?
    (a) 8.0 m
    (b) 16 m
    (c) 32 m
    (d) 50 m
    (e) 98 m

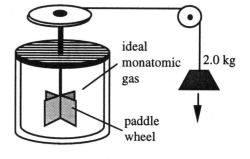

25. A paddle wheel frictionally adds thermal energy to an ideal monatomic gas in a sealed, insulated container. The paddle wheel is driven by a cord connected to a falling object as shown in the figure. In this experiment, a 5.0-kg object falls through a total distance of 2.0 m and the temperature of the gas is found to increase by 4 C°. Assume that all of the mechanical energy lost by the falling object goes into the gas. How many moles of gas must be present in this container?

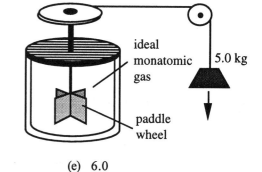

    (a) 2.0
    (b) 3.0
    (c) 4.0
    (d) 5.0
    (e) 6.0

26. One mole of a monatomic gas at 400 K is reversibly taken to half of its original volume by an isobaric process. How much work is done by the gas?
    (a) +1700 J
    (b) −1700 J
    (c) +3300 J
    (d) −3300 J
    (e) −8300 J

27. An ideal gas is taken from state **A** to state **B** through process shown on the pressure-volume graph.

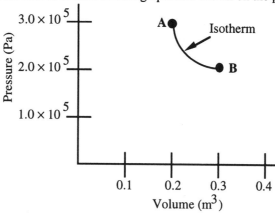

How much heat is added to the gas in this process?
    (a) zero
    (b) $1.0 \times 10^4$ J
    (c) $2.4 \times 10^4$ J
    (d) $6.0 \times 10^4$ J
    (e) This cannot be determined since $n$ and $T$ are not specified.

28. Two moles of a confined ideal monatomic gas begin at state **A** in the pressure-volume graph and follow the path shown to state **D**. If the temperature of the gas at **A** is 54 K, what is the temperature of the gas at **D**?
   (a) 32 K
   (b) 46 K
   (c) 54 K
   (d) 60 K
   (e) 78 K

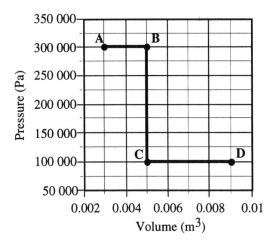

*Questions 29 through 32 pertain to the situation described below:*

An ideal monatomic gas originally in state **A** is taken reversibly to state **B** along the straight line path shown in the pressure-volume graph.

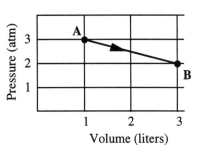

29. What is the change in the internal energy of the gas for this process?
   (a) zero
   (b) +12 cal
   (c) −110 cal
   (d) +110 cal
   (e) +122 cal

30. How much work was done by the gas?
   (a) zero
   (b) +12 cal
   (c) −110 cal
   (d) +110 cal
   (e) +122 cal

31. How much heat was exchanged during this process?
   (a) −110 cal
   (b) −12 cal
   (c) zero
   (d) +122 cal
   (e) +232 cal

32. Suppose that the same gas is originally in state **A** as described above, but its volume is increased *isothermally* until a new volume of 3.0 liters is reached. Which one of the following statements for this isothermal process is false?
   (a) The change in the internal energy is zero.
   (b) The final state of the system will still be **B**.
   (c) The work done will be smaller for the isothermal process.
   (d) The heat added will be smaller for the isothermal process.
   (e) The heat added for the isothermal process will be equal to the work done.

## 146  THERMODYNAMICS

*Questions 33 through 35 pertain to the situation described below:*

An ideal monatomic gas expands isobarically from state **A** to state **B**. It is then compressed isothermally from state **B** to state **C** and finally cooled at constant volume until it returns to its initial state **A**.

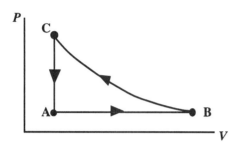

$V_A = 4 \times 10^{-3}$ m$^3$
$V_B = 8 \times 10^{-3}$ m$^3$
$P_A = P_B = 1 \times 10^6$ Pa
$P_C = 2 \times 10^6$ Pa
$T_A = 600$ K

33. What is the temperature of the gas when it is in state **B**?
    (a) 437 K
    (b) 573 K
    (c) 927 K
    (d) 1200 K
    (e) 1473 K

34. How much work is done by the gas in expanding isobarically from **A** to **B**?
    (a) $1 \times 10^3$ J
    (b) $2 \times 10^3$ J
    (c) $3 \times 10^3$ J
    (d) $4 \times 10^3$ J
    (e) $5 \times 10^3$ J

35. How much work is done on the gas in going from **B** to **C**?
    (a) $2.5 \times 10^6$ J
    (b) $5.6 \times 10^6$ J
    (c) $4.5 \times 10^6$ J
    (d) $6.5 \times 10^6$ J
    (e) $8.0 \times 10^6$ J

*Questions 36 through 38 pertain to the following situation:*

2.0 moles of an ideal gas have an initial Kelvin temperature $T_0$ and absolute pressure $P_0$. The gas undergoes a reversible isothermal compression from an initial volume $V_0$ to a final volume $V_0/2$.

36. Which one of the following expressions is equivalent to the final volume of the gas after the units have been properly considered?
    (a) $P_0/2$
    (b) $P_0/31.7$
    (c) $2P_0$
    (d) $3.17P_0$
    (e) $4P_0$

37. How much heat is exchanged with the environment? If heat is exchanged, is it absorbed or released?
    (a) $P_0V_0 \ln 2$, released
    (b) $P_0V_0 \ln 2$, absorbed
    (c) $P_0V_0/2$, released
    (d) $P_0V_0/2$, absorbed
    (e) zero

38. What is the change in entropy of the sample?
    (a) $+8.31 \ln 2$
    (b) $-8.31 \ln 2$
    (c) $+16.62 \ln 2$
    (d) $-16.62 \ln 2$
    (e) zero

## Section 15.6 Specific Heat Capacities and the First Law of Thermodynamics
## Section 15.7 The Second Law of Thermodynamics

39. Which one of the following statements is true concerning the *ratio* of the molar heat capacities $C_p/C_v$ for an ideal gas?
    (a) The ratio is always 1.
    (b) The ratio is always less than 1.
    (c) The ratio is always greater than 1.
    (d) The ratio is sometimes less than 1.
    (e) The ratio is sometimes greater than 1.

40. The ratio of the molar specific heat capacity at constant pressure to that at constant volume, $\gamma$, for diatomic hydrogen gas is 7/5. In an adiabatic compression, the gas, originally at atmospheric pressure, is compressed from an original volume of 0.30 m³ to 0.15 m³. What is the final pressure of the gas?
    (a) $2.0 \times 10^5$ Pa
    (b) $2.7 \times 10^5$ Pa
    (c) $3.0 \times 10^5$ Pa
    (d) $3.7 \times 10^5$ Pa
    (e) $4.0 \times 10^5$ Pa

41. Determine the quantity of heat added to 3.5 moles of the ideal gas Argon if the temperature increases from 75 °C to 225 °C during an *isobaric* process. **Note**: The molar specific heats of Argon are $C_V = 3.0$ cal/K•mol and $C_P = 5.0$ cal/K•mol.
    (a) 2600 cal
    (b) 2100 cal
    (c) 1600 cal
    (d) 1100 cal
    (e) 750 cal

## Section 15.8 Heat Engines

42. Which one of the following statements is *not* a consequence of the second law of thermodynamics?
    (a) The efficiency of any engine is less than 100%.
    (b) The natural direction of heat flow is from hot to cold.
    (c) A motor can operate using the atmosphere as a heat reservoir.
    (d) Only reversible processes have a zero net entropy change for the universe.
    (e) There is zero net entropy change for the universe in the operation of a real refrigerator.

43. An engine is used to lift a 2700 kg truck to a height of 3.0 m at constant speed. In the lifting process, the engine received $3.3 \times 10^5$ J of heat from the fuel burned in its interior. What is the efficiency of the engine?
    (a) 0.19
    (b) 0.24
    (c) 0.29
    (d) 0.34
    (e) 0.39

44. A gasoline engine with an efficiency of 0.40 generates 1500 W of power. If a liter of gasoline has an energy content of $3.7 \times 10^7$ J, how many liters of gasoline does the engine consume each hour?
    (a) 0.36
    (b) 0.48
    (c) 1.4
    (d) 2.8
    (e) 6.9

45. In a reversible heat engine, one mole of an ideal gas is carried through a circular cycle beginning and ending at point **A** as shown in the figure. Which one of the following statements concerning this system is false?
    (a) The entropy must increase in one cycle.
    (b) The heat added in one cycle must be 314 J.
    (c) The work done in completing one cycle is 314 J.
    (d) The change in internal energy for one cycle is zero.
    (e) The internal energy for this system is dependent on its state.

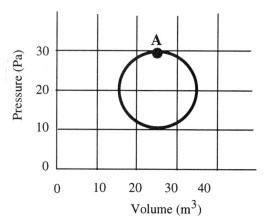

46. Which one of the following statements best describes the operation of a heat engine?
    (a) A heat engine uses input heat to perform work and rejects excess heat to a lower temperature reservoir.
    (b) A heat engine performs work and generates an equal amount of heat in a cyclic process.
    (c) A heat engine transfers heat from a lower temperature reservoir to a higher temperature reservoir through work performed on the system.
    (d) A heat engine transfers heat from a higher temperature reservoir to a lower temperature reservoir through work performed on the system.
    (e) A heat engine decreases the entropy of the universe by generating an equal amount of heat and work.

## Section 15.9 Carnot's Principle and the Carnot Engine

47. Under which one of the following conditions would a Carnot heat engine be 100% efficient?
    (a) The engine does no work.
    (b) The engine uses no heat.
    (c) The engine discharges heat at 0 K
    (d) The engine can be operated in reverse.
    (e) The engine uses an ideal gas.

48. A heat engine operates in a Carnot cycle between reservoirs of temperatures 1000 K and 400 K. It is found to discharge 20 J of heat per cycle to the cold reservoir. What is the work output per cycle?
    (a) 10 J
    (b) 20 J
    (c) 30 J
    (d) 40 J
    (e) 50 J

49. A Carnot engine operates between hot and cold reservoirs with temperatures 527 °C and –73.0 °C, respectively. If the engine performs 1000.0 J of work per cycle, how much heat is extracted per cycle from the hot reservoir?
    (a) 878 J
    (b) 1333 J
    (c) 1163 J
    (d) 1527 J
    (e) 2010 J

50. A Carnot heat engine is to be designed with an efficiency of 85%. If the low temperature reservoir is 25 °C, what is the temperature of the "hot" reservoir?
    (a) 45 °C
    (b) 430 °C
    (c) 850 °C
    (d) 1300 °C
    (e) 1700 °C

51. What is the maximum possible efficiency of an engine operating between the boiling and freezing points of water at sea level?
    (a) 0.27
    (b) 0.49
    (c) 0.61
    (d) 0.85
    (e) 1.0

*Questions 52 through 54 pertain to the following situation:*

A heat engine operates between a hot reservoir at 1500 K and a cold reservoir at 500 K. During each cycle, $1.0 \times 10^5$ J of heat is removed from the hot reservoir and $5.0 \times 10^4$ J of work is performed.

52. What is the actual efficiency of this engine?
    (a) 0.15
    (b) 0.34
    (c) 0.50
    (d) 0.67
    (e) 0.81

53. Determine the Carnot efficiency of this engine.
    (a) 0.15
    (b) 0.34
    (c) 0.50
    (d) 0.67
    (e) 0.81

54. Determine the change in entropy of the cold reservoir.
    (a) $-1.0 \times 10^2$ J/K
    (b) $+1.0 \times 10^2$ J/K
    (c) $-1.8 \times 10^2$ J/K
    (d) $+1.8 \times 10^2$ J/K
    (e) $-2.0 \times 10^2$ J/K

## Section 15.10 Refrigerators, Air Conditioners, and Heat Pumps

55. If the coefficient of performance for a refrigerator is 5.0 and 65 J of work are done on the system, how much heat is rejected to the room?
    (a) 390 J
    (b) 330 J
    (c) 260 J
    (d) 210 J
    (e) 130 J

56. A heat pump is found to remove 2400 J of heat from the exterior of a house and deliver 3500 J of heat to the interior of the house. What is the coefficient of performance of the heat pump?
    (a) 1.5
    (b) 3.2
    (c) 0.31
    (d) 3.9
    (e) 0.69

57. A heat pump extracts $7.0 \times 10^6$ J of heat per hour from a well at 280 K and delivers its output heat into a house at 320 K. If the heat pump uses an ideal Carnot cycle in its operation, what minimum work must be supplied to the heat pump *per hour*?
    (a) $8.6 \times 10^5$ J
    (b) $1.0 \times 10^6$ J
    (c) $7.7 \times 10^6$ J
    (d) $8.0 \times 10^6$ J
    (e) $4.7 \times 10^7$ J

*Questions 58 through 60 pertain to the situation described below:*

A container holding 1.2 kg of water at 20 °C is placed in a freezer which is kept at –20 °C. The water freezes and comes into thermal equilibrium with the interior of the freezer.

58. How much heat is extracted from the water in this process?
    (a) 48 000 J
    (b) 100 000 J
    (c) 400 000 J
    (d) 549 000 J
    (e) 348 000 J

59. What is the *minimum* amount of electrical energy required by the refrigerator to do this if it operates between reservoirs at temperatures of 20 °C and –20 °C?
    (a) 63 000 J
    (b) 77 000 J
    (c) 87 000 J
    (d) 348 000 J
    (e) 549 000 J

60. Which one of the following statements is true concerning this process?
    (a) The water gains entropy in accord with the second law of thermodynamics.
    (b) The water loses entropy so the process violates the second law of thermodynamics.
    (c) The water gains entropy, but the air outside the refrigerator loses entropy in accord with the second law of thermodynamics.
    (d) Both the water and the air outside the refrigerator lose entropy, but the universe gains entropy in accord with the second law of thermodynamics.
    (e) The water loses entropy, but the air outside the refrigerator gains entropy in accord with the second law of thermodynamics.

*Section 15.11 Entropy and the Second Law of Thermodynamics*

61. Which one of the following processes represents a *decrease* in entropy?
    (a) the melting of ice
    (b) the sublimation of carbon dioxide
    (c) the evaporation of perfume
    (d) the vaporization of liquid helium
    (e) the condensation of steam on a kitchen window

62. Which of the following samples exhibits the highest degree of entropy?
    (a) a diamond crystal
    (b) ammonia vapor
    (c) a block of graphite
    (d) a block of paraffin
    (e) liquid oxygen

63. In which one of these processes will there be *no net change* in the entropy of the system?
    (a) the growth of a microorganism
    (b) the fusion of a crystalline solid
    (c) the heating of water in an open container
    (d) the combustion of fuel in a machine engine
    (e) the evaporation and condensation of benzene in a closed vessel

64. A block that slides on a rough surface slows down and eventually stops. The reverse process never occurs. That is, a block at rest never begins to move and accelerate on a rough surface without the action of an external agent. The second situation is forbidden because it would violate
    (a) conservation of total energy.
    (b) conservation of momentum.
    (c) the first law of thermodynamics.
    (d) the second law of thermodynamics.
    (e) both the first and second laws of thermodynamics.

65. A 1.00-kg sample of steam at 100.0 °C condenses to water at 100.0 °C. What is the entropy change of the sample if he heat of vaporization of water is $2.26 \times 10^6$ J/kg?
    (a) $-6.05 \times 10^3$ J/K
    (b) $+6.05 \times 10^3$ J/K
    (c) $-2.26 \times 10^4$ J/K
    (d) $+2.26 \times 10^4$ J/K
    (e) zero

66. In an isothermal and reversible process, 945 J of heat is removed from a system and transferred to the surroundings. The temperature is 314 K. What is the change in entropy of the system?
    (a) −3.01 J/K
    (b) +3.01 J/K
    (c) −0.332 J/K
    (d) +0.332 J/K
    (e) $+2.97 \times 10^5$ J/K

## Additional Problems

*Questions 67 through 72 pertain to the situation described below:*

An ideal monatomic gas expands isothermally from state **A** to state **B**. It then cools at constant volume to state **C**. The gas is then compressed isobarically to **D** where it is then heated until it returns to state **A**.

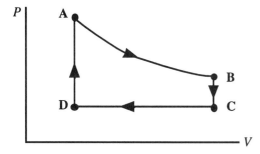

$P_A = 10$ atm

$P_C = 2$ atm

$V_A = V_D = 2$ liters

$V_B = V_C = 4$ liters

$T_A = 327$ °C

67. What is the internal energy of the gas at point **B**?
    (a) $1 \times 10^3$ J
    (b) $2 \times 10^3$ J
    (c) $3 \times 10^3$ J
    (d) $4 \times 10^3$ J
    (e) $5 \times 10^3$ J

68. What is the pressure of the gas when it is in state **B**?
    (a) 5 atm
    (b) 10 atm
    (c) 20 atm
    (d) 25 atm
    (e) 30 atm

69. What is the temperature of the gas when it is in state **C**?
    (a) –33 °C
    (b) 130 °C
    (c) 327 °C
    (d) 817 °C
    (e) 1500 °C

70. What is the ratio of the internal energy of the gas in state **C** to that in state **A**?
    (a) 2/1
    (b) 1/2
    (c) 130/327
    (d) 240/600
    (e) 600/240

71. How much work is done on the gas as it is compressed isobarically from state **C** to state **D**?
    (a) zero
    (b) 50 J
    (c) 100 J
    (d) 200 J
    (e) 400 J

72. What is the net amount of work done after one complete cycle?
    (a) zero
    (b) 20 J
    (c) 40 J
    (d) 1000 J
    (e) 1340 J

# CHAPTER 16   Waves and Sound

## Section 16.1 The Nature of Waves
## Section 16.2 Periodic Waves

1. Which wave is purely *longitudinal*?
   (a) sound waves in air
   (b) radio waves traveling through air
   (c) light waves traveling through vacuum
   (d) waves on a plucked guitar string
   (e) surface waves in a shallow pan of water

2. A periodic wave is produced on a stretched string. Which one of the following properties is *not* related to the speed of the wave?
   (a) frequency
   (b) amplitude
   (c) period
   (d) wavelength
   (e) tension in the string

3. A periodic wave travels along a stretched string in the direction shown by the arrow. The sketch shows a "snapshot" of the pulse at a certain instant. Points **A**, **B**, and **C** are on the string. Which entry in the table below correctly describes how the particles of the string between **A** and **B** and between **B** and **C** are moving?

   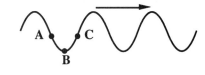

   |     | between **A** and **B** | between **B** and **C** |
   |-----|-------------------------|-------------------------|
   | (a) | down                    | down                    |
   | (b) | up                      | up                      |
   | (c) | left                    | right                   |
   | (d) | up                      | down                    |
   | (e) | down                    | up                      |

4. The speed of sound in a certain metal block is $3.00 \times 10^3$ m/s. The graph shows the amplitude (in meters) of a wave traveling through the block versus time (in milliseconds). What is the wavelength of this wave?
   (a) 0.5 m
   (b) 1.5 m
   (c) 3.0 m
   (d) 4.0 m
   (e) 6.0 m

   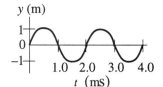

5. Two canoes are 10 m apart on a lake. Each bobs up and down with a period of 4.0 seconds. When one canoe is at its highest point, the other canoe is at its lowest point. Both canoes are always within a single cycle of the waves. Determine the speed of the waves.
   (a) 2.5 m/s
   (b) 5.0 m/s
   (c) 14 m/s
   (d) 40 m/s
   (e) 80 m/s

6. Which one of the following statements concerning waves is false?
   (a) A wave can have both transverse and longitudinal components.
   (b) A wave carries energy from one place to another.
   (c) A wave does not result in the bulk flow of the material of its medium.
   (d) A wave is a traveling disturbance.
   (e) A transverse wave is one in which the disturbance is parallel to the direction of travel.

7. What is the wavelength of a wave with a speed of 12 m/s and a period of 0.25 s?
   (a) 0.25 m
   (b) 1.5 m
   (c) 3.0 m
   (d) 24 m
   (e) 48 m

*Questions 8 through 11 pertain to the situation described below:*

The displacement of a vibrating string versus position along the string is shown in the figure.

The periodic waves have a speed of 10 cm/s.

**A** and **B** are two points on the string.

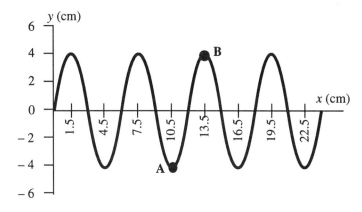

8. What is the amplitude of the wave?
   (a) 2 cm
   (b) 4 cm
   (c) 8 cm
   (d) 9 cm
   (e) 16 cm

9. What is the wavelength of the wave?
   (a) 3 cm
   (b) 6 cm
   (c) 9 cm
   (d) 12 cm
   (e) 15 cm

10. What is the frequency of the wave?
    (a) 0.60 Hz
    (b) 0.90 Hz
    (c) 1.11 Hz
    (d) 1.25 Hz
    (e) 1.67 Hz

11. What is the difference in phase between the points A and B?
    (a) ($\pi/4$) radians
    (b) ($\pi/2$) radians
    (c) $\pi$ radians
    (d) ($3\pi/4$) radians
    (e) $2\pi$ radians

## Section 16.3 The Speed of a Wave on a String

12. The tension in a taut rope is increased by a factor of 9. How does the speed of wave pulses on the rope change, if at all?
    (a) The speed remains the same.
    (b) The speed is reduced by a factor of 3.
    (c) The speed is reduced by a factor of 9.
    (d) The speed is increased by a factor of 3.
    (e) The speed is increased by a factor of 9.

13. A steel wire of mass 0.400 kg and length 0.640 m supports a 102-kg block. The wire is struck exactly at its midpoint causing a small displacement. How long does it take the peak of this displacement to reach the top of the wire?
    (a) $2.00 \times 10^{-3}$ s
    (b) $4.00 \times 10^{-3}$ s
    (c) $6.00 \times 10^{-3}$ s
    (d) $8.00 \times 10^{-3}$ s
    (e) $1.60 \times 10^{-2}$ s

102 kg

154  WAVES AND SOUND

14. A certain string on a piano is tuned to produce middle C ($f = 261.63$ Hz) by carefully adjusting the tension in the string. For a fixed wavelength, what is the frequency when this tension is doubled?
    (a) 130.08 Hz
    (b) 185.00 Hz
    (c) 370.00 Hz
    (d) 446.63 Hz
    (e) 523.26 Hz

15. A wave moves at a constant speed along a string. Which one of the following statements is false concerning the motion of particles in the string?
    (a) The particle speed is constant.
    (b) The particle speed is not the same as the wave speed.
    (c) The particle speed depends on the amplitude of the periodic motion of the source.
    (d) The particle speed depends on the frequency of the periodic motion of the source.
    (e) The particle speed is independent of the tension and linear density of the string.

16. A wave with a wavelength of 0.10 m is traveling at 5.5 m/s on a string with a linear density of 0.082 kg/m. What is the tension in the string?
    (a) 0.20 N
    (b) 0.45 N
    (c) 2.5 N
    (d) 4.4 N
    (e) 6.3 N

## Section 16.4 The Mathematical Description of a Wave

17. A transverse periodic wave described by the expression
$$y = \sin\left[2\pi\left(\frac{x}{2} + \frac{t}{10}\right)\right]$$
(where $y$ and $x$ are in meters and $t$ is in seconds) is established on a string. Which one of the following statements concerning this wave is false?
    (a) The wave is traveling in the negative $x$ direction.
    (b) The amplitude is 1.0 m.
    (c) The frequency of the wave is 0.10 Hz.
    (d) The wavelength of this wave is 2.0 m.
    (e) The wave travels with speed 5.0 m/s.

18. Of the three traveling waves listed below, which one(s) is(are) traveling *in the +x direction*?
    (1) $y = +3.2 \sin [4.1t + 2.3x]$
    (2) $y = -6.8 \sin [-3.0t + 1.5x]$
    (3) $y = +4.9 \sin [12.0t + 18x]$
    (a) *1* only
    (b) *2* only
    (c) *3* only
    (d) both *1* and *2*
    (e) both *2* and *3*

19. A wave has an amplitude of 0.35 m, a frequency of $1.05 \times 10^6$ Hz, and travels in the positive $x$ direction at the speed of light, $3.00 \times 10^8$ m/s. Which one of the following equations correctly represents this wave?
    (a) $y = 0.35 \sin (6.60 \times 10^6 t - 0.022x)$
    (b) $y = 0.35 \sin (6.60 \times 10^6 t + 0.022x)$
    (c) $y = 0.35 \sin (286t - 1.05 \times 10^6 x)$
    (d) $y = 0.35 \sin (286t + 1.05 \times 10^6 x)$
    (e) $y = 0.35 \sin (1.05 \times 10^6 t + 3.00 \times 10^8 x)$

20. A transverse periodic wave is established on a string. It is described by the expression
$$y = 0.005 \sin(20x - 2\pi ft)$$
where $y$ is in meters when $x$ and $t$ are in meters and seconds, respectively. If the wave travels with a speed of 20 m/s, what is its frequency, $f$?
(a) 0.16 Hz
(b) 0.64 Hz
(c) 31.9 Hz
(d) 63.7 Hz
(e) 400 Hz

21. A transverse periodic wave on a string with a linear density of 0.200 kg/m is described by the following equation: $y = 0.005 \sin(419t - 21.0x)$, where $x$ and $y$ are in meters and $t$ is in seconds. What is the tension in the string?
(a) 3.99 N
(b) 32.5 N
(c) 42.1 N
(d) 65.8 N
(e) 79.6 N

## Section 16.5 The Nature of Sound
## Section 16.6 The Speed of Sound

22. A guitar string is plucked and set into vibration. The vibrating string disturbs the surrounding air, resulting in a sound wave. Which entry in the table below is correct?

|     |                                                  | wave in the string | sound wave in air |
| --- | ------------------------------------------------ | ------------------ | ----------------- |
| (a) | The wave is transverse.                          | yes                | yes               |
| (b) | The wave speed increases if the temperature rises. | no               | yes               |
| (c) | The wave is longitudinal.                        | yes                | yes               |
| (d) | The wave is transmitted by particle vibrations.  | no                 | yes               |
| (e) | The wave transports energy.                      | yes                | no                |

23. A bell is ringing inside of a sealed glass jar which is connected to a vacuum pump. Initially, the jar is filled with air. What does one hear as the air is slowly removed from the jar by the pump?
(a) The sound intensity gradually decreases.
(b) The frequency of the sound gradually increases.
(c) The frequency of the sound gradually decreases.
(d) The speed of the sound gradually increases.
(e) The sound intensity of the bell does not change.

24. A railroad whistle is sounded. An echo is heard 5.0 seconds later. If the speed of sound is 343 m/s, how far away is the reflecting surface?
(a) 68 m
(b) 140 m
(c) 860 m
(d) 1700 m
(e) 2000 m

25. Two fans are watching a baseball game from different positions. One fan is located directly behind home plate, 18.3 m from the batter. The other fan is located in the centerfield bleachers, 127 m from the batter. Both fans observe the batter strike the ball at the same time, but the fan behind home plate hears the sound first. What is the time difference between hearing the sound at the two locations? Use 345 m/s as the speed of sound.
(a) 0.315 s
(b) 0.368 s
(c) 3.17 s
(d) 1.89 s
(e) 0.053 s

26. Ethanol has a density of 659 kg/m$^3$. If the speed of sound in ethanol is 1162 m/s, what is its adiabatic bulk modulus?
(a) $1.7 \times 10^8$ N/m$^2$
(b) $2.2 \times 10^8$ N/m$^2$
(c) $7.7 \times 10^8$ N/m$^2$
(d) $8.9 \times 10^8$ N/m$^2$
(e) $6.1 \times 10^9$ N/m$^2$

# 156 WAVES AND SOUND

27. A physics student is asked to determine the length of a long, slender, copper rod by measuring the time required for a sound pulse to travel the length of the rod. The Young's modulus of copper is $1.1 \times 10^{11}$ N/m$^2$; and its density is 8890 kg/m$^3$. The student finds that the time for the pulse to travel from one end to the other is $1.56 \times 10^{-3}$ s. How long is the rod?
   (a) 0.18 m
   (b) 1.2 m
   (c) 5.5 m
   (d) 19 m
   (e) 31 m

28. The speaker and two microphones shown in the figure are arranged inside a sealed container filled with neon gas. The signal from the microphones is monitored beginning at time $t = 0$ when a sound pulse is emitted from the speaker. The pulse is picked up by microphone 1 at $t_1 = 1.150 \times 10^{-2}$ s and by microphone 2 at $t_2 = 1.610 \times 10^{-2}$ s. What is the speed of sound in neon gas?
   (a) 124 m/s
   (b) 174 m/s
   (c) 362 m/s
   (d) 435 m/s
   (e) 724 m/s

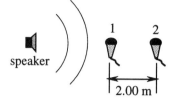

## Section 16.7 Sound Intensity

29. The intensity of a spherical wave at a distance of 4.0 m from the source is 120 W/m$^2$. What is the intensity at a point 9.0 m away from the source?
   (a) 11 W/m$^2$
   (b) 24 W/m$^2$
   (c) 53 W/m$^2$
   (d) 80 W/m$^2$
   (e) 270 W/m$^2$

30. A bell produces sound energy at a rate of $4.0 \times 10^{-3}$ W and radiates it uniformly in all directions. What is the intensity of the wave at a distance of 100 m from the bell?
   (a) $3.18 \times 10^{-8}$ W/m$^2$
   (b) $3.14 \times 10^{-7}$ W/m$^2$
   (c) $5.02 \times 10^{-2}$ W/m$^2$
   (d) $5.02 \times 10^{2}$ W/m$^2$
   (e) $6.28 \times 10^{7}$ W/m$^2$

31. How far must one stand from a 0.005 W point sound source if the intensity is at the hearing threshold? Assume the sound waves travel without being disturbed to the listener.
   (a) 500 m
   (b) 1 km
   (c) 2 km
   (d) 4 km
   (e) 20 km

32. Two boys are whispering in the library. The radiated sound power from one boy's mouth is $1.2 \times 10^{-9}$ W and it spreads out uniformly in all directions. What is the minimum distance the boys must be away from the librarian so that she will not be able to hear them? The threshold of hearing for the librarian is $1.00 \times 10^{-12}$ W/m$^2$.
   (a) 100 m
   (b) 35 m
   (c) 23 m
   (d) 16 m
   (e) 9.8 m

## Section 16.8 Decibels

33. The decibel level of a jackhammer is 130 dB relative to the threshold of hearing. Determine the sound intensity produced by the jackhammer.
   (a) 1.0 W/m$^2$
   (b) 10 W/m$^2$
   (c) 13 W/m$^2$
   (d) 130 W/m$^2$
   (e) $10^{13}$ W/m$^2$

34. The decibel level of a jackhammer is 130 dB relative to the threshold of hearing. Determine the decibel level if two jackhammers operate side by side.
    (a) 65 dB
    (b) 130 dB
    (c) 133 dB
    (d) 144 dB
    (e) 260 dB

35. A person is talking in a small room; and the sound intensity level is 60 dB everywhere within the room. If there are eight people talking simultaneously in the room, what is the sound intensity level?
    (a) 60 dB
    (b) 79 dB
    (c) 74 dB
    (d) 64 dB
    (e) 69 dB

36. At a distance of 5.0 m from a point sound source, the sound intensity level is 110 dB. At what distance is the intensity level 95 dB?
    (a) 5.0 m
    (b) 7.1 m
    (c) 14 m
    (d) 28 m
    (e) 42 m

*Section 16.9  Applications of Sound*
*Section 16.10  The Doppler Effect*
*Section 16.11  The Sensitivity of the Human Ear*

37. A train moving at a constant speed is passing a stationary observer on a platform. On one of the train cars, a flute player is continually playing the note known as concert A ($f = 440$ Hz). After the flute has passed, the observer hears the sound as a G which has a frequency of 392 Hz. What is the speed of the train? The speed of sound in air is 343 m/s.
    (a) 26 m/s
    (b) 12 m/s
    (c) 42 m/s
    (d) 7.3 m/s
    (e) 37 m/s

38. A car moving at 35 m/s approaches a stationary whistle that emits a 220 Hz sound. The speed of sound is 343 m/s. What is the speed of the sound *relative* to the driver of the car?
    (a) 300 m/s
    (b) 305 m/s
    (c) 340 m/s
    (d) 365 m/s
    (e) 378 m/s

39. A car moving at 35 m/s approaches a stationary whistle that emits a 220 Hz sound. The speed of sound is 343 m/s. What is the frequency of sound heard by the driver of the car?
    (a) 198 Hz
    (b) 220 Hz
    (c) 242 Hz
    (d) 282 Hz
    (e) 340 Hz

40. A source moving through water at 10 m/s generates water waves with a frequency of 5.0 Hz. The speed of these water waves relative to the water surface is 20.0 m/s. The source approaches an observer who is at rest in the water. What wavelength would be measured for these waves by the stationary observer?
    (a) 1.0 m
    (b) 2.0 m
    (c) 4.0 m
    (d) 6.0 m
    (e) 8.0 m

41. Two golf carts have horns which emit sound with a frequency of 390 Hz. The golf carts are traveling toward one another, each traveling with a speed of 9.0 m/s with respect to the ground. What frequency do the drivers of the carts hear? The speed of sound at the golf course is 343 m/s.
    (a) 390 Hz
    (b) 400 Hz
    (c) 410 Hz
    (d) 420 Hz
    (e) 430 Hz

## 158 WAVES AND SOUND

42. A loudspeaker at the base of a cliff emits a pure tone of frequency 3000 Hz. A man jumps from rest from the top of the cliff and safely falls into a net below. How far has the man fallen at the instant he hears the frequency of the tone as 3218 Hz? The speed of sound is 343 m/s.
    (a) 12.2 m
    (b) 15.3 m
    (c) 31.7 m
    (d) 46.8 m
    (e) 61.0 m

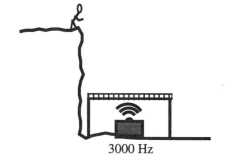

*Questions 43 through 45 pertain to the situation described below:*

The diagram shows the various positions of a child in motion on a swing. Somewhere in front of the child a stationary whistle is blowing.

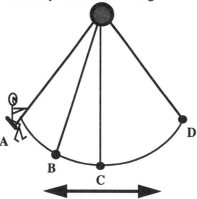

Stationary Whistle

43. At which position(s) will the child hear the highest frequency for the sound from the whistle?
    (a) at both **A** and **D**
    (b) at **B** when moving toward **A**
    (c) at **B** when moving toward **C**
    (d) at **C** when moving toward **B**
    (e) at **C** when moving toward **D**

44. At which position(s) will the child hear the lowest frequency for the sound from the whistle?
    (a) at both **A** and **D**
    (b) at **B** when moving toward **A**
    (c) at **B** when moving toward **C**
    (d) at **C** when moving toward **B**
    (e) at **C** when moving toward **D**

45. At which position(s) will the child hear the same frequency as that heard by a stationary observer standing next to the whistle?
    (a) at both **A** and **D**
    (b) at **B** when moving toward **A**
    (c) at **B** when moving toward **C**
    (d) at **C** when moving toward **B**
    (e) at **C** when moving toward **D**

*Questions 46 through 49 pertain to the following situation:*

The car in the figure is moving to the left at 35 m/s. The car's horn continuously emits a 220 Hz sound. The figure also shows the first two regions of compression of the emitted sound waves. The speed of sound is 343 m/s.

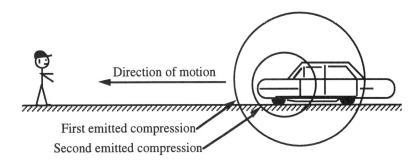

First emitted compression
Second emitted compression

46. How far does the car move in one period of the sound emitted from the horn?
    (a) 0.08 m
    (b) 0.16 m
    (c) 8 m
    (d) 16 m
    (e) 35 m

47. How far has the initial compression traveled when the second compression is emitted?
    (a) 0.77 m
    (b) 1.56 m
    (c) 7.7 m
    (d) 15.5 m
    (e) 35 m

48. What is the wavelength of the sound in the direction of motion of the car?
    (a) 1.40 m
    (b) 1.56 m
    (c) 1.70 m
    (d) 1.93 m
    (e) 35 m

49. What is the frequency heard by the observer?
    (a) 9.7 Hz
    (b) 176 Hz
    (c) 200 Hz
    (d) 219 Hz
    (e) 245 Hz

## Additional Problems

50. An unstretched spring has a length of 0.30 m. When the spring is stretched to a total length of 0.60 m, it supports traveling waves moving at 4.5 m/s. How fast will waves travel on this spring if it is stretched to 0.90 m?
    (a) 2.3 m/s
    (b) 4.5 m/s
    (c) 6.4 m/s
    (d) 9.0 m/s
    (e) 10.8 m/s

*Questions 51 through 54 pertain to the situation described below:*

A periodic transverse wave is established on a string such that there are exactly two cycles traveling along a 3.0 m section of the string. The crests move at 20 m/s.

51. What is the frequency of the wave?
    (a) 0.67 Hz
    (b) 1.33 Hz
    (c) 13 Hz
    (d) 30 Hz
    (e) 57 Hz

52. What is the shortest horizontal distance from a crest to a point of zero acceleration?
    (a) 0.38 m
    (b) 0.75 m
    (c) 1.5 m
    (d) 3.0 m
    (e) 6.0 m

53. How long does it take a particle at the top of a crest to reach the bottom of an adjacent trough?
    (a) 0.018 s
    (b) 0.038 s
    (c) 0.075 s
    (d) 0.150 s
    (e) 0.30 s

54. How could the speed of the wave be increased?
    (a) by increasing the period
    (b) by decreasing the amplitude
    (c) by decreasing the frequency
    (d) by increasing the tension in the string
    (e) by increasing amplitude

*Questions 55 through 59 pertain to the situation described below:*

A periodic traveling wave is generated on a string of linear density $8.0 \times 10^{-4}$ kg/m. Figure **A** shows the displacements of the particles in the string as a function of the position $x$ along the string at $t = 0$. Figure **B** shows the displacement of the particle at $x = 0$ as a function of time. The particle positions are measured from the left end of the string ($x = 0$) and the wave pulses move to the right.

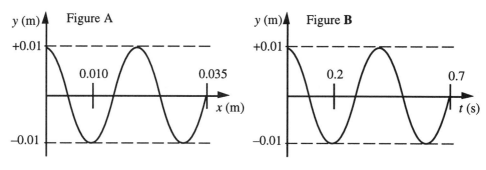

55. What is the wavelength of the wave?
    (a) 0.005 m
    (b) 0.010 m
    (c) 0.015 m
    (d) 0.020 m
    (e) 0.025 m

56. What is the minimum time required for the particles to return to their respective positions at $t = 0$?
    (a) 0.1 s
    (b) 0.2 s
    (c) 0.3 s
    (d) 0.4 s
    (e) 0.7 s

57. What is the amplitude of the wave?
    (a) 0.01 m
    (b) 0.02 m
    (c) 0.035 m
    (d) 0.070 m
    (e) 0.70 m

58. Determine the speed of the wave.
    (a) 0.01 m/s
    (b) 0.02 m/s
    (c) 0.035 m/s
    (d) 0.040 m/s
    (e) 0.050 m/s

59. What is the tension in the string?
    (a) $2 \times 10^{-6}$ N
    (b) $4 \times 10^{-6}$ N
    (c) $4 \times 10^{-5}$ N
    (d) $5 \times 10^{-5}$ N
    (e) 62.5 N

# CHAPTER 17

# The Principle of Linear Superposition and Interference Phenomena

*Section 17.1 The Principle of Linear Superposition*
*Section 17.2 Constructive and Destructive Interference of Sound Waves*

1. Two pulses of identical shape travel toward each other in opposite direction on a string as shown in the figure.

   Which one of the following statements concerning this situation is true?
   (a) The pulses will reflect from each other.
   (b) The pulses will diffract from each other.
   (c) The pulses will interfere to produce a standing wave.
   (d) The pulses will pass through each other and produce beats.
   (e) As the pulses pass through each other they will interfere destructively.

2. Two traveling waves on a rope move toward each other at a speed of 1.0 m/s. The waves have the same amplitude. The drawing shows the position of the waves at time $t = 0$. Which one of the following drawings depicts the waves on the rope at $t = 2.0$ s?

   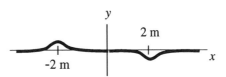

   (a)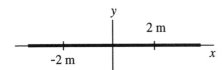

   (d)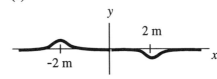

   (b)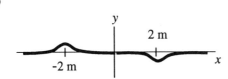

   (e)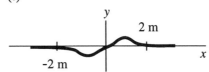

   (c)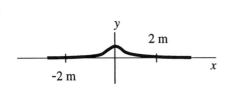

## 162 LINEAR SUPERPOSITION AND INTERFERENCE

*Questions 3 through 5 pertain to the statement and diagrams below:*

Consider the following figures depicting rectangular pulses on a string.

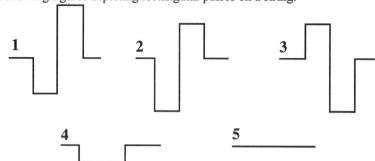

3. If pulse **1** were reflected from a wall, which one of the patterns above would represent the reflected pulse?
   (a) 1
   (b) 2
   (c) 3
   (d) 4
   (e) 5

4. Complete the following statement: If pulse **2** were superimposed on pulse **3**,
   (a) constructive interference would occur.
   (b) the resulting pattern would be represented by **1**.
   (c) the resulting pattern would be represented by **4**.
   (d) the resulting pattern would be represented by **5**.
   (e) the resulting pattern would be different than **1**, **4**, and **5**.

5. Which pulses must be superimposed to give the situation shown in **5**.
   (a) 1 and 2
   (b) 1 and 3
   (c) 2 and 4
   (d) 1, 2, and 4
   (5) 2, 3, and 4.

6. Sound waves are emitted from two speakers. Which one of the following statements about sound wave interference is false?
   (a) In a region where both destructive and constructive interference occur, energy is not conserved.
   (b) Destructive interference occurs when two waves are exactly out of phase when they meet.
   (c) Interference redistributes the energy carried by the individual waves.
   (d) Constructive interference occurs when two waves are exactly in phase when they meet.
   (e) Sound waves undergo diffraction as they exit each speaker.

7. A pebble is dropped in a lake and produces ripples with a frequency of 0.25 Hz. When should a second pebble be dropped at the same place to produce destructive interference?
   (a) 0.50 s after the first
   (b) 0.75 s after the first
   (c) 1.0 s after the first
   (d) 1.5 s after the first
   (e) 2.0 s after the first

8. Two loudspeakers, **A** and **B**, are separated by a distance of 2.0 m. The speakers emit sound waves at a frequency of 680 Hz that are exactly out of phase. The speed of sound is 343 m/s. How far from speaker **A** along the +x axis will a point of constructive interference occur?
   (a) 0.25 m
   (b) 0.30 m
   (c) 0.46 m
   (d) 0.88 m
   (e) 0.98 m

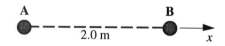

9. A speaker generates a continuous tone of 440 Hz. In the drawing, sound travels into a tube which splits into two segments, one longer than the other. The sound waves recombine before being detected by a microphone. The speed of sound in air is 339 m/s. What is the minimum difference in the lengths of the two paths for sound travel if the waves arrive in phase at the microphone?

   (a) 0.10 m
   (b) 0.39 m
   (c) 0.77 m
   (d) 1.11 m
   (e) 1.54 m

*Questions 10 and 11 pertain to the situation described below:*

Two loudspeakers are located 3 m apart on the stage of an auditorium. A listener at point **P** is seated 29 m from one speaker and 25 m from the other. A signal generator drives the speakers in phase with the same amplitude and frequency. The wave amplitude at **P** due to each speaker alone is $A$. The frequency is then varied between 20 Hz and 300 Hz. The speed of sound is 343 m/s.

10. At what frequency or frequencies will the listener at **P** hear a maximum intensity?
   (a) 170 Hz only
   (b) 113 Hz and 226 Hz
   (c) 85 Hz, 170 Hz, 257 Hz
   (d) 57 Hz, 113 Hz, 170 Hz, 227 Hz, and 284 Hz
   (e) 43 Hz, 85 Hz, 128 Hz, 170 Hz, 213 Hz, 257 Hz, and 298 Hz

11. Determine the value of the maximum amplitude in terms of $A$.
   (a) $2.0A$
   (b) $2.5A$
   (c) $3.0A$
   (d) $4.0A$
   (e) $5.0A$

## Section 17.3 Diffraction

12. Water waves approach an aperture. The resulting patterns are shown for two different cases, **A** and **B**, in which the wavelength and aperture size are varied.

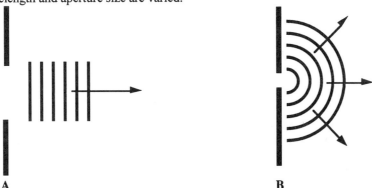

   Which one of the following statements concerning these cases is true?
   (a) Neither figure shows diffraction; and in both cases, the wavelength is much smaller than the aperture.
   (b) Diffraction occurs in **A**, but not in **B** because the wavelength in **A** is much smaller than the aperture.
   (c) Diffraction occurs in **B**, but not in **A** because the wavelength in **B** is much smaller than the aperture.
   (d) Both figures show diffraction; and in both cases the wavelengths are approximately the same size as the aperture.
   (e) Diffraction occurs in **B**, but not in **A** because the wavelength in **B** is approximately the same size as the aperture.

164   LINEAR SUPERPOSITION AND INTERFERENCE

13. A speaker is designed for wide dispersion for a high frequency sound. What should the diameter of the circular opening be in a speaker where the desired diffraction angle is 75° and an 9100 Hz sound is generated? The speed of sound is 343 m/s.
    (a) 0.039 m
    (b) 0.077 m
    (c) 0.21 m
    (d) 0.13 m
    (e) 0.048 m

14. The opening of a diffraction horn has a width of 0.15 m and a height of 0.75 m. If the speaker emits a continuous tone with a wavelength of 0.11 m, at what angle does the first minimum occur?
    (a) 47°
    (b) 39°
    (c) 23°
    (d) 12°
    (e) 8.4°

## Section 17.4 Beats

15. Which one of the following superpositions will result in *beats*?
    (a) the superposition of waves that travel with different speeds
    (b) the superposition of identical waves that travel in the same direction
    (c) the superposition of identical waves that travel in opposite directions
    (d) the superposition of waves that are identical except for slightly different amplitudes.
    (e) the superposition of waves that are identical except for slightly different frequencies.

16. Two identical tuning forks vibrate at 587 Hz. After a small piece of clay is placed on one them, 8 beats per second are heard. What is the period of the tuning fork that holds the clay?
    (a) $1.68 \times 10^{-3}$ s
    (b) $1.70 \times 10^{-3}$ s
    (c) $1.73 \times 10^{-3}$ s
    (d) $1.76 \times 10^{-3}$ s
    (e) $1.80 \times 10^{-3}$ s

17. A guitar string produces 4 beats/s when sounded with a 250 Hz tuning fork and 9 beats per second when sounded with a 255 Hz tuning fork. What is the vibrational frequency of the string?
    (a) 240 Hz
    (b) 246 Hz
    (c) 254 Hz
    (d) 259 Hz
    (e) 263 Hz

18. A guitar string has a linear density of $8.30 \times 10^{-4}$ kg/m. The length of the string is 0.660 m. The tension in the string is 52.0 N. When the fundamental frequency of the string is sounded with a tuning fork of frequency 196.0 Hz, what beat frequency is heard?
    (a) 6 Hz
    (b) 4 Hz
    (c) 12 Hz
    (d) 8 Hz
    (e) 2 Hz

19. The beat period occurring when two tuning forks are vibrating is 0.333 s. One of the forks is known to vibrate at 588.0 Hz. What are the possible vibration frequencies of the second tuning fork?
    (a) 587.7 or 588.3 Hz
    (b) 586.0 or 592.0 Hz
    (c) 580.3 or 596.7 Hz
    (d) 585.0 or 591.0 Hz
    (e) 584.5 Hz or 591.5 Hz

20. One string on a guitar is exactly in tune. The guitarist uses this string to produce a tone with a frequency of 196 Hz by pressing down at the proper fret. An adjacent string can also be used to produce this tone without being pressed against a fret. However, this adjacent string is out of tune and produces a tone that sounds lower in frequency than the other tone, When the tones are produced simultaneously, the beat frequency is 5.0 Hz. What frequency does the adjacent string produce?
    (a) 196 Hz
    (b) 191 Hz
    (c) 171 Hz
    (d) 201 Hz
    (e) 186 Hz

*Section 17.5 Transverse Standing Waves*

21. Which one of the following will result in *standing waves* ?
    (a) the superposition of waves that travel with different speeds
    (b) the superposition of identical waves that travel in the same direction
    (c) the superposition of identical waves that travel in opposite directions
    (d) the superposition of nearly identical waves of slightly different amplitudes
    (e) the superposition of nearly identical waves of slightly different frequencies

22. Which one of the following statements is true concerning the points on a string that sustain a standing wave pattern?
    (a) All points vibrate with the same energy.
    (b) All points undergo the same displacements.
    (c) All points vibrate with different frequencies.
    (d) All points vibrate with different amplitudes.
    (e) All points undergo motion that is purely longitudinal.

23. A rope of length $L$ is clamped at both ends. Which one of the following is not a possible wavelength for standing waves on this rope?
    (a) $L/2$
    (b) $2L/3$
    (c) $L$
    (d) $2L$
    (e) $4L$

24. What is the distance from the fixed end of a guitar string to the nearest antinode?
    (a) $\lambda$
    (b) $2\lambda$
    (c) $\lambda/2$
    (d) $\lambda/4$
    (e) $3\lambda/4$

25. A 4-m long string, clamped at both ends, vibrates at 200 Hz. If the string resonates in six segments, what is the speed of transverse waves on the string?
    (a) 100 m/s
    (b) 133 m/s
    (c) 267 m/s
    (d) 328 m/s
    (e) 400 m/s

26. A certain string, clamped at both ends, vibrates in 7 segments at a frequency of 240 Hz. What frequency will cause it to vibrate in 4 segments?
    (a) 89 Hz
    (b) 137 Hz
    (c) 274 Hz
    (d) 411 Hz
    (e) 420 Hz

27. Four standing wave segments, or loops, are observed on a string fixed at both ends as it vibrates at a frequency of 140 Hz. What is the fundamental frequency of the string?
    (a) 23 Hz
    (b) 28 Hz
    (c) 35 Hz
    (d) 47 Hz
    (e) 70 Hz

28. A string with a linear density of 0.035 kg/m and a mass of 0.014 kg is clamped at both ends. Under what tension in the string will it have a fundamental frequency of 110 Hz?
    (a) 270 N
    (b) 410 N
    (c) 550 N
    (d) 680 N
    (e) 790 N

# LINEAR SUPERPOSITION AND INTERFERENCE

*Questions 29 and 30 pertain to the situation described below:*

Vibrations with frequency 600 Hz are established on a 1.33-m length of string that is clamped at both ends. The speed of waves on the string is 400 m/s.

29. How many antinodes are contained in the resulting standing wave pattern?
    (a) 2
    (b) 3
    (c) 4
    (d) 5
    (e) 6

30. How far from either end of the string does the first node occur?
    (a) 0.17 m
    (b) 0.33 m
    (c) 0.49 m
    (d) 0.66 m
    (e) 0.75 m

*Questions 31 and 32 pertain to the situation described below:*

A 3-m long string sustains a three loop standing wave pattern as shown. The wave speed is 100 m/s.

31. What is the frequency of vibration?
    (a) 25 Hz
    (b) 33 Hz
    (c) 50 Hz
    (d) 75 Hz
    (e) 100 Hz

32. What is the *lowest* possible frequency for standing waves on this string?
    (a) 50.0 Hz
    (b) 33.3 Hz
    (c) 25.0 Hz
    (d) 16.7 Hz
    (e) 8.33 Hz

*Questions 33 through 37 pertain to the following situation:*

A string with a length of 2.5 m has two adjacent resonances at frequencies 112 Hz and 140 Hz.

33. Determine the fundamental frequency of the string.
    (a) 14 Hz
    (b) 28 Hz
    (c) 42 Hz
    (d) 56 Hz
    (e) 70 Hz

34. Determine the wavelength of the 140 Hz resonance.
    (a) 0.5 m
    (b) 0.75 m
    (c) 1.0 m
    (d) 1.5 m
    (e) 2.0 m

35. Determine the speed of the waves on the string.
    (a) 28 m/s
    (b) 42 m/s
    (c) 56 m/s
    (d) 112 m/s
    (e) 140 m/s

36. How many antinodes will be found on the 112 Hz resonance?
    (a) 1
    (b) 2
    (c) 3
    (d) 4
    (e) 5

37. What is the wavelength of the first harmonic?
    (a) 1.25 m
    (b) 2.50 m
    (c) 3.75 m
    (d) 5.00 m
    (e) 28 m

## Section 17.6  Longitudinal Standing Waves
## Section 17.7  Complex Sound Waves

*Questions 38 through 42 pertain to the statement and diagrams below:*

The figures show standing waves of sound in six organ pipes of the same length. Each pipe has one end open and the other end closed. *Note: some of the figures show situations that are not possible.*

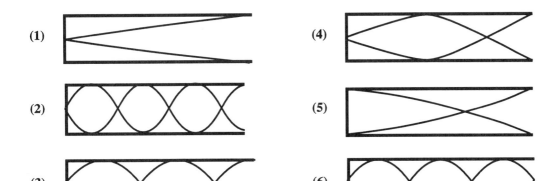

38. Which figures *do not* illustrate possible resonant situations?
    (a) 1 and 4
    (b) 2 and 3
    (c) 4 and 5
    (d) 5 and 6
    (e) 4, 5, and 6

39. Which one of the pipes emits sound with the lowest frequency?
    (a) 1
    (b) 2
    (c) 3
    (d) 4
    (e) 6

40. Which one of the pipes emits sound with the highest frequency?
    (a) 1
    (b) 2
    (c) 3
    (d) 4
    (e) 6

41. Which one of the pipes is resonating in its *third* harmonic?
    (a) 1
    (b) 2
    (c) 3
    (d) 4
    (e) 6

42. If the length of the pipes is 0.500 m, what is the frequency of the sound emitted from pipe 3?
    The speed of sound is 343 m/s.
    (a) 172 Hz
    (b) 344 Hz
    (c) 429 Hz
    (d) 515 Hz
    (e) 858 Hz

168  LINEAR SUPERPOSITION AND INTERFERENCE

43. Pipe **A** is 0.50 m long and open at both ends. Pipe **B** is open at one end and closed at the other end. Determine the length of **B** so that it has the same fundamental frequency as **A**.
    (a) 0.25 m
    (b) 0.50 m
    (c) 0.75 m
    (d) 1.0 m
    (e) 2.0 m

44. Determine the shortest length of pipe, open at both ends, that will resonate at 256 Hz. The speed of sound is 343 m/s.
    (a) 0.33 m
    (b) 0.67 m
    (c) 0.99 m
    (d) 1.32 m
    (e) 1.67 m

45. When a tuba is played, the player blows into one end of a tube which has an effective length of 3.50 m. The other end of the tube is open. If the speed of sound in air is 343 m/s, what is the lowest frequency the tuba can produce?
    (a) 8.00 Hz
    (b) 12.0 Hz
    (c) 16.0 Hz
    (d) 24.0 Hz
    (e) 49.0 Hz

46. Some of the lowest pitches attainable on a musical instrument are achieved on the worldest large pipe organs. What is the length of an organ pipe which is open on both ends and has a fundamental frequency of 8.75 Hz when the speed of sound in air is 341 m/s?
    (a) 9.83 m
    (b) 19.5 m
    (c) 21.2 m
    (d) 29.3 m
    (e) 32.4 m

# CHAPTER 18 — Electric Forces and Electric Fields

*Section 18.1 The Origin of Electricity*
*Section 18.2 Charged Objects and the Electric Force*
*Section 18.3 Conductors and Insulators*
*Section 18.4 Charging by Contact and by Induction*

1. Which one of the following statements *best* explains why tiny bits of paper are attracted to a charged rubber rod?
   (a) Paper is naturally a positive material.
   (b) Paper is naturally a negative material.
   (c) The paper becomes polarized by induction.
   (d) Rubber and paper always attract each other.
   (e) The paper acquires a net positive charge by induction.

2. Five styrofoam balls are suspended from insulating threads. Several experiments are performed on the balls and the following observations are made:

   I. Ball A attracts B and repels C.
   II. Ball D attracts B and has no effect on E.
   III. A negatively charged rod attracts both A and E.

   What are the charges, *if any*, on *each* ball?

   |     | A | B | C | D | E |
   |-----|---|---|---|---|---|
   | (a) | + | − | + | 0 | + |
   | (b) | + | − | + | + | 0 |
   | (c) | + | − | + | 0 | 0 |
   | (d) | − | + | − | 0 | 0 |
   | (e) | + | 0 | − | + | 0 |

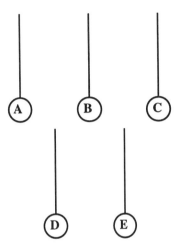

3. Two uncharged conducting spheres, **A** and **B**, are suspended from insulating threads so that they touch each other. While a negatively charged rod is held *near, but not touching* sphere **A**, the two spheres are separated. How will the spheres be charged, *if at all*?

   |     | Sphere A | Sphere B |
   |-----|----------|----------|
   | (a) | 0        | +        |
   | (b) | −        | +        |
   | (c) | 0        | 0        |
   | (d) | −        | 0        |
   | (e) | +        | −        |

4. Each of three objects has a net charge. Objects **A** and **B** attract one another. Objects **B** and **C** also attract one another, but objects **A** and **C** repel one another. Which one of the following table entries is a possible combination of the signs of the net charges on these three objects?

   |     | A | B | C |
   |-----|---|---|---|
   | (a) | + | + | − |
   | (b) | − | + | + |
   | (c) | + | − | − |
   | (d) | − | + | − |
   | (e) | − | − | + |

5. A conducting sphere has a net charge of $-4.8 \times 10^{-17}$ C. What is the approximate number of excess electrons on the sphere?
   (a) 100
   (b) 200
   (c) 300
   (d) 400
   (e) 500

6. Complete the following statement: When an ebonite rod is rubbed with animal fur, the rod becomes negatively charged as
   (a) positive charges are transferred from the fur to the rod.
   (b) negative charges are transferred from the rod to the fur.
   (c) negative charges are created on the surface of the rod.
   (d) negative charges are transferred from the fur to the rod.
   (e) positive charges are transferred from the rod to the fur.

7. Complete the following statement: When a glass rod is rubbed with silk cloth, the rod becomes positively charged as
   (a) positive charges are transferred from the silk to the rod.
   (b) negative charges are transferred from the rod to the silk.
   (c) positive charges are created on the surface of the rod.
   (d) negative charges are transferred from the silk to the rod.
   (e) positive charges are transferred from the rod to the silk.

8. A charged conductor is brought near an uncharged insulator. Which one of the following statements is true?
   (a) Both objects will repel each other.
   (b) Both objects will attract each other.
   (c) Neither object exerts an electrical force on the other.
   (d) The objects will repel each other only if the conductor has a negative charge.
   (e) The objects will attract each other only if the conductor has a positive charge.

9. An aluminum nail has an excess charge of $+3.2$ μC. How many electrons must be added to the nail to make it electrically neutral?
   (a) $2.0 \times 10^{13}$
   (b) $2.0 \times 10^{19}$
   (c) $3.2 \times 10^{-6}$
   (d) $3.2 \times 10^{6}$
   (e) $5.0 \times 10^{-14}$

10. Two uncharged, conducting spheres, **A** and **B**, are held at rest on insulating stands and are in contact. A positively charged rod is brought near sphere **A** as suggested in the figure. While the rod is in place, the two spheres are separated. How will the spheres be charged, *if at all*?

    | | Sphere **A** | Sphere **B** |
    |---|---|---|
    | (a) | positive | positive |
    | (b) | positive | negative |
    | (c) | negative | positive |
    | (d) | negative | negative |
    | (e) | zero | zero |

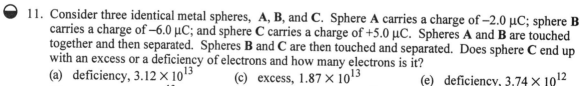

11. Consider three identical metal spheres, **A**, **B**, and **C**. Sphere A carries a charge of $-2.0$ μC; sphere B carries a charge of $-6.0$ μC; and sphere C carries a charge of $+5.0$ μC. Spheres **A** and **B** are touched together and then separated. Spheres **B** and **C** are then touched and separated. Does sphere **C** end up with an excess or a deficiency of electrons and how many electrons is it?
    (a) deficiency, $3.12 \times 10^{13}$
    (b) excess, $3.12 \times 10^{13}$
    (c) excess, $1.87 \times 10^{13}$
    (d) deficiency, $6.24 \times 10^{12}$
    (e) deficiency, $3.74 \times 10^{12}$

## Section 18.5 Coulomb's Law

12. Three charged particles **A**, **B**, and **C** are located near one another. Both the *magnitude* and *direction* of the force that particle **A** exerts on particle **B** is *independent* of
    (a) the sign of charge **B**.
    (b) the sign of charge **A**.
    (c) the distance between **C** and **B**.
    (d) the distance between **A** and **B**.
    (e) the magnitude of the charge on **B**.

13. Four point charges, each of the same magnitude, with varying signs are arranged at the corners of a square as shown. Which of the arrows labeled **A**, **B**, **C**, and **D** gives the correct direction of the net force that acts on the charge at the upper right corner?
    (a) A
    (b) B
    (c) C
    (d) D
    (e) The net force on that charge is zero.

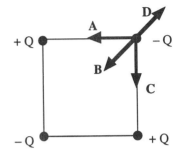

14. Two positive point charges $Q$ and $2Q$ are separated by a distance $R$. If the charge $Q$ experiences a force of magnitude $F$ when the separation is $R$, what is the magnitude of the force on the charge $2Q$ when the separation is $2R$?
    (a) $F/4$
    (b) $F/2$
    (c) $F$
    (d) $2F$
    (e) $4F$

15. A charge $Q$ exerts a 12 N force on another charge $q$. If the distance between the charges is doubled, what is the magnitude of the force exerted on $Q$ by $q$?
    (a) 3 N
    (b) 6 N
    (c) 24 N
    (d) 36 N
    (e) 48 N

16. At what separation will two charges, each of magnitude 6 μC, exert a force of 1.4 N on each other?
    (a) $5.1 \times 10^{-6}$ m
    (b) 0.23 m
    (c) 0.48 m
    (d) 2.0 m
    (e) 40 m

17. One mole of a substance contains $6.02 \times 10^{23}$ protons and an equal number of electrons. If the protons could somehow be separated from the electrons and placed in separate containers separated by $1.00 \times 10^3$ m, what would be the magnitude of the electrostatic force exerted by one box on the other?
    (a) $8.7 \times 10^8$ N
    (b) $9.5 \times 10^9$ N
    (c) $2.2 \times 10^{10}$ N
    (d) $8.3 \times 10^{13}$ N
    (e) $1.6 \times 10^{19}$ N

18. Three charges are positioned as indicated in the figure. What are the horizontal and vertical components of the net force exerted on the +15 μC charge by the +11 μC and +13 μC charges?

    |   | horizontal | vertical |
    |---|---|---|
    | (a) | 95 N | 310 N |
    | (b) | 76 N | 310 N |
    | (c) | 250 N | 130 N |
    | (d) | 95 N | 130 N |
    | (e) | 76 N | 370 N |

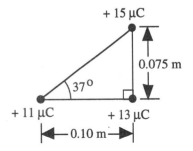

19. In Frame **1**, two identical conducting spheres, **A** and **B**, carry equal amounts of excess charge that have the same sign. The spheres are separated by a distance $d$; and sphere **A** exerts an electrostatic force on sphere **B** which has a magnitude $F$. A third sphere, **C**, which is handled only by an insulating rod, is introduced in Frame **2**. Sphere **C** is identical to **A** and **B** except that it is *initially uncharged*. Sphere **C** is touched first to sphere **A**, in Frame **2**, and then to sphere **B**, in Frame **3**, and is finally removed in Frame **4**.

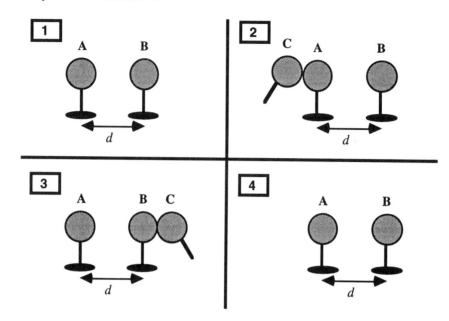

Determine the magnitude of the electrostatic force that sphere **A** exerts on sphere **B** in Frame **4**.
(a) $F/2$
(b) $F/3$
(c) $3F/4$
(d) $3F/8$
(e) zero

20. Three identical point charges, $Q$, are placed at the vertices of an equilateral triangle as shown in the figure. The length of each side of the triangle is $d$. Determine the magnitude and direction of the total electrostatic force on the charge at the top of the triangle.

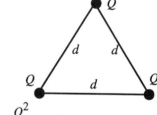

(a) $\dfrac{Q^2\sqrt{3}}{4\pi\varepsilon_0 d^2}$, directed upward

(b) $\dfrac{Q^2\sqrt{3}}{4\pi\varepsilon_0 d^2}$, directed downward

(c) $\dfrac{Q^2}{2\pi\varepsilon_0 d^2}$, directed upward

(d) $\dfrac{Q^2}{2\pi\varepsilon_0 d^2}$, directed downward

(e) zero

21. A $-4.0\ \mu C$ charge is located 0.30 m to the left of a $+6.0\ \mu C$ charge. What is the magnitude and direction of the electrostatic force on the positive charge?
(a) 2.4 N, to the right
(b) 2.4 N, to the left
(c) 4.8 N, to the right
(d) 4.8 N, to the left
(e) 7.2 N, to the right

22. Four point charges are held fixed at the corners of a square as shown in the figure. Which of the five arrows shown below most accurately shows the direction of the net force on the charge −Q due to the presence of the three other charges?

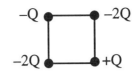

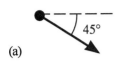

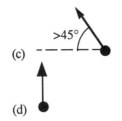

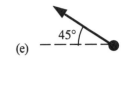

(a)  (c)  (e)

(b)  (d)

*Questions 23 and 24 pertain to the situation described below:*

The figure shows an equilateral triangle **ABC**. A positive point charge $+q$ is located at each of the three vertices **A**, **B**, and **C**. Each side of the triangle is of length $a$.

A point charge $Q$ (that may be positive or negative) is placed at the mid-point between **B** and **C**.

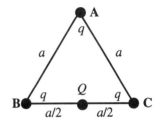

23. Is it possible to choose the value of $Q$ *(that is non-zero)* such that the force on $Q$ is zero? Explain why or why not.
    (a) Yes, because the forces on $Q$ are vectors and three vectors can add to zero.
    (b) No, because the forces on $Q$ are vectors and three vectors can never add to zero.
    (c) Yes, because the electric force at the mid-point between **B** and **C** is zero whether a charge is placed there or not.
    (d) No, because the forces on $Q$ due to the charges at **B** and **C** point in the same direction.
    (e) No, because a fourth charge would be needed to cancel the force on $Q$ due to the charge at **A**.

24. Determine an expression for the magnitude and sign of $Q$ so that the net force on the charge at **A** is zero.
    (a) $Q = +q\left(\dfrac{3\sqrt{3}}{4}\right)$
    (b) $Q = -q\left(\dfrac{3\sqrt{3}}{4}\right)$
    (c) $Q = -q\left(\dfrac{4\sqrt{3}}{3}\right)$
    (d) $Q = +q\left(\dfrac{3}{4\sqrt{3}}\right)$
    (e) $Q = +q\left(\dfrac{4\sqrt{3}}{3}\right)$

25. Determine the ratio of the electrostatic force to the gravitational force between a proton and an electron, $F_E/F_G$. **Note:** $k = 8.99 \times 10^9$ N·m$^2$/C$^2$; $G = 6.672 \times 10^{-11}$ N·m$^2$/kg$^2$; $m_e = 9.109 \times 10^{-31}$ kg; and $m_p = 1.672 \times 10^{-27}$ kg.
    (a) $1.24 \times 10^{23}$
    (b) $2.52 \times 10^{29}$
    (c) $1.15 \times 10^{31}$
    (d) $2.27 \times 10^{39}$
    (e) $1.42 \times 10^{58}$

*Section 18.6 The Electric Field*
*Section 18.7 Electric Field Lines*
*Section 18.8 The Electric Field Inside a Conductor: Shielding*

26. Which one of the following statements is true concerning the magnitude of the electric field at a point in space?
    (a) It is a measure of the total charge on the object.
    (b) It is a measure of the electric force on any charged object.
    (c) It is a measure of the ratio of the charge on an object to its mass.
    (d) It is a measure of the electric force per unit mass on a test charge.
    (e) It is a measure of the electric force per unit charge on a test charge.

27. In the figure, point **A** is a distance $L$ away from a point charge $Q$. Point **B** is a distance $4L$ away from $Q$. What is the ratio of the electric field at **B** to that at **A**, $E_B/E_A$?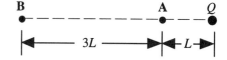
    (a) 1/16
    (b) 1/9
    (c) 1/4
    (d) 1/3
    (e) This cannot be determined since neither the value of $Q$ nor the length $L$ is specified.

28. At which point (or points) is the electric field zero for the two point charges shown on the $x$ axis?

    (a) The electric field is never zero in the vicinity of these charges.
    (b) The electric field is zero somewhere on the $x$ axis to the left of the $+4q$ charge.
    (c) The electric field is zero somewhere on the $x$ axis to the right of the $-2q$ charge.
    (d) The electric field is zero somewhere on the $x$ axis between the two charges, but this point is nearer to the $-2q$ charge.
    (e) The electric field is zero at two points along the $x$ axis; one such point is to the right of the $-2q$ charge and the other is to the left of the $+4q$ charge.

29. An electron traveling horizontally enters a region where a uniform electric field is directed upward. What is the direction of the force exerted on the electron once it has entered the field?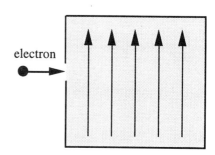
    (a) to the left
    (b) to the right
    (c) upward
    (d) downward
    (e) out of the page, toward the reader

30. Which one of the following statements is true concerning the strength of the electric field between two oppositely charged parallel plates?
    (a) It is zero midway between the plates.
    (b) It is a maximum midway between the plates.
    (c) It is a maximum near the positively charged plate.
    (d) It is a maximum near the negatively charged plate.
    (e) It is constant between the plates except near the edges.

31. Two particles of the same mass carry charges $+3Q$ and $-2Q$, respectively. They are shot into a region that contains a uniform electric field as shown. The particles have the same initial velocities in the positive $x$ direction. The lines, numbered 1 through 5, indicate possible paths for the particles. If the electric field points in the negative $y$ direction, what will be the resulting paths for these particles?

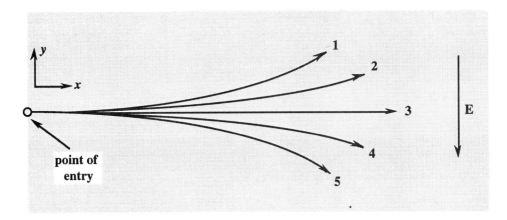

(a) path **1** for $+3Q$ and path **4** for $-2Q$
(b) path **3** for $+3Q$ and path **2** for $-2Q$
(c) path **4** for $+3Q$ and path **3** for $-2Q$
(d) path **2** for $+3Q$ and path **5** for $-2Q$
(e) path **5** for $+3Q$ and path **2** for $-2Q$

*Questions 32 through 34 pertain to the following situation:*

Five particles are shot from the left into a region that contains a uniform electric field. The numbered lines show the paths taken by the five particles. A negatively charged particle with a charge $-3Q$ follows path **2** while it moves through this field.

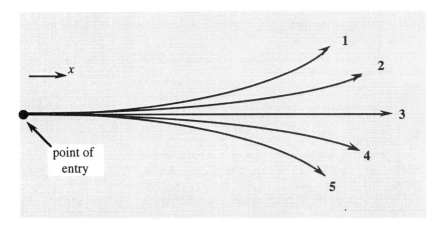

32. In which direction does the electric field point?
    (a) toward the top of the page
    (b) toward the left of the page
    (c) toward the right of the page
    (d) toward the bottom of the page
    (e) out of the page, toward the reader

33. Which path would be followed by a helium atom (an electrically neutral particle)?
    (a) path **1**
    (b) path **2**
    (c) path **3**
    (d) path **4**
    (e) path **5**

34. Which path would be followed by a charge $+6Q$?
    (a) path **1**
    (b) path **2**
    (c) path **3**
    (d) path **4**
    (e) path **5**

35. What is the magnitude of the electric field due to a $4.0 \times 10^{-9}$ C charge at a point 0.020 m away?
    (a) $1.8 \times 10^3$ N/C
    (b) $9.0 \times 10^4$ N/C
    (c) $1.0 \times 10^5$ N/C
    (d) $3.6 \times 10^6$ N/C
    (e) $7.2 \times 10^7$ N/C

*Questions 36 and 38 pertain to the statement and diagram below:*

The figure shows a parallel plate capacitor. The surface charge density on each plate is $8.8 \times 10^{-8}$ C/m$^2$. The point **P** is located $1.0 \times 10^{-5}$ m away from the positive plate.

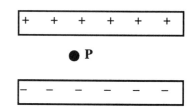

36. Which one of the following statements concerning the direction of the electric field between the plates is true?
    (a) It points to the left.
    (b) It points to the right.
    (c) It points toward the negative plate.
    (d) It points toward the positive plate.
    (e) It points up out of the plane of the page.

37. What is the magnitude of the electric field at the point **P**.
    (a) 8.8 N/C
    (b) 88 N/C
    (c) 100 N/C
    (d) 880 N/C
    (e) 10 000 N/C

38. If a $+2.0 \times 10^{-5}$ C point charge is placed at **P**, what is the force exerted on it?
    (a) 0.2 N, toward the negative plate
    (b) 0.2 N, toward the positive plate
    (c) $5 \times 10^4$ N, toward the positive plate
    (d) $5 \times 10^4$ N, toward the negative plate
    (e) $5 \times 10^4$ N, into the plane of the page

39. A small sphere of mass $1.0 \times 10^{-6}$ kg carries a total charge of $2.0 \times 10^{-8}$ C. The sphere hangs from a silk thread between two large parallel conducting plates. The excess charge on each plate is equal in magnitude, but opposite in sign. If the thread makes an angle of 30° with the positive plate as shown, what is the magnitude of the charge density on each plate?
    (a) $2.5 \times 10^{-9}$ C/m$^2$
    (b) $5.2 \times 10^{-9}$ C/m$^2$
    (c) $1.0 \times 10^{-9}$ C/m$^2$
    (d) $2.1 \times 10^{-8}$ C/m$^2$
    (e) $4.2 \times 10^{-8}$ C/m$^2$

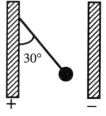

40. Complete the following statement: The magnitude of the electric field at a point in space does *not* depend upon
    (a) the distance from the charge causing the field.
    (b) the sign of the charge causing the field.
    (c) the magnitude of the charge causing the field.
    (d) the force that a unit positive charge will experience at that point.
    (e) the force that a unit negative charge will experience at that point.

41. Four point charges are placed at the corners of a square as shown in the figure. Each side of the square has length 2.0 m. Determine the magnitude of the electric field at the point **P**, the center of the square.
    (a) $2.0 \times 10^{-6}$ N/C
    (b) $3.0 \times 10^{-6}$ N/C
    (c) $9.0 \times 10^{3}$ N/C
    (d) $1.8 \times 10^{4}$ N/C
    (e) $2.7 \times 10^{4}$ N/C

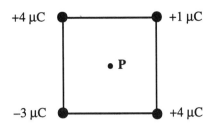

42. The figure shows the electric field lines in the vicinity of two point charges. Which one of the following statements concerning this situation is true?
    (a) $q_1$ is negative and $q_2$ is positive.
    (b) The magnitude of the ratio $(q_2/q_1)$ is 0.70.
    (c) Both $q_1$ and $q_2$ have the same sign of charge.
    (d) The magnitude of the electric field is the same everywhere.
    (e) The electric field is strongest midway between the charges.

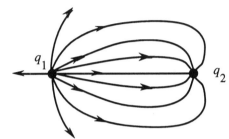

43. A rigid electric dipole is free to move in the electric field represented in the figure.

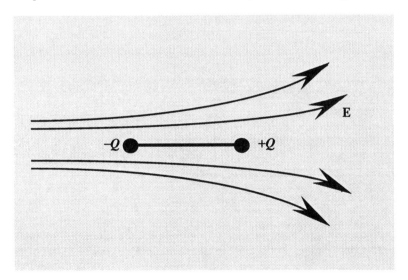

Which one of the following phrases most accurately describes the initial motion of the dipole if it is released from rest in the position shown?
(a) It moves to the left.
(b) It moves to the right.
(c) It does not move at all.
(d) It moves toward to the top of the page.
(e) It moves toward the bottom of the page.

## 178  ELECTRIC FORCES AND ELECTRIC FIELDS

44. A electric dipole is released from rest in a uniform electric field with the orientation shown. Which entry in the table below correctly describes the net torque and the net force on the dipole?

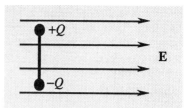

| | net torque | net force |
|---|---|---|
| (a) | zero | zero |
| (b) | clockwise | zero |
| (c) | counterclockwise | zero |
| (d) | clockwise | non-zero |
| (e) | counterclockwise | non-zero |

45. Which one of the following statements is true concerning the *spacing* of the electric field lines in the vicinity of two point charges of equal magnitude and opposite sign?
    (a) It indicates the direction of the electric field.
    (b) It does not depend on the magnitude of the charges.
    (c) It is large when the magnitude of the charges is large.
    (d) It indicates the relative magnitude of the electric field.
    (e) It is small when the magnitude of the charges is small.

46. The magnitude of the electric field at a distance of 10 meters from a negative point charge is $E$. What is the magnitude of the electric field at the same location if the magnitude of the charge is doubled?
    (a) $E/4$
    (b) $E/2$
    (c) $E$
    (d) $2E$
    (e) $4E$

47. What is the magnitude and direction of the electric force on a $-1.2$ μC charge at a point where the electric field is 2500 N/C and is directed along the $+y$ axis.
    (a) 0.15 N, $-y$ direction
    (b) 0.15 N, $+y$ direction
    (c) 0.0030 N, $-y$ direction
    (d) 0.0030 N, $+y$ direction
    (e) 4.3 N, $+x$ direction

### Section 18.9  Gauss' Law

48. A straight, copper wire has a length of 0.50 m and an excess charge of $-1.0 \times 10^{-5}$ C distributed uniformly along its length. Find the magnitude of the electric field at a point that is located $7.5 \times 10^{-3}$ m away from the midpoint of the wire.
    (a) $1.9 \times 10^{10}$ N/C
    (b) $1.5 \times 10^{6}$ N/C
    (c) $6.1 \times 10^{13}$ N/C
    (d) $7.3 \times 10^{8}$ N/C
    (e) $4.8 \times 10^{7}$ N/C

49. What is the electric flux passing through a Gaussian surface that surrounds a +0.075 C point charge?
    (a) $8.5 \times 10^{9}$ N·m²/C
    (b) $6.8 \times 10^{8}$ N·m²/C
    (c) $1.3 \times 10^{7}$ N·m²/C
    (d) $4.9 \times 10^{6}$ N·m²/C
    (e) $7.2 \times 10^{5}$ N·m²/C

50. A uniform electric field with a magnitude of 125 000 N/C passes through a rectangle with sides of 2.50 m and 5.00 m. The angle between the electric field vector and the vector normal to the rectangular plane is 65.0°. What is the electric flux through the rectangle?
    (a) $1.56 \times 10^{6}$ N·m²/C
    (b) $6.60 \times 10^{5}$ N·m²/C
    (c) $1.42 \times 10^{5}$ N·m²/C
    (d) $5.49 \times 10^{4}$ N·m²/C
    (e) $4.23 \times 10^{4}$ N·m²/C

51. A cubical Gaussian surface is placed in a uniform electric field as shown in the figure to the right. The length of each edge of the cube is 1.0 m. The uniform electric field has a magnitude of $5.0 \times 10^8$ N/C and passes through the left and right sides of the cube perpendicular to the surface. What is the total electric flux which passes through the cubical Gaussian surface?

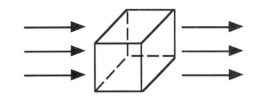

(a) $5.0 \times 10^8$ N·m²/C
(b) $3.0 \times 10^9$ N·m²/C
(c) $2.5 \times 10^6$ N·m²/C
(d) $1.5 \times 10^7$ N·m²/C
(e) zero

## Additional Problems

52. A helium nucleus is located between the plates of a parallel-plate capacitor as shown. The nucleus has a charge of +2e and a mass of $6.6 \times 10^{-27}$ kg. What is the magnitude of the electric field such that the electric force exactly balances the weight of the helium nucleus so that it remains stationary?

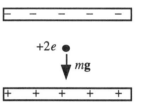

(a) $4.0 \times 10^{-7}$ N/C
(b) $6.6 \times 10^{-26}$ N/C
(c) $2.0 \times 10^{-7}$ N/C
(d) $5.0 \times 10^{-3}$ N/C
(e) $1.4 \times 10^8$ N/C

53. Two identical conducting spheres carry charges of +5.0 μC and −1.0 μC respectively. The centers of the spheres are initially separated by a distance $L$. The two spheres are brought together so that they are in contact. The spheres are then returned to their original separation $L$. What is the ratio of the magnitude of the force on either charge *after* the spheres are touched to that *before* they were touched?
(a) 1/1
(b) 4/5
(c) 9/5
(d) 5/1
(e) 4/9

*Questions 54 and 55 pertain to the situation described below:*

Two point charges, $A$ and $B$, lie along a line separated by a distance $L$. The point x is the midpoint of their separation.

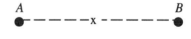

54. Which combination of charges would yield the greatest *repulsive* force between the charges?
(a) −2q and −4q
(b) +1q and −3q
(c) −1q and −4q
(d) −2q and +4q
(e) +1q and +7q

55. Which combination of charges will yield zero electric field at the point x ?
(a) +1q and −1q
(b) +2q and −3q
(c) +1q and −4q
(d) −1q and +4q
(e) +4q and +4q

*Questions 56 through 58 pertain to the situation described below:*

A solid, conducting sphere of radius $a$ carries an excess charge of +6 μC. This sphere is located at the center of a hollow, conducting sphere with an inner radius of $b$ and an outer radius of $c$ as shown. The hollow sphere also carries a total excess charge of +6 μC.

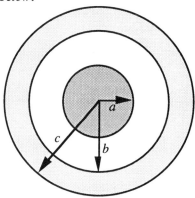

56. Determine the excess charge on the *inner surface* of the outer sphere (a distance $b$ from the center of the system).
    (a) zero
    (b) −6 μC
    (c) +6 μC
    (d) +12 μC
    (e) −12 μC

57. Determine the excess charge on the *outer surface* of the outer sphere (a distance $c$ from the center of the system).
    (a) zero
    (b) −6 μC
    (c) +6 μC
    (d) +12 μC
    (e) −12 μC

58. Which one of the following figures shows a qualitatively accurate sketch of the electric field lines in and around this system?

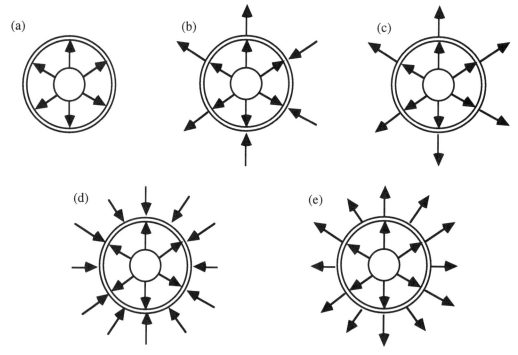

# CHAPTER 19

# Electric Potential Energy and the Electric Potential

## Section 19.1 Potential Energy
## Section 19.2 The Electric Potential Difference

1. Which one of the following statements is true concerning the work done by an external force in moving an electron at constant speed between two points in an electrostatic field?
   (a) It is always zero
   (b) It is always positive.
   (c) It is always negative.
   (d) It depends on the total distance covered.
   (e) It depends only on the displacement of the electron.

2. Which one of the following statements best explains why it is possible to define an *electrostatic potential* in a region of space that contains an electrostatic field?
   (a) Work must be done to bring two positive charges closer together.
   (b) Like charges repel one another and unlike charges attract one another.
   (c) A positive charge will gain kinetic energy as it approaches a negative charge.
   (d) The work required to bring two charges together is independent of the path taken.
   (e) A negative charge will gain kinetic energy as it moves away from another negative charge.

3. Complete the following statement: The *electron volt* is a unit of
   (a) energy.
   (b) electric field strength.
   (c) electric charge.
   (d) electric potential difference.
   (e) electric power.

4. The electric potential at a certain point is space is 12 V. What is the electric potential energy of a $-3.0$ μC charge placed at that point?
   (a) $+4$ μJ
   (b) $-4$ μJ
   (c) $+36$ μJ
   (d) $-36$ μJ
   (e) zero

5. A completely ionized beryllium atom (net charge = $+4e$) is accelerated through a potential difference of 6.0 V. What is the atom's gain in kinetic energy?
   (a) zero
   (b) 0.67 eV
   (c) 4 eV
   (d) 6 eV
   (e) 24 eV

6. Two positive point charges are separated by a distance $R$. If the distance between the charges is reduced to $R/2$, what happens to the total electric potential energy of the system?
   (a) It is doubled.
   (b) It remains the same.
   (c) It increases by a factor of 4.
   (d) It is reduced to one-half of its original value.
   (e) It is reduced to one-fourth of its original value.

7. A charge $q = -4.0$ μC is moved 0.25 m horizontally to point **P** in a region where an electric field is 150 V/m and directed vertically as shown. What is the change in the electric potential energy of the charge?
   (a) $-2.4 \times 10^{-3}$ J
   (b) $-1.5 \times 10^{-4}$ J
   (c) zero
   (d) $+1.5 \times 10^{-4}$ J
   (e) $+2.4 \times 10^{-3}$ J

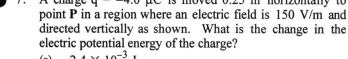

# 182 ELECTRIC POTENTIAL ENERGY AND THE ELECTRIC POTENTIAL

## Section 19.3 The Electric Potential Difference Created by Point Charges

8. A +1.0 μC point charge is moved from point **A** to **B** in the uniform electric field as shown. Which one of the following statements is necessarily true concerning the potential energy of the point charge?
   (a) It increases by $6.0 \times 10^{-6}$ J.
   (b) It decreases by $6.0 \times 10^{-6}$ J.
   (c) It decreases by $9.0 \times 10^{-6}$ J.
   (d) It increases by $10.8 \times 10^{-6}$ J.
   (e) It decreases by $10.8 \times 10^{-6}$ J.

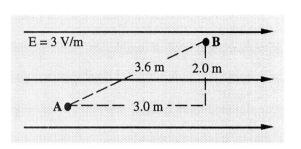

9. Three point charges $-Q$, $-Q$, and $+3Q$ are arranged along a line as shown in the sketch.

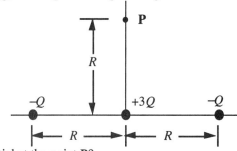

   What is the electric potential at the point **P**?
   (a) $+kQ/R$
   (b) $-2kQ/R$
   (c) $-1.6kQ/R$
   (d) $+1.6kQ/R$
   (e) $+4.4kQ/R$

10. Two point charges are arranged along the $x$ axis as shown in the figure. At which of the following values of $x$ is the electric potential equal to zero?
    Note: At infinity, the electric potential is zero.
    (a) +0.05 m
    (b) +0.29 m
    (c) +0.40 m
    (d) +0.54 m
    (e) +0.71 m

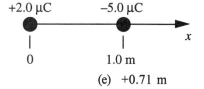

11. Two point charges are located at two of the vertices of a right triangle, as shown in the figure. If a third charge $-2q$ is brought from infinity and placed at the third vertex, what will its electric potential energy be? Use the following values: $a = 0.15$ m; $b = 0.45$ m, and $q = 2.0 \times 10^{-5}$ C.
    (a) −17 J
    (b) −12 J
    (c) −2.8 J
    (d) +8.5 J
    (e) +14 J

12. If the work required to move a +0.35 C charge from point **A** to point **B** is +125 J, what is the potential difference between the two points?
    (a) zero
    (b) 44 V
    (c) 88 V
    (d) 180 V
    (e) 360 V

*Questions 13 through 15 pertain to the electrostatic system described below:*

Two point charges are held at the corners of a rectangle as shown in the figure. The lengths of sides of the rectangle are 0.050 m and 0.150 m. Assume that the electric potential is defined to be zero at infinity.

13. Determine the electric potential at corner **A**.
    (a) $+6.0 \times 10^4$ V
    (b) $-2.4 \times 10^5$ V
    (c) $+4.6 \times 10^5$ V
    (d) $-7.8 \times 10^5$ V
    (e) zero

14. What is the potential difference, $V_B - V_A$, between corners **A** and **B**?
    (a) $-8.4 \times 10^5$ V
    (b) $-7.8 \times 10^5$ V
    (c) $-7.2 \times 10^5$ V
    (d) $-6.0 \times 10^5$ V
    (e) zero

15. What is the electric potential energy of a $+3$ µC charge placed at corner **A** ?
    (a) 0.10 J
    (b) 0.18 J
    (c) 2.34 J
    (d) 2.50 J
    (e) zero


*Questions 16 and 17 pertain to the situation described below:*

Four point charges are placed at the corners of a square as shown in the figure.

Each charge has the identical value $+Q$. The length of the diagonal of the square is $2a$.

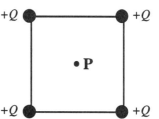

16. What is the magnitude of the electric field at **P**, the center of the square?
    (a) $kQ/a^2$
    (b) $2kQ/a^2$
    (c) $4kQ/a^2$
    (d) $kQ/4a^2$
    (e) zero

17. What is the electric potential at **P**, the center of the square?
    (a) $kQ/a$
    (b) $2kQ/a$
    (c) $4kQ/a$
    (d) $kQ/4a$
    (e) zero


*Questions 18 and 19 refer to the following statement and figure:*

**P** and **Q** are points within a uniform electric field that are separated by a distance of 0.1 m as shown. The potential difference between **P** and **Q** is 50 V.

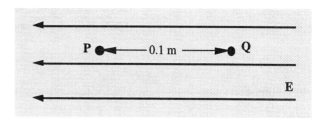

## 184  ELECTRIC POTENTIAL ENERGY AND THE ELECTRIC POTENTIAL

18. Determine the magnitude of this electric field.
    (a) 0.5 V/m
    (b) 5.0 V/m
    (c) 50 V/m
    (d) 500 V/m
    (e) 5000 V/m

19. How much work is required to move a +1000 µC point charge from **P** to **Q**?
    (a) 0.02 J
    (b) 0.05 J
    (c) 200 J
    (d) 1000 J
    (e) 5000 J

*Questions 20 and 21 pertain to the following situation:*

Two point charges are separated by $1.00 \times 10^{-2}$ m. One charge is $-2.8 \times 10^{-8}$ C; and the other is $+2.8 \times 10^{-8}$ C. The points **A** and **B** are located $2.5 \times 10^{-3}$ m from the lower and upper point charges as shown.

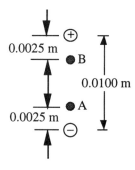

20. If an electron, which has a charge of $1.60 \times 10^{-19}$ C, is moved from rest at **A** to rest at **B**, what is the change in electric potential energy of the electron?
    (a) $+4.3 \times 10^{-15}$ J
    (b) $+5.4 \times 10^{-15}$ J
    (c) $-2.1 \times 10^{-14}$ J
    (d) $-3.2 \times 10^{-14}$ J
    (e) zero

21. If a proton which has a charge of $+1.60 \times 10^{-19}$ C is moved from rest at **A** to rest at **B**, what is change in electrical potential energy of the proton?
    (a) $+2.1 \times 10^{-14}$ J
    (b) $+3.2 \times 10^{-14}$ J
    (c) $-4.3 \times 10^{-15}$ J
    (d) $-5.4 \times 10^{-15}$ J
    (e) zero

*Questions 22 through 24 pertain to the situation described below:*

Two charges of opposite sign and equal magnitude $Q = 2.0$ C are held 2.0 m apart as shown in the figure.

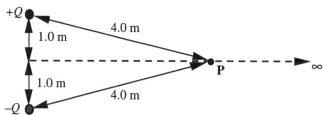

22. Determine the magnitude of the electric field at the point **P**.
    (a) $2.2 \times 10^9$ V/m
    (b) $5.6 \times 10^8$ V/m
    (c) $4.4 \times 10^8$ V/m
    (d) $2.8 \times 10^8$ V/m
    (e) zero

23. Determine the electric potential at the point **P**.
    (a) $1.1 \times 10^9$ V
    (b) $2.2 \times 10^9$ V
    (c) $4.5 \times 10^9$ V
    (d) $9.0 \times 10^9$ V
    (e) zero

24. How much work is required to move a 1.0 C charge from infinity to the point **P**?
    (a) zero
    (b) $2.2 \times 10^9$ J
    (c) $4.5 \times 10^9$ J
    (d) $9.0 \times 10^9$ J
    (e) infinite

## Section 19.4 Equipotential Surfaces and Their Relation to the Electric Field

25. Which one of the following statements concerning electrostatic situations is false?
    (a) **E** is zero everywhere inside a conductor.
    (b) Equipotential surfaces are always perpendicular to **E**.
    (c) It takes zero work to move a charge along an equipotential surface.
    (d) If $V$ is constant throughout a *region of space* then $E$ must be zero in that region.
    (e) No force component acts along the path of a charge as it is moved along an equipotential surface.

26. Which one of the following statements best describes the equipotential surfaces surrounding a point charge?
    (a) The equipotential surfaces are planes extending radially outward from the charge.
    (b) The equipotential surfaces are curved planes surrounding the charge, but only one passes through the charge.
    (c) The equipotential surfaces are concentric cubes with the charge at the center.
    (d) The equipotential surfaces are concentric spheres with the charge at the center.
    (e) The equipotential surfaces are concentric cylinders with the charge on the axis at the center.

*Questions 27 through 35 refer to the statement and figure below:*

The sketch below shows cross sections of equipotential surfaces between two charged conductors that are shown in solid black. Various points on the equipotential surfaces near the conductors are labeled **A, B, C, ..., I**.

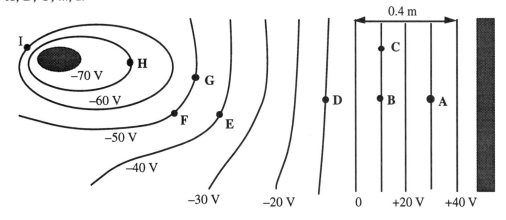

27. At which of the labeled points will the electric field have the greatest magnitude?
    (a) **G**
    (b) **I**
    (c) **A**
    (d) **H**
    (e) **D**

28. At which of the labeled points will an electron have the greatest potential energy?
    (a) **A**
    (b) **D**
    (c) **G**
    (d) **H**
    (e) **I**

186  ELECTRIC POTENTIAL ENERGY AND THE ELECTRIC POTENTIAL

○ 29. What is the potential difference between points **B** and **E**?
   (a) 10 V
   (b) 30 V
   (c) 40 V
   (d) 50 V
   (e) 60 V

◐ 30. What is the direction of the electric field at **B**?
   (a) toward **A**
   (b) toward **D**
   (c) toward **C**
   (d) into the page
   (e) up and out of the page

◐ 31. How much work is required to move a $-1$ µC charge from **A** to **E**?
   (a) $+3.0 \times 10^{-5}$ J
   (b) $-4.0 \times 10^{-5}$ J
   (c) $+7.0 \times 10^{-5}$ J
   (d) $-7.0 \times 10^{-5}$ J
   (e) zero

○ 32. How much work is required to move a $-1$ µC charge from **B** to **D** to **C**?
   (a) $+2.0 \times 10^{-5}$ J
   (b) $-2.0 \times 10^{-5}$ J
   (c) $+4.0 \times 10^{-5}$ J
   (d) $-4.0 \times 10^{-5}$ J
   (e) zero

○ 33. A positive point charge is placed at **F**. Complete the following statement: When it is released,
   (a) no force will be exerted on it.
   (b) a force will cause it to move toward **E**.
   (c) a force will cause it to move toward **G**.
   (d) a force will cause it to move away from **E**.
   (e) it would subsequently lose kinetic energy.

◐ 34. What is the magnitude of the electric field at point **A**?
   (a) 10 V/m
   (b) 25 V/m
   (c) 30 V/m
   (d) 75 V/m
   (e) 100 V/m

◐ 35. A point charge gains 50 µJ of electric potential energy when it is moved from point **D** to point **G**. Determine the magnitude of the charge.
   (a) 1.0 µC
   (b) 1.3 µC
   (c) 25 µC
   (d) 50 µC
   (e) 130 µC

*Questions 36 through 38 pertain to the following situation:*

The sketch shows cross sections of equipotential surfaces between two charged conductors shown in solid black. Points on the equipotential surfaces near the conductors are labeled **A, B, C, ..., H**.

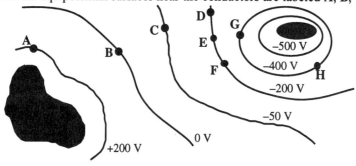

○ 36. What is the magnitude of the potential difference between points **A** and **H**?
   (a) 100 V
   (b) 200 V
   (c) 400 V
   (d) 600 V
   (e) 700 V

37. What is the direction of the electric field at point **E**?
    (a) toward **G**
    (b) toward **B**
    (c) toward **H**
    (d) toward **C**
    (e) toward **F**

38. How much work is required to move a +6 µC point charge from **B** to **F** to **D** to **A**?
    (a) $+1.2 \times 10^{-3}$ J
    (b) $-1.2 \times 10^{-3}$ J
    (c) $+3.6 \times 10^{-3}$ J
    (d) $-3.6 \times 10^{-3}$ J
    (e) zero

## Section 19.5 Capacitors and Dielectrics
## Section 19.6 Biomedical Applications of Electric Potential Differences

39. The magnitude of the charge on the plates of an *isolated* parallel plate capacitor is doubled. Which one of the following statements is true concerning the capacitance of this parallel-plate system?
    (a) The capacitance is decreased to one half of its original value.
    (b) The capacitance is increased to twice its original value.
    (c) The capacitance remains unchanged.
    (d) The capacitance depends on the electric field between the plates.
    (e) The capacitance depends on the potential difference across the plates.

40. A parallel plate capacitor with plates of area $A$ and plate separation $d$ is charged so that the potential difference between its plates is $V$. If the capacitor is then isolated and its plate separation is decreased to $d/2$, what happens to the potential difference between the plates?
    (a) The potential difference is increased by a factor of four.
    (b) The potential difference is twice it original value.
    (c) The potential difference is one half of its original value.
    (d) The potential difference is one forth of its original value.
    (e) The potential difference is unchanged.

41. A parallel plate capacitor with plates of area $A$ and plate separation $d$ is charged so that the potential difference between its plates is $V$. If the capacitor is then isolated and its plate separation is decreased to $d/2$, what happens to its capacitance?
    (a) The capacitance is twice its original value.
    (b) The capacitance is four times its original value.
    (c) The capacitance is eight times its original value.
    (d) The capacitance is one half of its original value.
    (e) The capacitance is unchanged.

42. A parallel plate capacitor is fully charged at a potential $V$. A dielectric with constant $\kappa = 4$ is inserted between the plates of the capacitor while the potential difference between the plates remains constant. Which one of the following statements is false concerning this situation?
    (a) The energy density remains unchanged.
    (b) The capacitance increases by a factor of four.
    (c) The stored energy increases by a factor of four.
    (d) The charge on the capacitor increases by a factor of four.
    (e) The electric field between the plates increases by a factor of four.

43. Which one of the following changes will necessarily increase the capacitance of a capacitor?
    (a) decreasing the charge on the plates
    (b) increasing the charge on the plates
    (c) placing a dielectric between the plates
    (d) increasing the potential difference between the plates
    (e) decreasing the potential difference between the plates

## 188 ELECTRIC POTENTIAL ENERGY AND THE ELECTRIC POTENTIAL

44. When a dielectric with constant κ is inserted between the plates of a charged *isolated* capacitor
    (a) the capacitance is reduced by a factor κ.
    (b) the charge on the plates is reduced by a factor of κ.
    (c) the charge on the plates is increased by a factor of κ.
    (d) the electric field between the plates is reduced by a factor of κ.
    (e) the potential difference between the plates is increased by a factor of κ.

45. A parallel plate capacitor has a potential difference between its plates of 1.2 V and a plate separation distance of 2.0 mm. What is the magnitude of the electric field if a material that has a dielectric constant of 3.3 is inserted between the plates?
    (a) 75 V/m
    (b) 180 V/m
    (c) 250 V/m
    (d) 400 V/m
    (e) 500 V/m

46. A capacitor has a very large capacitance of 10 F. The capacitor is charged by placing a potential difference of 2 V between its plates. How much energy is stored in the capacitor?
    (a) 2000 J
    (b) 500 J
    (c) 100 J
    (d) 40 J
    (e) 20 J

47. A uniform electric field of 8 V/m exists between the plates of a parallel plate capacitor. How much work is required to move a +20 µC point charge from the negative plate to the positive plate if the plate separation is 0.050 m?
    (a) 0.4 J
    (b) 1.6 J
    (c) $8 \times 10^{-4}$ J
    (d) $8 \times 10^{-5}$ J
    (e) $8 \times 10^{-6}$ J

48. A capacitor is initially charged to 2 V. It is then connected to a 4 V battery. What is the ratio of the final to the initial energy stored in the capacitor?
    (a) 2
    (b) 4
    (c) 6
    (d) 8
    (e) 10

49. A parallel plate capacitor has plates of area $2.0 \times 10^{-3}$ m$^2$ and plate separation $1.0 \times 10^{-4}$ m. Determine the capacitance of this system if air fills the volume between the plates.
    (a) $1.1 \times 10^{-10}$ F
    (b) $1.8 \times 10^{-10}$ F
    (c) $3.2 \times 10^{-10}$ F
    (d) $4.4 \times 10^{-10}$ F
    (e) $5.3 \times 10^{-10}$ F

50. A parallel plate capacitor has plates of area $2.0 \times 10^{-3}$ m$^2$ and plate separation $1.0 \times 10^{-4}$ m. Air fills the volume between the plates. What potential difference is required to establish a charge of 3.0 µC on the plates?
    (a) $9.3 \times 10^{2}$ V
    (b) $2.4 \times 10^{4}$ V
    (c) $1.7 \times 10^{4}$ V
    (d) $6.9 \times 10^{3}$ V
    (e) $3.7 \times 10^{5}$ V

51. A potential difference of 120 V is established between two parallel metal plates. The magnitude of the charge on each plate is 0.020 C. What is the capacitance of this capacitor?
    (a) 170 µC
    (b) 24 µC
    (c) 7.2 µC
    (d) 0.12 C
    (e) 2.4 C

*Questions 52 through 57 pertain to the situation described below:*

The plates of a parallel plate capacitor each have an area of 0.40 m$^2$ and are separated by a distance of 0.02 m. They are charged until the potential difference between the plates is 3000 V. The charged capacitor is then isolated.

52. Determine the magnitude of the electric field between the capacitor plates.
    (a) 60 V/m
    (b) 120 V/m
    (c) $1.0 \times 10^5$ V/m
    (d) $1.5 \times 10^5$ V/m
    (e) $3.0 \times 10^5$ V/m

53. Determine the value of the capacitance.
    (a) $9.0 \times 10^{-11}$ F
    (b) $1.8 \times 10^{-10}$ F
    (c) $3.6 \times 10^{-10}$ F
    (d) $4.8 \times 10^{-10}$ F
    (e) $6.4 \times 10^{-10}$ F

54. Determine the magnitude of the charge on either capacitor plate.
    (a) $1.8 \times 10^{-7}$ C
    (b) $2.7 \times 10^{-7}$ C
    (c) $4.9 \times 10^{-7}$ C
    (d) $5.4 \times 10^{-7}$ C
    (e) $6.8 \times 10^{-7}$ C

55. How much work is required to move a –4 µC charge from the negative plate to the positive plate of this system?
    (a) $-1.2 \times 10^{-2}$ J
    (b) $+1.2 \times 10^{-2}$ J
    (c) $-2.4 \times 10^{-2}$ J
    (d) $+2.4 \times 10^{-2}$ J
    (e) $-5.4 \times 10^{-2}$ J

56. Suppose that a dielectric sheet is inserted to completely fill the space between the plates and the potential difference between the plates drops to 1000 V. What is the capacitance of the system after the dielectric is inserted?
    (a) $1.8 \times 10^{-10}$ F
    (b) $2.7 \times 10^{-10}$ F
    (c) $5.4 \times 10^{-10}$ F
    (d) $6.2 \times 10^{-10}$ F
    (e) $6.8 \times 10^{-10}$ F

57. Suppose that a dielectric sheet is inserted to completely fill the space between the plates and the potential difference between the plates drops to 1000 V. Determine the dielectric constant.
    (a) 0.333
    (b) 0.666
    (c) 3.0
    (d) 6.0
    (e) 2000

*Questions 58 through 60 pertain to the situation described below:*

The figure below shows four parallel plate capacitors: **A**, **B**, **C**, and **D**. Each capacitor carries the same charge $q$ and has the same plate area $A$. As suggested by the figure, the plates of capacitors **A** and **C** are separated by a distance $d$ while those of **B** and **D** are separated by a distance $2d$. Capacitors **A** and **B** are maintained in vacuum while capacitors **C** and **D** contain dielectrics with constant κ = 5.

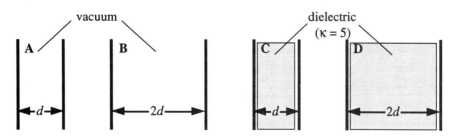

# 190 ELECTRIC POTENTIAL ENERGY AND THE ELECTRIC POTENTIAL

58. Which list below places the capacitors in order of *increasing* capacitance?
    (a) A, B, C, D
    (b) B, A, C, D
    (c) A, B, D, C
    (d) B, A, D, C
    (e) D, C, B, A

59. Which capacitor has the largest potential difference between its plates?
    (a) A
    (b) B
    (c) C
    (d) D
    (e) A and D are the same and larger than B or C.

60. Which capacitor is storing the greatest amount of electric potential energy?
    (a) A
    (b) B
    (c) C
    (d) D
    (e) Since all four carry the same charge, they will store the same amount of energy.

## Additional Problems

*Questions 61 through 64 pertain to the situation described below:*

Two positive charges are located at points **A** and **B** as shown in the figure.

The distance from each charge to the point **P** is $a = 2.0$ m.

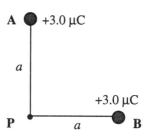

61. Determine the magnitude of the electric field at the point **P**.
    (a) $3.38 \times 10^3$ V/m
    (b) $6.75 \times 10^3$ V/m
    (c) $9.55 \times 10^3$ V/m
    (d) $1.35 \times 10^4$ V/m
    (e) $2.70 \times 10^4$ V/m

62. Which statement is true concerning the direction of the electric field at **P**?
    (a) The direction is toward **A**.
    (b) The direction is toward **B**.
    (c) The direction is directly away from **A**.
    (d) The direction makes a 45° angle above the horizontal direction.
    (e) The direction makes a 135° angle below the horizontal direction.

63. Determine the electric potential at the point **P**.
    (a) $1.35 \times 10^4$ V
    (b) $1.89 \times 10^4$ V
    (c) $2.30 \times 10^4$ V
    (d) $2.70 \times 10^4$ V
    (e) $3.68 \times 10^4$ V

64. Suppose that the charges are rearranged as shown in this figure.

    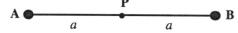

    Which one of the following statements is true for this new arrangement?
    (a) The electric field will be zero, but the electric potential remains unchanged.
    (b) Both the electric field and the electric potential are zero at **P**.
    (c) The electric field will remain unchanged, but the electric potential will be zero.
    (d) The electric field will remain unchanged, but the electric potential will decrease.
    (e) Both the electric field and the electric potential will be changed and will be non-zero.

*Questions 65 and 66 pertain to the situation described below:*

An isolated system consists of two conducting spheres **A** and **B**. Sphere **A** has five times the radius of sphere **B**. Initially, the spheres are given equal amounts of positive charge and are isolated from each other. The two spheres are then connected by a conducting wire.

**Note:** The potential of a sphere of radius $R$ that carries a charge $Q$ is $V = kQ/R$, if the potential at infinity is zero.

65. Which one of the following statements is true *after* the spheres are connected by the wire?
    (a) The electric potential of **A** is 1/25 as large as that of **B**.
    (b) Both spheres are at the same electric potential.
    (c) The electric potential of **A** is 25 times larger than that of **B**.
    (d) The electric potential of **A** is 1/5 as large as that of **B**.
    (e) The electric potential of **A** is five times larger than that of **B**.

66. Determine the ratio of the charge on sphere **A** to that on sphere **B**, $q_A/q_B$, after the spheres are connected by the wire.
    (a) 1
    (b) 1/5
    (c) 5
    (d) 25
    (e) 1/25

# CHAPTER 20    Electric Circuits

*Section 20.1 Electromotive Force and Current*
*Section 20.2 Ohm's Law*

1. Which one of the following situations results in a conventional electric current that flows westward?
   (a) a beam of protons moves eastward
   (b) an electric dipole moves westward
   (c) a beam of electrons moves westward
   (d) a beam of electrons moves eastward
   (e) a beam of neutral atoms moves westward

2. The electromotive force is
   (a) the maximum potential difference between the terminals of a battery.
   (b) the force that accelerates electrons through a wire when a battery is connected to it.
   (c) the force that accelerates protons through a wire when a battery is connected to it.
   (d) the maximum capacitance between the terminals of a battery.
   (e) the maximum electric potential energy stored within a battery.

3. How many electrons flow through a battery that delivers a current of 3.0 A for 12 s?
   (a) 4
   (b) 36
   (c) $4.8 \times 10^{15}$
   (d) $6.4 \times 10^{18}$
   (e) $2.2 \times 10^{20}$

4. A 10-A current is maintained in a simple circuit with a total resistance of 200 Ω. What net charge passes through any point in the circuit during a 1 minute interval?
   (a) 200 C
   (b) 400 C
   (c) 500 C
   (d) 600 C
   (e) 1200 C

5. Which one of the following combinations of units is equivalent to the ohm?
   (a) V/C
   (b) A/J
   (c) J/s
   (d) J·s/C$^2$
   (e) W/A

6. The potential difference across the ends of a wire is doubled in magnitude. If Ohm's law is obeyed, which one of the following statements concerning the resistance of the wire is true?
   (a) The resistance is one half of its original value.
   (b) The resistance is twice its original value.
   (c) The resistance is not changed.
   (d) The resistance increases by a factor of four.
   (e) The resistance decreases by a factor of four.

7. Which one of the following circuits has the largest resistance?

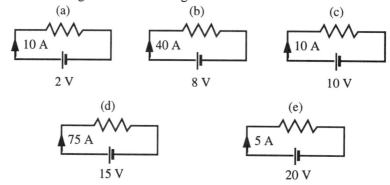

8. A physics student performed an experiment in which the potential difference $V$ between the ends of a long straight wire was varied. The current $I$ in the wire was measured at each value of the potential difference with an ammeter and the results of the experiment are shown in the table.

| Trial | V (volts) | I (amperes) |
|---|---|---|
| 1 | 5.0 | 0.25 |
| 2 | 10.0 | 0.50 |
| 3 | 15.0 | 0.75 |
| 4 | 20.0 | 1.00 |
| 5 | 25.0 | 1.50 |
| 6 | 30.0 | 1.65 |
| 7 | 35.0 | 1.55 |
| 8 | 40.0 | 1.53 |

Which one of the following statements is the *best* conclusion based on the data?
(a) The resistance of the wire is 20 Ω.
(b) The wire does not obey Ohm's law.
(c) The current in the wire is directly proportional to the applied potential difference.
(d) The wire obeys Ohm's law over the range of potential differences between 5 and 30 V.
(e) The wire obeys Ohm's law over the range of potential differences between 5 and 20 V.

## Section 20.3 Resistance and Resistivity

9. Which one of the following statements concerning resistance is true?
   (a) The resistance of a semiconductor increases with temperature.
   (b) Resistance is a property of resistors, but not conductors.
   (c) The resistance of a metal wire changes with temperature.
   (d) The resistance is the same for all samples of the same material.
   (e) The resistance of a wire is inversely proportional to the length of the wire.

10. Which one of the following statements concerning superconductors is false?
    (a) Below its critical temperature, the resistivity of a superconductor is zero.
    (b) Critical temperatures for some superconductors exceed 100 K.
    (c) All materials are superconducting at temperatures near absolute zero.
    (d) A constant current can be maintained in a superconducting ring for several years without an emf.
    (e) Superconductors are perfect conductors.

11. Determine the length of a copper wire that has a resistance of 0.172 Ω and a cross-sectional area of $1 \times 10^{-4}$ m$^2$. The resistivity of copper is $1.72 \times 10^{-8}$ Ω · m.
    (a) 0.1 m        (c) 100 m         (e) 10 000 m
    (b) 10 m         (d) 1000 m

*Questions 12 through 14 pertain to the following situation:*

The characteristics of five wires are given in the table.

| Wire | Material | Length | Gauge |
|---|---|---|---|
| A | iron | 2.0 m | #22 |
| B | copper | 2.0 m | #22 |
| C | copper | 2.0 m | #18 |
| D | copper | 1.0 m | #18 |
| E | iron | 2.0 m | #18 |

## 194 ELECTRIC CIRCUITS

The gauge is a measure of the diameter of the wire; and #18 corresponds to a diameter of $1.2 \times 10^{-3}$ m; and #22 corresponds to a diameter of $6.4 \times 10^{-4}$ m. The resistivity of iron is $9.7 \times 10^{-8}$ $\Omega \cdot$ m; and the value for copper is $1.72 \times 10^{-8}$ $\Omega \cdot$ m.

12. Which one of the wires carries the smallest current when they are connected to identical batteries?
    (a) wire E
    (b) wire D
    (c) wire C
    (d) wire B
    (e) wire A

13. Of the five wires, which one has the smallest resistance?
    (a) wire **A**
    (b) wire **B**
    (c) wire **C**
    (d) wire **D**
    (e) wire **E**

14. Which one of the five wires have the largest resistance?
    (a) wire **A**
    (b) wire **B**
    (c) wire **C**
    (d) wire **D**
    (e) wire **E**

### Section 20.4 Electric Power

15. Complete the following statement: The unit *kilowatt • hour* measures
    (a) current.
    (b) energy.
    (c) power.
    (d) potential drop.
    (e) voltage.

16. Which one of the following quantities can be converted to kilowatt • hours (kWh)?
    (a) 2.0 A
    (b) 8.3 V
    (c) 5.8 J
    (d) 9.6 W
    (e) 6.2 C/V

17. The current through a certain heater wire is found to be fairly independent of its temperature. If the current through the heater wire is doubled, the amount of energy delivered by the heater in a given time interval will
    (a) increase by a factor of two.
    (b) decrease by a factor of two.
    (c) increase by a factor of four.
    (d) decrease by a factor of four.
    (e) increase by a factor of eight.

18. A 4-A current is maintained in a simple circuit with a total resistance of 2 $\Omega$. How much energy is dissipated in 3 seconds?
    (a) 3 J
    (b) 6 J
    (c) 12 J
    (d) 24 J
    (e) 96 J

19. A 40-W and a 60-W light bulb are designed for use with the same voltage. What is the ratio of the resistance of the 60-W bulb to the resistance of the 40-W bulb?
    (a) 1.5
    (b) 0.67
    (c) 2.3
    (d) 0.44
    (e) 3.0

20. A 5-A current is maintained in a simple circuit that consists of a resistor between the terminals of an ideal battery. If the battery supplies energy at a rate of 20 W, how large is the resistance?
    (a) 0.4 $\Omega$
    (b) 0.8 $\Omega$
    (c) 2 $\Omega$
    (d) 4 $\Omega$
    (e) 8 $\Omega$

21. A resistor dissipates 1.5 W when it is connected to a battery with a potential difference of 12 V. What is the resistance of the resistor?
   (a) 0.13 Ω
   (b) 220 Ω
   (c) 18 Ω
   (d) 8.0 Ω
   (e) 96 Ω

## Section 20.5 Alternating Current

22. A 220-Ω resistor is connected across an ac voltage source $V = (150 \text{ V}) \sin [2\pi(60 \text{ Hz})t]$. What is the average power delivered to this circuit?
   (a) 51 W
   (b) 110 W
   (c) 280 W
   (d) 320 W
   (e) 550 W

23. A lamp uses an average power of 55 W when it is connected to an *rms* voltage of 120 V. Which entry in the following table is correct for this circuit?

   |     | lamp resistance $R$ (Ω) | $I_{rms}$ (A) |
   | --- | --- | --- |
   | (a) | 260 | 0.46 |
   | (b) | 22  | 3.8  |
   | (c) | 130 | 0.65 |
   | (d) | 170 | 0.57 |
   | (e) | 38  | 1.2  |

24. When a 1500-W hair dryer is in use, the current passing through the dryer may be represented as $I = (17.7 \text{ A}) \sin (120\pi t)$. What is the rms current for this circuit?
   (a) 17.7 A
   (b) 12.5 A
   (c) 85.7 A
   (d) 25.0 A
   (e) 8.85 A

*Questions 25 through 29 pertain to the situation described below:*

The figure shows variation of the current through the heating element with time in an iron when it is plugged into a standard 120 V, 60 Hz outlet.

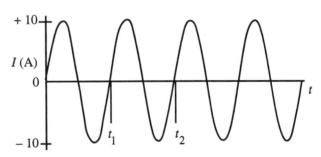

25. What is the peak voltage?
   (a) 10 V
   (b) 60 V
   (c) 120 V
   (d) 170 V
   (e) 240 V

26. What is the *rms* value of the current in this circuit?
   (a) 1.4 A
   (b) 7.1 A
   (c) 11 A
   (d) 14 A
   (e) 18 A

27. What is the resistance of the iron?
   (a) 24 Ω
   (b) 7.1 Ω
   (c) 17 Ω
   (d) 12 Ω
   (e) 1.8 Ω

196  ELECTRIC CIRCUITS

28. If $t_1 = 0.050$ s, what is the value of $t_2$? **Note**: The origin for the graph is not necessarily at $t = 0$.
    (a) 0.067 s
    (b) 0.079 s
    (c) 0.10 s
    (d) 0.60 s
    (e) 61 s

29. What is the approximate average power dissipated in the iron?
    (a) 450 W
    (b) 600 W
    (c) 850 W
    (d) 1200 W
    (e) 1700 W

*Section 20.6  Series Wiring*

30. Which one of the following statements concerning resistors in series is true?
    (a) The voltage across each resistor is the same.
    (b) The current through each resistor is the same.
    (c) The power dissipated by each resistor is the same.
    (d) The rate at which charge flows through each resistor depends on its resistance.
    (e) The total current through the resistors is the sum of the current through each resistor.

31. Two wires, **A** and **B**, and a variable resistor, **R**, are connected in series to a battery. Which one of the following results will occur if the resistance of **R** is increased?
    (a) The current through **A** and **B** will increase.
    (b) The voltage across **A** and **B** will increase.
    (c) The voltage across the entire circuit will increase.
    (d) The power used by the entire circuit will increase.
    (e) The current through the entire circuit will decrease.

32. Three resistors, 50-$\Omega$, 100-$\Omega$, 200-$\Omega$, are connected in series in a circuit. What is the equivalent resistance of this combination of resistors?
    (a) 350 $\Omega$
    (b) 250 $\Omega$
    (c) 200 $\Omega$
    (d) 120 $\Omega$
    (e) 29 $\Omega$

33. Two 15-$\Omega$ and three 25-$\Omega$ light bulbs and a 24 V battery are connected in a series circuit. What is the current that passes through each bulb?
    (a) 0.23 A
    (b) 0.51 A
    (c) 0.96 A
    (d) 1.6 A
    (e) The current will be 1.6 A in the 15-$\Omega$ bulbs and 0.96 A in the 25-$\Omega$ bulbs.

*Section 20.7  Parallel Wiring*

34. Complete the following statement: A simple series circuit contains a resistance $R$ and an ideal battery. If a second resistor is connected in parallel with $R$,
    (a) the voltage across $R$ will decrease.
    (b) the current through $R$ will decrease.
    (c) the total current through the battery will increase.
    (d) the rate of energy dissipation in $R$ will increase.
    (e) the equivalent resistance of the circuit will increase.

35. Some light bulbs are connected in parallel to a 120 V source as shown in the figure. Each bulb dissipates an average power of 60 W. The circuit has a fuse **F** that burns out when the current in the circuit exceeds 9 A. Determine the largest number of bulbs, that can be used in this circuit without burning out the fuse.

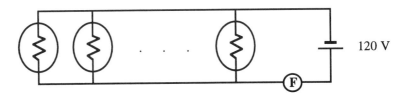

(a) 9
(b) 17
(c) 25
(d) 34
(e) 36

36. Two resistors are arranged in a circuit that carries a total current of 15 A as shown in the figure. Which one of the entries in the following table is correct?

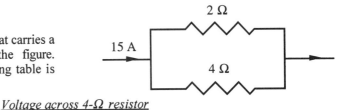

|     | Current through 2-Ω resistor | Voltage across 4-Ω resistor |
|-----|------------------------------|-----------------------------|
| (a) | 5 A                          | 10 V                        |
| (b) | 5 A                          | 20 V                        |
| (c) | 10 A                         | 20 V                        |
| (d) | 15 A                         | 15 V                        |
| (e) | 10 A                         | 10 V                        |

37. What is the total power dissipated in the two resistors in the circuit shown?

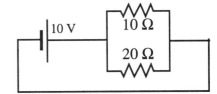

(a) 10 W
(b) 15 W
(c) 33 W
(d) 67 W
(e) 670 W

38. Three resistors, 6.0-Ω, 9.0-Ω, 15-Ω, are connected in parallel in a circuit. What is the equivalent resistance of this combination of resistors?
(a) 30 Ω
(b) 10 Ω
(c) 3.8 Ω
(d) 2.9 Ω
(e) 0.34 Ω

## Section 20.8 Circuits Wired Partially in Series and Partially in Parallel

39. Five resistors are connected as shown. What is the equivalent resistance between points **A** and **B**?
(a) 6.8 Ω
(b) 9.2 Ω
(c) 3.4 Ω
(d) 2.1 Ω
(e) 16 Ω

*Questions 40 through 42 pertain to the statement and diagram below:*

Three resistors are connected as shown in the figure. The potential difference between points **A** and **B** is 26 V.

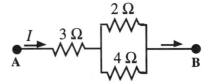

198   ELECTRIC CIRCUITS

40. What is the equivalent resistance between the points A and B?
    (a) 3.8 Ω
    (b) 4.3 Ω
    (c) 5.1 Ω
    (d) 6.8 Ω
    (e) 9.0 Ω

41. How much current flows through the 3-Ω resistor?
    (a) 2.0 A
    (b) 4.0 A
    (c) 6.0 A
    (d) 8.7 A
    (e) 10.0 A

42. How much current flows through the 2-Ω resistor?
    (a) 2.0 A
    (b) 4.0 A
    (c) 6.0 A
    (d) 8.7 A
    (e) 10.0 A

*Questions 43 through 45 pertain to the statement and diagram below:*

Four resistors and a 6-V battery are arranged as shown in the circuit diagram.

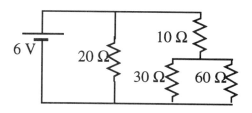

43. Determine the equivalent resistance in for this circuit.
    (a) 50 Ω
    (b) 120 Ω
    (c) 29 Ω
    (d) 5 Ω
    (e) 12 Ω

44. The smallest current passes through which resistor(s)?
    (a) the 10-Ω resistor
    (b) the 20-Ω resistor
    (c) the 30-Ω resistor
    (d) the 60-Ω resistor
    (e) It is the same and the smallest in the 30-Ω and 60-Ω resistors.

45. The largest potential difference is across which resistor(s)?
    (a) the 10-Ω resistor
    (b) the 20-Ω resistor
    (c) the 30-Ω resistor
    (d) the 60-Ω resistor
    (e) It is the same and the largest for the 30-Ω and 60-Ω resistors.

*Questions 46 through 49 pertain to the statement and diagram below:*

Three resistors are placed in a circuit as shown. The potential difference between points A and B is 30 V.

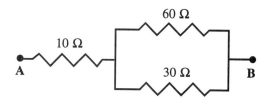

46. What is the equivalent resistance between the points A and B?
    (a) 10 Ω
    (b) 20 Ω
    (c) 30 Ω
    (d) 50 Ω
    (e) 100 Ω

47. What is the potential drop across the 10-Ω resistor?
    (a) 10 V
    (b) 20 V
    (c) 30 V
    (d) 60 V
    (e) 100 V

48. What is the potential drop across the 30-Ω resistor?
    (a) 10 V
    (b) 20 V
    (c) 30 V
    (d) 60 V
    (e) 100 V

49. What is the current through the 30-Ω resistor?
    (a) 0.33 A
    (b) 0.50 A
    (c) 0.67 A
    (d) 1.0 A
    (e) 2.0 A

*Section 20.9 Internal Resistance*

50. A non-ideal battery has a 6.0-V *emf* and an internal resistance of 0.6 Ω. Determine the terminal voltage when the current drawn from the battery is 1.0 A.
    (a) 5.0 V
    (b) 6.0 V
    (c) 5.4 V
    (d) 6.6 V
    (e) 5.8 V

51. A battery has a terminal voltage of 12 V when no current flows and an internal resistance of 2 Ω. The battery is placed in series with a 1-Ω resistor. Which one of the entries in the following table is correct?

|     | Terminal voltage | Current through the 1-Ω resistor |
|-----|------------------|----------------------------------|
| (a) | 4 V              | 4 A                              |
| (b) | 4 V              | 12 A                             |
| (c) | 12 V             | 4 A                              |
| (d) | 12 V             | 12 A                             |
| (e) | 18 V             | 3 A                              |

52. A battery is manufactured to have an emf of 24.0 V, but the terminal voltage is only 22.0 V when the battery is connected across a 7.5-Ω resistor. What is the internal resistance of the battery?
    (a) 3.2 Ω
    (b) 0.27 Ω
    (c) 1.2 Ω
    (d) 0.75 Ω
    (e) 0.68 Ω

*Section 20.10 Kirchhoff's Rules*

53. Three resistors and two 10.0-V batteries are arranged as shown in the circuit diagram. Which one of the following entries in the table is correct?

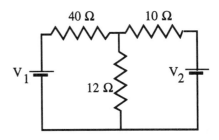

|     | Power Delivered by Battery 1 | Power Delivered by Battery 2 |
|-----|------------------------------|------------------------------|
| (a) | 2.5 W                        | 2.5 W                        |
| (b) | 4.0 W                        | 1.0 W                        |
| (c) | 1.0 W                        | 1.0 W                        |
| (d) | 1.0 W                        | 4.0 W                        |
| (e) | 4.0 W                        | 4.0 W                        |

## 200 ELECTRIC CIRCUITS

54. Three resistors and two batteries are connected as shown in the circuit diagram. What is the magnitude of the current through the 12-V battery?
   (a) 0.15 A
   (b) 0.82 A
   (c) 0.30 A
   (d) 0.67 A
   (e) 0.52 A

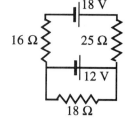

55. Determine the power dissipated by the 40-Ω resistor in the circuit shown.
   (a) 3.6 W
   (b) 4.5 W
   (c) 9.0 W
   (d) 14 W
   (e) 27 W

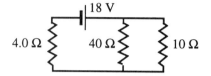

*Questions 56 through 60 pertain to the statement and diagram below:*

Five resistors are connected as shown in the diagram. The potential difference between points **A** and **B** is 25 V.

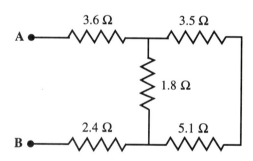

56. What is the equivalent resistance between the points **A** and **B**?
   (a) 1.5 Ω
   (b) 4.8 Ω
   (c) 7.5 Ω
   (d) 9.4 Ω
   (e) 11 Ω

57. What is the current through the 3.6-Ω resistor?
   (a) 1.3 A
   (b) 3.3 A
   (c) 6.9 A
   (d) 7.5 A
   (e) 25 A

58. What is the current through the 1.8-Ω resistor?
   (a) 2.8 A
   (b) 3.3 A
   (c) 5.6 A
   (d) 6.9 A
   (e) 14 A

59. How much energy is dissipated in the 1.8-Ω resistor in 4.0 seconds?
   (a) 18 J
   (b) 28 J
   (c) 55 J
   (d) 64 J
   (e) 93 J

60. What is the potential drop across the 3.5-Ω resistor?
   (a) 2.0 V
   (b) 5.0 V
   (c) 8.0 V
   (d) 17 V
   (e) 25 V

*Questions 61 through 64 pertain to the statement and diagram below:*

Five resistors are connected as shown in the diagram. The potential difference between points **A** and **B** is 15 V.

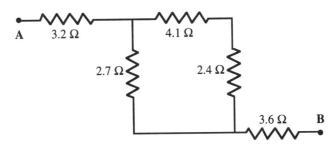

61. What is the equivalent resistance between the points **A** and **B**?
    (a) 1.5 Ω
    (b) 4.8 Ω
    (c) 8.7 Ω
    (d) 10.4 Ω
    (e) 11.1 Ω

62. What is the current in the 3.6-Ω resistor?
    (a) 1.3 A
    (b) 1.7 A
    (c) 2.9 A
    (d) 3.5 A
    (e) 15 A

63. What is the current in the 2.7-Ω resistor?
    (a) 12 A
    (b) 0.8 A
    (c) 2.2 A
    (d) 0.4 A
    (e) 1.2 A

64. What is the amount energy is dissipated in the 2.7-Ω resistor in 9.0 seconds?
    (a) 15 J
    (b) 24 J
    (c) 29 J
    (d) 36 J
    (e) 52 J

## Section 20.12 Capacitors in Series and Parallel

65. Which one of the following statements is true concerning capacitors of *unequal capacitance* connected in series?
    (a) Each capacitor holds a different amount of charge.
    (b) The equivalent capacitance of the circuit is the sum of the individual capacitances.
    (c) The total voltage supplied by the battery is the sum of the voltages across each capacitor.
    (d) The total positive charge in the circuit is the sum of the positive charges on each capacitor.
    (e) The total voltage supplied by the battery is equal to the average voltage across all the capacitors.

66. Three parallel plate capacitors, each having a capacitance of 1.0 μF are connected in parallel. The potential difference across the combination is 100 V. What is the equivalent capacitance of this combination?
    (a) 0.3 μF
    (b) 1 μF
    (c) 3 μF
    (d) 6 μF
    (e) 30 μF

67. Three parallel plate capacitors, each having a capacitance of 1.0 μF are connected in parallel. The potential difference across the combination is 100 V. What is the charge on any one of the capacitors.
    (a) 30 μC
    (b) 100 μC
    (c) 300 μC
    (d) 1000 μC
    (e) 3000 μC

## 202 ELECTRIC CIRCUITS

68. What is the equivalent capacitance of the combination of capacitors shown in the circuit?

    (a) 0.37 µF
    (b) 3.3 µF
    (c) 4.6 µF
    (d) 0.67 µF
    (e) 2.1 µF

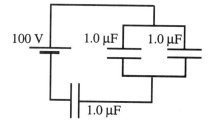

69. How much energy is stored in the combination of capacitors shown?

    (a) 0.01 J
    (b) 0.02 J
    (c) 0.03 J
    (d) 0.04 J
    (e) 0.05 J

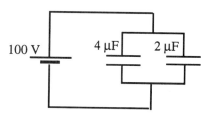

70. A battery supplies a total charge of 5.0 µC to a circuit which consists of a series combination of two identical capacitors, each with capacitance $C$. Determine the charge on either capacitor.
    (a) 5.0 µC
    (b) 2.5 µC
    (c) 1.5 µC
    (d) 1.0 µC
    (e) 0.50 µC

71. When two capacitors are connected in series, the equivalent capacitance of the combination is 100 µF. When the two are connected in parallel, however, the equivalent capacitance of is 450 µF. What are the capacitances of the individual capacitors?
    (a) 200 µF and 250 µF
    (b) 125 µF and 325 µF
    (c) 175 µF and 275 µF
    (d) 150 µF and 300 µF
    (e) 80 µF and 370 µF

*Questions 72 through 74 pertain to the situation described below:*

A 10.0-µF capacitor is charged so that the potential difference between its plates is 10.0 V. A 5.0-µF capacitor is similarly charged so that the potential difference between its plates is 5.0 V. The two charged capacitors are then connected to each other in parallel with positive plate connected to positive plate and negative plate connected to negative plate.

72. How much charge flows from one capacitor to the other when the capacitors are connected?
    (a) 17 µC
    (b) 33 µC
    (c) 67 µC
    (d) 83 µC
    (e) zero

73. What is the final potential difference across the plates of the capacitors when they are connected in parallel?
    (a) 5.0 V
    (b) 6.7 V
    (c) 7.5 V
    (d) 8.3 V
    (e) 10 V

74. How much energy is "lost" when the two capacitors are connected together?
    (a) 33 µJ
    (b) 42 µJ
    (c) 63 µJ
    (d) 130 µJ
    (e) 560 µJ

## Section 20.13 RC Circuits

75. A simple RC circuit consists of a 1-μF capacitor in series with a 3000-Ω resistor, a 6-V battery, and an open switch. Initially, the capacitor is uncharged. How long after the switch is closed will the voltage across the capacitor be 3.8 V?
    (a) $3 \times 10^9$ s
    (b) 3 s
    (c) $3 \times 10^{-9}$ s
    (d) $3 \times 10^{-8}$ s
    (e) 0.003 s

*Questions 76 and 77 pertain to the statement and diagram below:*

The figure shows a simple RC circuit consisting of a 100.0-V battery in series with a 10.0-μF capacitor and a resistor. Initially, the switch S is open and the capacitor is uncharged. Two seconds after the switch is closed, the voltage across the resistor is 37 V.

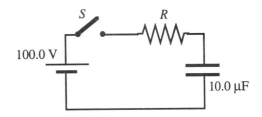

76. Determine the numerical value of the resistance R.
    (a) 0.37 Ω
    (b) 2.70 Ω
    (c) $5.0 \times 10^4$ Ω
    (d) $2.0 \times 10^5$ Ω
    (e) $4.3 \times 10^5$ Ω

77. How much charge is on the capacitor 2.0 s after the switch is closed?
    (a) $1.1 \times 10^{-3}$ C
    (b) $2.9 \times 10^{-3}$ C
    (c) $3.7 \times 10^{-4}$ C
    (d) $5.2 \times 10^{-4}$ C
    (e) $6.6 \times 10^{-4}$ C

*Questions 78 through 80 pertain to the situation described below:*

An uncharged 5.0-μF capacitor and a resistor are connected in series to a 12-V battery to form a simple RC circuit as shown below at the left. The switch is closed at $t = 0$. The graph at the right shows the time variation of the current through the resistor. The time constant of the circuit is 4.0 s.

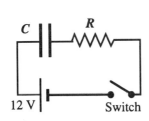 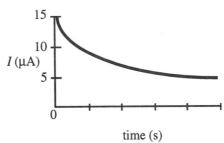

78. Determine the value of the resistance R.
    (a) 15 Ω
    (b) 60 Ω
    (c) $8.0 \times 10^5$ Ω
    (d) $8.0 \times 10^7$ Ω
    (e) $8.0 \times 10^8$ Ω

79. What is the current through the resistor after one time constant has elapsed?
    (a) 5.6 μA
    (b) 9.5 μA
    (c) 22 μA
    (d) 38 μA
    (e) 40 μA

80. Determine the maximum charge on the capacitor.
    (a) 5.5 μC
    (b) 9.5 μC
    (c) 15 μC
    (d) 48 μC
    (e) 60 μC

*Questions 81 through 83 pertain to the situation described below:*

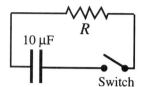

The figure shows a simple RC circuit consisting of a 10-μF capacitor in series with a resistor. Initially, the switch is open as suggested in the figure. The capacitor has been charged. The potential difference between its plates is 100 V. At $t = 0$, the switch is closed. The capacitor then discharges exponentially so that 2.0 s after the switch is closed, the potential difference between the capacitor plates is 37 V. In other words, in 2.0 s the potential difference between the capacitor plates is reduced to 37 % of its original value.

81. Calculate the electric potential energy stored in the capacitor before the switch is closed.
    (a) 0.01 J
    (b) 0.02 J
    (c) 0.03 J
    (d) 0.04 J
    (e) 0.05 J

82. Determine the potential drop across the resistor $R$ at $t = 2.0$ s (i.e., two seconds after the switch is closed).
    (a) zero
    (b) 37 V
    (c) 63 V
    (d) 87 V
    (e) 100 V

83. Determine the numerical value of the resistance $R$.
    (a) $1.0 \times 10^5 \, \Omega$
    (b) $2.0 \times 10^5 \, \Omega$
    (c) $5.0 \times 10^5 \, \Omega$
    (d) $1.0 \times 10^6 \, \Omega$
    (e) $2.5 \times 10^6 \, \Omega$

*Questions 84 through 87 pertain to the situation described below:*

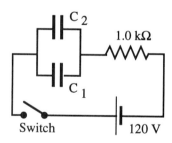

An RC circuit consists of a resistor with resistance 1.0 kΩ, a 120-V battery, and two capacitors, $C_1$ and $C_2$, with capacitances of 20 μF and 60 μF, respectively. Initially, the capacitors are uncharged; and the switch is closed at $t = 0$.

84. What is the current through the resistor *a long time* after the switch is closed? Recall that current is the charge per unit time that *flows* in a circuit.
    (a) 0.60 A
    (b) 0.12 A
    (c) 0.24 A
    (d) 0.48 A
    (e) zero

85. What is the time constant of the circuit?
    (a) $1.0 \times 10^{-2}$ s
    (b) $2.0 \times 10^{-2}$ s
    (c) $6.0 \times 10^{-2}$ s
    (d) $8.0 \times 10^{-2}$ s
    (e) $3.0 \times 10^{-1}$ s

86. How much charge will be stored in each capacitor after a long time has elapsed?

|  | Charge on $C_1$ | Charge on $C_2$ |
|---|---|---|
| (a) | $2.4 \times 10^{-3}$ C | $7.2 \times 10^{-3}$ C |
| (b) | $1.8 \times 10^{-3}$ C | $1.8 \times 10^{-3}$ C |
| (c) | $6.0 \times 10^{-3}$ C | $2.0 \times 10^{-3}$ C |
| (d) | $9.6 \times 10^{-3}$ C | $9.6 \times 10^{-3}$ C |
| (e) | zero | zero |

87. Determine the total charge on both capacitors two time constants after the switch is closed.
    (a) $1.3 \times 10^{-3}$ C
    (b) $2.2 \times 10^{-3}$ C
    (c) $4.7 \times 10^{-3}$ C
    (d) $6.1 \times 10^{-3}$ C
    (e) $8.3 \times 10^{-3}$ C

## Additional Problems

88. Two wires **A** and **B** are made of the same material and have the same diameter. Wire **A** is twice as long as wire **B**. If each wire has the same potential difference across its ends, which one of the following statements is true concerning the current in wire **A**?
    (a) The current is one-fourth that in **B**.
    (b) The current is four times that in **B**.
    (c) The current is equal to the current in **B**.
    (d) The current is half as much as that in **B**.
    (e) It is twice as much as that in **B**.

*Questions 89 through 92 pertain to the situation described below:*

The figure shows a circuit. The switch S can be closed on either point **A** or **C**, but not both at the same time. Use the following quantities:
$V_1 = V_2 = 12$ V
$R_1 = R_4 = 1.0$ Ω
$R_2 = R_3 = 2.0$ Ω

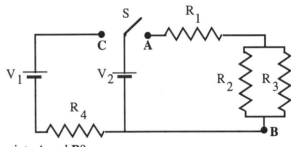

89. What is the equivalent resistance between the points **A** and **B**?
    (a) 1 Ω
    (b) 2 Ω
    (c) 3 Ω
    (d) 4 Ω
    (e) 5 Ω

90. Determine the current through $R_1$ when the switch S is closed on **A**.
    (a) 1 A
    (b) 2 A
    (c) 3 A
    (d) 6 A
    (e) 12 A

91. At what rate is energy dissipated by $R_1$ when the switch S is closed on **A**?
    (a) 1 W
    (b) 4 W
    (c) 9 W
    (d) 36 W
    (e) 144 W

92. Determine the current through $R_4$ when the switch S is closed on **C**.
    (a) zero
    (b) 2 A
    (c) 3 A
    (d) 6 A
    (e) 24 A

# CHAPTER 21
## Magnetic Forces and Magnetic Fields

*Section 21.1 Permanent Magnets*
*Section 21.2 The Force that a Magnetic Field Exerts on a Moving Charge*

1. Which one of the following statements concerning permanent magnets is false?
   (a) The north pole of a permanent magnet is attracted to a south pole.
   (b) All permanent magnets are surrounded by a magnetic field.
   (c) The direction of a magnetic field is indicated by the north pole of a compass.
   (d) Magnetic field lines outside a permanent magnet originate from the north pole and end on the south pole.
   (e) When a permanent magnet is cut in half, one piece will be a north pole and one piece will be a south pole.

2. Which combination of units can be used to express the *magnetic field*?
   (a) kg • m$^2$/C
   (b) kg • s/C$^2$
   (c) N • m$^2$/C
   (d) kg/(C • s)
   (e) kg • m/(C • s$^2$)

3. Which one of the following statements concerning the magnetic force on a charged particle in a magnetic field is true?
   (a) It is a maximum if the particle is stationary.
   (b) It is zero if the particle moves perpendicular to the field.
   (c) It is a maximum if the particle moves parallel to the field.
   (d) It acts in the direction of motion for a positively charged particle.
   (e) It depends on the component of the particle's velocity that is perpendicular to the field.

4. Complete the following statement: The magnitude of the magnetic force that acts on a charged particle in a magnetic field is independent of
   (a) the sign of the charge.
   (b) the magnitude of the charge.
   (c) the magnitude of the magnetic field.
   (d) the direction of motion of the particle.
   (e) the velocity components of the particle.

5. A charged particle is moving in a uniform, constant magnetic field. Which one of the following statements concerning the magnetic force exerted on the particle is false?
   (a) It does no work on the particle.
   (b) It increases the speed of the particle.
   (c) It changes the velocity of the particle.
   (d) It can act only on a particle in motion.
   (e) It does not change the kinetic energy of the particle.

6. A proton traveling due east in a region that contains only a magnetic field experiences a vertically *upward force* away from the surface of the earth. What is the direction of the magnetic field?
   (a) north
   (b) east
   (c) south
   (d) west
   (e) down

*TEST BANK*  Chapter 21   207

7.  Which one of the following statements best explains why a *constant* magnetic field can do *no work* on a moving charged particle?
    (a) The magnetic field is conservative.
    (b) The magnetic force is a velocity dependent force.
    (c) The magnetic field is a vector and work is a scalar quantity.
    (d) The magnetic force is always perpendicular to the velocity of the particle.
    (e) The electric field associated with the particle cancels the effect of the magnetic field on the particle.

8.  An electron traveling due north enters a region that contains a uniform magnetic field that points due east. In which direction will the electron be deflected?
    (a) east
    (b) west
    (c) down
    (d) up
    (e) south

9.  Two electrons are located in a region of space where the magnetic field is zero. Electron **A** is at rest; and electron **B** is moving westward with a constant velocity. A non-zero magnetic field directed eastward is then applied to the region. In what direction, if any, will each electron be moving after the field is applied?

    |  | electron **A** | electron **B** |
    |---|---|---|
    | (a) | at rest | westward |
    | (b) | northward | eastward |
    | (c) | at rest | eastward |
    | (d) | southward | downward, toward the earth |
    | (e) | upward, away from earth | westward |

10. An electron is moving with a speed of $3.5 \times 10^5$ m/s when it encounters a magnetic field of 0.60 T. The direction of the magnetic field makes an angle of 60.0° with respect to the electron's velocity. What is the magnitude of the magnetic force on the electron?
    (a) $4.9 \times 10^{-13}$ N
    (b) $3.2 \times 10^{-13}$ N
    (c) $1.7 \times 10^{-13}$ N
    (d) $3.4 \times 10^{-14}$ N
    (e) $2.9 \times 10^{-14}$ N

## Section 21.3 The Motion of a Charged Particle in a Magnetic Field
## Section 21.4 The Mass Spectrometer

11. A proton traveling due north enters a region that contains both a magnetic field and an electric field. The electric field lines point due west. It is observed that the proton continues to travel in a straight line due north. In which direction must the magnetic field lines point?
    (a) up
    (b) down
    (c) east
    (d) west
    (e) south

12. An electron travels through a region of space with no acceleration. Which one of the following statements is the best conclusion?
    (a) Both **E** and **B** must be zero in that region.
    (b) **E** must be zero, but **B** might be non-zero in that region.
    (c) **E** and **B** might both be non-zero, but they must be mutually perpendicular.
    (d) **B** must be zero, but **E** might be non-zero in that region.
    (e) **E** and **B** might both be non-zero, but they must point in opposite directions.

13. An electron traveling horizontally enters a region where a uniform magnetic field is directed into the plane of the paper as shown. Which one of the following phrases most accurately describes the motion of the electron once it has entered the field?
    (a) upward and parabolic
    (b) upward and circular
    (c) downward and circular
    (d) upward, along a straight line
    (e) downward and parabolic

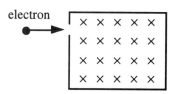

14. An electron enters a region that contains a magnetic field directed into the page as shown. The velocity vector of the electron makes an angle of 30° with the +y axis. What is the direction of the magnetic force on the electron when it enters the field?
    (a) up, out of the page
    (b) at an angle of 30° below the positive $x$ axis
    (c) at an angle of 30° above the positive $x$ axis
    (d) at an angle of 60° below the positive $x$ axis
    (e) at an angle of 60° above the positive $x$ axis

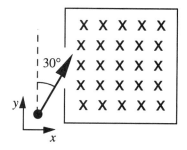

15. Two particles move through a uniform magnetic field that is directed out of the plane of the page. The figure shows the paths taken by the two particles as they move through the field. The particles are not subject to any other forces or fields. Which one of the following statements concerning these particles is true?
    (a) The particles may both be neutral.
    (b) Particle **1** is positively charged; **2** is negative.
    (c) Particle **1** is positively charged; **2** is positive.
    (d) Particle **1** is negatively charged; **2** is negative.
    (e) Particle **1** is negatively charged; **2** is positive.

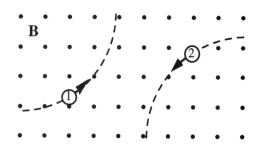

16. Two charged particles of equal mass are traveling in circular orbits in a region of uniform, constant magnetic field as shown. The particles are observed to move in circular paths of radii $R_1$ and $R_2$ with speeds $v_1$ and $v_2$, respectively.

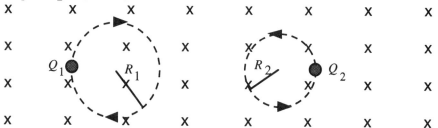

    As the figure shows, the path of particle 2 has a smaller radius than that of particle 1. Which one of the following statements about this system is false?
    (a) $|v_1/Q_1| < |v_2/Q_2|$
    (b) Particle 2 carries a positive charge.
    (c) Particle 1 carries a negative charge.
    (d) Neither particle gains energy from the magnetic field.
    (e) The particle velocities have no components parallel to the magnetic field.

17. A proton is traveling south as it enters a region that contains a magnetic field. The proton is deflected downward toward the earth. What is the direction of the magnetic field?
    (a) downward toward the earth
    (b) west
    (c) north
    (d) east
    (e) south

18. Two charged particles are traveling in circular orbits with the same speed in a region of uniform magnetic field that is directed into the page, as shown. The magnitude of the charge on each particle is identical, but the signs of the charges are unequal.

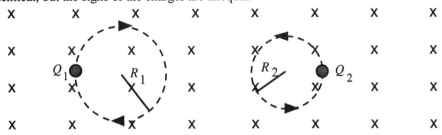

   Which one of the entries in the table below is correct?

|     | Mass Relationship | Sign of charge $Q_1$ | Sign of charge $Q_2$ |
|-----|-------------------|----------------------|----------------------|
| (a) | $m_1 = m_2$       | +                    | −                    |
| (b) | $m_1 > m_2$       | −                    | +                    |
| (c) | $m_1 < m_2$       | −                    | +                    |
| (d) | $m_1 > m_2$       | +                    | −                    |
| (e) | $m_1 < m_2$       | +                    | −                    |

19. A mass spectrometer is used to separate two isotopes of uranium with masses $m_1$ and $m_2$ where $m_2 > m_1$. The two types of uranium atom exit an ion source S with the same charge of $+e$ and are accelerated through a potential difference $V$. The charged atoms then enter a constant, uniform magnetic field $B$ as shown. If $r_1 = 0.5049$ m and $r_2 = 0.5081$ m, what is the value of the ratio $m_1/m_2$?

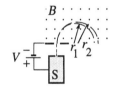

    (a) 0.9984
    (b) 0.9929
    (c) 0.9874
    (d) 0.9812
    (e) 0.9749

*Questions 20 through 22 pertain to the situation described below:*

A beam consisting of five types of ions labeled **A**, **B**, **C**, **D**, and **E** enters a region that contains a uniform magnetic field as shown in the figure below. The field is perpendicular to the plane of the paper, but its precise direction is not given. All ions in the beam travel with the same speed. The table below gives the masses and charges of the ions.

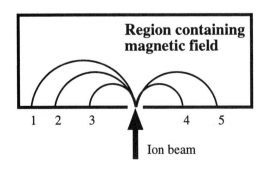

| Ion | Mass    | Charge |
|-----|---------|--------|
| A   | 2 units | $+e$   |
| B   | 4 units | $+e$   |
| C   | 6 units | $+e$   |
| D   | 2 units | $-e$   |
| E   | 4 units | $-e$   |

where
1 mass unit = $1.67 \times 10^{-27}$ kg
and $e = 1.6 \times 10^{-19}$ C

210  MAGNETIC FORCES AND MAGNETIC FIELDS

20. Which ion falls at position **2** ?
    (a) A
    (b) B
    (c) C
    (d) D
    (e) E

21. What is the direction of the magnetic field?
    (a) toward the right
    (b) toward the left
    (c) into the page
    (d) out of the page
    (e) toward the bottom of the page

22. Determine the magnitude of the magnetic field if ion **A** travels in a semicircular path of radius 0.50 m at a speed of $5.0 \times 10^6$ m/s.
    (a) 1.0 T
    (b) 0.84 T
    (c) 0.42 T
    (d) 0.21 T
    (e) 0.11 T

## Section 21.5  The Force on a Current in a Magnetic Field

23. A 0.150-m wire oriented horizontally between the poles of an electromagnet carries a direct current of 12.5 A. The angle between the direction of the current and that of the magnetic field is 25.0°. If the magnetic field strength is 0.625 T, what is the magnitude and direction of the magnetic force on the wire between the poles?

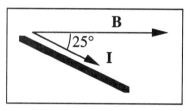

    (a) 1.17 N, upward
    (b) 3.30 N, downward
    (c) 0.792 N, upward
    (d) 1.44 N, downward
    (e) 0.495 N, upward

24. A loop of wire with a weight of 1.47N is oriented vertically and carries a current $I = 1.75$ A. A segment of the wire passes through a magnetic field directed into the plane of the page as shown. The net force on the wire is measured using a balance and found to be zero. What is the magnitude of the magnetic field?

    (a) zero
    (b) 0.51 T
    (c) 0.84 T
    (d) 1.5 T
    (e) 4.2 T

*Questions 25 and 26 pertain to the statement and figure below:*

A long straight vertical segment of wire traverses a magnetic field of magnitude 2.0 T in the direction shown in the diagram. The length of the wire that lies in the magnetic field is 0.060 m. When the switch is closed, a current of 4.0 A flows through the wire from point **P** to point **Q**.

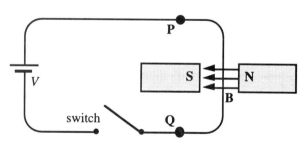

25. Which one of the following statements concerning the effect of the magnetic force on the wire is true?
    (a) The wire will be pushed to the left.
    (b) The wire will be pushed to the right.
    (c) The wire will have no net force acting on it.
    (d) The wire will be pushed downward, into the plane of the paper.
    (e) The wire will be pushed upward, out of the plane of the paper.

26. What is the magnitude of the magnetic force acting on the wire?
   (a) 0.12 N
   (b) 0.24 N
   (c) 0.48 N
   (d) 67 N
   (e) zero

## Section 21.6  The Torque on a Current-Carrying Coil

27. A circular coil consists of 5 loops each of diameter 1.0 m. The coil is placed in an external magnetic field of 0.5 T. When the coil carries a current of 4.0 A, a torque of magnitude 3.93 N•m acts on it. Determine the angle between the normal to the plane of the coil and the direction of the magnetic field.
   (a) 0°
   (b) 30°
   (c) 45°
   (d) 60°
   (e) 90°

28. A single circular loop of 1.0 m radius carries a current of 10.0 mA. It is placed in a uniform magnetic field of magnitude 0.50 T that is directed parallel to the plane of the loop as suggested in the figure. What is the magnitude of the torque exerted on the loop by the magnetic field?
   (a) $1.57 \times 10^{-3}$ N•m
   (b) $3.14 \times 10^{-3}$ N•m
   (c) $6.28 \times 10^{-3}$ N•m
   (d) $9.28 \times 10^{-3}$ N•m
   (e) zero

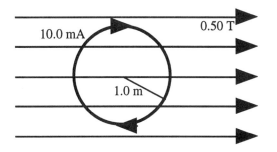

29. A rectangular loop has sides of length 0.06 m and 0.08 m. The wire carries a current of 10 A in the direction shown. The loop is in a uniform magnetic field of magnitude 0.2 T and directed in the positive $x$ direction. What is the magnitude of the torque on the loop?
   (a) $4.2 \times 10^{-2}$ N•m
   (b) $4.8 \times 10^{-2}$ N•m
   (c) $4.8 \times 10^{-3}$ N•m
   (d) $8.3 \times 10^{-3}$ N•m
   (e) $9.6 \times 10^{-3}$ N•m

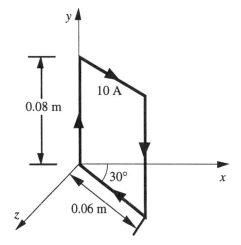

30. A single circular loop of wire of radius 0.75 m carries a constant current of 3.0 A. The loop may be rotated about an axis that passes through the center and lies in the plane of the loop. When the orientation of the normal to the loop with respect to the direction of the magnetic field is 25°, the torque on the coil is 1.8 N•m. What is the magnitude of the uniform magnetic field exerting this torque on the loop?
   (a) 0.37 T
   (b) 1.7 T
   (c) 3.0 T
   (d) 0.46 T
   (e) 0.80 T

# 212 MAGNETIC FORCES AND MAGNETIC FIELDS

31. A coil consists of 240 circular loops, each of radius 0.044 m, and carries a current of 2.2 A. Determine the magnetic moment of the coil.
    (a) 0.21 A·m$^2$
    (b) 0.65 A·m$^2$
    (c) 3.2 A·m$^2$
    (d) 15 A·m$^2$
    (e) 23 A·m$^2$

## Section 21.7 Magnetic Fields Produced by Currents

32. Which one of the following statements concerning the magnetic field well inside a long, current-carrying solenoid is true?
    (a) The magnetic field is zero.
    (b) The magnetic field is non-zero and nearly uniform.
    (c) The magnetic field is independent of the number of windings.
    (d) The magnetic field is independent of the current in the solenoid.
    (e) The magnetic field varies as 1/r as measured from the solenoid axis.

33. Complete the following statement: The magnetic field around a current-carrying, circular loop is most like that of
    (a) the earth.
    (b) a short bar magnet.
    (c) a current-carrying, rectangular loop.
    (d) a long straight wire that carries a current.
    (e) two long straight wires that carry currents in opposite directions.

34. A solenoid of length 0.25 m and radius 0.02 m is comprised of 120 turns of wire. Determine the magnitude of the magnetic field at the center of the solenoid when it carries a current of 15 A.
    (a) 2.26 × 10$^{-3}$ T
    (b) 4.52 × 10$^{-3}$ T
    (c) 9.05 × 10$^{-3}$ T
    (d) 7.50 × 10$^{-3}$ T
    (e) zero

35. The drawing shows two long, thin wires that carry currents in the positive z direction. Both wires are parallel to the z axis. The 50-A wire is in the x-z plane and is 5 m from the z axis. The 40-A wire is in the y-z plane and is 4 m from the z axis. What is the magnitude of the magnetic field at the origin?
    (a) zero
    (b) 1.6 × 10$^{-6}$ T
    (c) 2.8 × 10$^{-6}$ T
    (d) 3.2 × 10$^{-6}$ T
    (e) 4.0 × 10$^{-6}$ T

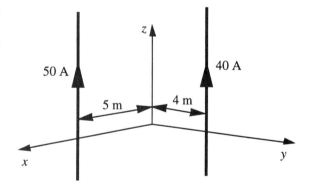

36. A long, straight wire carries a current $I$. If the magnetic field at a distance $d$ from the wire has magnitude $B$, what is the magnitude of the magnetic field at a distance $2d$ from the wire?
    (a) $B/2$
    (b) $B/4$
    (c) $2B$
    (d) $4B$
    (e) $8B$

37. An overhead electric power line carries a maximum current of 125 A. What is the magnitude of the maximum magnetic field at a point 4.50 m directly below the power line?
    (a) 5.56 × 10$^{-6}$ T
    (b) 1.75 × 10$^{-5}$ T
    (c) 3.49 × 10$^{-5}$ T
    (d) 4.69 × 10$^{-4}$ T
    (e) 7.95 × 10$^{-3}$ T

38. A long wire that carries a current *I* is bent into five loops as shown in the figure.

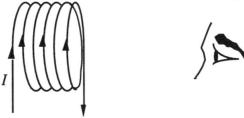

If the observer could "see" the magnetic field inside this arrangement of loops, how would it appear?

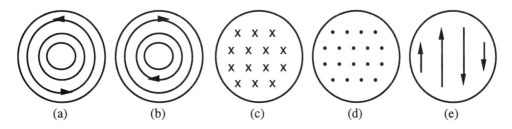

39. A wire, connected to a battery and switch, passes through the center of a long current-carrying solenoid as shown in the drawing.

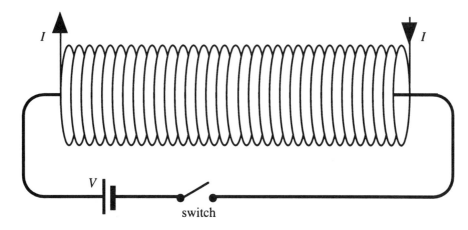

When the switch is closed and there is a current in the wire, what happens to the portion of the wire that runs inside of the solenoid?
(a) There is no effect on the wire.
(b) The wire is pushed downward.
(c) The wire is pushed upward.
(d) The wire is pushed into the plane of the paper.
(e) The wire is pushed out of the plane of the paper.

40. Two loops carry equal currents *I* in the same direction. The loops are held in the positions shown in the figure and are then released. Which one of the following statements correctly describes the subsequent behavior of the loops?
(a) Both loops move to the left.
(b) The loops remain in the positions shown.
(c) The top loop moves to the right; the bottom loop moves to the right.
(d) The loops repel each other.
(e) The loops attract each other.

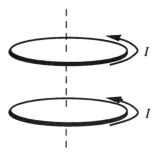

41. Four long, straight wires are parallel to each other; and their cross-section forms a square. Each side of the square is 0.02 m as shown in the figure. If each wire carries a current of 8.0 A in the direction shown in the figure, determine the magnitude of the total magnetic field at **P**, the center of the square.

(a) $5.1 \times 10^{-5}$ T
(b) $1.1 \times 10^{-4}$ T
(c) $1.7 \times 10^{-4}$ T
(d) $2.3 \times 10^{-4}$ T
(e) zero

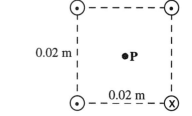

42. The figure shows two concentric metal loops, each carrying a current. The larger loop carries a current of 8 A and has a radius of 0.06 m. The smaller loop has a radius of 0.04 m. What is the value of a current in the smaller loop that will result in zero total magnetic field at the center of the system?

(a) 5.3 A
(b) 6.0 A
(c) 8.8 A
(d) 12 A
(e) 24 A

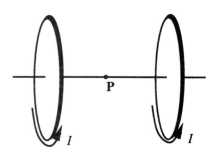

43. Two loops carry equal currents $I$ in the same direction. They are held in the positions shown in the figure and project above and below the plane of the paper. The point **P** lies exactly halfway between them on the line that joins their centers. The centers of the loops and the point **P** lie in the plane of the paper. Which one of the figures below shows the position of a compass needle if the compass were placed in the plane of the paper at **P**?

(a)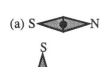

(b)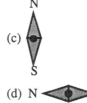

(c)

(d)

(e)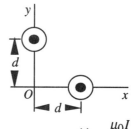

44. Two long, straight wires are perpendicular to the plane of the paper as shown in the drawing. Each wire carries a current of magnitude $I$. The currents are directed out of the paper toward you. Which one of the following expressions correctly gives the magnitude of the total magnetic field at the origin of the $x$, $y$ coordinate system?

(a) $\dfrac{\mu_0 I}{2d}$
(b) $\dfrac{\mu_0 I}{\sqrt{2}d}$
(c) $\dfrac{\mu_0 I}{2\pi d}$
(d) $\dfrac{\mu_0 I}{\pi d}$
(e) $\dfrac{\mu_0 I}{\sqrt{2}\pi d}$

*Questions 45 and 46 pertain to the statement and figure below:*

A long, straight wire is carrying a current of 5 A in the direction shown in the figure. The point **P** is 0.04 m from the wire.

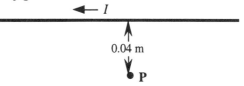

45. What is the direction of the magnetic field at point **P** due to the current in the wire?
    (a) to the right of page
    (b) to the left of the page
    (c) toward the bottom of the page
    (d) into the plane of the page
    (e) out of the plane of the page

46. What is the magnitude of the magnetic field at the point **P**?
    (a) $1.3 \times 10^{-5}$ T
    (b) $1.9 \times 10^{-5}$ T
    (c) $2.5 \times 10^{-5}$ T
    (d) $7.9 \times 10^{-5}$ T
    (e) $9.4 \times 10^{-5}$ T

*Questions 47 and 48 pertain to the statement and figure below:*

A single circular loop of wire with radius 0.02 m carries a current of 8 A. It is placed at the center of a solenoid that has length 0.65 m, radius 0.08 m, and 1400 turns.

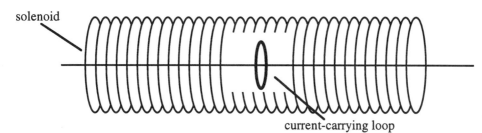

47. Determine the value of the current in the solenoid so that the magnetic field at the center of the loop is zero.
    (a) $1.4 \times 10^{-1}$ A
    (b) $2.5 \times 10^{-4}$ A
    (c) $4.4 \times 10^{-2}$ A
    (d) $5.0 \times 10^{-3}$ A
    (e) $9.3 \times 10^{-2}$ A

48. Determine the magnitude of the total magnetic field at the center of the loop (due both to the loop and the solenoid) if the current in the loop is reversed in direction from that needed to make the total field zero.
    (a) $2.5 \times 10^{-4}$ T
    (b) $5.0 \times 10^{-4}$ T
    (c) $6.4 \times 10^{-4}$ T
    (d) $8.7 \times 10^{-4}$ T
    (e) $9.2 \times 10^{-4}$ T

*Questions 49 and 50 pertain to the statement and figure below:*

Two long, straight, parallel wires separated by a distance $d$ carry currents in opposite directions as shown in the figure. The bottom wire carries a current of 6 A. Point **C** is at the midpoint between the wires and point **O** is a distance $0.5d$ below the 6-A wire as suggested in the figure. The total magnetic field at point **O** is zero.

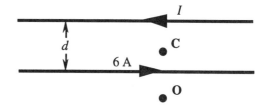

216   MAGNETIC FORCES AND MAGNETIC FIELDS

49. Determine the value of the current, $I$, in the top wire.
(a) 2 A
(b) 3 A
(c) 6 A
(d) 18 A
(e) This cannot be determined since the value of $d$ is not specified.

50. Determine the magnitude of the magnetic field at point **C** if $d = 0.10$ m.
(a) $2.4 \times 10^{-5}$ T
(b) $4.8 \times 10^{-5}$ T
(c) $9.6 \times 10^{-5}$ T
(d) $1.1 \times 10^{-4}$ T
(e) $1.4 \times 10^{-4}$ T

*Questions 51 and 52 pertain to the two wire system described below:*

Two long, straight wires separated by 0.10 m carry currents of 18 A and 6 A in the same direction as shown.

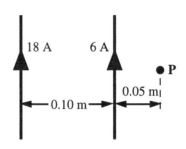

51. Determine the magnitude of the magnetic field at the point **P**.
(a) $2.4 \times 10^{-5}$ T
(b) $4.8 \times 10^{-5}$ T
(c) $7.2 \times 10^{-5}$ T
(d) $9.6 \times 10^{-5}$ T
(e) zero

52. What is the direction of the magnetic field at the point **P**?
(a) to the left of the page
(b) to the right of the page
(c) toward the bottom of the page
(d) out of the plane of the page
(e) into the plane of the page

## Section 21.8 Ampere's Law
## Section 21.9 Magnetic Materials

*Questions 53 and 54 pertain to the following:*

A long, coaxial cable, shown in cross-section in the drawing, is made using two conductors that share a common central axis, labeled **C**. The conductors are separated by an electrically insulating material that is also used as the outer cover of the cable. The current in the *inner* conductor is 2.0 A directed *into the page* and that in the *outer* conductor is 2.5 A directed *out of the page*.

The distance from point **C** to point **A** is 0.0015 m; and the distance from **C** to **B** is 0.0030 m. The radii $a$ and $b$ of the conductors are $6.0 \times 10^{-4}$ m and $1.9 \times 10^{-3}$ m, respectively.

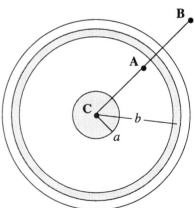

53. What is the magnitude and direction of the magnetic field at point **A**?
(a) $3.3 \times 10^{-5}$ T, clockwise
(b) $3.3 \times 10^{-5}$ T, counterclockwise
(c) $6.8 \times 10^{-5}$ T, clockwise
(d) $6.8 \times 10^{-5}$ T, counterclockwise
(e) $2.7 \times 10^{-4}$ T, clockwise

54. What is the magnitude and direction of the magnetic field at point **B**?
    (a) $3.3 \times 10^{-5}$ T, clockwise
    (b) $3.3 \times 10^{-5}$ T, counterclockwise
    (c) $6.8 \times 10^{-5}$ T, clockwise
    (d) $6.8 \times 10^{-5}$ T, counterclockwise
    (e) $2.7 \times 10^{-4}$ T, clockwise

55. Which one of the following materials is not ferromagnetic?
    (a) iron
    (b) chromium dioxide
    (c) nickel
    (d) aluminum
    (e) cobalt

## Additional Problems

56. A long straight wire carries a 40.0 A current in the +x direction. At a particular instant, an electron moving at $1.0 \times 10^7$ m/s in the +y direction is 0.10 m from the wire. The charge on the electron is $-1.6 \times 10^{-19}$ C. What is the force on the electron at this instant?

    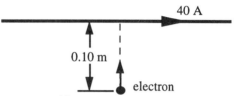

    (a) $1.3 \times 10^{-16}$ N in the +x direction
    (b) $1.3 \times 10^{-16}$ N in the −x direction
    (c) $6.5 \times 10^{-10}$ N in the +y direction
    (d) $6.5 \times 10^{-10}$ N in the −y direction
    (e) $6.5 \times 10^{-16}$ N in the −y direction

*Questions 57 through 59 pertain to the statement and diagram below:*

A wire is bent into the shape a circle of radius $r$ = 0.1 m and carries a 20 A current in the direction shown.

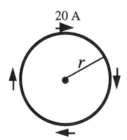

57. What is the direction of the magnetic field at the center of the loop?
    (a) to the right of the page
    (b) to the left of the page
    (c) toward the top of the page
    (d) into the plane of the paper
    (e) out of the plane of the paper

58. What is the magnitude of the magnetic field at the center of the loop?
    (a) $2.0 \times 10^{-5}$ T
    (b) $1.3 \times 10^{-5}$ T
    (c) $2.0 \times 10^{-4}$ T
    (d) $1.3 \times 10^{-4}$ T
    (e) zero

59. Determine the magnetic moment of the loop.
    (a) 0.20 A•m²
    (b) 0.40 A•m²
    (c) 0.63 A•m²
    (d) 0.84 A•m²
    (e) 1.3 A•m²

# MAGNETIC FORCES AND MAGNETIC FIELDS

*Questions 60 and 61 pertain to the following statement and figure:*

A long, straight wire carries a 10.0 A current in the $+y$ direction as shown in the figure.

Next to the wire is a square copper loop that carries a 2.0 A current as shown. The length of each side of the square is 1.0 m.

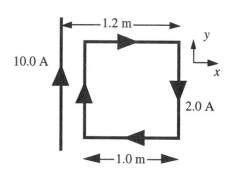

60. What is the magnitude of the net magnetic force that acts on the loop?
    (a) $8.0 \times 10^{-6}$ N
    (b) $1.1 \times 10^{-5}$ N
    (c) $1.4 \times 10^{-5}$ N
    (d) $1.7 \times 10^{-5}$ N
    (e) $2.3 \times 10^{-5}$ N

61. What is the direction of the net magnetic force that acts on the loop?
    (a) $+x$ direction
    (b) $-x$ direction
    (c) $+y$ direction
    (d) $-y$ direction
    (e) 30° with respect to the $+x$ direction

## CHAPTER 22 — Electromagnetic Induction

*Section 22.1 Induced EMF and Induced Current*
*Section 22.2 Motional EMF*

1. A conducting loop of wire is placed in a magnetic field that is normal to the plane of the loop. Which one of the following actions will not result in an induced current in the loop?
   (a) Rotate the loop about an axis that is parallel to the field and passes through the center of the loop.
   (b) Increase the strength of the magnetic field.
   (c) Decrease the area of the loop.
   (d) Decrease the strength of the magnetic field.
   (e) Rotate the loop about an axis that is perpendicular to the field and passes through the center of the loop.

2. The units of motional emf may be written
   (a) T • m/s
   (b) V • m$^2$/s
   (c) J/s
   (d) kg • m$^2$/(C • s$^2$)
   (e) T • m

3. The figure shows a uniform magnetic field which is normal to the plane of a conducting loop which has a resistance $R$.
   Which of the following changes will cause an induced current to flow through the resistor?

   (a) decreasing the area of the loop
   (b) decreasing the magnitude of the magnetic field
   (c) increasing the magnitude of the magnetic field
   (d) rotating the loop through 90° into the plane of the paper
   (e) all of the above

4. A conducting bar moves to the left at a constant speed $v$ on two conducting rails joined at the left as shown. As a result of the bar moving through a constant magnetic field, a current $I$ is induced in the indicated direction. Which one of the following directions is that of the magnetic field?
   (a) toward the right
   (b) toward the left
   (c) parallel to the long axis of the bar
   (d) into the page
   (e) out of the page

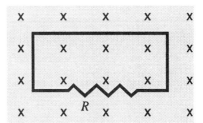

5. A 2.0-kg rod has a length of 1.0 m and a resistance of 4.0 Ω. It slides with *constant speed* down a pair of frictionless vertical conducting rails that are joined at the bottom. A uniform magnetic field of magnitude 3.0 T is perpendicular to the plane formed by the rod and the rails as shown.

   Determine the speed of the rod.
   (a) 0.38 m/s
   (b) 0.90 m/s
   (c) 2.6 m/s
   (d) 5.6 m/s
   (e) 8.7 m/s

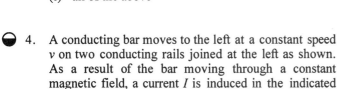

## Section 22.3 Magnetic Flux

6. A 0.50-T magnetic field is directed perpendicular to the plane of a circular loop of radius 0.25 m. What is the magnitude of the magnetic flux through the loop?
   (a) 0.049 Wb
   (b) 0.098 Wb
   (c) 0.20 Wb
   (d) 0.39 Wb
   (e) zero

7. The Earth's magnetic field passes through a square tabletop with a magnitude of $4.95 \times 10^{-5}$ T and directed at an angle of 165° relative to the normal of the tabletop. If the tabletop has 1.50-m sides, what is the magnitude of the magnetic flux through it?
   (a) $1.08 \times 10^{-4}$ Wb
   (b) $7.11 \times 10^{-5}$ Wb
   (c) $2.88 \times 10^{-5}$ Wb
   (d) $1.92 \times 10^{-5}$ Wb
   (e) $3.30 \times 10^{-6}$ Wb

8. A circular copper loop is placed perpendicular to a uniform magnetic field of 0.50 T. Due to external forces, the area of the loop decreases at a rate of $1.26 \times 10^{-3}$ m²/s. Determine the induced emf in the loop.
   (a) $3.1 \times 10^{-4}$ V
   (b) $6.3 \times 10^{-4}$ V
   (c) $1.2 \times 10^{-3}$ V
   (d) $7.9 \times 10^{-3}$ V
   (e) 3.1 V

9. A conducting loop has an area of 0.065 m² and is positioned such that a uniform magnetic field is perpendicular to the plane of the loop. When the magnitude of the magnetic field *decreases* to 0.30 T in 0.087 s, the average induced emf in the loop is 1.2 V. What is the initial value of the magnetic field?
   (a) 0.42 T
   (b) 0.75 T
   (c) 0.87 T
   (d) 1.2 T
   (e) 1.9 T

10. A uniform magnetic field passes through two areas, $A_1$ and $A_2$. The angles between the magnetic field and the normals of areas $A_1$ and $A_2$ are 30.0° and 60.0°, respectively. If the magnetic flux through the two areas is the same, what is the ratio $A_1/A_2$?
    (a) 0.577
    (b) 0.816
    (c) 1.00
    (d) 1.23
    (e) 1.73

## Section 22.4 Faraday's Law of Electromagnetic Induction

11. A circular coil of wire has 25 turns and has a radius of 0.075 m. The coil is located in a variable magnetic field whose behavior is shown on the graph. At all times, the magnetic field is directed at an angle of 75° relative to the normal to the plane of a loop. What is the average emf induced in the coil in the time interval from $t$ = 5.00 s to 7.50 s?
    (a) −18 mV
    (b) −49 mV
    (c) −92 mV
    (d) −140 mV
    (e) −180 mV

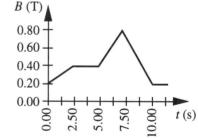

12. A magnetic field is directed perpendicular to the plane of the 0.15-m × 0.30-m rectangular coil comprised of 120 loops of wire. To induce an average emf of −1.2 V in the coil, the magnetic field is increased from 0.1 T to 1.5 T during a time interval $\Delta t$. Determine $\Delta t$.
    (a) 0.053 s
    (b) 0.13 s
    (c) 1.6 s
    (d) 6.3 s
    (e) 7.6 s

*Questions 13 through 16 pertain to the situation described below:*

The figure shows a uniform, 3.0-T magnetic field that is normal to the plane of a conducting, circular loop with a resistance of 1.5 Ω and a radius of 0.024 m. The magnetic field is directed out of the paper as shown.
Note: The area of the non-circular portion of the wire is considered negligible compared to that of the circular loop.

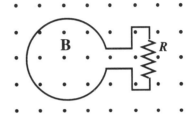

13. What is the magnitude of the average induced emf in the loop if the magnitude of the magnetic field is doubled in 0.4 s?
    (a) 0.43 V
    (b) 0.65 V
    (c) 0.014 V
    (d) 0.027 V
    (e) 0.038 V

14. What is the average current around the loop if the magnitude of the magnetic field is doubled in 0.4 s?
    (a) $2.8 \times 10^{-3}$ A, clockwise
    (b) $4.5 \times 10^{-3}$ A, clockwise
    (c) $4.5 \times 10^{-3}$ A, counterclockwise
    (d) $9.0 \times 10^{-3}$ A, clockwise
    (e) $9.0 \times 10^{-3}$ A, counterclockwise

15. If the magnetic field is held constant at 3.0 T and the loop is pulled out of the region that contains the field in 0.2 s, what is the magnitude of the average induced emf in the loop?
    (a) $8.6 \times 10^{-3}$ V
    (b) $9.8 \times 10^{-3}$ V
    (c) $2.7 \times 10^{-2}$ V
    (d) $5.4 \times 10^{-2}$ V
    (e) $6.4 \times 10^{-2}$ V

16. If the magnetic field is held constant at 3.0 T and the loop is pulled out of the region that contains the field in 0.2 s, at what rate is energy dissipated in $R$?
    (a) $1.8 \times 10^{-2}$ W
    (b) $3.6 \times 10^{-2}$ W
    (c) $3.8 \times 10^{-3}$ W
    (d) $2.7 \times 10^{-4}$ W
    (e) $4.9 \times 10^{-4}$ W

## Section 22.5 Lenz's Law
## Section 22.6 Applications of Electromagnetic Induction to the Reproduction of Sound

17. A long, straight wire is in the same plane as a wooden, *non-conducting* loop. The wire carries an increasing current $I$ in the direction shown in the figure.

    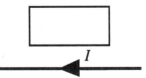

    (a) There will be no induced emf.
    (b) There will be a *counterclockwise* induced emf, but no induced current.
    (c) There will be a *clockwise* induced emf, but no induced current.
    (d) There will be an induced current which is *clockwise* around the loop.
    (e) There will be an induced current which is *counterclockwise* around the loop.

18. A long, straight wire is in the same plane as a rectangular, *conducting* loop. The wire carries a constant current $I$ as shown in the figure. Which one of the following statements is true if the wire is suddenly moved *toward* the loop?
    (a) There will be no induced emf.
    (b) There will be an induced emf, but no induced current.
    (c) There will be an induced current which is clockwise around the loop.
    (d) There will be an induced current which is counterclockwise around the loop.
    (e) There will be an induced electric field which is clockwise around the loop.

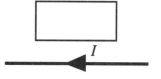

19. A metal ring is dropped from rest below a bar magnet that is fixed in position as suggested in the figure. An observer views the ring from below. Which one of the following statements concerning this situation is true?
    (a) As the ring falls, an induced current will flow *counterclockwise* as viewed by the observer.
    (b) As the ring falls, an induced current will flow *clockwise* as viewed by the observer.
    (c) As the ring falls, there will be an induced magnetic field around the ring which appears *counterclockwise* as viewed by the observer.
    (d) As the ring falls, there will be an induced magnetic field around the ring which appears *clockwise* as viewed by the observer.
    (d) Since the magnet is stationary, there will be no induced current in the ring.

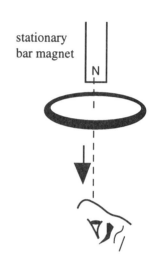

20. Two conducting loops carry equal currents $I$ in the same direction as shown in the figure. If the current in the upper loop suddenly drops to zero, what will happen to the current in the bottom loop according to Lenz's law?
    (a) The current will decrease.
    (b) The current will increase.
    (c) The current will not change.
    (d) The current will also drop to zero.
    (e) The current will reverse its direction.

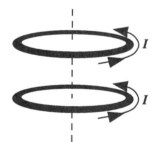

21. A sheet of copper is pulled at constant velocity **v** from a region that contains a uniform magnetic field. At the instant shown in the figure, the sheet is partially in and partially out of the field. The induced emf in the sheet leads to the eddy current shown. Which one of the following statements concerning the direction of the magnetic field is true?
    (a) The magnetic field points to the right.
    (b) The magnetic field points to the left.
    (c) The magnetic field points into the paper.
    (d) The magnetic field points out of the paper.
    (e) The direction of the magnetic field cannot be determined from the information given.

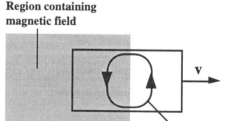

*Questions 22 and 23 pertain to the situation described below:*

A circuit is pulled with a 16-N force toward the right to maintain a constant speed v. At the instant shown, the loop is partially in and partially out of a uniform magnetic field that is directed into the paper. As the circuit moves, a 6.0-A current flows through a 4.0-Ω resistor.

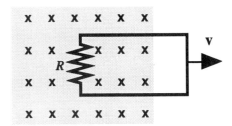

22. Which one of the following statements concerning this situation is true?
    (a) The temperature of the circuit remains constant.
    (b) The induced current flows clockwise around the circuit.
    (c) Since the circuit moves with constant speed, the force **F** does zero work.
    (d) If the circuit were replaced with a wooden loop, there would be no induced emf.
    (e) As the circuit moves through the field, the field does work to produce the current.

23. With what speed does the circuit move?
    (a) 1.5 m/s
    (b) 3.0 m/s
    (c) 6.4 m/s
    (d) 9.0 m/s
    (e) 12 m/s

## Section 22.7 The Electric Generator

24. A circular coil has 275 turns and a radius of 0.045 m. The coil is used as an ac generator by rotating it in a 0.500 T magnetic field, as shown in the figure. At what angular speed should the coil be rotated so that the maximum emf is 175 V?
    (a) 28 rad/s
    (b) 50 rad/s
    (c) 130 rad/s
    (d) 200 rad/s
    (e) 490 rad/s

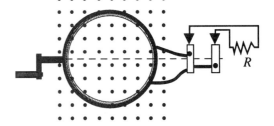

*Questions 25 through 27 pertain to the situation described below:*

A single conducting loop with an area of 2.0 m² rotates in a uniform magnetic field so that the induced emf has a sinusoidal time dependence as shown.

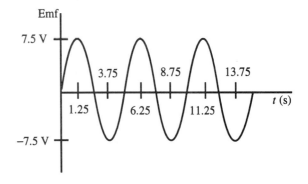

25. Determine the strength of the magnetic field in which the loop rotates.
    (a) 0.5 T
    (b) 2.4 T
    (c) 3.0 T
    (d) 7.5 T
    (e) 18.8 T

224  ELECTROMAGNETIC INDUCTION

26. With what angular frequency does the loop rotate?
    (a) 0.16 rad/s
    (b) 0.30 rad/s
    (c) 0.52 rad/s
    (d) 0.80 rad/s
    (e) 1.26 rad/s

27. What is the period of the induced current?
    (a) 1.25 s
    (b) 2.50 s
    (c) 3.75 s
    (d) 5.00 s
    (e) 6.25 s

*Section 22.8 Mutual Inductance and Self-Inductance*

28. Which one of the following combinations of units is equivalent to one henry?
    (a) N • m • s/C
    (b) N • m • $s^2$/C
    (c) N • m • $C^2/s^2$
    (d) N • m • C/$s^2$
    (e) N • m • $s^2/C^2$

29. The current in a solenoid is decreased to one-half of its original value. Which one of the following statements is true concerning the self-inductance of the solenoid?
    (a) The self-inductance does not change.
    (b) The self-inductance increases by a factor of two.
    (c) The self-inductance decreases by a factor of two.
    (d) The self-inductance increases by a factor of four.
    (e) The self-inductance decreases by a factor of four.

30. A solenoid with 1000 turns has a cross-sectional area of 7.0 $cm^2$ and length of 25 cm. How much energy is stored in the magnetic field of the solenoid when it carries a current of 10.0 A?
    (a) 0.10 J
    (b) 2.8 J
    (c) 0.18 J
    (d) 28 J
    (e) 0.36 J

31. A closed loop carries a current that increases with time. Which one of the quantities listed below relates the emf induced in the loop to the rate at which the current is increasing?
    (a) resistance of the loop
    (b) capacitance of the loop
    (c) self-inductance of the loop
    (d) power dissipated by the loop
    (e) mutual inductance of the loop

32. The figure shows a circular conducting loop that is connected to a 5.0-V battery and a switch *S*. Immediately after the switch *S* is closed, the current through the loop changes at a rate of 15 A/s and the emf induced in the loop has a magnitude of 5.0 V. Determine the *self-inductance* of the coil.
    (a) 0.33 H
    (b) 0.60 H
    (c) 1.5 H
    (d) 3.0 H
    (e) 5.0 H

33. Two coils share a common axis as shown in the figure. The mutual inductance of this pair of coils is 6.0 mH. If the current in coil **1** is changing at the rate of 3.5 A/s, what is the magnitude of the emf generated in coil **2**?
    (a) 5.8 × $10^{-4}$ V
    (b) 1.7 × $10^{-3}$ V
    (c) 3.5 × $10^{-3}$ V
    (d) 1.5 × $10^{-2}$ V
    (e) 2.1 × $10^{-2}$ V

coil **1**   coil **2**

34. A coil of wire with a resistance of 0.15 Ω has a self-inductance of 0.083 H. If a 5.0-V battery is connected across the ends of the coil and the current in the circuit reaches an equilibrium value, what is the stored energy in the inductor?
    (a) 92 J
    (b) 46 J
    (c) 16 J
    (d) 4.1 J
    (e) 1.4 J

35. In the drawing, a coil of wire is wrapped around a cylinder from which an iron core extends upward. The ends of the coil are connected to an ac voltage source. After the alternating current is established in the coil, an aluminum ring of resistance $R$ is placed onto the iron core and released. Which one of the following statements concerning this situation is false?
    (a) The induced current in the ring is an alternating current.
    (b) The temperature of the ring will increase.
    (c) At any instant, the direction of the induced current in the ring is in the same direction as that in the coil.
    (d) The induced magnetic field in the ring may by directed either upward or downward at an instant when the direction of the magnetic field generated by the current in the coil is upward.
    (e) The ring may remain suspended at the position shown with no vertical movement of its center of mass.

*Questions 36 through 40 pertain to the situation described below:*

A 0.10-m long solenoid has a radius of 0.05 m and $1.5 \times 10^4$ turns. The current in the solenoid changes at a rate of 6.0 A/s. A conducting loop of radius 0.02 m is placed at the center of the solenoid with its axis the same as that of the solenoid as shown.

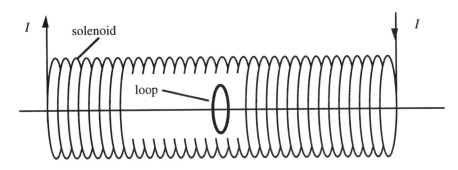

36. What is the magnetic flux through the small loop when the current through the solenoid is 2.5 A?
    (a) $2.95 \times 10^{-2}$ Wb
    (b) $7.28 \times 10^{-2}$ Wb
    (c) $2.95 \times 10^{-4}$ Wb
    (d) $4.38 \times 10^{-4}$ Wb
    (e) $5.92 \times 10^{-4}$ Wb

37. Determine the mutual inductance of this combination.
    (a) $1.8 \times 10^{-4}$ H
    (b) $2.4 \times 10^{-4}$ H
    (c) $3.6 \times 10^{-4}$ H
    (d) $4.4 \times 10^{-4}$ H
    (e) $5.9 \times 10^{-4}$ H

226  ELECTROMAGNETIC INDUCTION

⊖ 38. Determine the induced emf in the loop.
   (a) $0.7 \times 10^{-3}$ V
   (b) $1.4 \times 10^{-3}$ V
   (c) $2.8 \times 10^{-3}$ V
   (d) $5.6 \times 10^{-3}$ V
   (e) zero

○ 39. Determine the induced emf in the loop if the loop is oriented so that its axis is perpendicular to the axis of the solenoid instead of parallel.
   (a) $0.7 \times 10^{-4}$ V
   (b) $1.4 \times 10^{-4}$ V
   (c) $2.8 \times 10^{-4}$ V
   (d) $5.6 \times 10^{-4}$ V
   (e) zero

⊖ 40. Determine the self-induced emf in the solenoid due to the changing current.
   (a) 60 V
   (b) 98 V
   (c) 130 V
   (d) 180 V
   (e) 250 V

*Questions 41 through 47 pertain to the situation described below:*

Two coils, **1** and **2**, with iron cores are positioned as shown in the figure. Coil **1** is part of a circuit with a battery and a switch.

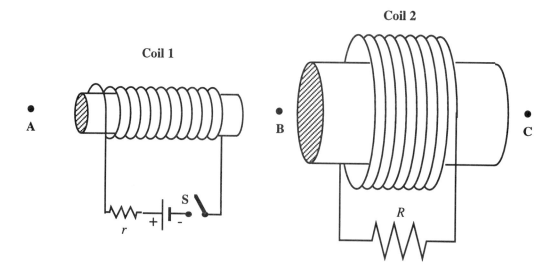

● 41. Immediately after the switch S is closed, which one of the following statements is true?
   (a) An induced current will flow from right to left in R.
   (b) An induced current will flow from left to right in r.
   (c) A magnetic field that points toward **B** appears inside coil **1**.
   (d) An induced magnetic field that points toward **B** appears inside coil **2**.
   (e) A current will pass through r, but there will be no current through R.

⊖ 42. Assume the switch S has been closed for a long time. Which one of the following statements is true?
   (a) An induced current will flow from right to left in R.
   (b) An induced current will flow from left to right in r.
   (c) A magnetic field that points toward **B** appears inside coil **1**.
   (d) An induced magnetic field that points toward **B** appears inside coil **2**.
   (e) A current will pass through r, but there will be no current through R.

43. Assume S has been closed for a long time. Which one of the following statements is true when coil **1** and its core are moved toward point **B**?
    (a) There is no induced current in *r*.
    (b) There is a magnetic field in coil **1** that points toward **B**.
    (c) There is an induced current in *R* that flows from left to right.
    (d) There is an induced current in *R* that flows from right to left.
    (e) There is an induced magnetic field in coil **2** that points toward **B**.

44. Assume that **S** has been closed for a long time. Which one of the following changes will *not* result in an induced current in coil **2** that flows from left to right through *R*.
    (a) Coil **1** and its core are moved toward **A**.
    (b) Coil **2** and its core are moved toward **B**.
    (c) Coil **2** and its core are moved toward **C**.
    (d) The switch **S** is opened.
    (e) The iron core is removed from coil **1**.

45. Assume that **S** has been closed for a long time. Which one of the following changes will result in an induced magnetic field in coil **2** that points toward **C**?
    (a) The switch **S** is opened.
    (b) The iron core is removed from coil **1**.
    (c) Coil **1** and its core are moved toward **A**.
    (d) Coil **1** and its core are moved toward **B**.
    (e) Coil **2** and its core are moved toward **C**.

46. Assume that **S** has been closed for a long time. Which one of the following statements is true if **S** is suddenly opened?
    (a) There is no induced current through *R*.
    (b) There is no induced magnetic field in coil **2**.
    (c) There is an induced current in *R* that flows from right to left.
    (d) There is an induced magnetic field in coil **2** that points toward **C**.
    (e) There is an induced magnetic field in coil **2** that points toward **B**.

47. Assume that **S** has been closed for a long time. Which one of the following statements is true if coil **2** is moved toward **C**?
    (a) There is an induced magnetic field in coil **2** that points toward **B**.
    (b) There is an induced magnetic field in coil **2** that points toward **C**.
    (c) There is an induced current in *R* that flows from right to left.
    (d) There is an induced north pole at the right end of coil **2**.
    (e) There is no induced current in *R*.

*Section 22.9 Transformers*

48. Which one of the following statements concerning transformers is false?
    (a) Their operation makes use of mutual induction.
    (b) They are an application of Faraday's and Lenz's laws.
    (c) A transformer can function with either an ac current or a steady dc current.
    (d) A transformer that steps down the voltage, steps up the current.
    (e) A transformer that steps up the voltage, steps down the current.

49. A transformer changes 120 V across the primary to 1200 V across the secondary. If the secondary coil has 800 turns, how many turns does the primary coil have?
    (a) 40
    (b) 80
    (c) 100
    (d) 400
    (e) 4000

## 228 ELECTROMAGNETIC INDUCTION

50. A transformer has 450 turns in its primary coil and 30 turns in its secondary coil. Which one of the following statements concerning this transformer is true?
    (a) This is a *step-up* transformer.
    (b) The *turns ratio* is 15 for this transformer.
    (c) The ratio of the voltages $V_s/V_p$ is 15 for this transformer.
    (d) The ratio of the currents $I_s/I_p$ is 0.067 for this transformer.
    (e) The power delivered to the secondary must be the same as that delivered to the primary.

**Questions 51 and 52 relate to the following situation:**

A power plant produces a voltage of 6.0 kV and 150 A. The voltage is stepped up to 240 kV by a transformer before it is transmitted to a substation. The resistance of the transmission line between the power plant and the substation is 75 Ω.

51. What is the current in the transmission line from the plant to the substation?
    (a) 3.8 A    (c) 6.4 A    (e) 7.5 A
    (b) 5.2 A    (d) 7.0 A

52. What percentage of the power produced at the power plant is lost in transmission to the substation?
    (a) 0.47 %   (c) 0.34 %   (e) 0.12 %
    (b) 0.41 %   (d) 0.23 %

## *Additional Problems*

53. A circular loop of copper wire with an area of 2.0 m² lies in a plane perpendicular to a time-dependent magnetic field oriented as shown. The time-dependence of the field is shown in the graph.

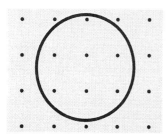
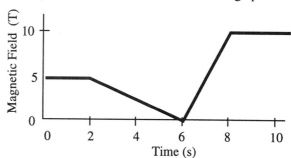

Which one of the entries in the table below is incorrect?

| | Time | Induced emf |
|---|---|---|
| (a) | 1 s | zero |
| (b) | 4 s | 2.5 V, counterclockwise |
| (c) | 5 s | 2.5 V, counterclockwise |
| (d) | 7 s | 10 V, clockwise |
| (e) | 9 s | 10 V, counterclockwise |

**Questions 54 and 55 pertain to the statement and diagram below:**

The figure shows a uniform magnetic field which is normal to the plane of the conducting loop of resistance $R$.
Note: the area of the non-circular portion of the circuit is negligible compared to that of the loop.

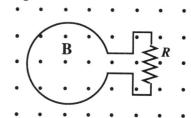

54. Which entry in the table below correctly pairs the change in the system with the direction of the induced current through R?

| change in the system | direction of current through R |
|---|---|
| (a) decrease the area of the loop | from top toward bottom |
| (b) rotate loop into the paper | no induced current |
| (c) increase the area of the loop | from bottom toward top |
| (d) decrease the magnitude of **B** | from bottom toward top |
| (e) pull loop to the right | from top toward bottom |

55. Suppose that the radius of the loop is 0.50 m. At what *rate* must **B** change with time if the emf induced in the loop is $3\pi$ volts?
    (a) 12.0 T/s
    (b) 18.8 T/s
    (c) 24.0 T/s
    (d) 37.7 T/s
    (e) 49.2 T/s

*Questions 56 through 58 pertain to the situation described below:*

A loop is pulled with a force **F** to the right to maintain a constant speed of 8.0 m/s. The loop has a length of 0.15 m, a width of 0.08 m, and a resistance of 200.0 Ω. At the instant shown, the loop is partially in and partially out of a uniform magnetic field that is directed into the paper. The magnitude of the field is 1.2 T.

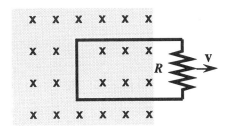

56. What is the magnitude of the emf induced in the loop?
    (a) zero
    (b) 0.77 V
    (c) 1.4 V
    (d) 4.9 V
    (e) 9.6 V

57. What is the induced current in the loop?
    (a) 0.048 A
    (b) 0.024 A
    (c) $7.2 \times 10^{-3}$ A
    (d) $3.8 \times 10^{-3}$ A
    (e) zero

58. Determine the magnitude of the force **F** required to pull the loop.
    (a) $1.3 \times 10^{-4}$ N
    (b) $2.1 \times 10^{-4}$ N
    (c) $3.7 \times 10^{-4}$ N
    (d) $6.8 \times 10^{-4}$ N
    (e) $9.0 \times 10^{-4}$ N

*Questions 59 through 63 pertain to the situation described below:*

A loop with a resistance of 2.0 Ω is pushed to the left at a constant speed of 4.0 m/s by a 32 N force. At the instant shown in the figure, the loop is partially in and partially out of a uniform magnetic field. An induced current flows from left to right through the resistor. The length and width of the loop are 2.0 m and 1.0 m, respectively.

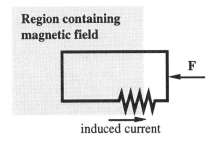

230  **ELECTROMAGNETIC INDUCTION**

59. What is the direction of the magnetic field?
    (a) to the left
    (b) to the right
    (c) out of the paper
    (d) into the paper
    (e) toward the top of the page

60. Determine the magnitude of the induced current through the resistor.
    (a) 2.0 A
    (b) 4.0 A
    (c) 8.0 A
    (d) 16 A
    (e) 32 A

61. Determine the magnitude of the induced emf in the loop.
    (a) 2.0 V
    (b) 4.0 V
    (c) 8.0 V
    (d) 12 V
    (e) 16 V

62. Determine the magnitude of the uniform magnetic field.
    (a) 2 T
    (b) 4 T
    (c) 6 T
    (d) 8 T
    (e) 12 T

63. At what rate is energy dissipated by the resistor?
    (a) 128 W
    (b) 96 W
    (c) 32 W
    (d) 16 W
    (e) 8.0 W

*Questions 64 through 66 pertain to the situation described below:*

A flexible, circular conducting loop of radius 0.15 m and resistance 4.0 Ω lies in a uniform magnetic field of 0.25 T. The loop is pulled on opposite sides by equal forces and stretched until its enclosed area is essentially zero, as suggested in the drawings. It takes 0.30 s to close the loop.

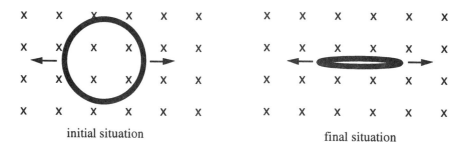

initial situation        final situation

64. Determine the magnitude of the emf induced in the loop.
    (a) $1.2 \times 10^{-1}$ V
    (b) $1.8 \times 10^{-2}$ V
    (c) $1.8 \times 10^{2}$ V
    (d) $5.9 \times 10^{-2}$ V
    (e) $5.9 \times 10^{2}$ V

65. At what rate is heat generated in the loop?
    (a) $3.6 \times 10^{-3}$ W
    (b) $8.7 \times 10^{-4}$ W
    (c) $8.7 \times 10^{4}$ W
    (d) $8.1 \times 10^{5}$ W
    (e) $8.1 \times 10^{-5}$ W

66. Which one of the following phrases best describes the direction of the *induced* magnetic field generated by the current induced in the loop while the loop is being stretched?
    (a) clockwise
    (b) counterclockwise
    (c) into the page
    (d) out of the page
    (e) The induced field is zero.

# CHAPTER 23 — Alternating Current Circuits

## Section 23.1 Capacitors and Capacitive Reactance

1. An ac voltage source that has a frequency $f$ is connected across the terminals of a capacitor. Which one of the following statements correctly indicates the effect on the capacitive reactance when the frequency is increased to $4f$?
   (a) The capacitive reactance increases by a factor of four.
   (b) The capacitive reactance increases by a factor of eight.
   (c) The capacitive reactance is unchanged.
   (d) The capacitive reactance decreases by a factor of eight.
   (e) The capacitive reactance decreases by a factor of four.

2. Which of the following units results when calculating the following quantity: $\dfrac{1}{2\pi f C}$?
   (a) F/s
   (b) F·s
   (c) Ω
   (d) V
   (e) Wb

3. The current in an ac circuit is found to be independent of the frequency at a given voltage. Which combination of elements is most likely to comprise the circuit?
   (a) resistors only
   (b) inductors only
   (c) capacitors only
   (d) a combination of inductors and resistors
   (e) a combination of inductors and capacitors

4. A battery is used to drive a circuit, but after a certain amount of time, the current is zero. When the same circuit is driven by an ac generator the current is non-zero and alternates. Which combination of elements is most likely to comprise the circuit?
   (a) resistors only
   (b) inductors only
   (c) capacitors only
   (d) a combination of inductors and resistors
   (e) a combination of inductors and capacitors

5. Which one of the following graphs illustrates how capacitive reactance varies with frequency?

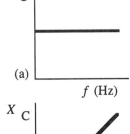
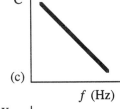
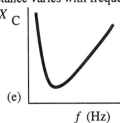
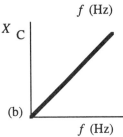
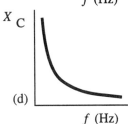

## 232 ALTERNATING CURRENT CIRCUITS

6. A variable capacitor is connected to an ac source. What effect does *decreasing* the capacitance have on the reactance and current in this circuit?

   | | Reactance | Current |
   |---|---|---|
   | (a) | no change | no change |
   | (b) | decreases | no change |
   | (c) | increases | increases |
   | (d) | decreases | increases |
   | (e) | increases | decreases |

7. The reactance of a capacitor at 110 Hz is 35 Ω. Determine the capacitance.
   (a) 41 μF
   (b) 260 μF
   (c) 51 mF
   (d) 0.10 F
   (e) 0.31 F

8. What is the capacitive reactance of a circuit comprised of a 65.0-μF capacitor and a 50.0-Hz generator?
   (a) 49.0 Ω
   (b) 72.5 Ω
   (c) 97.6 Ω
   (d) 145 Ω
   (e) 308 Ω

9. An ac generator is connected across the terminals of a 3.25-μF capacitor. Determine the frequency at which the capacitive reactance is 495 Ω.
   (a) 60.0 Hz
   (b) 72.4 Hz
   (c) 85.7 Hz
   (d) 98.9 Hz
   (e) 152 Hz

*Questions 10 through 14 pertain to the situation described below:*

The graph shows the voltage across and the current through a single circuit element connected to an ac generator.

10. Determine the frequency of the generator.
    (a) 0.14 Hz
    (b) 7.14 Hz
    (c) 12.5 Hz
    (d) 25.0 Hz
    (e) 50.0 Hz

11. Determine the *rms* voltage across this element.
    (a) 49.5 V
    (b) 70.0 V
    (c) 112 V
    (d) 140 V
    (e) 170 V

12. Determine the *rms* current through this element.
    (a) 1.4 A
    (b) 2.0 A
    (c) 3.4 A
    (d) 3.9 A
    (e) 5.6 A

13. What is the reactance of this element?
    (a) 20 Ω
    (b) 25 Ω
    (c) 30 Ω
    (d) 35 Ω
    (e) 40 Ω

14. Identify the circuit element.
    (a) The element is a 25-Ω resistor.
    (b) The element is a 35-Ω resistor.
    (c) The element is a 0.45-H inductor.
    (d) The element is a 360-μF capacitor.
    (e) The element is a 510-μF capacitor.

## Section 23.2 Inductors and Inductive Reactance

15. When the frequency of an ac circuit is decreased, the current in the circuit increases. Which combination of elements is most likely to comprise the circuit?
    (a) resistors only
    (b) inductors only
    (c) capacitors only
    (d) capacitors and resistors
    (e) inductors and capacitors

16. Which circuit elements act to oppose *changes* in the current in an ac circuit?
    (a) resistors only
    (b) capacitors only
    (c) inductors only
    (d) both resistors and inductors
    (e) both capacitors and resistors

17. Which one of the following graphs shows how the inductive reactance varies with frequency?

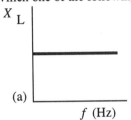
(a)

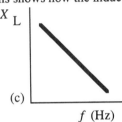
(c)

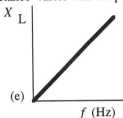
(e)

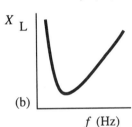
(b)

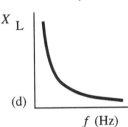
(d)

18. A variable inductor is connected to an ac source. What effect does *increasing* the inductance have on the reactance and current in this circuit?

    | | Reactance | Current |
    |---|---|---|
    | (a) | no change | no change |
    | (b) | decreases | no change |
    | (c) | increases | increases |
    | (d) | decreases | increases |
    | (e) | increases | decreases |

19. An ac voltage source that has a frequency $f$ is connected across the ends of an inductor. Which one of the following statements correctly indicates the effect on the inductive reactance when the frequency is increased to $2f$?
    (a) The inductive reactance increases by a factor of two.
    (b) The inductive reactance increases by a factor of four.
    (c) The inductive reactance is unchanged.
    (d) The inductive reactance decreases by a factor of four.
    (e) The inductive reactance decreases by a factor of two.

20. In an ac circuit, a 0.025-H inductor is connected to a generator which has an rms voltage of 25 V and operates at 50.0 Hz. What is the rms current through the inductor?
    (a) 0.62 A
    (b) 2.0 A
    (c) 3.2 A
    (d) 7.1 A
    (e) 14 A

## 234  ALTERNATING CURRENT CIRCUITS

21. Which one of the following phasor models correctly represents a circuit comprised of only an inductor and an ac generator?

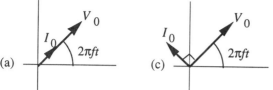
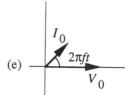
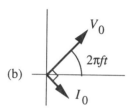
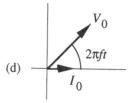

*Questions 22 through 25 pertain to the situation described below:*

The voltage across and the current through a single circuit element connected to an ac generator is shown in the graph.

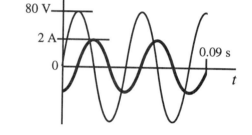

22. Which one of the following statements concerning this circuit element is true?
    (a) The element is a resistor.
    (b) The element is a capacitor.
    (c) The element is an inductor.
    (d) The element could be a resistor or an inductor.
    (e) The element could be an inductor or a capacitor.

23. At what frequency do charges oscillate in this circuit?
    (a) 0.01 Hz
    (b) 0.04 Hz
    (c) 0.09 Hz
    (d) 11 Hz
    (e) 25 Hz

24. Determine the *rms* current through this circuit element.
    (a) 1.0 A
    (b) 1.4 A
    (c) 2.0 A
    (d) 2.8 A
    (e) 4.0 A

25. Determine the reactance of this circuit element.
    (a) 20 Ω
    (b) 25 Ω
    (c) 40 Ω
    (d) 57 Ω
    (e) 80 Ω

### Section 23.3  Circuits Containing Resistance, Capacitance, and Inductance

26. Which one of the following statements concerning an ac circuit is true?
    (a) The current and voltage are *in phase* for a capacitor in an ac circuit.
    (b) On average the power dissipated by a resistor in an ac circuit is zero.
    (c) The current and voltage are 90° *out of phase* for a resistor in an ac circuit.
    (d) Inductors in an ac circuit offer little opposition to current at high frequencies.
    (e) When only resistance is present in an ac circuit, voltage and current are *in phase*.

TEST BANK  Chapter 23  235

○ 27. Note the following circuit elements: (*1*) resistors, (*2*) capacitors, and (*3*) inductors. Which of these elements uses no energy, on average, in an ac circuit?
 (a) *1* only
 (b) *2* only
 (c) *3* only
 (d) both *2* and *3*
 (e) both *1* and *3*

○ 28. Which one of the following statements concerning the *impedance* of an *RCL* circuit is true?
 (a) The impedance is dominated by the capacitance at low frequencies.
 (b) The impedance is dominated by the resistance at high frequencies.
 (c) The impedance depends only on the values of $C$ and $L$.
 (d) The impedance depends only on the resistance.
 (e) The impedance is independent of frequency.

◐ 29. When the frequency of an ac circuit is increased at constant voltage, the current increases and then decreases. Which combination of elements is most likely to comprise this circuit?
 (a) resistors only
 (b) inductors only
 (c) capacitors only
 (d) a combination of inductors and resistors
 (e) a combination of inductors and capacitors

○ 30. For which one of the following circuit arrangements will the *power factor* be *non-zero*?
 (a) a capacitor in series with an ac generator
 (b) an inductor in series with an ac generator
 (c) two capacitors in series with an ac generator
 (d) a capacitor and resistor in series with an ac generator
 (e) a capacitor and inductor in series with an ac generator

● 31. The table below shows the values of the resistance, capacitive reactance and inductive reactance for five *RCL* circuits. In which circuit will the voltage lead the current?

| | Resistance | Capacitive reactance | Inductive reactance |
|---|---|---|---|
| (a) | 30 Ω | 219 Ω | 180 Ω |
| (b) | 50 Ω | 288 Ω | 244 Ω |
| (c) | 120 Ω | 58 Ω | 18 Ω |
| (d) | 150 Ω | 79 Ω | 212 Ω |
| (e) | 212 Ω | 314 Ω | 78 Ω |

● 32. Complete the following statement: When the current in an oscillating *LC* circuit is zero,
 (a) the charge on the capacitor is zero.
 (b) the energy in the electric field is maximum.
 (c) the energy in the magnetic field is a maximum.
 (d) the charge is moving through the inductor.
 (e) the energy is equally shared between the electric and magnetic fields.

◐ 33. The graph shows the impedance as a function of frequency for a series RCL circuit. At what frequency does the capacitor make the largest contribution?
 (a) $f_1$
 (b) $f_2$
 (c) $f_3$
 (d) $f_1$ or $f_3$
 (e) $f_1$ or $f_2$ or $f_3$

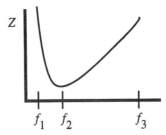

## 236 ALTERNATING CURRENT CIRCUITS

● 34. The table below shows the values of the resistance, capacitive reactance, and inductive reactance for five *RCL* circuits. Which circuit will have a negative phase angle?

| | Resistance | Capacitive Reactance | Inductive Reactance |
|---|---|---|---|
| (a) | 78 Ω | 306 Ω | 346 Ω |
| (b) | 86 Ω | 49 Ω | 86 Ω |
| (c) | 120 Ω | 314 Ω | 314 Ω |
| (d) | 127 Ω | 218 Ω | 306 Ω |
| (e) | 148 Ω | 219 Ω | 180 Ω |

○ 35. An ac circuit consists of a series combination of an inductor and a capacitor. What is the phase angle between the voltages across these two circuit elements?
  (a) 0°
  (b) 90°
  (c) 180°
  (d) 270°
  (e) 360°

◐ 36. A series RCL circuit operating at 60.0 Hz contains a 35-Ω resistor and a 8.2-μF capacitor. If the power factor of the circuit is +1.00, what is the inductance of the inductor in this circuit?
  (a) 0.86 H
  (b) 1.1 H
  (c) 2.3 H
  (d) 57 H
  (e) 320 H

◐ 37. Use the information given in the figure for the series RCL circuit to determine its total impedance.
  (a) 300 Ω
  (b) 500 Ω
  (c) 1500 Ω
  (d) 1700 Ω
  (e) 1900 Ω

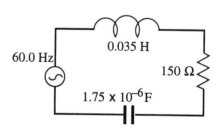

● 38. Use the information given in the figure for the series RCL circuit to determine the phase angle between the current and the voltage.
  (a) zero
  (b) +5.3°
  (c) −9.6°
  (d) −84°
  (e) +90°

◐ 39. A 7.70-μF capacitor and a 1250-Ω resistor are connected in series to a generator operating at 50.0 Hz and producing an *rms* voltage of 208 V. What is the average power dissipated in this circuit?
  (a) 346 W
  (b) 31.2 W
  (c) 19.7 W
  (d) 1.66 W
  (e) zero

*Questions 40 through 46 pertain to the ac circuit described below:*

An ac generator supplies a peak (not *rms*) voltage of 150 V at 60.0 Hz. It is connected in series with a 35-mH inductor, a 45-μF capacitor and an 85-Ω resistor.

◐ 40. Determine the *rms* voltage of the generator.
  (a) 220 V
  (b) 110 V
  (c) 82 V
  (d) 75 V
  (e) 53 V

◐ 41. What is the capacitive reactance for this circuit?
  (a) 58.9 Ω
  (b) 85.0 Ω
  (c) 99.1 Ω
  (d) 124 Ω
  (e) 301 Ω

42. What is the inductive reactance for this circuit?
    (a) 13.2 Ω
    (b) 23.4 Ω
    (c) 58.9 Ω
    (d) 85.0 Ω
    (e) 99.1 Ω

43. What is the impedance of the circuit?
    (a) 13.2 Ω
    (b) 23.4 Ω
    (c) 58.9 Ω
    (d) 85.0 Ω
    (e) 96.5 Ω

44. Determine the *rms* current in the circuit.
    (a) 0.28 A
    (b) 0.40 A
    (c) 0.75 A
    (d) 1.1 A
    (e) 1.6 A

45. What is the *power factor* for this circuit?
    (a) 0.47
    (b) 0.54
    (c) 0.64
    (d) 0.88
    (e) 28.3

46. Which one of the following statements concerning this circuit is true?
    (a) The voltage leads the current.
    (b) The circuit is more capacitive than inductive.
    (c) The voltage and current are exactly out of phase.
    (d) The voltage and current are in phase.
    (e) The phase angle for this circuit is positive.


*Questions 47 through 51 pertain to the ac circuit described below:*

The following table gives the reactance and rms voltage across the elements of a series RCL circuit:

| Circuit element | Reactance | Voltage across element |
|---|---|---|
| resistor | 200 Ω | 86 V |
| capacitor | 663 Ω | 285 V |
| inductor | 377 Ω | 162 V |

47. What is the *rms* current in the circuit?
    (a) 0.25 A
    (b) 0.43 A
    (c) 0.50 A
    (d) 0.86 A
    (e) 1.08 A

48. What is the impedance of the circuit?
    (a) 11 Ω
    (b) 22 Ω
    (c) 349 Ω
    (d) 486 Ω
    (e) 1240 Ω

49. Determine the peak (*not rms*) voltage of the ac generator.
    (a) 150 V
    (b) 212 V
    (c) 300 V
    (d) 414 V
    (e) 533 V

50. What is the power factor for this circuit?
    (a) 0.40
    (b) 0.57
    (c) 0.81
    (d) 1.4
    (e) 5.5

51. What is the average power consumed by the circuit?
    (a) 37.0 W
    (b) 64.5 W
    (c) 73.3 W
    (d) 95.4 W
    (e) 129 W

## ALTERNATING CURRENT CIRCUITS

*Questions 52 through 61 pertain to the ac circuit described below:*

An ac generator supplies a rms (*not peak*) voltage of 180 V at 60 Hz. It is connected in series with a 0.5-H inductor, a 6.0-µF capacitor and an 300-Ω resistor.

52. Determine the peak voltage of the generator.
    (a) 180 V
    (b) 255 V
    (c) 300 V
    (d) 360 V
    (e) 480 V

53. What is the capacitive reactance?
    (a) 150 Ω
    (b) 190 Ω
    (c) 360 Ω
    (d) 390 Ω
    (e) 440 Ω

54. What is the inductive reactance?
    (a) 150 Ω
    (b) 190 Ω
    (c) 360 Ω
    (d) 390 Ω
    (e) 440 Ω

55. What is the impedance of the circuit?
    (a) 150 Ω
    (b) 190 Ω
    (c) 360 Ω
    (d) 390 Ω
    (e) 440 Ω

56. What is the peak (*not rms*) current through the resistor?
    (a) 0.46 A
    (b) 0.52 A
    (c) 0.65 A
    (d) 0.78 A
    (e) 0.85 A

57. What is the phase angle for this circuit?
    (a) 26°
    (b) –32°
    (c) 38°
    (d) –40°
    (e) 50°

58. What is the average power supplied to the circuit?
    (a) 48 W
    (b) 54 W
    (c) 63 W
    (d) 83 W
    (e) 112 W

59. What is the peak (*not rms*) voltage across the inductor?
    (a) 122 V
    (b) 180 V
    (c) 194 V
    (d) 255 V
    (e) 286 V

60. What is the peak (*not rms*) voltage across the resistor?
    (a) 122 V
    (b) 138 V
    (c) 194 V
    (d) 255 V
    (e) 286 V

61. What is the peak (*not rms*) voltage across the capacitor?
    (a) 122 V
    (b) 180 V
    (c) 194 V
    (d) 203 V
    (e) 286 V

## Section 23.4 Resonance in Electric Circuits

62. The graph shows the impedance as a function of frequency for a series RCL circuit. At what frequency will this circuit resonate?
    (a) $f_1$
    (b) $f_2$
    (c) $f_3$
    (d) $f_1$ or $f_3$
    (e) $f_1$ or $f_2$ or $f_3$

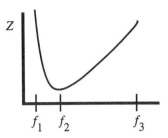

63. A series RCL circuit is comprised of a 6.00-mH inductor and a 2.50-µF capacitor. What is the resonant frequency of this circuit?
    (a) 11.9 Hz
    (b) 69.4 Hz
    (c) $1.06 \times 10^2$ Hz
    (d) $8.27 \times 10^2$ Hz
    (e) $1.30 \times 10^3$ Hz

64. What is the inductance of the inductor in a series RCL circuit with a resonant frequency of 640 Hz if the only other element in the circuit is a 1.2-µF capacitor?
    (a) 52 mH
    (b) 74 mH
    (c) 0.12 H
    (d) 2.1 H
    (e) 13 H

*Questions 65 through 67 pertain to the LC circuit described below:*

An oscillating LC circuit has a resonant frequency of $1.2 \times 10^6$ Hz. The diagram shows the system at $t = 0$ when there is no charge on the capacitor and the current is in the direction shown. At this time, the energy stored in the inductor is $2.5 \times 10^{-7}$ J.

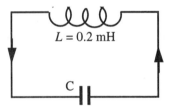

65. Determine the numerical value of the capacitance.
    (a) $2.4 \times 10^{-11}$ F
    (b) $3.6 \times 10^{-11}$ F
    (c) $4.4 \times 10^{-11}$ F
    (d) $6.8 \times 10^{-11}$ F
    (e) $8.8 \times 10^{-11}$ F

66. Determine the value of the current at the instant shown in the drawing.
    (a) 0.02 A
    (b) 0.03 A
    (c) 0.05 A
    (d) 0.09 A
    (e) 0.12 A

67. Which one of the following statements is true for this circuit?
    (a) When the current in the inductor is zero, $\Delta I/\Delta t = 0$.
    (b) The circuit will allow *dc* as well as *ac* current to flow.
    (c) The magnetic field in the inductor is a maximum at $t = 0$.
    (d) The charge on the capacitor will be a maximum at $t = 4.2 \times 10^{-7}$ s.
    (e) As the current through the inductor *decreases*, the induced emf in the coil is directed *opposite* to the direction of the current.

## Section 23.5 Semiconductor Devices

68. Which one of the following statements best explains why semiconductors in transistors are doped?
    (a) Doping makes the charge carriers mobile.
    (b) Doping makes the charge carriers immobile.
    (c) Doping makes the semiconducting material non-conducting.
    (d) Doping changes the conductive properties of the semiconductors.
    (e) Doping converts the semiconducting material into a metallic material like copper.

69. Complete the following statement: The main difference between an *npn* transistor and a *pnp* transistor is that
    (a) the bias voltages are reversed.
    (b) they are made from different materials.
    (c) one is a bipolar transistor and the other is not.
    (d) one conducts electricity and the other does not.
    (e) the current directions are the same while the voltages are reversed.

70. Which one of the following materials would be in an *n*-type semiconductor?
    (a) silicon doped with boron
    (b) silicon doped with gallium
    (c) germanium doped with boron
    (d) silicon doped with phosphorus
    (e) germanium doped with gallium

71. Which one of the following statements concerning diodes is false?
    (a) A diode is fabricated by forming junctions between one *n*-type and two *p*-type semiconductors or one *p*-type and two *n*-type semiconductors.
    (b) At an *n-p* junction in a diode which is not connected to a battery, an electric field exists across the junction.
    (c) A battery can be connected to a diode in such a way so as to produce a continual flow of charge.
    (d) A diode connected to an ac source allows current to flow only in one direction.
    (e) Holes move in a diode under a forward-bias condition.

## Additional Problems

72. Two 250-µF capacitors, with equal capacitances, and a 1.5-mH inductor are connected as shown in the figure. It is desired to "drive" the circuit by placing an ac generator in series with the inductor. Which output voltage for the generator will produce the largest current in the above circuit? Assume that $V$ is in volts when $t$ is in seconds.
    (a) $V = 0.6 \sin[(1.1 \times 10^6 \text{ rad/s})t]$
    (b) $V = 0.6 \sin[(1.2 \times 10^3 \text{ rad/s})t]$
    (c) $V = 1.2 \sin[(5.3 \times 10^6 \text{ rad/s})t]$
    (d) $V = 1.2 \sin[(2.3 \times 10^6 \text{ rad/s})t]$
    (e) $V = 2.8 \sin[(4.6 \times 10^6 \text{ rad/s})t]$

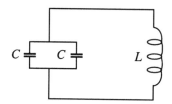

73. In the circuit shown in the drawing, the generator supplies the same amount of rms current at either very small or very large frequencies. What can be deduced about the resistances $R_1$, $R_2$, and $R_3$ from this information?
    (a) $R_1 = R_2 = R_3$
    (b) $R_1 = 2R_2$
    (c) $R_1 = R_2$
    (d) $R_1 = R_3$
    (e) $R_2 = R_3$

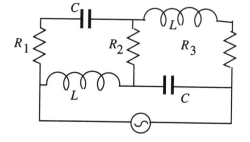

*Questions 74 through 79 refer to the ac circuit described below:*

An ac generator supplies a peak (not *rms*)voltage of 120 V at 60.0 Hz. It is connected in series with a 50.0-Ω resistor and a 0.5-H inductor.

74. What is the *rms* voltage of the generator?
    (a) 84.9 V  (c) 151 V  (e) 240 V
    (b) 138 V  (d) 169 V

75. What is the inductive reactance of the circuit?
    (a) 60 Ω  (c) 377 Ω  (e) 754 Ω
    (b) 188 Ω  (c) 514 Ω

76. Determine the impedance of the circuit.
    (a) 50 Ω  (c) 200 Ω  (e) 210 Ω
    (b) 190 Ω  (d) 240 Ω

77. What is the *rms* current through the resistor?
    (a) 2.4 A  (c) 1.3 A  (e) 0.44 A
    (b) 1.7 A  (d) 0.62 A

78. What is the peak current through the resistor?
    (a) 2.4 A  (c) 1.3 A  (e) 0.44 A
    (b) 1.7 A  (d) 0.62 A

79. What is the average power dissipated in the circuit?
    (a) 9.5 W  (c) 74 W  (e) 140 W
    (b) 21 W  (d) 110 W

# CHAPTER 24    Electromagnetic Waves

## Section 24.1 The Nature of Electromagnetic Waves

1. Which one of the following statements concerning electromagnetic waves is false?
   (a) Electromagnetic waves are longitudinal waves.
   (b) Electromagnetic waves transfer energy through space.
   (c) The existence of electromagnetic waves was predicted by Maxwell.
   (d) Electromagnetic waves can propagate through a material substance.
   (e) Electromagnetic waves do not require a physical medium for propagation.

2. Which one of the following will *not* generate electromagnetic waves or pulses?
   (a) a steady direct current
   (b) an accelerating electron
   (c) a proton in simple harmonic motion
   (d) an alternating current
   (e) charged particles traveling in a circular path in a mass spectrometer

3. Complete the following sentence: When electrons from a heated filament accelerate through vacuum toward a positive plate,
   (a) only an electric field will be produced.
   (b) only a magnetic field will be produced.
   (c) electromagnetic waves will be produced.
   (d) longitudinal waves will be produced.
   (e) neither electric nor magnetic fields will be produced.

4. The electric field **E** of an electromagnetic wave traveling the positive *x* direction is illustrated in the figure. This is the wave of the radiation field of an antenna. What is the direction and the phase relative to the electric field of the magnetic field at a point where the electric field is in the negative *y* direction?
   Note: The wave is shown in a region of space that is a large distance from its source.
   (a) +*y* direction, in phase
   (b) −*z* direction, 90° out of phase
   (c) +*z* direction, 90° out of phase
   (d) −*z* direction, in phase
   (e) +*z* direction, in phase

## Section 24.2 The Electromagnetic Spectrum

5. Which one of the following statements concerning electromagnetic waves is false?
   (a) Electromagnetic waves carry energy.
   (b) X-rays have longer wavelengths than radio waves.
   (c) In vacuum, all electromagnetic waves travel at the same speed.
   (d) Lower frequency electromagnetic waves can be produced by oscillating circuits.
   (e) They consist of mutually perpendicular electric and magnetic fields that oscillate perpendicular to the direction of propagation.

6. Which one of the following types of wave is intrinsically different from the other four?
   (a) radio waves      (c) gamma rays      (e) visible light
   (b) sound waves      (d) ultraviolet radiation

7. Which one of the following statements concerning the wavelength of an electromagnetic wave in a vacuum is true?
   (a) The wavelength is independent of the speed of the wave for a fixed frequency.
   (b) The wavelength is inversely proportional to the speed of the wave.
   (c) The wavelength is the same for all types of electromagnetic waves.
   (d) The wavelength is directly proportional to the frequency of the wave.
   (e) The wavelength is inversely proportional to the frequency of the wave.

8. Complete the following sentence: The various colors of visible light differ in
   (a) frequency only.
   (b) wavelength only.
   (c) their speeds in a vacuum.
   (d) frequency and wavelength.
   (e) frequency and their speed in a vacuum.

9. Note the different types of electromagnetic radiation:
   (1) X-rays
   (2) radio waves
   (3) gamma rays
   (4) visible light
   (5) infrared radiation
   (6) ultraviolet radiation
   Which list correctly ranks the electromagnetic waves in order of *increasing* frequency?
   (a) 2, 3, 4, 5, 6, 1
   (b) 2, 5, 4, 1, 6, 3
   (c) 2, 5, 4, 6, 1, 3
   (d) 3, 1, 6, 4, 5, 2
   (e) 3, 6, 1, 4, 5, 2

10. Which one of the following colors of visible light has the highest frequency?
    (a) yellow
    (b) red
    (c) green
    (d) blue
    (e) violet

11. What is the correct order, beginning with shortest wavelength and extending to the longest wavelength, of the following colors in the visible light spectrum: blue, green, red, violet, and yellow?
    (a) red, yellow, blue, green, violet
    (b) violet, blue, yellow, red, green
    (c) red, yellow, green, blue, violet
    (d) violet, blue, green, yellow, red
    (e) red, blue, violet, green, yellow

12. When a radio telescope observes a region of space between two stars, it detects electromagnetic radiation which has a wavelength of 0.21 m. This radiation was emitted by hydrogen atoms in the gas and dust located in that region. What is the frequency of this radiation?
    (a) $7.1 \times 10^{10}$ Hz
    (b) $2.1 \times 10^{14}$ Hz
    (c) $3.0 \times 10^{8}$ Hz
    (d) $6.9 \times 10^{11}$ Hz
    (e) $1.4 \times 10^{9}$ Hz

13. An FM radio station generates radio waves that have a frequency of 95.5 MHz. The frequency of the waves from a competing station have a frequency of 102.7 MHz. What is the difference in wavelength between the waves emitted from the two stations?
    (a) 0.220 m
    (b) 0.454 m
    (c) 0.844 m
    (d) 2.39 m
    (e) 41.7 m

## Section 24.3 The Speed of Light

14. Which one of the following scientists *did not* attempt to measure the speed of light?
    (a) Galileo
    (b) Newton
    (c) Fizeau
    (d) Foucault
    (e) Michelson

244   ELECTROMAGNETIC WAVES

15. A radio wave sent from the earth's surface reflects from the surface of the moon and returns to the earth. The elapsed time between the generation of the wave and the detection of the reflected wave is 2.6444 s. Determine the distance from the surface of the earth to the surface of the moon.
    Note: The speed of light is $2.9979 \times 10^8$ m/s.
    (a) $3.76882 \times 10^8$ m
    (b) $3.8445 \times 10^8$ m
    (c) $3.9638 \times 10^8$ m
    (d) $4.0551 \times 10^8$ m
    (e) $7.9276 \times 10^8$ m

16. The average distance between the surface of the earth and the surface of the sun is $1.49 \times 10^{11}$ m. How much time, in minutes, does it take for light leaving the surface of the sun to reach the earth?
    (a) zero
    (b) $2.9 \times 10^{-3}$ min
    (c) 8.3 min
    (d) 74 min
    (e) 500 min

## Section 24.4 The Energy Carried by Electromagnetic Waves

17. Which one of the following statements concerning the energy carried of an electromagnetic wave is true?
    (a) The energy is carried only by the electric field.
    (b) More energy is carried by the electric field than by the magnetic field.
    (c) The energy is carried equally by the electric and magnetic fields.
    (d) More energy is carried by the magnetic field than by the electric field.
    (e) The energy is carried only by the magnetic field.

18. The peak value of the electric field component of an electromagnetic wave is $E$. At a particular instant, the intensity of the wave is of 0.020 W/m$^2$. If electric field were increased to $5E$, what would be the intensity of the wave?
    (a) 0.020 W/m$^2$
    (b) 0.10 W/m$^2$
    (c) 0.25 W/m$^2$
    (d) 0.50 W/m$^2$
    (e) 1.0 W/m$^2$

19. An electromagnetic wave has an electric field with peak value 250 N/C. What is the average intensity of the wave?
    (a) 0.66 W/m$^2$
    (b) 0.89 W/m$^2$
    (c) 83 W/m$^2$
    (d) 120 W/m$^2$
    (e) 170 W/m$^2$

20. An electromagnetic wave has an electric field with peak value 250.0 N/C. What is the average energy delivered to a surface with area 2.00 m$^2$ by this wave in one minute?
    (a) 83.1 J
    (b) 166 J
    (c) 2490 J
    (d) 4980 J
    (e) 9960 J

21. An incandescent light bulb radiates uniformly in all directions with a total average power of $1.0 \times 10^2$ W. What is the maximum value of the magnetic field at a distance of 0.50 m from the bulb?
    (a) $8.4 \times 10^{-7}$ T
    (b) $5.2 \times 10^{-7}$ T
    (c) $3.1 \times 10^{-7}$ T
    (d) $1.6 \times 10^{-7}$ T
    (e) zero

22. A local radio station transmits radio waves uniformly in all directions with a total power of $1.50 \times 10^5$ W. What is the intensity of these waves when they reach a receiving antenna located 40.0 km from the transmitting antenna?
    (a) $2.98 \times 10^{-5}$ W/m$^2$
    (b) $7.46 \times 10^{-6}$ W/m$^2$
    (c) $9.25 \times 10^{-7}$ W/m$^2$
    (d) $1.17 \times 10^{-8}$ W/m$^2$
    (e) $5.60 \times 10^{-10}$ W/m$^2$

## Section 24.5 The Doppler Effect and Electromagnetic Waves

23. An astronomer observes electromagnetic waves emitted by oxygen atoms in a distant galaxy that have a frequency of $5.710 \times 10^{14}$ Hz. In the laboratory on earth, oxygen atoms emit waves that have a frequency of $5.841 \times 10^{14}$ Hz. Determine the relative velocity of the galaxy with respect to the astronomer in earth.
    (a) $6.724 \times 10^6$ m/s, away from earth
    (b) $6.724 \times 10^6$ m/s, toward earth
    (c) $2.931 \times 10^8$ m/s, away from earth
    (d) $4.369 \times 10^4$ m/s, toward earth
    (e) $4.369 \times 10^4$ m/s, away from earth

24. What would the speed of an observer be if a red ($4.688 \times 10^{14}$ Hz) traffic light appeared green ($5.555 \times 10^{14}$ Hz) to the observer?
    (a) $4.445 \times 10^8$ m/s
    (b) $2.219 \times 10^8$ m/s
    (c) $8.438 \times 10^7$ m/s
    (d) $5.548 \times 10^7$ m/s
    (e) $2.890 \times 10^6$ m/s

25. A minivan moving at 38 m/s passes an unmarked state police car moving at 24 m/s. The electromagnetic waves produced by the radar gun in the police car have a frequency of $8.25 \times 10^9$ Hz. What is the magnitude of the difference in frequency between the waves emitted by the gun and those that are reflected back from the speeding minivan?
    (a) 180 Hz
    (b) 390 Hz
    (c) 770 Hz
    (d) 1440 Hz
    (e) 2100 Hz

## Section 24.6 Polarization

26. The most convincing evidence that electromagnetic waves are *transverse* waves is that
    (a) they can be polarized.
    (b) they carry energy through space.
    (c) they can travel through a material substance.
    (d) they do not require a physical medium for propagation.
    (e) all electromagnetic waves travel with the same speed through vacuum.

27. The magnitude of the magnetic field component of a plane polarized electromagnetic wave traveling in vacuum is given by $B_y = B_0 \sin[kz - \omega t]$. Which one of the following statements concerning this electromagnetic wave is true?
    (a) The wavelength is equal to $k/\omega$.
    (b) The wave propagates in the $y$ direction.
    (c) The wave is polarized in the $x$ direction.
    (d) The electric field component vibrates in the $z$ direction.
    (e) The electric field component has a magnitude of $E = cB_0 \cos[kz - \omega t]$.

28. Light emerges from a polarizer that has its transmission axis located along the $x$ axis. The light then passes through two additional sheets of polarizing material. It is desired to orient the two sheets so that, after passing through both of them, the electromagnetic wave has the maximum possible intensity and is polarized 90° with respect to the $x$ axis. How should the transmission axes of the sheets be oriented? Note: the following answers give the angles that the transmission axes make with respect to the $x$ axis.

    | | First polarizing sheet | Second polarizing sheet |
    |---|---|---|
    | (a) | 45° with respect to the $x$ axis | 45° with respect to the $x$ axis |
    | (b) | 45° with respect to the $x$ axis | 90° with respect to the $x$ axis |
    | (c) | 90° with respect to the $x$ axis | 45° with respect to the $x$ axis |
    | (d) | 30° with respect to the $x$ axis | 60° with respect to the $x$ axis |
    | (e) | 30° with respect to the $x$ axis | 90° with respect to the $x$ axis |

## 246 ELECTROMAGNETIC WAVES

29. Unpolarized light of intensity $S_o$ passes through two sheets of polarizing material whose transmission axes make an angle of 60° with each other as shown in the figure. What is the intensity of the transmitted beam, $S_t$?

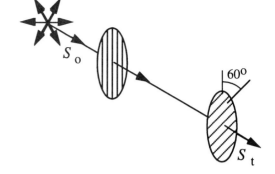

(a) $S_o/4$
(b) $S_o/8$
(c) $3S_o/4$
(d) $S_o/16$
(e) zero

30. Linearly polarized light is incident of a sheet of polarizing material. The angle between the transmission axis and the incident electric field is 52°. What percentage of the incident intensity is transmitted?
(a) 38 %
(b) 43 %
(c) 52 %
(d) 62 %
(e) 79 %

31. A linearly polarized beam of light is incident upon a group of three polarizing sheets which are arranged so that the transmission axis of each sheet is rotated by 45° with respect to the preceding sheet as shown.

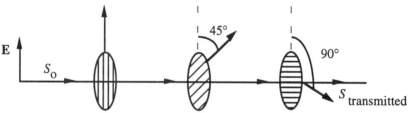

What fraction of the incident intensity is transmitted?
(a) 1/8
(b) 1/4
(c) 3/8
(d) 1/2
(e) 3/4

32. Unpolarized light with an average intensity of 750.0 W/m$^2$ enters a polarizer with a vertical transmission axis. The transmitted light then enters a second polarizer. The light which exits the second polarizer is found to have an average intensity of 125 W/m$^2$. What is the orientation angle of the second polarizer relative to the first one?
(a) 54.7°
(b) 19.5°
(c) 29.0°
(d) 70.5°
(e) zero

33. Two polarizing sheets have their transmission axes parallel so that the intensity of unpolarized light transmitted through both of them is a maximum. Through what angle must either sheet be rotated if the transmitted intensity is 25 % of the incident intensity?
(a) 15°
(b) 30°
(c) 45°
(d) 60°
(e) 75°

34. Linearly polarized light is incident upon a polarizing sheet. The sheet is rotated about its axis through 360° (one complete revolution). At how many positions, including the initial and final positions, will the transmitted intensity be a maximum?
(a) 6
(b) 5
(c) 4
(d) 3
(e) 2

*Questions 35 through 38 pertain to the situation described below:*

A linearly polarized electromagnetic wave is sent through two sheets of polarizing material. The first sheet, **A**, is oriented so that its transmission axis makes an angle of 30° with respect to the incident electric field of the wave. The second sheet, **B**, is oriented so that its transmission axis makes an angle of 90° with the incident electric field of the wave. The incident beam has an electric field of peak magnitude $E_0$ and average intensity $S_0$.

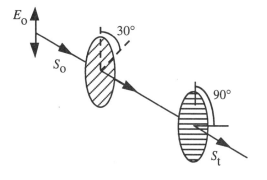

35. What is the peak value of the electric field amplitude after it goes through sheet **A**?
    (a) $0.30E_0$
    (b) $0.50E_0$
    (c) $0.60E_0$
    (d) $0.75E_0$
    (e) $0.87E_0$

36. What is the average intensity of the wave after it passes through **A**?
    (a) $0.30S_0$
    (b) $0.50S_0$
    (c) $0.60S_0$
    (d) $0.75S_0$
    (e) $0.86S_0$

37. What is the average intensity of the wave after it passes through **B**?
    (a) $0.19S_0$
    (b) $0.34S_0$
    (c) $0.43S_0$
    (d) $0.50S_0$
    (e) zero

38. Suppose that **A** and **B** are interchanged so that the wave is first incident upon **B**. What is the average wave intensity after passing through both polarizing sheets?
    (a) $0.19S_0$
    (b) $0.34S_0$
    (c) $0.43S_0$
    (d) $0.50S_0$
    (e) zero

## Additional Problems

39. An FM radio station emits an electromagnetic wave which is received by a circuit containing a $3.33 \times 10^{-7}$ H inductor and a variable capacitor set at $7.31 \times 10^{-12}$ F. What is the frequency of the radio wave?
    (a) $1.02 \times 10^8$ Hz
    (b) $8.80 \times 10^7$ Hz
    (c) $1.58 \times 10^8$ Hz
    (d) $9.40 \times 10^7$ Hz
    (e) $9.80 \times 10^7$ Hz

*Questions 40 through 48 pertain to the statement and figure below:*

The electromagnetic wave of the radiation field from a wire antenna travels toward the plane of the paper (which is in the $-z$ direction). At time $t = 0$, the wave strikes the paper at normal incidence. The magnetic field vector at point **O** in the figure points in the $-y$ direction and has a magnitude of $4.0 \times 10^{-8}$ T. The frequency of the wave is $1.0 \times 10^{16}$ Hz.

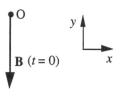

40. What is the magnitude of the associated electric field **E** at time $t = 0$?
    (a) $7.5 \times 10^{-16}$ N/C
    (b) 1.3 N/C
    (c) 7.5 N/C
    (d) 12 N/C
    (e) $7.5 \times 10^{15}$ N/C

248   ELECTROMAGNETIC WAVES

41. What is the direction of the electric field?
    (a) It points in the negative y direction.
    (b) It points in the positive y direction.
    (c) It points in the negative z direction.
    (d) It points in the negative x direction.
    (e) It points in the positive x direction.

42. What is the direction of polarization of this electromagnetic wave?
    (a) the x direction
    (b) the y direction
    (c) the z direction
    (d) 45° with respect to the x direction
    (e) the wave is not polarized

43. What is the wavelength of this electromagnetic wave?
    (a) 0.33 nm
    (b) 3.3 nm
    (c) 20 nm
    (d) 30 nm
    (e) 40 nm

44. What is the *rms* value of the electric field? Assume that the figure shows the peak value of **B**.
    (a) 7.5 N/C
    (b) 8.5 N/C
    (c) 17 N/C
    (d) $1.1 \times 10^{15}$ N/C
    (e) $8.5 \times 10^{-16}$ N/C

45. What is the *rms* value of the magnetic field? Assume that the figure shows the peak value of **B**.
    (a) $1.0 \times 10^{-8}$ T
    (b) $1.4 \times 10^{-8}$ T
    (c) $2.8 \times 10^{-8}$ T
    (d) $4.6 \times 10^{-8}$ T
    (e) $5.7 \times 10^{-8}$ T

46. What is the intensity of the electromagnetic wave at time $t = 0$?
    (a) 0.24 W/m$^2$
    (b) 0.38 W/m$^2$
    (c) 0.48 W/m$^2$
    (d) 0.76 W/m$^2$
    (e) 24 W/m$^2$

47. What is the direction of electromagnetic energy transport?
    (a) the positive x direction
    (b) the negative x direction
    (c) the positive y direction
    (d) the negative y direction
    (e) the negative z direction

48. What is the magnitude of the magnetic field at point **O** at time $t = 5.0 \times 10^{-17}$ s? **Note:** Assume that the figure shows the peak value of **B**.
    (a) $1.0 \times 10^{-8}$ T
    (b) $2.0 \times 10^{-8}$ T
    (c) $3.0 \times 10^{-8}$ T
    (d) $4.0 \times 10^{-8}$ T
    (e) $5.0 \times 10^{-8}$ T

*Questions 49 through 59 pertain to the statement and figure below:*

The figure shows the time variation of the magnitude of the electric field of an electromagnetic wave produced by a wire antenna.

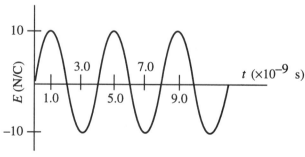

49. Determine the *rms* value of the electric field magnitude.
    (a) 7.1 N/C
    (b) 12 N/C
    (c) 14 N/C
    (d) 19 N/C
    (e) 28 N/C

50. What is the peak value of the magnetic field?
    (a) $1.4 \times 10^{-8}$ T
    (b) $2.3 \times 10^{-8}$ T
    (c) $3.3 \times 10^{-8}$ T
    (d) $4.6 \times 10^{-8}$ T
    (e) $5.4 \times 10^{-8}$ T

51. What is the *rms* value of the magnitude of the magnetic field?
    (a) $1.4 \times 10^{-8}$ T
    (b) $2.4 \times 10^{-8}$ T
    (c) $3.3 \times 10^{-8}$ T
    (d) $4.6 \times 10^{-8}$ T
    (e) $5.4 \times 10^{-8}$ T

52. Determine the frequency of the wave.
    (a) $1.0 \times 10^9$ Hz
    (b) $1.3 \times 10^8$ Hz
    (c) $2.5 \times 10^8$ Hz
    (d) $3.8 \times 10^8$ Hz
    (e) $5.0 \times 10^8$ Hz

53. Determine the wavelength of the wave.
    (a) 0.30 m
    (b) 0.60 m
    (c) 0.79 m
    (d) 1.20 m
    (e) 2.3 m

54. What is the average total energy density of this electromagnetic wave?
    (a) $6.2 \times 10^{-11}$ J/m$^3$
    (b) $8.6 \times 10^{-11}$ J/m$^3$
    (c) $1.1 \times 10^{-10}$ J/m$^3$
    (d) $1.8 \times 10^{-10}$ J/m$^3$
    (e) $4.4 \times 10^{-10}$ J/m$^3$

55. What is the average intensity of this electromagnetic wave?
    (a) 0.13 W/m$^2$
    (b) 0.26 W/m$^2$
    (c) 0.33 W/m$^2$
    (d) 0.36 W/m$^2$
    (e) 0.54 W/m$^2$

56. What is the magnitude of the electric field at $t = 3.34 \times 10^{-10}$ s?
    (a) 2.0 N/C
    (b) 5.0 N/C
    (c) 7.1 N/C
    (d) 10.0 N/C
    (e) zero

57. What is the magnitude of the magnetic field at $t = 6.67 \times 10^{-10}$ s?
    (a) $1.7 \times 10^{-8}$ T
    (b) $2.3 \times 10^{-8}$ T
    (c) $2.9 \times 10^{-8}$ T
    (d) $3.3 \times 10^{-8}$ T
    (e) zero

58. What is the magnitude of the magnetic field at $t = 6.0 \times 10^{-9}$ s?
    (a) $1.7 \times 10^{-8}$ T
    (b) $2.3 \times 10^{-8}$ T
    (c) $2.8 \times 10^{-8}$ T
    (d) $3.3 \times 10^{-8}$ T
    (e) zero

59. To which region of the electromagnetic spectrum does this wave belong?
    (a) X-rays
    (b) gamma rays
    (c) visible light
    (d) infrared radiation
    (e) radio waves

## CHAPTER 25 — The Reflection of Light: Mirrors

*Section 25.1 Wave Fronts and Rays*
*Section 25.2 The Reflection of Light*
*Section 25.3 The Formation of Images by a Plane Mirror*

1. Which one of the following phrases most accurately describes the term *wave front*?
   (a) the surface of a plane mirror
   (b) the surface of a convex mirror
   (c) a surface upon which a wave is incident
   (d) a surface of constant phase within a wave
   (e) a surface that is parallel to the direction of wave propagation

2. Which one of the following statements concerning rays is false?
   (a) Rays point in the direction of the wave velocity.
   (b) Rays point outward from the wave source.
   (c) Rays are parallel to the wave front.
   (d) Rays are radial lines that originate from a point source of waves.
   (e) Rays for a plane wave are parallel to each other.

3. Which one of the following statements is *not* a characteristic of a plane mirror?
   (a) The image is real.
   (b) The magnification is +1.
   (c) The image is always upright.
   (d) The image is reversed right to left.
   (e) The image and object distances are equal in magnitude.

4. The view in the figure is from above a plane mirror suspended by a thread connected to the center of the mirror at point **C**. A scale is located 0.65 m (the distance from point **C** to point **A**) to the right of the center of the mirror. Initially, the plane of the mirror is parallel to the side of the scale; and the angle of incidence of a light ray which is directed at the center of the mirror is 30°. A small torque applied to the thread causes the mirror to turn 12° away from its initial position. The reflected ray then intersects the scale at point **B**. What is the distance from point **A** to point **B** on the scale?
   (a) 0.37 m
   (b) 0.58 m
   (c) 0.76 m
   (d) 0.89 m
   (e) 1.0 m

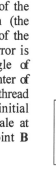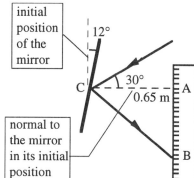

5. A 6.0-ft tall football player stands in front of a plane mirror. How tall must the mirror be so that the football player can see his full-sized image?
   (a) 2.0 ft
   (b) 2.5 ft
   (c) 3.0 ft
   (d) 3.5 ft
   (e) 6.0 ft

6. Daniel walks directly toward a plane mirror at a speed of 0.25 m/s. Determine the speed of the image *relative to him*.
   (a) 0.13 m/s
   (b) 0.25 m/s
   (c) 0.50 m/s
   (d) 0.75 m/s
   (e) 1.0 m/s

7. An object is placed near two perpendicular plane mirrors as shown in the figure. How many images will be formed?
   (a) 1
   (b) 2
   (c) 3
   (d) 4
   (e) 5

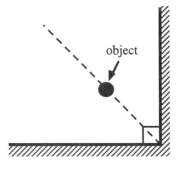

8. Five balls labeled **A**, **B**, **C**, **D**, and **E** are placed in front of a plane mirror as shown in the figure. Which ball(s) will the observer see reflected in the mirror?
   (a) A only
   (b) C only
   (c) A and B
   (d) A, B, D and E
   (e) A, B, C, D and E

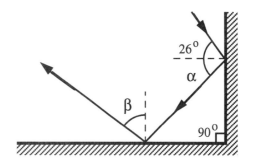

9. A ray of light is reflected from two plane mirror surfaces as shown in the figure. What are the correct values of α and β?

   |     | Value of α | Value of β |
   | --- | --- | --- |
   | (a) | 26° | 26° |
   | (b) | 26° | 64° |
   | (c) | 38° | 52° |
   | (d) | 52° | 26° |
   | (e) | 64° | 26° |

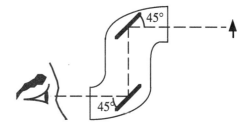

10. In the figure, a Joe is using a device known as a periscope to observe an object, represented by an upright arrow. The periscope contains two plane mirrors tilted at an angle of 45° relative to the observer's line of sight. Which one of the following pairs of characteristics of the final image is correct relative to those of the object?
    (a) inverted, real
    (b) upright, enlarged
    (c) inverted, smaller
    (d) inverted, same size
    (e) upright, virtual

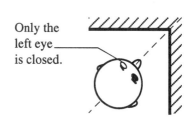

11. Two plane mirrors are connected perpendicular to each other as shown in the figure. If a person looks into the corner at the image *straight ahead of him* and closes his left eye as shown, which one of the following statements is true?
    (a) The left eye of the image is closed.
    (b) The right eye of the image is closed.
    (c) Both eyes of the image are open.
    (d) Both eyes of the image are closed.
    (e) There is no image straight ahead of him, so this question is meaningless.

252   THE REFLECTION OF LIGHT: MIRRORS

12. An object is placed 1 m in front of a plane mirror. An observer stands 3 m behind the object. For what distance must the observer focus his eyes in order to see the image of the object?
    (a) 1 m
    (b) 2 m
    (c) 3 m
    (d) 4 m
    (e) 5 m

*Section 25.4 Spherical Mirrors*
*Section 25.5 The Formation of Images by Spherical Mirrors*

13. The focal length of a spherical concave mirror is 20 cm. What is its radius of curvature?
    (a) 5 cm
    (b) 10 cm
    (c) 20 cm
    (d) 40 cm
    (e) 50 cm

14. The radius of curvature of a spherical convex mirror is 50 cm. What is its focal length?
    (a) +25 cm
    (b) −25 cm
    (c) +50 cm
    (d) −50 cm
    (e) +100 cm

15. Which one of the following statements concerning a convex mirror is true?
    (a) It can form a real image.
    (b) It must be spherical in shape.
    (c) The image will always be inverted relative to the object.
    (d) It produces a larger image than does a plane mirror for the same object distance.
    (e) The image it produces is closer to the mirror than it would be in a plane mirror for the same object distance.

16. A concave mirror has a radius of curvature of 30 cm. How close to the mirror should an object be placed so that the rays travel parallel to each other after reflection?
    (a) 10 cm
    (b) 15 cm
    (c) 30 cm
    (d) 45 cm
    (e) 60 cm

17. Which one of the following phrases most accurately describes *paraxial rays*?
    (a) rays that pass through the principal focus
    (b) any rays that are parallel to the principal axis
    (c) rays that come to a focus on the principal axis
    (d) rays close to the principal axis and parallel to it
    (e) rays close to the principal axis, but not necessarily parallel to it

18. Which one of the following statements concerning the image formed by a concave spherical mirror is true?
    (a) When the object distance is less than the focal length, the image is virtual.
    (b) When the object distance is larger than the focal length, the image is virtual.
    (c) When the object is at the center of curvature, the image is formed at infinity.
    (d) When the object distance is less than the focal length, the image is inverted relative to the object.
    (e) When the object distance is larger than the focal length, the image is upright relative to the object.

19. Santa Claus looks at his reflection in a spherical Christmas tree ornament. Which one of the following statements concerning Santa's image is true?
    (a) The image must be real.
    (b) The image is farther from the ornament than Santa is.
    (c) The image is larger than Santa.
    (d) The image must be inverted.
    (e) The image must be smaller than Santa.

*Questions 20 through 22 pertain to the system described below:*

An object is placed in front of a concave spherical mirror as shown below. The three rays **1**, **2**, and **3**, leave the top of the object and, after reflection, converge at a point on the top of the image. Ray **1** is parallel to the principal axis, ray **2** passes through $F$, and ray **3** passes through $C$.

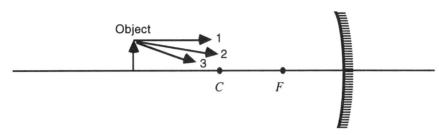

20. Which ray(s) will pass through $F$ after reflection?
    (a) **1** only
    (b) **2** only
    (c) **3** only
    (d) both **1** and **2**
    (e) **1**, **2**, and **3**

21. Which ray(s) will reflect back on itself (themselves)?
    (a) **1** only
    (b) **2** only
    (c) **3** only
    (d) both **1** and **2**
    (e) **1**, **2**, and **3**

22. Which one of the following groups of terms best describes the image?
    (a) real, upright, enlarged
    (b) real, inverted, reduced
    (c) virtual, upright, enlarged
    (d) real, inverted, enlarged
    (e) virtual, inverted, reduced

## Section 25.6 The Mirror Equation and the Magnification Equation

23. A concave mirror in an amusement park has a radius of curvature of 4.0 m. A child stands in front of the mirror so that she appears 2.5 times taller than her actual height. If the image is upright, how far is she standing from the mirror?
    (a) 1.2 m
    (b) 3.5 m
    (c) 2.8 m
    (d) 4.0 m
    (e) 7.0 m

24. A convex mirror in an amusement park has a radius of curvature of 3.00 m. A man stands in front of the mirror so that his image is half as tall as his actual height. At what distance must the man focus his eyes in order to see his image?
    (a) 2.25 m
    (b) 3.00 m
    (c) 4.50 m
    (d) 5.00 m
    (e) 6.75 m

25. A woman stands 2.0 m in front of a convex mirror and notices that her image height is 1/4 of her actual height. Determine the radius of curvature of the mirror.
    (a) 0.67 m
    (b) 1.33 m
    (c) 2.0 m
    (d) 4.0 m
    (e) 6.0 m

26. A concave mirror has a radius of curvature of 20 cm. For which one of the following object distances will the image be real, inverted and smaller than the object?
    (a) 5 cm
    (b) 10 cm
    (c) 15 cm
    (d) 18 cm
    (e) 25 cm

## 254 THE REFLECTION OF LIGHT: MIRRORS

27. The table below lists object and image distances for five objects in front of mirrors. For which one of the following cases is the image formed by a convex spherical mirror?

    |     | Object distance | Image distance |
    |-----|-----------------|----------------|
    | (a) | 7.10 cm         | 18.0 cm        |
    | (b) | 25.0 cm         | 16.7 cm        |
    | (c) | 5.0 cm          | −10.0 cm       |
    | (d) | 20.0 cm         | −5.71 cm       |
    | (e) | 40.0 cm         | −80.0 cm       |

28. An object is 1.0 m in front of a mirror. A virtual image is formed 10.0 m behind the mirror. What is the radius of curvature of the mirror?
    - (a) 0.56 m
    - (b) 1.1 m
    - (c) 2.2 m
    - (d) 4.4 m
    - (e) 10 m

29. An object is placed 30 cm in front of a concave spherical mirror which has a radius of curvature 40 cm. Which one of the following phrases best describes the image?
    - (a) virtual and located at infinity
    - (b) real and located 60 cm from the mirror
    - (c) real and located 120 cm from the mirror
    - (d) virtual and located 60 cm from the mirror
    - (e) virtual and located 120 cm from the mirror

30. An object is placed 30 cm from a convex spherical mirror with radius of curvature 40 cm. Which one of the following phrases best describes the image?
    - (a) virtual and located at infinity
    - (b) real and located 12 cm from the mirror
    - (c) real and located 17 cm from the mirror
    - (d) virtual and located 12 cm from the mirror
    - (e) virtual and located 17 cm from the mirror

31. A spherical concave mirror has a radius of curvature of 6.0 cm. At what distance from the mirror should a 6.0-cm object be placed to obtain an image that is 48 cm tall?
    - (a) 1.3 cm
    - (b) 3.4 cm
    - (c) 4.2 cm
    - (d) 5.3 cm
    - (e) 6.8 cm

32. A convex mirror has a radius of curvature of 0.50 m. Where must an object be placed in front of the mirror such that the image is formed 0.15 m behind the mirror?
    - (a) 0.38 m
    - (b) 0.19 m
    - (c) 0.77 m
    - (d) 0.093 m
    - (e) 0.57 m

33. A concave mirror is found to focus parallel rays at a distance of 9.0 cm. Where is the image formed when an object is placed 6.0 cm in front of the mirror?
    - (a) 11 cm in front of the mirror
    - (b) 18 cm behind the mirror
    - (c) 3.6 cm in front of the mirror
    - (d) 5.6 cm behind the mirror
    - (e) 9.2 cm in front of the mirror

34. A rubber ball is held 4.0 m above a concave spherical mirror with a radius of curvature of 1.5 m. At time equals zero, the ball is dropped from rest and falls along the principal axis of the mirror. How much time elapses before the ball and its image are at the same location?
    - (a) 0.30 s
    - (b) 0.55 s
    - (c) 0.63 s
    - (d) 0.71 s
    - (e) 0.90 s

35. A 0.127-m pencil is oriented perpendicular to the principal axis of a concave spherical mirror that has a radius of 0.300 m. What is the image distance and the image height if the pencil is 0.250 m from the mirror?

|     | Image Distance | Image Height |
|-----|----------------|--------------|
| (a) | 0.150 m        | −0.076 m     |
| (b) | 0.225 m        | −0.114 m     |
| (c) | 0.250 m        | −0.127 m     |
| (d) | 0.300 m        | −0.152 m     |
| (e) | 0.375 m        | −0.191 m     |

## Additional Problems

36. Which one of the following statements concerning a virtual image produced by a mirror is true?
    (a) A virtual image is always larger than the object.
    (b) A virtual image is always smaller than the object.
    (c) A virtual image is always upright relative to the object.
    (d) A virtual image is always inverted relative to the object.
    (e) A virtual image can be photographed or projected onto a screen.

37. A plane mirror is 8 ft tall. What is its focal length?
    (a) zero
    (b) 4 ft
    (c) 8 ft
    (d) 16 ft
    (e) infinity

*Questions 38 and 39 pertain to the situation described below:*

A 3.0-cm object is placed 8.0 cm in front of a mirror. The virtual image is 4.0 cm further from the mirror when the mirror is concave than when it is planar.

38. Determine the focal length of the concave mirror.
    (a) 6.0 cm
    (b) 12 cm
    (c) 24 cm
    (d) 48 cm
    (e) 96 cm

39. Determine the image height in the concave mirror.
    (a) 0.5
    (b) 1.5
    (c) 2.0
    (d) 3.0
    (e) 4.5

# CHAPTER 26

# The Refraction of Light: Lenses and Optical Instruments

## Section 26.1 The Index of Refraction

1. The table lists the index of refraction for various substances at 20 °C for light with a wavelength of 589 nm in a vacuum. Through which substance will light with a vacuum wavelength of 589 nm travel with the greatest speed?

   | Substance | n |
   |---|---|
   | fused quartz | 1.458 |
   | ethyl alcohol | 1.362 |
   | crown glass | 1.520 |
   | carbon tetrachloride | 1.461 |
   | crystalline quartz | 1.544 |

   (a) fused quartz
   (b) crown glass
   (c) ethyl alcohol
   (d) carbon tetrachloride
   (e) crystalline quartz

2. Which one of the following statements concerning the index of refraction for a given material is true?
   (a) It may be less than 1.
   (b) It may be measured in nanometers.
   (c) It does not depend on the frequency of the incident light.
   (d) For a given frequency, it is inversely proportional to the wavelength of light in vacuum.
   (e) For a given frequency, it is inversely proportional to the wavelength of light in the material.

3. The bending of light as it moves from one medium to another with differing indices of refraction is due to a change in what property of the light?
   (a) amplitude
   (b) period
   (c) frequency
   (d) speed
   (e) color

4. When certain light rays pass from a vacuum into a block of an unknown material, the measured index of refraction of the material is 3.50. What is the speed of light inside the block?
   (a) $1.0 \times 10^7$ m/s
   (b) $4.8 \times 10^7$ m/s
   (c) $8.6 \times 10^7$ m/s
   (d) $1.9 \times 10^8$ m/s
   (e) $2.9 \times 10^8$ m/s

5. What is the frequency of light that has a wavelength in water of 600 nm if the refractive index for this light is 1.33?
   (a) $3.76 \times 10^{14}$ Hz
   (b) $5.00 \times 10^{14}$ Hz
   (c) $6.65 \times 10^{14}$ Hz
   (d) $7.25 \times 10^{14}$ Hz
   (e) $9.52 \times 10^{14}$ Hz

6. Blue light with a wavelength of 425 nm passes from a vacuum into a glass lens; and the index of refraction is found to be 1.65. The glass lens is replaced with a plastic lens. The index of refraction for the plastic lens is 1.54. In which one of the two lenses does the light have the greatest speed and what is that speed?
   (a) glass, $2.28 \times 10^8$ m/s
   (b) plastic, $2.13 \times 10^8$ m/s
   (c) glass, $1.82 \times 10^8$ m/s
   (d) plastic, $1.95 \times 10^8$ m/s
   (e) The speed of the blue light is the same in the vacuum and both lenses; and it is $3.00 \times 10^8$ m/s.

## Section 26.2 Snell's Law and the Refraction of Light

7. A beam of light passes from air into water. Which is necessarily true?
   (a) The frequency is unchanged and the *wavelength increases.*
   (b) The frequency is unchanged and the *wavelength decreases.*
   (c) The wavelength is unchanged and the *frequency decreases.*
   (d) Both the wavelength and frequency *increase.*
   (e) Both the wavelength and frequency *decrease.*

8. A ray of light passes from air into a block of glass with a refractive index of 1.50 as shown in the figure. Note: *The drawing is not to scale.*

   What is the value of the distance *D*?
   (a) 1.42 cm
   (b) 1.66 cm
   (c) 1.90 cm
   (d) 2.14 cm
   (e) 2.38 cm

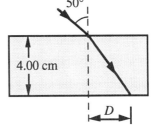

9. A fish swims 2.00 m below the surface of a pond. At what apparent depth does the fish appear to swim if viewed from directly above? The index of refraction of water is 1.33.
   (a) 1.33 m
   (b) 1.50 m
   (c) 2.00 m
   (d) 2.66 m
   (e) 3.00 m

10. A grizzly bear is sitting on a rock in the middle of a calm river when she observes a fish directly below. If the apparent depth of the fish is 0.60 m, what is the actual depth at which the fish is swimming? The index of refraction of water is 1.33.
    (a) 0.80 m
    (b) 0.71 m
    (c) 0.62 m
    (d) 0.53 m
    (e) 0.45 m

11. A scuba diver shines a flashlight from beneath the surface of water ($n = 1.33$) such that the light strikes the water-air boundary with an angle of incidence of 43°. At what angle is the beam refracted?
    (a) 31°
    (b) 43°
    (c) 48°
    (d) 65°
    (e) 90°

12. The figure shows the path of a portion of a ray of light as it passes through three different materials. *Note: The figure is drawn to scale.*

    What can be concluded concerning the refractive indices of these three materials?

    (a) $n_1 < n_2 < n_3$
    (b) $n_1 > n_2 > n_3$
    (c) $n_3 < n_1 < n_2$
    (d) $n_2 < n_1 < n_3$
    (e) $n_1 < n_3 < n_2$

13. Light with a wavelength of 589 nm in a vacuum strikes the surface of an unknown liquid at an angle of 31.2° with respect to the normal to the surface. If the light travels at a speed of $1.97 \times 10^8$ m/s through the liquid, what is the angle of refraction?
    (a) 19.9°
    (b) 26.1°
    (c) 34.2°
    (d) 39.3°
    (e) 51.9°

## Questions 14 and 15 pertain to the statement and diagram below:

The figure shows the path of a ray of light as it travels through air and crosses a boundary into water.

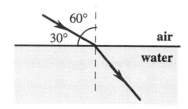

The index of refraction of water for this light is 1.33.

14. What is the speed of this ray of light as it travels through the water?
    (a) $1.54 \times 10^8$ m/s
    (b) $2.26 \times 10^8$ m/s
    (c) $2.86 \times 10^8$ m/s
    (d) $3.99 \times 10^8$ m/s
    (e) $4.43 \times 10^9$ m/s

15. What is the angle of refraction for this situation?
    (a) 0.37°
    (b) 0.65°
    (c) 22°
    (d) 41°
    (e) 60°

## Section 26.3 Total Internal Reflection

16. Complete the following statement: Fiber optics make use of
    (a) total internal reflection.
    (b) polarization.
    (c) chromatic aberration.
    (d) Brewster's angle.
    (e) dispersion.

17. Which one of the following expressions determines the critical angle for quartz ($n = 1.5$) immersed in oil ($n = 1.1$)?
    (a) $\theta_c = 1.5/1.1$
    (b) $\theta_c = 1.5/1.1$
    (c) $\theta_c = \sin^{-1}(1.1/1.5)$
    (d) $\theta_c = \sin(1.1/1.5)$
    (e) $\theta_c = \tan^{-1}(1.1/1.5)$

18. A ray of light originates in medium **A** and is incident upon medium **B**. For which one of the following pairs of indices of refraction for **A** and **B** is total internal reflection *not possible*?

    | | $n_A$ | $n_B$ |
    |---|---|---|
    | (a) | 1.36 | 1.00 |
    | (b) | 1.26 | 1.15 |
    | (c) | 2.54 | 1.63 |
    | (d) | 1.28 | 1.36 |
    | (e) | 1.12 | 1.06 |

19. A glass block with an index of refraction of 1.7 is immersed in an unknown liquid. A ray of light inside the block undergoes total internal reflection as shown in the figure. Which one of the following relations best indicates what may be concluded concerning the index of refraction of the liquid, $n_L$?
    (a) $n_L < 1.00$
    (b) $n_L \geq 1.09$
    (c) $n_L \geq 1.30$
    (d) $n_L \leq 1.09$
    (e) $n_L \leq 1.30$

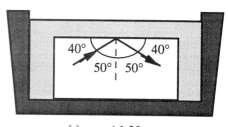

20. A light ray is traveling in a diamond (n = 2.419). If the ray approaches the diamond-air interface, what is the minimum angle of incidence that will result in all of the light being reflected back into the diamond? The index of refraction for air is 1.000.
    (a) 24.42°
    (b) 32.46°
    (c) 54.25°
    (d) 65.58°
    (e) 77.54°

21. A fiber optic line is composed of a core with an index of refraction of 1.47 and cladding with an index of 1.31. Which one of the following relations best describes angles of incidence θ that will result in total internal reflection within the fiber optic line?
   (a) θ < 63°
   (b) θ > 63°
   (c) θ < 27°
   (d) θ > 27°
   (e) 0 ≤ θ ≤ 90°

## Section 26.4 Polarization and the Reflection and Refraction of Light

22. A child is looking at a reflection of the sun in a pool of water. When she puts on a pair of Polaroid sunglasses with a vertical transmission axis, she can no longer see the reflection. At what angle is she looking at the pool of water?
   (a) 45.0°
   (b) 48.8°
   (c) 53.1°
   (d) 61.6°
   (e) 77.3°

23. A ray of light originating in oil (n = 1.21) is incident at the *Brewster angle* upon a flat surface of a quartz crystal (n = 1.458). Determine the angle of incidence for this ray.
   (a) 0.82°
   (b) 1.2°
   (c) 40°
   (d) 50°
   (e) 56°

24. What is the Brewster angle if light is reflected from a plastic plate (n = 1.575) submerged in ethyl alcohol (n = 1.362)?
   (a) 68.3°
   (b) 40.8°
   (c) 59.8°
   (d) 30.1°
   (e) 49.1°

## Section 26.5 The Dispersion of Light: Prisms and Rainbows

25. A ray of green light travels through air and is refracted as it enters a glass prism shown in the figure. An unknown liquid is in contact with the right side of the prism. The light then follows the path shown. Which one of the following statements concerning this situation is true?

    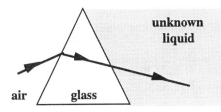

   (a) The frequency of the light changes inside the prism.
   (b) The index of refraction of the glass is smaller than that of air.
   (c) The index of refraction of the unknown liquid is the same as that of the glass.
   (d) The speed of light is larger in the liquid than in the glass.
   (e) The refractive index of the liquid is the same as that of air.

26. Complete the following statement: The term *dispersion* refers to the fact that the index of refraction of certain materials
   (a) depends on the Brewster angle.
   (b) depends on the wavelength of light.
   (c) depends on the angle of incidence.
   (d) depends on the intensity of light.
   (e) depends on the polarization of light.

27. White light enters a glass prism, but the color components of the light are observed to emerge from the prism. Which one of the following statements best explains this observation?
   (a) The separation of white light into its color components is due to the increase in the speed of light within the glass.
   (b) Some of the color components of the white light are absorbed by the glass and only the remaining components are observed.
   (c) The index of refraction of the glass depends on the wavelength, so the color components are refracted at different angles.
   (d) Only some of the color components are refracted by the glass; and these are the ones that are observed.
   (e) White light is separated into its color components by total internal reflection within the prism.

*Questions 28 and 29 pertain to the situation described below:*

A ray of light is normally incident on face **ab** of a plastic prism with an index of refraction $n = 1.20$ as shown.

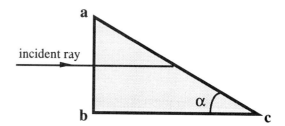

28. Determine the largest value of the angle $\alpha$ so that the ray is totally reflected at the face **ac** if the prism is immersed in air.
   (a) 28°
   (b) 34°
   (c) 45°
   (d) 56°
   (e) Total internal reflection will not occur for any value of $\alpha$.

29. Determine the largest value of the angle $\alpha$ so that the ray is totally reflected at the face **ac** if the prism is immersed in a liquid with refractive index 1.12.
   (a) 21°
   (b) 34°
   (c) 69°
   (d) 78°
   (e) Total internal reflection will not occur for any value of $\alpha$.

*Questions 30 through 33 pertain to the situation described below:*

A beam of light that consists of a *mixture* of **red**, **green** and **violet** light strikes a prism (surrounded by air) as shown. Indices of refraction for this prism for the various colors are indicated in the table. An observer is located to the right of the prism as shown.

| color | n |
|---|---|
| red | 1.43 |
| green | 1.40 |
| violet | 1.37 |

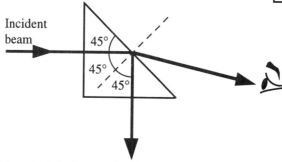

30. Which color(s) could, in principle, be seen by the observer?
   (a) only red light
   (b) only green light
   (c) only violet light
   (d) only green and red light
   (e) only green and violet light

31. Which physical phenomenon is illustrated by the fact that the observer will not see all three colors of light?
    (a) Doppler effect
    (b) dispersion
    (c) diffraction
    (d) total internal reflection
    (e) interference

32. Which physical phenomenon is illustrated by the fact that the emerging rays are spread into the component colors of the beam?
    (a) Doppler effect
    (b) dispersion
    (c) diffraction
    (d) total internal reflection
    (e) interference

33. Which physical phenomenon is illustrated by the fact that the prism has different refractive indices for different colors?
    (a) Doppler effect
    (b) dispersion
    (c) diffraction
    (d) total internal reflection
    (e) interference

*Section 26.6 Lenses*
*Section 26.7 The Formation of Images by Lenses*
*Section 26.8 The Thin-Lens Equation and the Magnification Equation*

34. Which one of the following statements is true concerning the *focal length* of a lens?
    (a) The focal length is the same for all colors.
    (b) The focal length is different for different colors because of reflection.
    (c) The focal length is different for different colors because of dispersion.
    (d) The focal length is different for different colors because of polarization.
    (e) The focal length is different for different colors because of spherical aberration.

35. An object is placed at the focal point of a converging lens of focal length $f$. What is the image distance?
    (a) $f$
    (b) $2f$
    (c) $1/f$
    (d) $2/f$
    (e) at an infinite distance

36. An object is placed at the focal point of a thin diverging lens of focal length $f$. What is the image distance?
    (a) $f$
    (b) $2f$
    (c) $1/f$
    (d) $f/2$
    (e) at an infinite distance

37. An object is placed 4.0 cm from a thin converging lens with a focal length of 12 cm. Which one of the following statements is true concerning the image?
    (a) The image is virtual and 6.0 cm from the lens.
    (b) The image is virtual and 12 cm from the lens.
    (c) The image is real and 3.0 cm from the lens.
    (d) The image is real and 6.0 cm from the lens.
    (e) The image is real and 12 cm from the lens.

38. A converging lens is used to focus light from a small bulb onto a book. The lens has a focal length of 10 cm and is located 40 cm from the book. Determine the distance from the lens to the light bulb.
    (a) 8 cm
    (b) 13 cm
    (c) 20 cm
    (d) 33 cm
    (e) 50 cm

39. When an object is placed 25 cm from a lens, a real image is formed. Which one of the following conclusions is *incorrect*?
    (a) The image is upright.
    (b) The lens is a converging lens.
    (c) The image may be reduced or enlarged.
    (d) The image distance can be less than 25 cm.
    (e) The focal length of the lens is less than 25 cm.

40. When an object is placed 15 cm from a lens, a virtual image is formed. Which one of the following conclusions is *incorrect*?
    (a) The lens may be a convex or concave.
    (b) If the image is upright the lens must be a diverging lens.
    (c) If the image is reduced, the lens must be a diverging lens.
    (d) If the lens is a diverging lens, the image distance must be less than 15 cm.
    (e) If the lens is a converging lens, the focal length must be greater than 15 cm.

41. When an object is placed 20 cm from a diverging lens, a reduced image is formed. Which one of the following statements is necessarily true?
    (a) The image is inverted.
    (b) The image could be real.
    (c) The image distance must be greater than 20 cm.
    (d) The focal length of the lens may be less than 20 cm.
    (e) The refractive power of the lens must be greater than 0.05 diopters.

42. A converging lens with focal length 12 cm produces a 3-cm high virtual image of a 1-cm high object. Which entry in the table below is correct?

    |     | image distance | location of image |
    | --- | --- | --- |
    | (a) | 8 cm  | same side of lens as object |
    | (b) | 8 cm  | opposite side of lens from object |
    | (c) | 12 cm | opposite side of lens from object |
    | (d) | 24 cm | opposite side of lens from object |
    | (e) | 24 cm | same side of lens as object |

43. A camera with a focal length of 0.050 m (a 50-mm lens) is focused for an object at infinity. To focus the camera on a subject which is 4.00 m away, how should the lens be moved?
    (a) 1.0 cm closer to the film
    (b) 0.06 cm closer to the film
    (c) 4.94 cm closer to the film
    (d) 0.06 cm farther from the film
    (e) 4.94 cm farther from the film

44. A 6.0-cm object is placed 30.0 cm from a lens. The resulting image height has a magnitude of 2.0 cm; and the image is inverted. What is the focal length of the lens?
    (a) 7.5 cm
    (b) 15.0 cm
    (c) 22.5 cm
    (d) 30.0 cm
    (e) 45.0 cm

45. A 4-cm object is placed in front of a converging lens of focal length 20 cm. The image is formed 60 cm on the other side of the lens. Which entry in the table below is correct?

    |     | object distance | magnitude of the image height |
    | --- | --- | --- |
    | (a) | 15 cm | 2 cm |
    | (b) | 15 cm | 4 cm |
    | (c) | 30 cm | 4 cm |
    | (d) | 30 cm | 8 cm |
    | (e) | 60 cm | 2 cm |

46. A 2.00-cm tall object is placed 40.0 cm from a lens. The resulting image is 8.00-cm tall and upright relative to the object. Determine the focal length of the lens.
   (a) 26.6 cm
   (b) 32.0 cm
   (c) 53.3 cm
   (d) 64.0 cm
   (e) 80.0 cm

47. In a slide projector, the slide is illuminated; and light passing through the slide then passes through a converging lens of focal length 0.10 m. If a screen is placed 5.0 m from the lens, a sharp image is observed. How far is the slide from the lens?
   (a) 0.082 m
   (b) 0.050 m
   (c) 0.50 m
   (d) 0.27 m
   (e) 0.10 m

*Questions 48 through 51 pertain to the statement and diagram below:*

The figure is a scaled diagram of a an object and a converging lens surrounded by air. Only one focal point, $F$, of the lens is shown.

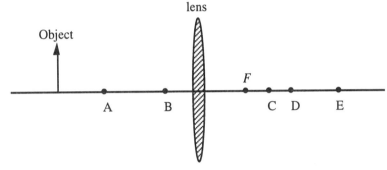

48. At which of the labeled points will the image be formed?
   (a) A
   (b) B
   (c) C
   (d) D
   (e) E

49. Which pair of terms most accurately describes the image?
   (a) real, upright
   (b) real, inverted
   (c) virtual, upright
   (d) virtual, inverted
   (e) virtual, reduced

50. The index of refraction of this lens is 1.51 for red light and 1.53 for blue light. Blue light is focused at the point $F$. Which one of the following statements is true concerning the focal point for red light?
   (a) It is also at $F$.
   (b) It is very close to D.
   (c) It is very close to the lens.
   (d) It is to the left of and close to $F$.
   (e) It is to the right of and close to $F$.

51. The system is immersed in a fluid other than air that has an index of refraction that is larger than that of the lens. Which one of the following statements is true concerning this new situation?
   (a) The image will be real.
   (b) The image will be inverted.
   (c) The image will be enlarged relative to the object.
   (d) The image will be formed on the same side of the lens as the object.
   (e) The lens may act as a diverging lens or a converging lens depending on the location of the object.

# 264  THE REFRACTION OF LIGHT: LENSES AND OPTICAL INSTRUMENTS

*Questions 52 through 54 refer to the statement below:*

A diverging lens has a focal length of −10 cm. A 3-cm object is placed 25 cm from the lens.

● 52. Determine the approximate distance between the object and the image.
   (a) 7 cm
   (b) 10 cm
   (c) 18 cm
   (d) 32 cm
   (e) 35 cm

◐ 53. What is the magnification of the image?
   (a) +0.29
   (b) −0.29
   (c) +0.72
   (d) −0.72
   (e) +0.84

◐ 54. Which pair of terms most accurately describes the image?
   (a) real, upright
   (b) virtual, upright
   (c) real, inverted
   (d) virtual, inverted
   (e) real, reduced

*Questions 55 through 57 pertain to the statement and diagram below:*

A 4.0-cm object is placed 30 cm from a converging lens that has a focal length of 10 cm as shown in the diagram.

Note: *The diagram is not drawn to scale.*

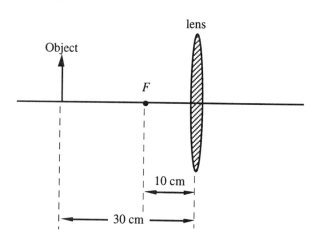

◐ 55. Where is the image located?
   (a) 15 cm to the left of the lens
   (b) 7.5 cm to the left of the lens
   (c) 7.5 cm to the right of the lens
   (d) 15 cm to the right of the lens
   (e) 30 cm to the right of the lens

◐ 56. Determine the height and orientation of the image.
   (a) 2 cm and upright
   (b) 1 cm and inverted
   (c) 2 cm and inverted
   (d) 8 cm and upright
   (e) 8 cm and inverted

● 57. A second converging lens is placed 20.0 cm to the right of the lens shown in the figure. Determine the focal length of the second lens if an inverted image (relative to the object in the diagram) is formed 13.3 cm to the right of the first lens.
   (a) 1.33 cm
   (b) 6.67 cm
   (c) 13.3 cm
   (d) 15.4 cm
   (e) 19.7 cm

*Questions 58 through 60 pertain to the statement and diagram below:*

The figure is a scaled diagram of a an object and a converging lens. The focal length of the lens is 5.0 units. An object is placed 3.0 units from the lens as shown.

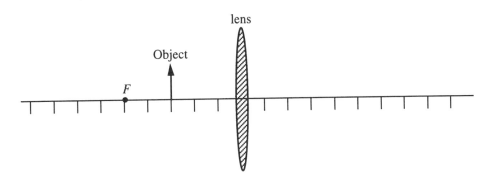

58. Approximately, what is the image distance?
    (a) −2.0 units
    (b) −4.0 units
    (c) +6.0 units
    (d) −7.5 units
    (e) +9.0 units

59. The object has a height of 1.5 units. What is the approximate height of the image?
    (a) 2.0 units
    (b) 1.2 units
    (c) 5.0 units
    (d) 3.8 units
    (e) 9.8 units

60. Which pair of terms most accurately describes the image?
    (a) real, upright
    (b) real, enlarged
    (c) real, inverted
    (d) virtual, inverted
    (e) virtual, upright

## Section 26.9 Lenses in Combination

61. Two converging lenses, each with a focal length of 0.12 m, are used in combination to form an image of an object located 0.36 m to the left of the left lens in the pair. If the distance between the lenses is 0.24 m, where is the final image located relative to the lens on the right?
    (a) 0.06 m to the left of the lens
    (b) 0.12 m to the left of the lens
    (c) 0.18 m to the left of the lens
    (d) 0.12 m to the right of the lens
    (e) 0.36 m to the right of the lens

*Questions 62 and 63 pertain to the following statement:*

A 1.5-cm tall object is placed 0.50 m to the left of a diverging lens with a focal length of 0.20 m. A converging lens with a focal length of 0.17 m is located 0.08 m to the right of the diverging lens.

62. What is the location of the final image with respect to the object?
    (a) The final image is located 0.14 m to the left of the object.
    (b) The final image is located 0.32 m to the right of the object.
    (c) The final image is located 0.40 m to the right of the object.
    (d) The final image is located 0.83 m to the right of the object.
    (e) The final image is located 1.33 m to the right of the object.

266  THE REFRACTION OF LIGHT: LENSES AND OPTICAL INSTRUMENTS

63. What is the height and orientation with respect to the original object of the final image?
    (a) 1.4 cm, inverted
    (b) 1.4 cm, upright
    (c) 0.95 cm, inverted
    (d) 0.95 cm, upright
    (e) 0.28 cm, inverted

*Questions 64 through 66 pertain to the statement and diagram below:*

An object is placed 20 cm from a converging lens with focal length 15 cm. A concave mirror with focal length 10 cm is located 75 cm to the right of the lens as shown in the figure.
Note: *The figure is not drawn to scale.*

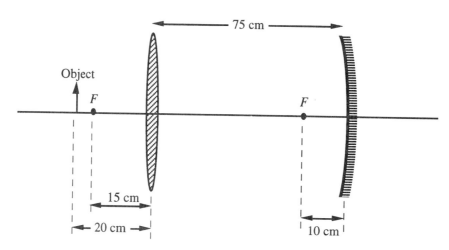

64. Determine the location of the final image.
    (a) 48 cm to the right of the lens
    (b) 96 cm to the right of the lens
    (c) 30 cm to the left of the mirror
    (d) 0.225 cm to the left of the mirror
    (e) 0.225 cm to the right of the mirror

65. If the height of the object is 1.0 cm, what is the height of the image?
    (a) 1.2 cm
    (b) 2.4 cm
    (c) 6.0 cm
    (d) 12 cm
    (e) 24 cm

66. Which pair of terms most accurately describes the final image?
    (a) real, upright
    (b) virtual, upright
    (c) real, inverted
    (d) virtual, inverted
    (e) inverted, enlarged

## Section 26.10 The Human Eye

67. An object is placed 15 cm from a converging lens of refractive power 5 diopters. How far from the object will the image be located?
    (a) 15 cm
    (b) 20 cm
    (c) 45 cm
    (d) 60 cm
    (e) 75 cm

68. Rachel has a far point of 5 m. Which statement below concerning Rachel's vision is true?
    (a) She has normal vision.
    (b) She is myopic and requires diverging lenses to correct her vision.
    (c) She is myopic and requires converging lenses to correct her vision.
    (d) She is hyperopic and requires diverging lenses to correct her vision.
    (e) She is hyperopic and requires converging lenses to correct her vision.

69. Without his contact lenses, Mr. Liu can focus from 0.80 m to infinity. What refractive power of the lenses does he require for normal reading (0.25 m from the eyes)?
    (a) 1.25 diopters
    (b) 2.75 diopters
    (c) 4.00 diopters
    (d) 5.25 diopters
    (e) −5.25 diopters

70. The right lens of Frank's contact lenses is a converging lens of +2.5 diopters. He can read a book held as close as 25 cm from his eyes. Without his lenses, Frank's right eye has
    (a) a far point of 15.4 cm.
    (b) a far point of 40.0 cm.
    (c) a far point of 66.7 cm.
    (d) a near point of 15.4 cm.
    (e) a near point of 66.7 cm.

71. Mrs. York has been prescribed eyeglasses with lenses that have a refractive power of +3.2 diopters. The glasses are worn 2.0 cm from her eyes. With the lenses, she can read a magazine held 25 cm from her eyes. Which one of the following statements is necessarily true? **Note:** The near points and far points given in the following answers are measured relative to her eye.
    (a) She has a far point of 3.2 m.
    (b) She has a far point of 0.25 m.
    (c) She has a near point of 3.2 m
    (d) She has a near point of 6.4 m.
    (e) She has a near point of 0.87 m.

72. George's near point is 20 cm and his far point is 2 m. His contact lenses are designed so that he can see objects that are infinitely far away. What is the closest distance that he can see an object clearly when he wears his contacts?
    (a) 18 cm
    (b) 22 cm
    (c) 25 cm
    (d) 75 cm
    (e) 180 cm

73. In a scene from a movie, a nearsighted character removes his eyeglasses and uses them to focus the nearly parallel rays of the sun to start a fire. What is physically wrong with this scene?
    (a) The eyeglasses have diverging lenses and cannot be used to focus parallel rays.
    (b) The eyeglasses have converging lenses and cannot be used to focus parallel rays.
    (c) Sunlight cannot be used to start a fire.
    (d) A fire can only be started if the image is virtual.
    (e) Parallel rays cannot be focused.

74. Light which is incident upon the eye is refracted several times before it reaches the retina. As light passes through the eye, at which boundary does the majority of the overall refraction occur?
    (a) lens/aqueous humor
    (b) air/cornea
    (c) lens/vitreous humor
    (d) aqueous humor/iris
    (e) vitreous humor/retina

## Section 26.11 Angular Magnification and the Magnifying Glass

75. Which one of the following statements is true concerning a magnifying glass?
    (a) It produces a virtual image.
    (b) It produces an inverted image.
    (c) It can be made from a diverging lens.
    (d) The image distance will be less than the object distance.
    (e) The object must be placed outside the focal point of the lens.

76. A stamp collector observed his favorite stamp with a magnifying glass with a focal length of 24.0 cm. The image of the stamp was located at his near point distance of 30.0 cm. What was the approximate angular magnification of the magnifying glass when the collector observed his stamp?
    (a) 1.00
    (b) 1.25
    (c) 1.75
    (d) 2.00
    (e) 2.25

## Section 26.12 The Compound Microscope

77. The compound microscope discussed in *Cutnell and Johnson's* text is made from two lenses. Which one of the following statements is true concerning the operation of this microscope?
    (a) Both lenses form real images.
    (b) Both lenses form virtual images.
    (c) Only the lens closest to the eye forms an image.
    (d) The lens closest to the object forms a real image; the other lens forms a virtual image.
    (e) The lens closest to the object forms a virtual image; the other lens forms a real image.

78. In her biology class, Chris examines an insect wing under a compound microscope that has an objective lens with a focal length of 0.70 cm, an eyepiece with a focal length of 3.0 cm, and a lens separation distance of 16.00 cm. Chris has a near point distance of 22.5 cm. What is the approximate angular magnification of the microscope as Chris views the insect wing?
    (a) −75
    (b) −110
    (c) −140
    (d) −190
    (e) −250

## Section 26.13 The Telescope
## Section 26.14 Lens Aberrations

79. The moon is observed using a telescope that has an objective lens with a focal length of 3.0 m and an eyepiece with a focal length of 7.5 cm. What is the angular diameter of the moon if the earth-moon distance is $3.85 \times 10^8$ m and the diameter of the moon is $3.48 \times 10^6$ m?
    (a) 0.36 rad
    (b) 4.7 rad
    (c) 9.0 rad
    (d) 22 rad
    (e) 40 rad

80. Which one of the following statements best explains why chromatic aberration occurs in lenses, but not in mirrors?
    (a) The shape of the mirror prevents chromatic aberration.
    (b) The thickness of a lens varies from top to bottom.
    (c) The frequency of light changes when it passes through glass.
    (d) The angle of incidence varies over the surface of a lens for incident parallel rays of light.
    (e) Different colors of light are refracted by different amounts as the light passes through a lens.

## Additional Problems

81. A child sitting at the edge of a swimming pool sees a coin resting on the bottom of the pool. The coin appears to be 2.0 ft directly below the water's surface. How deep is the pool at the location of the coin? The index of refraction of water is 1.33.
    (a) 1.5 ft
    (b) 2.0 ft
    (c) 2.7 ft
    (d) 3.2 ft
    (e) 4.0 ft

82. A physics student desires to create a beam of light that consists of parallel rays. Which one of the following arrangements would allow her to accomplish this task?
    (a) A light bulb is placed at the focal point of a convex mirror.
    (b) A light bulb is placed at the focal point of a diverging lens.
    (c) A light bulb is placed at the focal point of a converging lens.
    (d) A light bulb is located at twice the focal length from a concave mirror.
    (e) A light bulb is located at twice the focal length from a converging lens.

83. The length of the wing of an insect is 1 mm. When viewed through a microscope, the image is 1 m long and located 5 m away. Determine the angular magnification if the observer has a near point of 25 cm.
   (a) 50
   (b) 100
   (c) 200
   (d) 500
   (e) 1000

84. The leg of a spider is 0.2 cm long. When viewed through a microscope, a person with a near point of 25 cm sees an image 2 m long located 10 m away. What is the angular magnification?
   (a) 25
   (b) 50
   (c) 100
   (d) 250
   (e) 1000

*Questions 85 through 88 pertain to the statement and figure below:*

The figure shows a point source of unpolarized light at **A** inside a uniform transparent crystal. The ray **AO** in the crystal strikes the plane surface **SS'** making an angle of 30° with the normal. This angle is the critical angle for transmission into air.

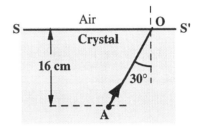

85. Determine the index of refraction of the crystal.
   (a) 0.58
   (b) 1.2
   (c) 1.7
   (d) 2.0
   (e) 2.4

86. If point **A** is 16 cm below the plane **SS'**, what is the radius of the largest circle at the air-crystal interface through which light can emerge from the crystal?
   (a) 9.2 cm
   (b) 16.0 cm
   (c) 18.4 cm
   (d) 27.0 cm
   (e) 32.0 cm

87. For what angle of incidence will the reflected rays of light be completely polarized?
   (a) 15°
   (b) 27°
   (c) 30°
   (d) 60°
   (e) 63°

88. Which one of the following statements is true concerning this situation?
   (a) When the angle of incidence is equal to the Brewster angle, the angle of refraction is 13°.
   (b) When the angle of incidence is equal to the Brewster angle, the angle of refraction is 63°.
   (c) When the angle of incidence is equal to the Brewster angle, the angle of refraction is 42°.
   (d) Since the Brewster angle is less than the critical angle, there is no refraction when the angle of incidence is equal to $\theta_B$.
   (e) Since the Brewster angle is greater than the critical angle, there is no refraction when the angle of incidence is equal to $\theta_B$.

# THE REFRACTION OF LIGHT: LENSES AND OPTICAL INSTRUMENTS

*Questions 89 through 91 pertain to the statement and diagram below:*

A fish swims 4.0 m below the surface of a still lake as shown in the figure. When an archer attempts to shoot the fish, the arrow enters the water at point **P** which is a horizontal distance 1.2 m from the fish.

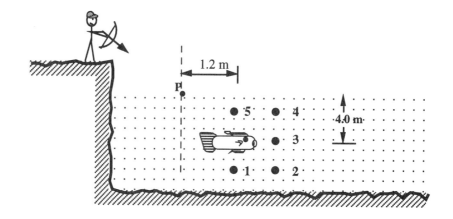

89. At which of the numbered positions should the archer aim to hit the fish?
    (a) 1
    (b) 2
    (c) 3
    (d) 4
    (e) 5

90. Which one of the following phrases most accurately describes the image of the fish as seen by the archer?
    (a) real and inverted
    (b) virtual and inverted
    (c) real and reversed right to left
    (d) virtual with its orientation unaltered
    (e) real with its orientation unaltered

91. If the archer is successful in shooting the fish, what angle does the arrow make *with the horizontal* as it enters the water?
    (a) 12°
    (b) 17°
    (c) 23°
    (d) 67°
    (e) 73°

*Questions 92 and 93 pertain to the situation described below:*

The figure shows a ray of light that originates in an aquarium. It travels through water, is incident on the glass side, and emerges into the air. Ignore any partial reflections.

Note: *The figure is not drawn to scale.*

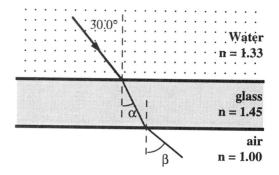

92. Which entry in the table below gives the correct values for the angles shown in the figure?

|  | α | β |
|---|---|---|
| (a) | 0.55° | 2.66° |
| (b) | 27.3° | 18.4° |
| (c) | 27.3° | 30.0° |
| (d) | 27.3° | 41.7° |
| (e) | 33.0° | 52.2° |

93. Which one of the following figures shows the smallest angle of incidence in the water for which no light emerges into air?
    Note: *Only one figure is physically possible.*

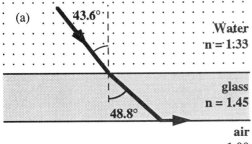

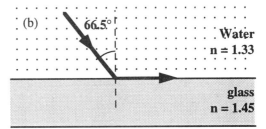

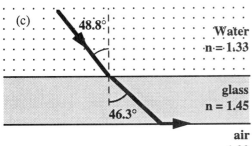

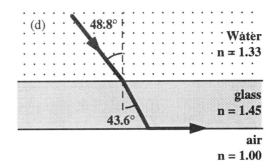

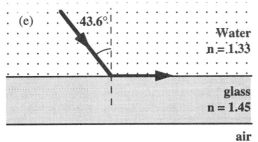

# CHAPTER 27
# Interference and The Wave Nature of Light

*Section 27.1 The Principle of Linear Superposition*
*Section 27.2 Young's Double Slit Experiment*

1. Complete the following sentence: The term *coherence* relates to
   (a) the phase relationship between two waves.
   (b) the polarization state of two waves.
   (c) the diffraction of two waves.
   (d) the amplitude of two waves.
   (e) the frequency of two waves.

2. Two identical light waves, **A** and **B**, are emitted from different sources and meet at a point **P**. The distance from the source of **A** to the point **P** is $\ell_A$; and the source of **B** is a distance $\ell_B$ from P. Which of the following statements is necessarily true concerning the interference of the two waves?
   (a) **A** and **B** will interfere constructively because their amplitudes are the same.
   (b) **A** and **B** will interfere constructively if $\ell_A = \ell_B$.
   (c) **A** and **B** will interfere destructively if $\ell_A - \ell_B = m\lambda$ where m = 0, 1, 2, 3, ...
   (d) **A** and **B** will interfere destructively if $\ell_A$ is not equal to $\ell_B$.
   (e) **A** and **B** will interfere constructively because their wavelengths are the same.

3. Which one of the following statements provides the most convincing evidence that *visible light* is a form of electromagnetic radiation?
   (a) Two light sources can be coherent.
   (b) Light can be reflected from a surface.
   (c) Light can be diffracted through an aperture.
   (d) Light can form a double-slit interference pattern.
   (e) Light travels through vacuum at the same speed as X-rays.

4. Which one of the following statements best explains why interference patterns are not usually observed for light from two ordinary light bulbs?
   (a) Diffraction effects predominate.
   (b) The two sources are out of phase.
   (c) The two sources are not coherent.
   (d) The interference pattern is too small to observe.
   (e) Light from ordinary light bulbs is not polarized.

5. What does one observe on the screen in a Young's experiment if white light illuminates the double slit instead of light of a single wavelength?
   (a) a white central fringe and no other fringes
   (b) a dark central fringe and a series of alternating white and dark fringes on each side of the center
   (c) a white central fringe and a series of colored and dark fringes on each side of the center
   (d) a continuous band of colors with no dark fringes anywhere
   (e) a dark screen since no constructive interference can occur

6. A double slit is illuminated with monochromatic light of wavelength 600 nm. The $m = 0$ and $m = 1$ bright fringes are separated by 3.0 cm on a screen which is located 4.0 m from the slits. What is the separation between the slits?
   (a) $4.0 \times 10^{-5}$ m
   (b) $8.0 \times 10^{-5}$ m
   (c) $1.2 \times 10^{-4}$ m
   (d) $1.6 \times 10^{-4}$ m
   (e) $2.4 \times 10^{-4}$ m

## TEST BANK — Chapter 27

7. Two slits separated by $2.00 \times 10^{-5}$ m are illuminated by light of wavelength 500 nm. If the screen is 8.00 m from the slits, what is the distance between the $m = 0$ and $m = 1$ bright fringes?
   (a) 1.25 cm
   (b) 2.50 cm
   (c) 5.00 cm
   (d) 10.0 cm
   (e) 20.0 cm

8. Two slits are separated by $2.00 \times 10^{-5}$ m. They are illuminated by light of wavelength $5.60 \times 10^{-7}$ m. If the distance from the slits to the screen is 6.00 m, what is the separation between the central bright fringe and the third dark fringe?
   (a) 0.420 m
   (b) 0.224 m
   (c) 0.168 m
   (d) 0.084 m
   (e) 0.070 m

9. In a Young's double slit experiment, the separation between the slits is $1.20 \times 10^{-4}$ m; and the screen is located 3.50 m from the slits. The distance between the central bright fringe and the second-order bright fringe is 0.0415 m. What is the wavelength of the light used in this experiment?
   (a) 428 nm
   (b) 474 nm
   (c) 517 nm
   (d) 642 nm
   (e) 711 nm

10. Light of wavelength 530 nm is incident on two slits that are spaced 1.0 mm apart. How far from the slits should the screen be placed so that the distance between the $m = 0$ and $m = 1$ bright fringes is 1.0 cm?
    (a) 7.9 m
    (b) 9.5 m
    (c) 16 m
    (d) 19 m
    (e) 36 m

11. Light is incident on two slits that are separated by 0.2 mm. The figure shows the resulting interference pattern observed on a screen 1.0 m from the slits.

    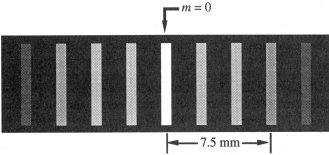

    Determine the wavelength of light used in this experiment.
    (a) 0.05 nm
    (b) 0.50 nm
    (c) 50 nm
    (d) 500 nm
    (e) 5000 nm

12. The figure shows the interference pattern produced when light of wavelength 500 nm is incident on two slits. Fringe **A** is equally distant from each slit. By what distance is fringe **B** closer to one slit than the other?

    (a) 250 nm
    (b) 500 nm
    (c) 750 nm
    (d) 1000 nm
    (e) 1500 nm

## 274 INTERFERENCE AND THE WAVE NATURE OF LIGHT

13. In a Young's double slit experiment, green light is incident on the two slits. The inference pattern is observed on a screen. Which one of the following changes would cause the fringes to be more closely spaced?
    (a) Reduce the slit separation distance.
    (b) Use red light instead of green light.
    (c) Use blue light instead of green light.
    (d) Move the screen farther away from the slits.
    (e) Move the light source farther away from the slits.

*Questions 14 through 19 pertain to the interference pattern shown below:*

The figure shows the interference pattern obtained in a double-slit experiment using light of wavelength 600 nm.

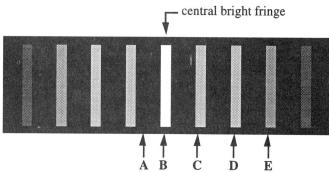

14. Which fringe is the same distance from both slits?
    (a) A
    (b) B
    (c) C
    (d) D
    (e) E

15. Which fringe is the *third order maximum*?
    (a) A
    (b) B
    (c) C
    (d) D
    (e) E

16. Which fringe is 300 nm closer to one slit than to the other?
    (a) A
    (b) B
    (c) C
    (d) D
    (e) E

17. Which fringe results from a phase difference of $4\pi$?
    (a) A
    (b) B
    (c) C
    (d) D
    (e) E

18. Which one of the following phenomena would be observed if the wavelength of light were *increased*?
    (a) The fringes would be brighter.
    (b) More bright fringes would appear on the screen.
    (c) The distance between dark fringes would decrease.
    (d) Single-slit diffraction effects would become non-negligible.
    (e) The angular separation between bright fringes would increase.

19. Which one of the following phenomena would be observed if the distance between the slits were *increased*?
    (a) The fringes would become brighter.
    (b) The central bright fringe would change position.
    (c) The distance between dark fringes would increase.
    (d) The distance between bright fringes would increase.
    (e) The angular separation between the dark fringes would decrease.

## Section 27.3 Thin-film Interference
## Section 27.4 The Michelson Interferometer

20. A $4.0 \times 10^2$-nm thick film of kerosene ($n = 1.2$) is floating on water. White light is normally incident on the film. What is the visible wavelength *in air* that has a maximum intensity after the light is reflected? **Note**: the visible wavelength range is 380 nm to 750 nm.
    (a) 380 nm
    (b) 430 nm
    (c) 480 nm
    (d) 530 nm
    (e) 580 nm

21. A portion of a soap bubble appears green ($\lambda = 500$ nm in vacuum) when viewed at normal incidence in white light. Determine the two smallest, non-zero thicknesses for the soap film if its index of refraction is 1.40.
    (a) 89 nm and 179 nm
    (b) 89 nm and 268 nm
    (c) 125 nm and 250 nm
    (d) 125 nm and 375 nm.
    (e) 170 nm and 536 nm

22. Light of wavelength $\lambda$ in vacuum strikes a lens which is made of glass with index of refraction 1.6. The lens has been coated with a film of thickness $t$ and index of refraction 1.3. For which condition will there be no reflection?
    (a) $2t = \dfrac{\lambda}{2}$
    (b) $2t = \dfrac{\lambda}{1.33}$
    (c) $2t = \dfrac{\lambda}{1.6}$
    (d) $2t = \dfrac{1}{2}\left(\dfrac{\lambda}{1.6}\right)$
    (e) $2t = \dfrac{1}{2}\left(\dfrac{\lambda}{1.3}\right)$

23. Light of wavelength 650 nm is incident normally upon a glass plate. The glass plate rests on top of a second plate so that they touch at one end and are separated by 0.0325 mm at the other end as shown in the figure.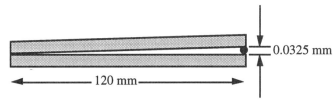
    Which range of values contains the horizontal separation between adjacent bright fringes?
    (a) 1.1 mm to 1.4 mm
    (b) 1.4 mm to 2.8 mm
    (c) 2.8 mm to 4.2 mm
    (d) 4.2 mm to 5.6 mm
    (e) 5.6 mm to 7.0 mm

24. A transparent film ($n = 1.4$) is deposited on a glass lens ($n = 1.5$) to form a non-reflective coating. What thickness would prevent reflection of light with wavelength 500 nm in air?
    (a) 89 nm
    (b) 125 nm
    (c) 170 nm
    (d) 250 nm
    (e) 357 nm

25. A lens which has an index of refraction of 1.61 is coated with a non-reflective coating which has an index of refraction of 1.45. Determine the minimum thickness for the film if it is to be non-reflecting for light of wavelength 560 nm.
    (a) $1.93 \times 10^{-7}$ m
    (b) $3.86 \times 10^{-7}$ m
    (c) $4.83 \times 10^{-8}$ m
    (d) $9.66 \times 10^{-8}$ m
    (e) $8.69 \times 10^{-8}$ m

# 276  INTERFERENCE AND THE WAVE NATURE OF LIGHT

26. Two glass plates, each with an index of refraction of 1.55, are separated by a small distance $D$. The space between them is filled with water ($n = 1.33$) as shown. For which one of the following conditions will the reflected light appear green? **Note**: The wavelength of green light is 460 nm in vacuum.

    (a) $D = \left(\dfrac{460 \text{ nm}}{2}\right)$

    (b) $2D = \left(\dfrac{460 \text{ nm}}{1.33}\right)$

    (c) $2D = \left(\dfrac{460 \text{ nm}}{1.55}\right)$

    (d) $2D = \dfrac{1}{2}\left(\dfrac{460 \text{ nm}}{1.33}\right)$

    (e) $2D = \dfrac{1}{2}\left(\dfrac{460 \text{ nm}}{1.55}\right)$

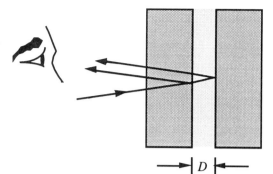

27. The table lists the range of wavelengths in vacuum corresponding to a given color. If one looks through a film which has a refractive index of 1.333 and thickness of 340 nm, which color will be 100% transmitted through the film?

    (a) red            (d) green
    (b) yellow         (e) white
    (c) blue

    | Color  | Wavelength (nm) |
    |--------|-----------------|
    | red    | 780 - 622       |
    | orange | 622 - 597       |
    | yellow | 597 - 577       |
    | green  | 577 - 492       |
    | blue   | 492 - 455       |
    | violet | 455 - 390       |

28. Monochromatic light of wavelength $\lambda_{film}$ is normally incident on a soap film in air. In terms of the wavelength, what is the thickness of the thinnest film for which the reflected light will be a maximum?

    (a) $\lambda_{film}/4$         (c) $3\lambda_{film}/4$         (e) $3\lambda_{film}/2$
    (b) $\lambda_{film}/2$         (d) $\lambda_{film}$

29. A Michelson interferometer is used to measure the wavelength of light emitted from a monochromatic source. As the adjustable mirror is slowly moved through a distance $D_A = 0.265$ mm, an observer counts 411 alternations between bright and dark fields. What is the wavelength of the monochromatic light used in this experiment?

    (a) 193 nm         (c) 258 nm         (e) 369 nm
    (b) 227 nm         (d) 346 nm

## Section 27.5  Diffraction

30. Which one of the following statements best explains why the diffraction of sound is more apparent than the diffraction of light under most circumstances?
    (a) Sound requires a physical medium for propagation.
    (b) Sound waves are longitudinal, and light waves are transverse.
    (c) Light waves can be represented by rays while sound waves cannot.
    (d) The speed of sound in air is six orders of magnitude smaller than that of light.
    (e) The wavelength of light is considerably smaller than the wavelength of sound.

31. Light of wavelength 700.0 nm passes through a diffraction grating. The $m = 0$ and $m = 1$ bright spots are 6.0 cm apart on a screen positioned 2.0 cm from the grating. What is the spacing between the slits in the grating?
   (a) 233 nm
   (b) 420 nm
   (c) 467 nm
   (d) 738 nm
   (e) 1240 nm

32. Light of wavelength 625 nm shines through a single slit of width 0.32 mm and forms a diffraction pattern on a flat screen located 8.0 m away. Determine the distance between the middle of the central bright fringe and the first dark fringe.
   (a) 0.156 cm
   (b) 0.516 cm
   (c) 1.56 cm
   (d) 5.16 cm
   (e) 6.51 cm

33. Light of 600.0 nm is incident on a single slit of width 6.5 μm. The resulting diffraction pattern is observed on a nearby screen and has a central maximum of width 3.5 m. What is the distance between the screen and the slit?
   (a) 9.50 m
   (b) 19.0 m
   (c) 38.0 m
   (d) 57.0 m
   (e) 76.0 m

34. Light from a red laser passes through a single slit to form a diffraction pattern. If the width of the slit is increased by a factor of two, what happens to the width of the central maximum?
   **Note**: *Assume that the angle θ is sufficiently small so that* (sin θ) *is nearly equal to* θ.
   (a) The width of the central maximum increases by a factor of 4.
   (b) The width of the central maximum decreases by a factor of 2.
   (c) The width of the central maximum decreases by a factor of 4.
   (d) The width of the central maximum increases by a factor of 2.
   (e) The width of the central maximum does not change.

35. Light of wavelength 600 nm is incident upon a single slit with width $4 \times 10^{-4}$ m. The figure shows the pattern observed on a screen positioned 2 m from the slits.

   Determine the distance $s$.
   (a) 0.002 m
   (b) 0.003 m
   (c) 0.004 m
   (d) 0.006 m
   (e) 0.008 m

36. Light of 600.0 nm is incident upon a single slit. The resulting diffraction pattern is observed on a screen which is 0.50 m from the slit. The distance between the first and third minima of the diffraction pattern is 0.80 mm. Which range of values listed below contains the width of the slit?
   (a) 0.1 mm to 0.4 mm
   (b) 0.4 mm to 0.8 mm
   (c) 0.8 mm to 1.2 mm
   (d) 1.2 mm to 1.6 mm
   (e) 1.6 mm to 2.0 mm

37. A monochromatic beam of microwaves with a wavelength of 0.052 m is directed at a rectangular opening of width 0.35 m. The resulting diffraction pattern is measured along a wall 8.0 m from the opening. What is the distance between the first- and second-order dark fringes?
   (a) 1.3 m
   (b) 1.8 m
   (c) 2.1 m
   (d) 2.5 m
   (e) 3.7 m

## 278  INTERFERENCE AND THE WAVE NATURE OF LIGHT

38. The table lists the range of wavelengths in vacuum corresponding to a given color. Which one of these colors will produce a diffraction pattern with the widest central maximum, assuming all other factors are equal?
    (a) red
    (b) yellow
    (c) green
    (d) blue
    (e) violet

| Color | Wavelength (nm) |
|---|---|
| red | 780 - 622 |
| orange | 622 - 597 |
| yellow | 597 - 577 |
| green | 577 - 492 |
| blue | 492 - 455 |
| violet | 455 - 390 |

39. Diffraction occurs when light passes through a single slit. Rank the following three choices in decreasing order, according to the extent of the diffraction that occurs (largest diffraction first):
    A - blue light, narrow slit
    B - red light, narrow slit
    C - blue light, wide slit

    **Note:** *The blue light referred to in choices A and C is the same. Also, the narrow slit referred to in choices A and B is the same.*
    (a) A, B, C
    (b) B, A, C
    (c) C, A, B
    (d) A, C, B
    (e) B, C, A

## Section 27.6  Resolving Power

40. The Hubble Space Telescope in orbit above the Earth has a 2.4 m circular aperture. The telescope has equipment for detecting ultraviolet light. What is the minimum angular separation between two objects that the Hubble Space Telescope can resolve in ultraviolet light of wavelength 95 nm?
    (a) $4.8 \times 10^{-8}$ rad
    (b) $7.0 \times 10^{-8}$ rad
    (c) $1.9 \times 10^{-7}$ rad
    (d) $1.5 \times 10^{-7}$ rad
    (e) $3.3 \times 10^{-9}$ rad

41. A spy satellite is in orbit at a distance of $1.0 \times 10^6$ m above the ground. It carries a telescope that can resolve the two rails of a railroad track that are 1.4 m apart using light of wavelength 600 nm. Which one of the following statements best describes the diameter of the lens in the telescope?
    (a) It is less than 0.14 m.
    (b) It is less than 0.35 m.
    (c) It is less than 0.52 m.
    (d) It is greater than 0.35 m.
    (e) It is greater than 0.52 m.

42. The headlights of a car are 1.6 m apart and produce light of wavelength 575 nm in vacuum. The pupil of the eye of the observer has a diameter of 4.0 mm and a refractive index of 1.4. What is the maximum distance from the observer that the two headlights can be distinguished?
    (a) 8.0 km
    (b) 9.1 km
    (c) 11.1 km
    (d) 12.8 km
    (e) 15.6 km

43. Two stars are just barely resolved by a telescope with a lens diameter of 0.500 m. Determine the angular separation of the two stars. Assume incident light of wavelength 500.0 nm.
    (a) $1.22 \times 10^{-6}$ rad
    (b) $5.66 \times 10^{-5}$ rad
    (c) $2.44 \times 10^{-7}$ rad
    (d) $4.88 \times 10^{-5}$ rad
    (e) $1.22 \times 10^7$ rad

44. The wavelength of light emitted from two distant objects is 715 nm. What is the minimum angle at which these objects can just be resolved when using binoculars with a 50-mm objective lens?
    (a) $10^{-2}$ degrees
    (b) $10^{-3}$ degrees
    (c) $10^{-4}$ degrees
    (d) $10^{-5}$ degrees
    (e) $10^{-6}$ degrees

45. Two candles are lit and separated by 0.10 m. If the diameter of the pupil of an observer's eye is 3.5 mm, what is the maximum distance that the candles can be away from the observer and be seen as two light sources? Use 545 nm for the wavelength of light in the eye.
  (a) 170 m
  (b) 340 m
  (c) 530 m
  (d) 680 m
  (e) 850 m

## Section 27.7 The Diffraction Grating
## Section 27.8 Compact Discs and the Use of Interference

46. A 30.0-mm wide diffraction grating produces a deviation of 30.0° in the second order principal maxima. The wavelength of light is 600.0 nm. What is the total number of slits on the grating?
  (a) 10 000
  (b) 11 500
  (c) 12 500
  (d) 14 000
  (e) 15 000

47. A beam of light which consists of a mixture of red light ($\lambda$ = 660 nm in vacuum) and violet light ($\lambda$ = 410 nm in vacuum) falls on a grating that contains $1.0 \times 10^4$ lines/cm. Find the angular separation between the first-order maxima of the two wavelengths if the experiment takes place in a vacuum.
  (a) 11°
  (b) 17°
  (c) 24°
  (d) 41°
  (e) 65°

48. Red light of wavelength 600.0 nm is incident on a grating. If the separation between the slits is $5.0 \times 10^{-5}$ m, at what angle does the first principal maximum occur?
  (a) $0.6 \times 10^{-2}$ rad
  (b) $0.8 \times 10^{-2}$ rad
  (c) $1.2 \times 10^{-2}$ rad
  (d) $3.6 \times 10^{-2}$ rad
  (e) $5.0 \times 10^{-2}$ rad

49. A diffraction grating which has 4500 lines/cm is illuminated by light which has a single wavelength. If a second order maximum is observed at an angle of 42° with respect to the central maximum, what is the wavelength of this light?
  (a) 1500 nm
  (b) 370 nm
  (c) 930 nm
  (d) 1100 nm
  (e) 740 nm

50. Light from a laser ($\lambda$ = 640 nm) passes through a diffraction grating and spreads out into three beams as shown in the figure.

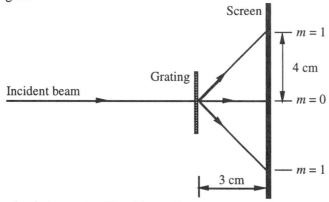

Determine the spacing between the slits of the grating.
  (a) 240 nm
  (b) 410 nm
  (c) 500 nm
  (d) 680 nm
  (e) 800 nm

280  INTERFERENCE AND THE WAVE NATURE OF LIGHT

51. Visible light of wavelength 589 nm is incident on a diffraction grating which has 3500 lines/cm. At what angle with respect to the central maximum is the fifth order maximum observed?
    (a) 17.9°
    (b) 23.8°
    (c) 35.7°
    (d) 71.3°
    (e) A fifth order maximum cannot be observed.

52. White light is passed through a diffraction grating that has $2.50 \times 10^5$ lines/m. On each side of the white central maximum, a spectrum of colors is observed. What is the wavelength of the light observed at an angle of 7.00° in the first-order bright fringes?
    (a) 487 nm
    (b) 589 nm
    (c) 632 nm
    (d) 668 nm
    (e) 731 nm

*Additional Problems*

53. Which one of the following statements provides the most convincing evidence that electromagnetic waves have a transverse character?
    (a) Electromagnetic waves can be refracted.
    (b) Electromagnetic waves can be reflected.
    (c) Electromagnetic waves can be polarized.
    (d) Electromagnetic waves can be diffracted.
    (e) Electromagnetic waves exhibit interference.

54. Light of wavelength 530 nm is incident on two slits that are spaced 1.0 mm apart. If each of the slits has a width of 0.10 mm, how many interference maxima lie within the central diffraction peak?
    (a) 1
    (b) 4
    (c) 12
    (d) 21
    (e) ∞

# CHAPTER 28    Special Relativity

**Section 28.1 Events and Inertial Reference Frames**
**Section 28.2 The Postulates of Special Relativity**
**Section 28.3 The Relativity of Time: Time Dilation**

1. At time $t = 2.3$ s, a 4-kg block that initially moves with a constant speed of 6 m/s undergoes an inelastic collision with another block. Any two inertial observers must agree that
   (a) the event took place at $t = 2.3$ s.
   (b) the initial speed of the block is 6 m/s.
   (c) the initial momentum of the block has magnitude 24 kg·m/s.
   (d) the second block is moving after the collision.
   (e) the momentum of the two block system is conserved during the collision.

2. Which one of the following systems would constitute an inertial reference frame?
   (a) a weather balloon descending at constant velocity
   (b) a rocket undergoing uniform acceleration
   (c) a train rounding a turn at constant speed
   (d) an orbiting space station
   (e) a rotating merry-go-round

3. Which one of the following is a consequence of the postulates of special relativity?
   (a) There is no such thing as an inertial reference frame.
   (b) Newton's laws of motion apply in every reference frame.
   (c) Coulomb's law of electrostatics applies in any reference frame.
   (d) The question of whether an object is at rest in the universe is meaningless.
   (e) The value of every physical quantity depends on the reference frame in which it is measured.

4. Complete the following statement: The Michelson-Morley experiment
   (a) confirmed that time dilation occurs.
   (b) proved that length contraction occurs.
   (c) verified the conservation of momentum in inertial reference frames.
   (d) supported the relationship between mass and energy.
   (e) indicated that the speed of light is the same in all inertial reference frames.

5. Danelle is moving in a spaceship at a constant velocity away from a group of stars. Which one of the following statements indicates a method by which she can determine her absolute velocity through space?
   (a) She can measure her increase in mass.
   (b) She can measure the contraction of her ship.
   (c) She can measure the vibration frequency of a quartz crystal.
   (d) She can measure the change in total energy of her ship.
   (e) She can perform no measurement to determine this quantity.

6. Which one of the following statements concerning the *proper time interval* for between two events is true?
   (a) It is the longest time interval that any inertial observer can measure for the event.
   (b) It is the shortest time interval that any inertial observer can measure for the event.
   (c) It is the time measured by an observer who is in motion with respect to the event.
   (d) Its value depends upon the speed of the observer.
   (e) Its value depends upon the choice of reference frame.

## SPECIAL RELATIVITY

7. Which one of the following statements concerning *time dilation* is true?
   (a) It is predicted by special relativity, but has never been observed.
   (b) It has been observed only in experiments involving radioactive decay processes.
   (c) It has been observed in experiments involving both atomic clocks and radioactive decay processes.
   (d) It was demonstrated by the Michelson-Morley experiment.
   (e) It has been disproved in experiments with atomic clocks.

8. Which one of the following statements is a consequence of Special Relativity?
   (a) Clocks that are moving run slower than when they are at rest.
   (b) The length of a moving object is larger than it was at rest.
   (c) Events occur at the same coordinates for observers in all inertial reference frames.
   (d) Events occur at the same time for observers in all inertial reference frames.
   (e) The speed of light has the same value for observers in all reference frames.

9. Two helium-filled balloons are released simultaneously at points **A** and **B** on the $x$ axis in an earth-based reference frame. Which one of the following statements is true for an observer moving in the $+x$ direction?

    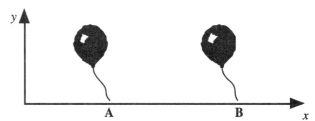

   (a) The observer always sees the balloons released simultaneously.
   (b) The observer could see either balloon released first depending on her speed and the distance between **A** and **B**.
   (c) The observer sees balloon **A** released before balloon **B**.
   (d) The observer sees balloon **B** released before balloon **A**.
   (e) The observer cannot determine whether they were released separately or simultaneously.

10. In the year 2098, an astronaut wears an antique, but accurate, "quartz" wristwatch on a journey at a speed of $2.0 \times 10^8$ m/s. According to mission control in Houston, the trip lasts 12 hours. How long was the trip as measured on the watch?
    (a) 6.7 hr          (c) 12.0 hr         (e) 21.6 hr
    (b) 8.9 hr          (d) 16.1 hr

11. In a science fiction novel, a starship takes 3 days to travel between two distant space stations according to its own clocks. Instruments on one of the space stations indicate that the trip took 4 days. How fast did the starship travel, relative to the space station?
    (a) $1.98 \times 10^8$ m/s      (c) $2.51 \times 10^8$ m/s      (e) $2.99 \times 10^8$ m/s
    (b) $2.24 \times 10^8$ m/s      (d) $2.83 \times 10^8$ m/s

12. The proper mean lifetime of a muon is $2.2 \times 10^{-6}$ s. A beam of muons is moving with speed $0.6c$ relative to an inertial observer. How far will a muon in the beam travel, on average, before it decays?
    (a) 288 m           (c) 500 m           (e) 800 m
    (b) 360 m           (d) 600 m

13. A bomb is designed to explode 2.00 s after it is armed. The bomb is launched from earth and accelerated to an unknown final speed. After reaching its final speed, however, the bomb is observed by people on earth to explode 4.25 s after it is armed. What is the final speed of the bomb just before it explodes?
    (a) $0.995c$        (c) $0.939c$        (e) $0.882c$
    (b) $0.971c$        (d) $0.904c$

14. During a baseball game, a batter hits a ball directly back to the pitcher who catches it. An observer flying over the stadium at a speed of 0.75c, measures 0.658 s as the time between the two events (hitting and catching the ball). What is the proper time interval between the two events?
   (a) 0.288 s
   (b) 0.435 s
   (c) 0.658 s
   (d) 0.715 s
   (e) 0.994 s

## Section 28.4 The Relativity of Length: Length Contraction

15. Which one of the following statements concerning the *proper length* of a meter stick is true?
   (a) The proper length is always one meter.
   (b) The proper length depends upon the speed of the observer.
   (c) The proper length depends upon the acceleration of the observer.
   (d) The proper length depends upon the reference frame in which it is measured.
   (e) The proper length is the length measured by an observer who is moving with respect to the meter stick.

16. A meter stick is observed to be only 0.9 meters long to an inertial observer. At what speed, relative to the observer, must the meter stick be moving?
   (a) $0.44 \times 10^8$ m/s
   (b) $0.57 \times 10^8$ m/s
   (c) $0.95 \times 10^8$ m/s
   (d) $1.31 \times 10^8$ m/s
   (e) $2.70 \times 10^8$ m/s

17. A UFO flies directly over a football stadium at a speed of 0.5c. If the proper length of the field is 100 yards, what field length is measured by the pilot of the UFO?
   (a) 59 yards
   (b) 75 yards
   (c) 87 yards
   (d) 113 yards
   (e) 121 yards

18. A spaceship leaves our solar system at a constant speed of 0.900c and travels to a point in the Andromeda galaxy. According to astronomers in an inertial reference frame on Earth, the distance to the galaxy is $2.081 \times 10^{22}$ m. What distance does the crew on the ship measure on its journey?
   (a) $9.07 \times 10^{21}$ m
   (b) $9.85 \times 10^{21}$ m
   (c) $1.91 \times 10^{22}$ m
   (d) $2.83 \times 10^{22}$ m
   (e) $4.77 \times 10^{22}$ m

19. Complete the following statement: To measure the proper length of an object moving relative to the surface of the earth, one must note the coordinates of points on the front and back ends
   (a) at the same time with respect to a clock at rest on the earth.
   (b) at different times with respect to a clock on the moving object.
   (c) at the same time with respect to a clock on the moving object.
   (d) at different times with respect to a clock at rest on the earth.
   (e) at the same time with respect to a clock moving at the same speed on the surface of the earth.

## Section 28.5 Relativistic Momentum

20. A proton has a mass of $1.673 \times 10^{-27}$ kg. If the proton is accelerated to a speed of 0.93c, what is the magnitude of the relativistic momentum of the proton?
   (a) $6.2 \times 10^{-17}$ kg·m/s
   (b) $1.3 \times 10^{-18}$ kg·m/s
   (c) $4.7 \times 10^{-19}$ kg·m/s
   (d) $5.9 \times 10^{-24}$ kg·m/s
   (e) $1.6 \times 10^{-27}$ kg·m/s

284   SPECIAL RELATIVITY

21. The momentum of an electron is 1.60 times larger than the value computed non-relativistically. What is the speed of the electron?
    (a) $2.94 \times 10^8$ m/s
    (b) $2.76 \times 10^8$ m/s
    (c) $2.61 \times 10^8$ m/s
    (d) $2.34 \times 10^8$ m/s
    (e) $1.83 \times 10^8$ m/s

22. An electron gun inside a computer monitor sends an electron toward the screen at a speed of $1.20 \times 10^8$ m/s. If the mass of the electron is $9.109 \times 10^{-31}$ kg, what is the magnitude of its relativistic momentum?
    (a) $9.88 \times 10^{-23}$ kg·m/s
    (b) $1.09 \times 10^{-22}$ kg·m/s
    (c) $1.20 \times 10^{-22}$ kg·m/s
    (d) $1.41 \times 10^{-22}$ kg·m/s
    (e) $3.25 \times 10^{-22}$ kg·m/s

## *Section 28.6 The Equivalence of Mass and Energy*

23. Determine the total energy of an electron traveling at $0.98c$.
    (a) 0.25 MeV
    (b) 0.51 MeV
    (c) 0.76 MeV
    (d) 1.8 MeV
    (e) 2.6 MeV

24. Determine the speed at which the kinetic energy of an electron is equal to twice its rest energy.
    (a) $0.45c$
    (b) $0.63c$
    (c) $0.87c$
    (d) $0.94c$
    (e) $0.99c$

25. A muon has rest energy 105 MeV. What is its kinetic energy when its speed is $0.95c$ ?
    (a) 37 MeV
    (b) 47 MeV
    (c) 231 MeV
    (d) 441 MeV
    (e) 741 MeV

26. A space ship at rest on a launching pad has a mass of $1.00 \times 10^5$ kg. How much will its energy have increased when the ship is moving at $0.600c$ ?
    (a) $1.12 \times 10^{21}$ J
    (b) $1.62 \times 10^{21}$ J
    (c) $2.25 \times 10^{21}$ J
    (d) $6.00 \times 10^{21}$ J
    (e) $9.00 \times 10^{21}$ J

27. The rest energies of three subatomic particles are:
    Particle **X**: 107 MeV;   Particle **Y**: 140 MeV;   Particle **Z**: 0.51 MeV.
    Which one of the following statements is necessarily true concerning these three particles?
    (a) Particle **Z**, at rest, could decay into particle **X** and give off electromagnetic radiation.
    (b) Particle **X**, at rest, could decay into particle **Y** and give off electromagnetic radiation.
    (c) Particle **X**, at rest, could decay into particles **Y** and **Z** and give off electromagnetic radiation.
    (d) Particle **Y**, at rest, could decay into particles **X** and **Z** and give off electromagnetic radiation.
    (e) Particle **Z**, at rest, could decay into particles **X** and **Y** and give off electromagnetic radiation.

28. The temperature of a 5.00-kg lead brick is increased by 225 C°. If the specific heat capacity of lead is 128 J/(kg·C°), what is the *increase* in the mass of the lead brick when it has reached its final temperature?
    (a) $4.33 \times 10^{-11}$ kg
    (b) $9.66 \times 10^{-11}$ kg
    (c) $1.60 \times 10^{-12}$ kg
    (d) $2.40 \times 10^{-12}$ kg
    (e) $4.80 \times 10^{-4}$ kg

29. How much energy would be released if 1.0 g of material were completely converted into energy?
    (a) $9 \times 10^8$ J
    (b) $9 \times 10^9$ J
    (c) $9 \times 10^{11}$ J
    (d) $9 \times 10^{13}$ J
    (e) $9 \times 10^{16}$ J

30. A particle travels at $0.60c$. Determine the ratio of its kinetic energy to its rest energy.
    (a) 0.25
    (b) 0.50
    (c) 0.60
    (d) 0.64
    (e) 0.80

31. The average power output of a nuclear power plant is 500 MW. In 1 minute, what is the change in the mass of the nuclear fuel due to the energy being taken from the reactor? *Assume 100% efficiency.*
    (a) $9.3 \times 10^{-17}$ kg
    (b) $9.3 \times 10^{-11}$ kg
    (c) $3.3 \times 10^{-13}$ kg
    (d) $3.3 \times 10^{-7}$ kg
    (e) 9.3 kg

32. At what speed is a particle traveling if its kinetic energy is three times its rest energy?
    (a) $0.879c$
    (b) $0.918c$
    (c) $0.943c$
    (d) $0.968c$
    (e) $0.989c$

33. How much energy is required to accelerate a golf ball of mass 0.046 kg initially at rest to a speed of $0.75c$?
    (a) $1.2 \times 10^{14}$ J
    (b) $2.1 \times 10^{15}$ J
    (c) $6.3 \times 10^{15}$ J
    (d) $3.6 \times 10^{16}$ J
    (e) $7.5 \times 10^{16}$ J

34. Calculate the ratio of the relativistic kinetic energy to the classical kinetic energy, $KE_{rel}/KE_{class}$, for an electron (mass = $9.109 \times 10^{-31}$ kg) moving with a constant speed of $0.75c$.
    (a) 1.8
    (b) 1.6
    (c) 1.4
    (d) 0.74
    (e) 0.56

*Questions 35 through 39 pertain to the situation described below:*

A subatomic particle **X** spontaneously decays into two particles, **A** and **B**, each of rest energy 140 MeV. The particles fly off in opposite directions, each with speed 0.827c relative to an inertial reference frame S.

35. Determine the total energy of particle **A**.
    (a) 109 MeV
    (b) 140 MeV
    (c) 200 MeV
    (d) 249 MeV
    (e) 314 MeV

36. Determine the kinetic energy of particle **B** (relative to frame *S*).
    (a) 109 MeV
    (b) 140 MeV
    (c) 206 MeV
    (d) 249 MeV
    (e) 314 MeV

37. Which one of the following statements concerning particle **X** is true?
    (a) Momentum conservation requires that it was moving in frame *S*.
    (b) Energy conservation requires that it must have been at rest in frame *S*.
    (c) Momentum conservation requires that it must have been at rest in frame *S*.
    (d) Energy conservation requires that its total energy was 280 MeV in frame *S*.
    (e) There is not enough information to determine its state of motion at the time of the decay.

38. Which expression gives the momentum of particle **A** (relative to frame *S*)?
    (a) 109 MeV/$c$
    (b) 140 MeV/$c$
    (c) 206 MeV/$c$
    (d) 249 MeV/$c$
    (e) 314 MeV/$c$

## 286 SPECIAL RELATIVITY

● 39. Use energy conservation to determine the rest energy of particle **X**.
(a) 206 MeV
(b) 249 MeV
(c) 280 MeV
(d) 392 MeV
(e) 498 MeV

## Section 28.7 The Relativistic Addition of Velocities

◐ 40. Two space ships are observed from earth to be approaching each other along a straight line. Ship **A** moves at $0.40c$ relative to the earth observer, while ship **B** moves at $0.50c$ relative to the same observer. What speed does the captain of ship **A** report for the speed of ship **B**?
(a) $0.10c$
(b) $0.75c$
(c) $0.85c$
(d) $0.95c$
(e) $0.99c$

● 41. Rocket ship **A** travels at $0.400c$ relative to an earth observer. According to the same observer, rocket ship **A** overtakes a slower moving rocket ship **B** that moves in the same direction. The captain of **B** sees **A** pass her ship at $0.114c$. Determine the speed of **B** relative to the earth observer.
(a) $0.100c$
(b) $0.214c$
(c) $0.300c$
(d) $0.625c$
(e) $0.700c$

◐ 42. An earth observer sees an alien ship pass overhead at $0.3c$. The ion gun of the alien ship shoots ions straight ahead of the ship at a speed $0.4c$ relative to the ship. What is the speed of the ions relative to the earth observer?
(a) $0.40c$
(b) $0.63c$
(c) $0.70c$
(d) $0.79c$
(e) $0.99c$

◐ 43. Two rockets, **A** and **B**, travel toward each other with speeds $0.5c$ relative to an inertial observer.

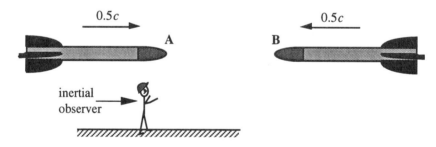

Determine the speed of rocket **A** relative to rocket **B**.
(a) $0.2c$
(b) $0.4c$
(c) $0.6c$
(d) $0.8c$
(e) $c$

◐ 44. The starship *Enterprise* approaches the Klingon home world with speed $0.6c$ relative to the planet. To announce its arrival, the *Enterprise* sends a message in a projectile that travels toward the planet with speed $0.4c$ relative to the *Enterprise*. At what speed does a Klingon at the surface of the planet see the projectile approach?
(a) $0.2c$
(b) $0.4c$
(c) $0.5c$
(d) $0.6c$
(e) $0.8c$

◐ 45. An astronomer on earth observes two galaxies moving away from each other along a line that passes through the earth. The astronomer finds that each is moving with a speed of $2.1 \times 10^8$ m/s relative to the earth. At what speed are the galaxies moving apart relative to each other?
(a) $2.8 \times 10^8$ m/s
(b) $2.4 \times 10^8$ m/s
(c) $2.1 \times 10^8$ m/s
(d) $1.8 \times 10^8$ m/s
(e) $1.4 \times 10^8$ m/s

46. The starship *Enterprise* approaches the planet Risa at a speed of 0.8c relative to the planet. On the way, it overtakes the intergalactic freighter *Astra*. The relative speed of the two ships as measured by the navigator on the *Enterprise* is 0.5c. At what speed is *Astra* approaching the planet?
    (a) 0.3c
    (b) 0.5c
    (c) 0.6c
    (d) 0.92c
    (e) 0.99c

47. Rocket **A** travels with speed 0.800c relative to the earth. Rocket **B** passes rocket **A** with speed 0.500c relative to rocket **A**.

    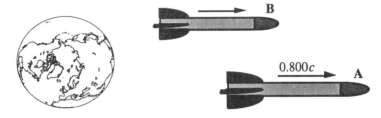

    Determine the speed of rocket **B** relative to the earth. Assume that the earth is an inertial reference frame.
    (a) 0.714c
    (b) 0.929c
    (c) 1.30c
    (d) 1.40c
    (e) 1.70c

## Additional Problems

48. Complete the following statement: The results of special relativity indicate that
    (a) Newtonian mechanics is a valid approximation at low speeds ($v \ll c$).
    (b) the laws of electromagnetism are invalid at speeds that are comparable to that of light.
    (c) Newtonian mechanics is an incorrect theory.
    (d) moving clocks run fast compared to when they are at rest.
    (e) moving objects appear to be longer than when they are at rest.

49. A cubic asteroid with proper side length 10.0 m is stationary in an inertial reference frame S. A rocket ship moves along one side of the asteroid as shown in the figure with speed 0.80c relative to frame S. An astronaut in the rocket ship measures the volume of the asteroid. What volume does the astronaut measure?

    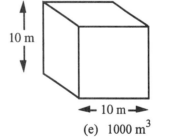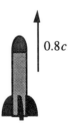

    (a) 220 m$^3$
    (b) 300 m$^3$
    (c) 360 m$^3$
    (d) 600 m$^3$
    (e) 1000 m$^3$

*Questions 50 and 51 pertain to the situation described below:*

The rest energy of a block is $E_0$. Relative to inertial observer **O**, the block is moving with speed $v$ so that $\sqrt{1-\dfrac{v^2}{c^2}} = \dfrac{1}{4}$.

288  SPECIAL RELATIVITY

● 50. Which one of the following table entries is correct?

| | Kinetic Energy of Block | Total Energy of Block |
|---|---|---|
| (a) | $4E_o$ | $5E_o$ |
| (b) | $0.75E_o$ | $0.25E_o$ |
| (c) | $4E_o$ | $4E_o$ |
| (d) | $5E_o$ | $4E_o$ |
| (e) | $3E_o$ | $4E_o$ |

◐ 51. Observer **O** finds that the block takes 12 s to go from **A** to **B**. How long would this time interval appear to be to an observer riding on the block?
  (a) 3 s  (c) 12 s  (e) 48 s
  (b) 6 s  (d) 24 s

*Questions 52 through 54 pertain to the situation described below:*

The figure shows a side view of a galaxy that is 50.0 light years in diameter. This value for the diameter is its proper length. A starship enters the galactic plane with speed 0.995c relative to the galaxy. Assume that the galaxy can be treated as an inertial reference frame. **Note:** *A light year is the distance that light travels through vacuum in one year; that is,* 1 light year = $c \times$ (1 year).

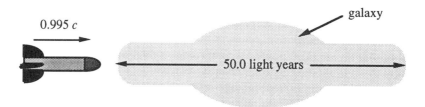

◐ 52. How long does it take the rocket to cross the galaxy according to an observer at rest in the galaxy?
  (a) 5.00 years  (c) 49.75 years  (e) 89.96 years
  (b) 12.5 years  (d) 50.25 years

◐ 53. How long does it take the rocket to cross the galaxy according to a clock on board the rocket?
  (a) 5.02 years  (c) 49.75 years  (e) 89.96 years
  (b) 12.5 years  (d) 50.25 years

◐ 54. Determine the diameter of the galaxy as perceived by a person in the rocket.
  (a) 2.49 light years  (c) 4.99 light years  (e) 36.3 light years
  (b) 3.63 light years  (d) 12.4 light years

*Questions 55 through 57 pertain to the situation described below:*

Two rockets, **A** and **B**, approach each other as shown in the figure. **A** travels to the right at 0.7c while **B** travels to the left at 0.8c. Both speed measurements are made relative to an inertial observer.

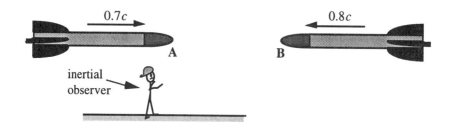

55. Determine the speed of **A** relative to **B**.
    (a) $c$
    (b) $0.96c$
    (c) $0.92c$
    (d) $0.89c$
    (e) $0.86c$

56. Both rockets have a supply of unstable mesons with a mean proper lifetime of $2.6 \times 10^{-8}$ s. Which one of the following is a correct observation for the inertial observer?
    (a) The mesons in **A** have a mean lifetime of $4.3 \times 10^{-8}$ s.
    (b) The mesons in **B** have a mean lifetime of $3.6 \times 10^{-8}$ s.
    (c) The mesons in both rockets have a mean lifetime of $2.6 \times 10^{-8}$ s.
    (d) The mean lifetime of the mesons is the same for both rockets, but less than $2.6 \times 10^{-8}$ s.
    (e) On average, the mesons in **A** will decay before the mesons in **B**.

57. Determine the ratio of the kinetic energy of rocket **B** to its rest energy.
    (a) 0.80
    (b) 0.67
    (c) 0.54
    (d) 0.33
    (e) 0.25

*Questions 58 through 62 pertain to the situation described below:*

A space ship traveling east flies directly over the head of an inertial observer who is at rest on the earth's surface. The speed of the space ship is such that $\sqrt{1 - \frac{v^2}{c^2}} = \frac{1}{2}$.

58. The observer is 5 feet tall. According to the navigator of the space ship, how tall is the observer?
    (a) 2.5 ft
    (b) 3.6 ft
    (c) 5 ft
    (d) 8 ft
    (e) 10 ft

59. The navigator of the space ship observes a neon sign on a storefront. If he measures the speed of light emitted from the sign *as he approaches* the sign, what value will he obtain?
    (a) $1.5 \times 10^8$ m/s
    (b) $1.8 \times 10^8$ m/s
    (c) $2.2 \times 10^8$ m/s
    (d) $2.8 \times 10^8$ m/s
    (e) $3.0 \times 10^8$ m/s

60. The navigator's on-board instruments indicate that the length of the space ship is 20 m. If the length of the ship is measured by the inertial earth-bound observer, what value will be obtained?
    (a) 5 m
    (b) 10 m
    (c) 20 m
    (d) 40 m
    (e) 80 m

61. The pilot fires an ion gun that propels ions from the space ship at $1.0 \times 10^8$ m/s relative to the ship. What is the speed of the ions as measured by the earth observer?
    (a) $1.0 \times 10^8$ m/s
    (b) $2.0 \times 10^8$ m/s
    (c) $2.4 \times 10^8$ m/s
    (d) $2.8 \times 10^8$ m/s
    (e) $3.0 \times 10^8$ m/s

62. An apple falls from a tree and takes 4 s to reach the ground as reported by the earth-bound observer. According to the navigator's instruments, how long did it take for the apple to fall?
    (a) 1 s
    (b) 2 s
    (c) 4 s
    (d) 6 s
    (e) 8 s

# CHAPTER 29    Particles and Waves

*Section 29.1 The Wave-Particle Duality*
*Section 29.2 Blackbody Radiation and Planck's Constant*
*Section 29.3 Photons and the Photoelectric Effect*

1. Light is usually thought of as wave-like in nature and electrons as particle-like. In which one of the following activities does light behave as a particle *or* does an electron behave as a wave?
   (a) A Young's double slit experiment is conducted using blue light.
   (b) X-rays are used to examine the crystal structure of sodium chloride.
   (c) Water is heated to its boiling point in a microwave oven.
   (d) An electron enters a parallel plate capacitor and is deflected downward.
   (e) A beam of electrons is diffracted as it passes through a narrow slit.

2. The energy of a photon depends upon which one of the following parameters?
   (a) mass
   (b) amplitude
   (c) polarization
   (d) frequency
   (e) phase relationships

3. For which one of the following problems did Max Planck make contributions which eventually led to the development of the "quantum" hypothesis?
   (a) photoelectric effect
   (b) uncertainty principle
   (c) blackbody radiation curves
   (d) the motion of the earth in the ether
   (e) the invariance of the speed of light through vacuum

4. Determine the energy of a single photon in a beam of light of wavelength 450 nm.
   (a) 2.0 eV
   (b) 2.5 eV
   (c) 2.8 eV
   (d) 4.2 eV
   (e) 4.5 eV

5. A laser emits a single, 2.0-ms pulse of light that has a frequency of $2.83 \times 10^{11}$ Hz and a total power of 75 000 W. How many photons are in the pulse?
   (a) $8.0 \times 10^{23}$
   (b) $1.6 \times 10^{24}$
   (c) $2.4 \times 10^{25}$
   (d) $3.2 \times 10^{25}$
   (e) $4.0 \times 10^{26}$

6. A laser emits a pulse of light with energy 5000 J. Determine the number of photons in the pulse if the wavelength of light is 480 nm.
   (a) $5.2 \times 10^{16}$
   (b) $2.5 \times 10^{19}$
   (c) $1.2 \times 10^{22}$
   (d) $3.1 \times 10^{22}$
   (e) $8.1 \times 10^{22}$

7. A laser emits photons of energy 2.5 eV with a power of $10^{-3}$ W. How many photons are emitted in one second?
   (a) $4.0 \times 10^{14}$
   (b) $2.5 \times 10^{15}$
   (c) $4.0 \times 10^{18}$
   (d) $1.0 \times 10^{21}$
   (e) $2.5 \times 10^{21}$

8. An X-ray generator produces photons with energy 49 600 eV or less. Which one of the following phrases most accurately describes the wavelength of these photons?
   (a) 0.025 nm or longer
   (b) 0.050 nm or longer
   (c) 0.75 nm or longer
   (d) 0.25 nm or shorter
   (e) 0.75 nm or shorter

9. A laser produces 3.0 W of light at wavelength 600 nm. How many photons per second are produced?
   (a) $7.3 \times 10^{15}$
   (b) $4.2 \times 10^{17}$
   (c) $1.0 \times 10^{17}$
   (d) $3.0 \times 10^{18}$
   (e) $9.1 \times 10^{18}$

10. Complete the following statement: The *photon* description of light is necessary to explain
    (a) polarization
    (b) photoelectric effect
    (c) diffraction of light
    (d) electron diffraction
    (e) interference of light

11. Which one of the following phrases best describes the term *work function*?
    (a) the minimum energy required to vaporize a metal
    (b) the work required to place a charged particle on a metal surface
    (c) the minimum energy required to remove electrons from the metal
    (d) the minimum energy required to remove an atom from a metal surface
    (e) the work done by electromagnetic radiation when it hits a metal surface

12. Which one of the following quantities is the same for all photons in vacuum?
    (a) speed
    (b) frequency
    (c) kinetic energy
    (d) wavelength
    (e) total energy

13. Complete the following statement: The term *photon* applies
    (a) only to X-rays.
    (b) only to visible light.
    (c) to any form of wave motion.
    (d) to any form of particle motion.
    (e) to any form of electromagnetic radiation.

14. Which type of wave motion *does not* involve photons?
    (a) gamma rays
    (b) microwaves
    (c) radio waves
    (d) infrared radiation
    (e) sound waves

15. Which one of the following statements concerning photons is false?
    (a) Photons have zero mass.
    (b) The rest energy of all photons is zero.
    (c) Photons travel at the speed of light in a vacuum.
    (d) Photons have been brought to rest by applying a strong magnetic field to them.
    (e) The energy of a photon is proportional to its frequency.

16. Photons of what minimum frequency are required to remove electrons from gold?
    **Note:** *The work function for gold is 4.8 eV.*
    (a) $7.24 \times 10^{14}$ Hz
    (b) $1.16 \times 10^{15}$ Hz
    (c) $3.84 \times 10^{17}$ Hz
    (d) $6.47 \times 10^{15}$ Hz
    (e) $4.64 \times 10^{14}$ Hz

17. Photons of energy 6 eV cause electrons to be emitted from a certain metal with a maximum kinetic energy of 2 eV. If photons of twice the wavelength are incident on this metal which one of the following statements is true?
    (a) No electrons will be emitted.
    (b) Electrons will be emitted with a maximum kinetic energy of 1 eV.
    (c) Electrons will be emitted with a maximum kinetic energy of 8 eV.
    (d) Electrons will be emitted with a maximum kinetic energy of 10 eV.
    (e) Electrons will be emitted with a maximum kinetic energy of 20 eV.

## 292 PARTICLES AND WAVES

18. When ultraviolet photons with a wavelength of $3.45 \times 10^{-7}$ m are incident on an unknown metal surface in a vacuum, electrons with a maximum kinetic energy of 1.52 eV are emitted from the surface. What is the work function of the metal?
    (a) 3.60 eV
    (b) 3.11 eV
    (c) 2.59 eV
    (d) 2.08 eV
    (e) 1.98 eV

19. Photons of energy 5.0 eV strike a metal whose work function is 3.5 eV. Determine which one of the following best describes the kinetic energy of the emitted electrons.
    (a) 1.5 eV or less
    (b) 1.5 eV or more
    (c) 2.5 eV or more
    (d) 3.5 eV or more
    (e) 3.5 eV or less

20. The work function for a particular metal is 4.0 eV. Which one of the following best describes the wavelength of electromagnetic radiation needed to eject electrons from this metal?
    (a) 310 nm or greater
    (b) 310 nm or smaller
    (c) 620 nm or greater
    (d) 620 nm or smaller
    (e) 800 nm or greater

*Questions 21 and 22 pertain to the situation described below:*

A physicist wishes to produce electrons by shining light on a metal surface. The light source emits light with a wavelength of 450 nm. The table lists the only available metals and their work functions.

| Metal | $W_0$ (eV) |
| --- | --- |
| barium | 2.5 |
| lithium | 2.3 |
| tantalum | 4.2 |
| tungsten | 4.5 |

21. Which metal(s) can be used to produce electrons by the photoelectric effect?
    (a) barium only
    (b) tungsten only
    (c) tungsten or tantalum
    (d) barium or lithium
    (e) lithium, tantalum, or tungsten

22. Which entry in the table below correctly identifies the metal that will produce the most energetic electrons and their energies?

| | Metal | Maximum electron energy observed |
| --- | --- | --- |
| (a) | lithium | 2.30 eV |
| (b) | lithium | 0.46 eV |
| (c) | tungsten | 1.75 eV |
| (d) | tungsten | 2.75 eV |
| (e) | tungsten | 4.50 eV |

### Section 29.4 The Momentum of a Photon and the Compton Effect

23. In the Compton effect, a photon of wavelength $\lambda$ and frequency $f$ hits an electron which is initially at rest. Which one of the following occurs as a result of the collision?
    (a) The photon is absorbed completely.
    (b) The photon gains energy, so the final photon has a frequency greater than $f$.
    (c) The photon gains energy, so the final photon has a wavelength greater than $\lambda$.
    (d) The photon loses energy, so the final photon has a frequency less than $f$.
    (e) The photon loses energy, so the final photon has a wavelength less than $\lambda$

24. Which one of the following is demonstrated by the Compton effect?
    (a) time dilation
    (b) length contraction
    (c) the uncertainty principle
    (d) electrons have wave properties
    (e) electromagnetic radiation has particle properties

25. Complete the following statement: The photon or "particle" theory of electromagnetic radiation is necessary to explain the
    (a) refraction of light by a prism.
    (b) diffraction of light by a grating.
    (c) reflection of light from a mirrored surface.
    (d) results of Compton scattering experiments.
    (e) interference of light in Young's double-slit experiment.

26. In the Compton scattering experiment shown in the figure, a monochromatic beam of X-rays strikes a target containing free electrons. Scattered X-rays are detected with a wavelength of $2.50 \times 10^{-12}$ m at an angle of 45° away from the original beam direction. What is the wavelength of the incident monochromatic X-rays? **Note:** *The mass of an electron is $9.11 \times 10^{-31}$ kg.*

    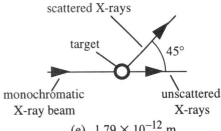

    (a) $3.21 \times 10^{-12}$ m
    (b) $2.86 \times 10^{-12}$ m
    (c) $2.50 \times 10^{-12}$ m
    (d) $2.03 \times 10^{-12}$ m
    (e) $1.79 \times 10^{-12}$ m

27. What is the speed of an electron which has the same momentum as a photon with a wavelength (in a vacuum) of 380 nm? The mass of an electron is $9.11 \times 10^{-31}$ kg.
    (a) $1.9 \times 10^3$ m/s
    (b) $2.1 \times 10^4$ m/s
    (c) $2.5 \times 10^5$ m/s
    (d) $2.7 \times 10^6$ m/s
    (e) $2.9 \times 10^8$ m/s

28. A photon of wavelength 200 nm is scattered by an electron that is initially at rest. Which one of the following statements concerning the wavelength of the scattered photon is true?
    (a) The wavelength is zero.
    (b) The wavelength is 200 nm.
    (c) The wavelength is 100 nm.
    (d) The wavelength is greater than 200 nm.
    (e) The wavelength is less than 200 nm, but greater than zero.

29. Incident X-rays in a Compton scattering experiment have a wavelength of 0.400 nm. Determine the wavelength of the scattered X-rays that are detected at an angle of 80.0°. The Compton wavelength of the electron is $2.43 \times 10^{-12}$ m.
    (a) 0.041 nm
    (b) 0.398 nm
    (c) 0.399 nm
    (d) 0.402 nm
    (e) 0.403 nm

30. A photon has a collision with a stationary electron ($h/mc = 2.43 \times 10^{-12}$ m) and loses 5.0% of its energy. The photon scattering angle is 180°. What is the wavelength of the incident photon in this scattering process?
    (a) $2.4 \times 10^{-12}$ m
    (b) $4.6 \times 10^{-11}$ m
    (c) $9.2 \times 10^{-11}$ m
    (d) $1.9 \times 10^{-10}$ m
    (e) $3.1 \times 10^{-12}$ m

## Section 29.5 The De Broglie Wavelength and the Wave Nature of Matter

31. What kinetic energy must a beam of neutrons have if their wavelength is 0.10 nm? The mass of a neutron is $1.67 \times 10^{-27}$ kg.
    (a) $6.6 \times 10^{-19}$ J
    (b) $1.3 \times 10^{-20}$ J
    (c) $2.6 \times 10^{-20}$ J
    (d) $6.3 \times 10^{-20}$ J
    (e) $7.1 \times 10^{-20}$ J

32. Approximately, what is the de Broglie wavelength of an electron that has been accelerated through a potential difference of 150 V? The mass of an electron is $9.11 \times 10^{-31}$ kg.
    (a) 0.1 nm
    (b) 1 nm
    (c) 10 nm
    (d) 100 nm
    (e) 1000 nm

33. Which experimental evidence confirms the hypothesis that matter exhibits wave properties?
    (a) photoelectric experiments
    (b) electron diffraction experiments
    (c) Young's double-slit experiments
    (d) Compton scattering experiments
    (e) Michelson-Morley experiment

34. Complete the following statement: According to the de Broglie relation, the wavelength of a "matter" wave is inversely proportional to
    (a) Planck's constant.
    (b) the mass of the particle.
    (c) the momentum of the particle.
    (d) the frequency of the wave.
    (e) the speed of the particle.

35. What is the kinetic energy that a beam of electrons must have if it is to produce a diffraction pattern of a crystal which is similar to that of a beam of 1.00 eV neutrons?
    Note: *The electron mass is $9.11 \times 10^{-31}$ kg; and the neutron mass is $1.67 \times 10^{-27}$ kg.*
    (a) 1830 eV
    (b) 42.8 eV
    (c) $3.35 \times 10^6$ eV
    (d) $5.46 \times 10^{-4}$ eV
    (e) $2.34 \times 10^{-2}$ eV

36. What is the de Broglie wavelength of an electron ($m = 9.11 \times 10^{-31}$ kg) in a $5.0 \times 10^3$-volt X-ray tube?
    (a) 0.007 nm
    (b) 0.014 nm
    (c) 0.017 nm
    (d) 0.028 nm
    (e) 0.034 nm

37. Determine the de Broglie wavelength of a neutron ($m = 1.67 \times 10^{-27}$ kg) which has a speed of 5.0 m/s.
    (a) 79 nm
    (b) 162 nm
    (c) 395 nm
    (d) 529 nm
    (e) 1975 nm

38. The de Broglie wavelength of an electron ($m = 9.11 \times 10^{-31}$ kg) is $1.2 \times 10^{-10}$ m. Determine the electron's kinetic energy.
    (a) $1.5 \times 10^{-15}$ J
    (b) $1.6 \times 10^{-16}$ J
    (c) $1.7 \times 10^{-17}$ J
    (d) $1.8 \times 10^{-18}$ J
    (e) $1.9 \times 10^{-19}$ J

39. What happens to the de Broglie wavelength of an electron if its momentum is doubled?
    (a) The wavelength decreases by a factor of 4.
    (b) The wavelength increases by a factor of 4.
    (c) The wavelength increases by a factor of 3.
    (d) The wavelength increases by a factor of 2.
    (e) The wavelength decreases by a factor of 2.

40. The Hubble Space Telescope has an orbital speed of $7.56 \times 10^3$ m/s and a mass of 11 600 kg. What is the de Broglie wavelength of the telescope?
    (a) $8.77 \times 10^7$ m
    (b) $5.81 \times 10^{-26}$ m
    (c) $6.63 \times 10^{-34}$ m
    (d) $3.78 \times 10^{-40}$ m
    (e) $7.56 \times 10^{-42}$ m

41. In a computer monitor, electrons approach the screen at $1.20 \times 10^8$ m/s. What is the de Broglie wavelength of these electrons? **Note:** *the mass of electrons is $9.109 \times 10^{-31}$ kg; and use the relativistic momentum in your calculation.*
    (a) $4.31 \times 10^{-12}$ m
    (b) $5.56 \times 10^{-12}$ m
    (c) $6.07 \times 10^{-12}$ m
    (d) $6.62 \times 10^{-12}$ m
    (e) $7.85 \times 10^{-12}$ m

*Questions 42 and 43 pertain to the situation described below:*

It is desired to obtain a diffraction pattern for electrons using a diffraction grating with lines separated by 10 nm. The mass of an electron is $9.11 \times 10^{-31}$ kg.

42. What is the approximate kinetic energy of electrons that would be diffracted by such a grating?
    (a) $1.5 \times 10^{-6}$ eV
    (b) $1.5 \times 10^{-4}$ eV
    (c) $1.5 \times 10^{-2}$ eV
    (d) $1.5 \times 10^2$ eV
    (e) $1.5 \times 10^8$ eV

43. Suppose it is desired to observe diffraction effects for a beam of electromagnetic radiation using the same grating. Roughly, what is the required energy of the individual photons in the beam?
    (a) $10^{-6}$ eV
    (b) $10^{-4}$ eV
    (c) $10^{-2}$ eV
    (d) $10^2$ eV
    (e) $10^4$ eV

## Section 29.6 The Heisenberg Uncertainty Principle

44. In an experiment to determine the speed and position of an electron (mass = $9.11 \times 10^{-31}$ kg), three researchers claim to have measured the position of the electron to within $\pm 10^{-9}$ m. They reported the following values for the speed of the electron:

    | Researcher A | $3 \times 10^6 \pm 2 \times 10^4$ m/s |
    | Researcher B | $4 \times 10^8 \pm 2 \times 10^7$ m/s |
    | Researcher C | $2 \times 10^7 \pm 5 \times 10^5$ m/s |

    Which of these measurements violates *one or more* basic laws of modern physics?
    (a) A only
    (b) B only
    (c) A and B
    (d) B and C
    (e) A, B, and C

45. The $x$ component of the velocity of an electron ($m = 9.11 \times 10^{-31}$ kg) is known to be between 100 m/s and 300 m/s. Which one of the following is a true statement concerning the uncertainty in the $x$ coordinate of the electron?
    (a) The maximum uncertainty is about $10^6$ m.
    (b) The minimum uncertainty is about $6 \times 10^{-7}$ m.
    (c) The maximum uncertainty is about $6 \times 10^{-7}$ m.
    (d) The minimum uncertainty is about $3 \times 10^{-36}$ m.
    (e) The maximum uncertainty is about $3 \times 10^{-36}$ m.

296  PARTICLES AND WAVES

46. The position of a 1-g object moving in the $x$ direction at 1 cm/s is known to within $\pm 10$ nm. In which range is the fractional uncertainty, $(\Delta p_x)/p_x$, in the $x$ component of its momentum?
   (a) $10^{-2}$ to $10^{-4}$
   (b) $10^{-12}$ to $10^{-14}$
   (c) $10^{-16}$ to $10^{-18}$
   (d) $10^{-20}$ to $10^{-22}$
   (e) less than $10^{-30}$

47. If Planck's constant were changed to 660 J·s, what would be the minimum uncertainty in the position of a 120-kg football player running at a speed of 3.5 m/s?
   (a) 0.032 m
   (b) 0.065 m
   (c) 0.13 m
   (d) 0.25 m
   (e) 0.50 m

48. The speed of a bullet with a mass of 0.050 kg is 420 m/s with an uncertainty of 0.010 %. What is the minimum uncertainty in the bullet's position if it is measured at the same time as the speed is measured?
   (a) $2.5 \times 10^{-31}$ m
   (b) $5.0 \times 10^{-32}$ m
   (c) $7.5 \times 10^{-33}$ m
   (d) $2.5 \times 10^{-34}$ m
   (e) $6.0 \times 10^{-36}$ m

49. The position of a hydrogen atom ($m = 1.7 \times 10^{-27}$ kg) is known to within $2.0 \times 10^{-6}$ m. What is the minimum uncertainty in the atom's velocity?
   (a) zero
   (b) 0.0085 m/s
   (c) 0.011 m/s
   (d) 0.016 m/s
   (e) 0.031 m/s

# CHAPTER 30    The Nature of the Atom

## Section 30.1 Rutherford Scattering and the Nuclear Atom

1. Which model of atomic structure was developed to explain the results of the experiment shown?
   (a) Bohr model
   (b) nuclear atom
   (c) billiard ball atom
   (d) plum-pudding model
   (e) quantum mechanical atom

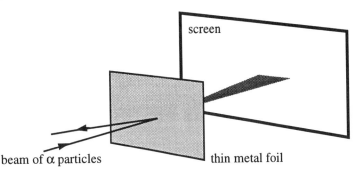

2. Which one of the following statements concerning the plum-pudding model of the atom is false?
   (a) There is no nucleus at the center of the plum-pudding model atom.
   (b) The plum-pudding model was proven to be correct in experiments by Ernest Rutherford.
   (c) The plum-pudding model was proposed by Joseph J. Thomson.
   (d) Positive charge is spread uniformly throughout the plum-pudding model atom.
   (e) Negative electrons are dispersed uniformly within the positively charged "pudding" within the plum-pudding model atom.

3. In the planetary model of the atom where electrons orbit a centralized nucleus, what is the approximate ratio of the radius of the nucleus to that of the electron orbits, $r_n/r_e$?
   (a) $10^5$
   (b) $10^{-3}$
   (c) $10^3$
   (d) $10^{-5}$
   (e) $10^{-7}$

## Section 30.2 Line Spectra
## Section 30.3 The Bohr Model of the Hydrogen Atom
## Section 30.4 De Broglie's Explanation of Bohr's Assumption about Angular Momentum

4. Each atom in the periodic table has unique set of spectral lines. Which one of the following statements is the best explanation for this observation?
   (a) Each atom has a dense central nucleus.
   (b) The electrons in atoms orbit the nucleus.
   (c) Each atom has a unique set of energy levels.
   (d) The electrons in atoms are in constant motion.
   (e) Each atom is composed of positive and negative charges.

5. Which one of the following pairs of characteristics of light is best explained by assuming that light can be described in terms of photons?
   (a) photoelectric effect and the effect observed in Young's experiment
   (b) diffraction and the formation of atomic spectra
   (c) polarization and the photoelectric effect
   (d) existence of line spectra and the photoelectric effect
   (e) polarization and the formation of line spectra

298  THE NATURE OF THE ATOM

6. Each atom in the periodic table has a unique set of spectral lines. The model of atomic structure that provides the best explanation for this observation was proposed by
   (a) Balmer.
   (b) Bohr.
   (c) Einstein.
   (d) Rutherford.
   (e) Thomson.

7. Which one of the following statements is the assumption that Niels Bohr made about the angular momentum of the electron in the hydrogen atom?
   (a) The angular momentum of the electron is zero.
   (b) The angular momentum can assume only certain discrete values.
   (c) Angular momentum is not quantized.
   (d) The angular momentum can assume any value greater than zero because it's proportional to the radius of the orbit.
   (e) The angular momentum is independent of the mass of the electron.

8. Why was it necessary to Bohr to require that electrons remain in *stationary orbits*?
   (a) An electron must travel in a circular path.
   (b) It was required by the Heisenberg uncertainty principle.
   (c) No two electrons can be in the same region in the atom.
   (d) It was required by the Pauli exclusion principle.
   (e) Classical physics predicts that the electron should spiral into the nucleus.

9. Complete the following statement: For the ground state of the hydrogen atom, the Bohr model correctly predicts
   (a) only the energy.
   (b) only the angular momentum.
   (c) only the angular momentum and the spin.
   (d) the angular momentum and the energy.
   (e) the energy, the angular momentum, and the spin.

10. Complete the following statement: An individual copper atom emits electromagnetic radiation with wavelengths that are
    (a) evenly spaced across the spectrum.
    (b) unique to that particular copper atom.
    (c) the same as other elements in the same column of the periodic table.
    (d) unique to all copper atoms.
    (e) the same as those of all elements.

11. Electrons have been removed from a beryllium atom ($Z = 4$) until only one remains. Determine the energy of the photon that can be emitted if the remaining electron is in the $n = 2$ level.
    (a) 13.6 eV
    (b) 54.4 eV
    (c) 122 eV
    (d) 163 eV
    (e) 218 eV

12. Determine the wavelength of incident electromagnetic radiation required to cause an electron transition from the $n = 6$ to the $n = 8$ level in a hydrogen atom.
    (a) $1.2 \times 10^3$ nm
    (b) $2.2 \times 10^3$ nm
    (c) $3.4 \times 10^3$ nm
    (d) $5.9 \times 10^3$ nm
    (e) $7.5 \times 10^3$ nm

13. The second ionization energy (the energy required to remove the second outermost electron) of calcium is 11.9 eV. Determine the maximum wavelength of incident radiation that can be used to remove the second electron from a calcium atom?
    (a) 16.6 nm
    (b) 52 nm
    (c) 104 nm
    (d) 208 nm
    (e) 416 nm

14. Which one of the following will result in an electron transition from the $n = 4$ level to the $n = 7$ level in a hydrogen atom?
    (a) emission of a 0.28 eV photon
    (b) emission of a 0.57 eV photon
    (c) emission of a 0.85 eV photon
    (d) absorption of a 0.28 eV photon
    (e) absorption of a 0.57 eV photon

15. Determine the maximum wavelength of incident radiation that can be used to remove the remaining electron from a singly ionized helium atom $He^+$ ($Z = 2$). Assume the electron is in its ground state.
    (a) 6.2 nm
    (b) 12.4 nm
    (c) 22.8 nm
    (d) 45.6 nm
    (e) 54.4 nm

16. Determine the energy of the photon emitted when the electron in a hydrogen atom undergoes a transition from the $n = 8$ level to the $n = 6$ level.
    (a) 0.17 eV
    (b) 0.21 eV
    (c) 0.36 eV
    (d) 0.57 eV
    (e) 13.4 eV

17. The kinetic energy of the ground state electron in hydrogen is +13.6 eV. What is its potential energy?
    (a) −13.6 eV
    (b) +27.2 eV
    (c) −27.2 eV
    (d) +56.2 eV
    (e) zero

18. An electron is in the ground state of a hydrogen atom. A photon is absorbed by the atom and the electron is excited to the $n = 2$ state. What is the energy in eV of the photon?
    (a) 13.6 eV
    (b) 10.2 eV
    (c) 3.40 eV
    (d) 1.51 eV
    (e) 0.54 eV

19. According to the Bohr model, what is the radius of a hydrogen atom when its electron is excited to the $n = 9$ state?
    (a) $5.87 \times 10^{-12}$ m
    (b) $5.29 \times 10^{-11}$ m
    (c) $4.76 \times 10^{-10}$ m
    (d) $4.28 \times 10^{-9}$ m
    (e) $1.51 \times 10^{-8}$ m

20. Determine the kinetic energy of an electron that has a de Broglie wavelength equal to twice the diameter of the hydrogen atom. Assume that the hydrogen atom is a sphere of radius $5.3 \times 10^{-11}$ m.
    (a) 13.6 eV
    (b) 27.2 eV
    (c) 33.6 eV
    (d) 48.9 eV
    (e) 65.2 eV

21. What is the shortest possible wavelength in the Lyman series for atomic hydrogen?
    (a) 91.2 nm
    (b) 104 nm
    (c) 122 nm
    (d) 364 nm
    (e) 820 nm

22. The electron in a hydrogen atom is in the $n = 3$ state. What is(are) the possible value(s) for an emitted photon?
    (a) 1.89 eV or 12.09 eV
    (b) 1.89 eV or 13.6 eV
    (c) 0.66 eV or 13.6 eV
    (d) 0.66 eV or 12.09 eV
    (e) 1.51 eV only

## Section 30.5 The Quantum Mechanical Picture of the Hydrogen Atom

23. According to the quantum mechanical picture of the atom, which one of the following is a true statement concerning the ground state electron in a hydrogen atom?
    (a) The ground state electron has zero kinetic energy.
    (b) The ground state electron has zero binding energy.
    (c) The ground state electron has zero ionization energy.
    (d) The ground state electron has zero spin angular momentum.
    (e) The ground state electron has zero orbital angular momentum.

24. A hydrogen atom is in a state for which the principle quantum number is $n = 3$. How many possible such states are there for which the magnetic quantum number is $m_\ell = 0$?
    (a) 2
    (b) 4
    (c) 6
    (d) 8
    (e) 10

25. According to the quantum mechanical picture of the atom, which one of the following statements is true concerning the magnitude of the angular momentum $L$ of an electron in the $n = 3$ level of the hydrogen atom?
    (a) $L$ is $0.318h$.
    (b) $L$ is $0.477h$.
    (c) $L$ could be $0.159h$ or $0.318h$.
    (d) $L$ could be $0.225h$ or $0.276h$.
    (e) $L$ could be $0.225h$ or $0.390h$.

26. An electron in a hydrogen atom is described by the quantum numbers: $n = 8$ and $m_\ell = 4$. What are the possible values for the orbital quantum number $\ell$?
    (a) only 0 or 4
    (b) only 4 or 7
    (c) only 5 or 8
    (d) only 4, 5, 6, or 7
    (e) only 5, 6, 7, or 8

27. Which one of the following sets of quantum numbers is *not* possible?

    |     | $n$ | $\ell$ | $m_\ell$ | $m_s$ |
    |-----|-----|--------|----------|-------|
    | (a) | 2   | 3      | −2       | +1/2  |
    | (b) | 4   | 3      | +2       | +1/2  |
    | (c) | 3   | 1      | 0        | −1/2  |
    | (d) | 6   | 2      | −1       | +1/2  |
    | (e) | 5   | 4      | −4       | −1/2  |

28. The principle quantum number for the electron in a hydrogen atom is $n = 4$. According to the quantum mechanical picture of the atom, what is the maximum possible value for the magnitude of the z-component of the angular momentum of the electron?
    (a) $3.17 \times 10^{-34}$ kg·m²/s
    (b) $4.22 \times 10^{-34}$ kg·m²/s
    (c) $1.99 \times 10^{-34}$ kg·m²/s
    (d) $1.06 \times 10^{-34}$ kg·m²/s
    (e) $2.11 \times 10^{-33}$ kg·m²/s

29. Which quantum number applies to most of the electrons in a collection of hydrogen atoms at room temperature?
    (a) $n = 1$
    (b) $n = 2$
    (c) $n = 3$
    (d) $n = 4$
    (e) $n = 5$

30. How many electron states (including spin states) are possible in a hydrogen atom if its energy is −3.4 eV?
    (a) 2
    (b) 4
    (c) 6
    (d) 8
    (e) 10

31. Determine the maximum number of electron states with principal quantum number $n = 3$?
    (a) 2
    (b) 3
    (c) 6
    (d) 9
    (e) 18

32. Which one of the following values of $m_\ell$ is not possible for $\ell = 2$?
    (a) zero
    (b) −1
    (c) +1
    (d) +2
    (e) +3

## Section 30.6 The Pauli Exclusion Principle and the Periodic Table of the Elements

33. To which model of atomic structure does the *Pauli exclusion principle* apply?
    (a) the nuclear atom
    (b) the quantum mechanical atom
    (c) the billiard ball atom
    (d) the plum-pudding model
    (e) the planetary model

34. Which one of the following factors best explains why the six electrons of a carbon atom are not all in the 1s state?
    (a) electron spin
    (b) Coulomb's law
    (c) Pauli exclusion principle
    (d) Heisenberg uncertainty principle
    (e) Rutherford model of atomic structure

35. Which one of the following statements concerning the electrons specified by the notation $3p^4$ is true?
    (a) The electrons are in the M shell.
    (b) The electrons are in the $\ell = 2$ subshell.
    (c) The electrons are necessarily in an excited state.
    (d) They have principal quantum number 4.
    (e) There are 3 electrons in the specified subshell.

36. How many electrons could be accommodated in a g subshell?
    (a) 4
    (b) 5
    (c) 8
    (d) 9
    (e) 18

37. Which one of the following subshells is not compatible with a principle quantum number of $n = 4$?
    (a) s
    (b) p
    (c) d
    (d) f
    (e) g

38. Which one of the following electronic configurations corresponds to an atomic ground state?
    (a) $1s^2 2s^1 2p^6$
    (b) $1s^1 2s^1 2p^1$
    (c) $1s^1 2s^2 3p^1$
    (d) $1s^2 2s^2 2p^1$
    (e) $1s^1 2s^2 2p^1$

39. An h subshell refers to orbital quantum number
    (a) $\ell = 1$
    (b) $\ell = 2$
    (c) $\ell = 3$
    (d) $\ell = 4$
    (e) $\ell = 5$

40. What is the total number of subshells in the $n = 3$ level?
    (a) 3
    (b) 6
    (c) 7
    (d) 9
    (e) 18

## 302  THE NATURE OF THE ATOM

41. Name the physicist credited with the following statement: No two electrons in an atom can have the same set of values for the four quantum numbers.
    (a) Werner Heisenberg
    (b) Wolfgang Pauli
    (c) Arthur Compton
    (d) Niels Bohr
    (e) Erwin Schrödinger

*Questions 42 through 44 pertain to the statement below:*

A neutral atom has the following electronic configuration: $1s^2 \, 2s^2 \, 2p^6 \, 3s^2 \, 3p^5$

42. How many electrons are in the M shell?
    (a) 2
    (b) 5
    (c) 6
    (d) 7
    (e) 8

43. How many protons are in the atomic nucleus?
    (a) 4
    (b) 7
    (c) 12
    (d) 17
    (e) 34

44. To which group of the periodic table does this element belong?
    (a) I
    (b) II
    (c) III
    (d) VI
    (e) VII

*Questions 45 through 47 pertain to the statement below:*

An electron in an atom has the following set of quantum numbers:
$$n = 3, \; \ell = 2, \; m_\ell = +1, \; m_s = +1/2.$$

45. What shell is this electron occupying?
    (a) K shell
    (b) L shell
    (c) M shell
    (d) N shell
    (e) O shell

46. In which subshell can the electron be found?
    (a) s
    (b) p
    (c) d
    (d) f
    (e) g

47. According to the quantum mechanical picture of the atom, which quantum number(s) *could* be different for electrons in this same atom that have exactly the same energy?
    (a) $n, \ell, m_\ell$ and $m_s$
    (b) only $\ell, m_\ell$ and $m_s$
    (c) only $\ell$ and $m_\ell$
    (d) only $m_\ell$ and $m_s$
    (e) $m_s$

*Questions 48 through 52 pertain to the statement below*

Consider the following list of electron configurations:
(1) $1s^2 \, 2s^2 \, 3s^2$
(2) $1s^2 \, 2s^2 \, 2p^6$
(3) $1s^2 \, 2s^2 \, 2p^6 \, 3s^1$
(4) $1s^2 \, 2s^2 \, 2p^6 \, 3s^2 \, 3p^6 \, 4s^2$
(5) $1s^2 \, 2s^2 \, 2p^6 \, 3s^2 \, 3p^6 \, 4s^2 \, 3d^6$

48. Which one of the above lists represents the electronic configuration for the ground state of the atom with $Z = 11$?
    (a) 1
    (b) 2
    (c) 3
    (d) 4
    (e) 5

49. Which electronic configuration is characteristic of noble gases?
    (a) 1
    (b) 2
    (c) 3
    (d) 4
    (e) 5

50. Which one of the above configurations represents a neutral atom that readily forms a singly charged positive ion?
    (a) 1
    (b) 2
    (c) 3
    (d) 4
    (e) 5

51. Which one of the above configurations represents an excited state of a neutral atom?
    (a) 1
    (b) 2
    (c) 3
    (d) 4
    (e) 5

52. Which one of the above configurations represents a transition element?
    (a) 1
    (b) 2
    (c) 3
    (d) 4
    (e) 5

## Section 30.7 X-rays

53. Which one of the following statements concerning the cutoff wavelength typically exhibited in X-ray spectra is true?
    (a) The cutoff wavelength depends on the target material.
    (b) The cutoff wavelength depends on the potential difference across the X-ray tube.
    (c) The cutoff wavelength is independent of the energy of the incident electrons.
    (d) The cutoff wavelength occurs because of the mutual shielding effects of K-shell electrons.
    (e) The cutoff wavelength occurs because an incident electron cannot give up all of its energy.

54. In an X-ray tube, electrons with energy 35 keV are incident on a cobalt ($Z = 27$) target. Determine the cutoff wavelength for X-ray production.
    (a) $1.4 \times 10^{-11}$ m
    (b) $1.8 \times 10^{-11}$ m
    (c) $2.8 \times 10^{-11}$ m
    (d) $3.2 \times 10^{-11}$ m
    (e) $3.6 \times 10^{-11}$ m

55. Which electron energy will produce the lowest cutoff wavelength for X-ray production from a nickel ($Z = 28$) surface?
    (a) 25 keV
    (b) 30 keV
    (c) 35 keV
    (d) 40 keV
    (e) 45 keV

56. Calculate the $K_\alpha$ X-ray wavelength for a gold atom ($Z = 79$).
    (a) $5.13 \times 10^{-10}$ m
    (b) $8.54 \times 10^{-10}$ m
    (c) $2.00 \times 10^{-11}$ m
    (d) $3.60 \times 10^{-11}$ m
    (e) $2.47 \times 10^{-13}$ m

57. Electrons in an X-ray tube are accelerated through a potential difference of 40 kV. The electrons then strike a zirconium ($Z = 40$) target. Determine the cutoff frequency for X-ray production.
    (a) $4.7 \times 10^{19}$ Hz
    (b) $9.7 \times 10^{18}$ Hz
    (c) $3.2 \times 10^{18}$ Hz
    (d) $6.7 \times 10^{17}$ Hz
    (e) $1.1 \times 10^{16}$ Hz

304   THE NATURE OF THE ATOM

*Section 30.8 The Laser*
*Section 30.9 Medical Applications of the Laser*
*Section 30.10 Holography*

58. Complete the following sentence: In the condition known as population inversion,
    (a) the amount of one type of gas atoms is larger than that of another in a mixture.
    (b) the number of energy levels that are populated is larger than that of unpopulated levels.
    (c) there are more electrons occupying lower energy levels than occupying higher energy levels.
    (d) there are more electrons occupying higher energy levels than occupying lower energy levels.
    (e) there are more photons than electrons in a given system.

59. An argon-ion laser emits a blue-green beam of light with a wavelength of 488 nm in a vacuum. What is the difference in energy in joules between the two energy states for the atomic transition which produces this light?
    (a) $4.08 \times 10^{-19}$ J
    (b) $1.05 \times 10^{-20}$ J
    (c) $6.18 \times 10^{-20}$ J
    (d) $4.76 \times 10^{-24}$ J
    (e) $5.10 \times 10^{-28}$ J

60. A pulsed laser has an average output power of 4 W. Each pulse consists of light at wavelength 500 nm and has a 25 ms duration. How many photons are emitted in a single pulse?
    (a) $1.0 \times 10^{17}$
    (b) $2.5 \times 10^{17}$
    (c) $3.7 \times 10^{17}$
    (d) $5.0 \times 10^{17}$
    (e) $7.4 \times 10^{17}$

61. An electron makes a transition from a higher energy state to a lower one without any external provocation. As a result of the transition, a photon is emitted and moves in a random direction. What is the name of this emission process?
    (a) stationary emission
    (b) stimulated emission
    (c) spectral emission
    (d) spontaneous emission
    (e) specular emission

62. Complete the following statement: In the laser-based medical procedure known as photorefractive keratectomy (PRK), nearsightedness and farsightedness can be treated using the laser to
    (a) remove small amounts of tissue from the lens and change its curvature.
    (b) remove small amounts of tissue from the cornea and change its curvature.
    (c) change the index of refraction of the aqueous humor.
    (d) alter the fluid pressure within the eye.
    (e) stimulate unused rods and cones on the retina.

63. Complete the following sentence: Holography is
    (a) the projection of an image produced by a combination of mirrors and lenses.
    (b) a photograph of the light produced by a laser.
    (c) a process for producing three dimensional images using the interference of laser light beams.
    (d) the name for an imaging process that occurs within a camera when a photograph is taken.
    (e) the production of a two dimensional image of the three dimensional object.

*Additional Problems*

64. An atom will emit photons when one of its electrons goes from
    (a) the K shell to the L shell.
    (b) the M shell to the N shell.
    (c) the K shell to the M shell.
    (d) the N shell to the L shell.
    (e) the K shell to the N shell.

65. Which one of the following statements best explains why a neon sign does not emit visible light after it is turned off?
   (a) All of the neon atoms are ionized.
   (b) Most of the neon atoms are in the ground state.
   (c) None of the neon atoms are in the $n = 2$ state.
   (d) All of the neon atoms have principle quantum number $n = 0$.
   (e) Only some of the neon atoms have returned to the $n = 1$ state.

*Questions 66 through 72 pertain to the statement and diagram below.*

The figure shows an energy level diagram for the hydrogen atom. Several transitions are shown and are labeled by letters.

Note: *The diagram is not drawn to scale.*

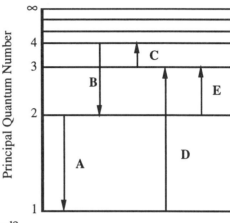

66. In which transition is a Balmer series photon absorbed?
   (a) A   (c) C   (e) E
   (b) B   (d) D

67. Which transition corresponds to the absorption of the photon with the longest wavelength?
   (a) A   (c) C   (e) E
   (b) B   (d) D

68. Determine the energy of the photon involved in transition **E**.
   (a) 1.5 eV   (c) 3.4 eV   (e) 12.1 eV
   (b) 1.9 eV   (d) 10.2 eV

69. Determine the wavelength of the radiation involved in transition **B**.
   (a) 291 nm   (c) 487 nm   (e) 1910 nm
   (b) 364 nm   (d) 652 nm

70. Which transition will occur when a hydrogen atom is irradiated with radiation of wavelength 103 nm?
   (a) A   (c) C   (e) E
   (b) B   (d) D

71. Which transition will occur when a hydrogen atom is irradiated with radiation of frequency $1.60 \times 10^{14}$ Hz?
   (a) A   (c) C   (e) E
   (b) B   (d) D

72. Which transition involves the longest wavelength line in the visible portion of the hydrogen spectrum?
   (a) A   (c) C   (e) E
   (b) B   (d) D

# CHAPTER 31
## Nuclear Physics and Radioactivity

*Section 31.1 Nuclear Structure*
*Section 31.2 The Strong Nuclear Force and the Stability of the Nucleus*

1. How many neutrons are there in the nucleus $^{205}_{82}Pb$?
   (a) 82
   (b) 123
   (c) 205
   (d) 246
   (e) 287

2. Which one of the following pairs of symbols represents two isotopes?
   (a) $^{16}_{8}O, ^{14}_{7}N$
   (b) $^{12}_{6}C, ^{14}_{6}C$
   (c) $^{16}_{8}O, ^{23}_{11}Na$
   (d) $^{14}_{7}N, ^{14}_{6}C$
   (e) $^{14}_{7}N, ^{13}_{6}C$

3. In which one of the following sets do the species have the same neutron number $N$?
   (a) $^{16}_{8}O, ^{14}_{7}N$
   (b) $^{12}_{6}C, ^{14}_{6}C$
   (c) $^{16}_{8}O, ^{23}_{11}Na$
   (d) $^{14}_{7}N, ^{14}_{6}C$
   (e) $^{14}_{7}N, ^{13}_{6}C$

4. In which one of the following sets do the species have the same number of nucleons?
   (a) $^{16}_{8}O, ^{14}_{7}N$
   (b) $^{12}_{6}C, ^{14}_{6}C$
   (c) $^{16}_{8}O, ^{23}_{11}Na$
   (d) $^{14}_{7}N, ^{14}_{6}C$
   (e) $^{14}_{7}N, ^{13}_{6}C$

5. Osmium has atomic number 76. A particular isotope of osmium has an atomic mass of 186.956 u. Which symbol correctly represents this isotope?
   (a) $^{187}_{111}Os$
   (b) $^{111}_{76}Os$
   (c) $^{76}_{187}Os$
   (d) $^{187}_{76}Os$
   (e) $^{111}_{187}Os$

6. A tellurium nucleus contains 73 neutrons. A particular isotope of tellurium has atomic mass 124.904418 u. Which symbol correctly represents this isotope?
   (a) $^{125}_{73}Te$
   (b) $^{125}_{52}Te$
   (c) $^{52}_{125}Te$
   (d) $^{73}_{125}Te$
   (e) $^{73}_{52}Te$

7. The nucleus of a certain isotope of tin contains 68 neutrons and 50 protons. Which symbol correctly represents this isotope?
   (a) $^{50}_{68}Sn$
   (b) $^{68}_{50}Sn$
   (c) $^{118}_{68}Sn$
   (d) $^{68}_{118}Sn$
   (e) $^{118}_{50}Sn$

8. A particular isotope of dysprosium has atomic number 66 and atomic mass 159.925 202 u. Identify the nucleus with a radius that is one half that of this isotope.
   (a) $^{20}_{10}Ne$
   (b) $^{16}_{8}O$
   (c) $^{50}_{25}Mn$
   (d) $^{60}_{28}Ni$
   (e) $^{180}_{74}W$

9. The nucleus of a particular isotope of beryllium contains 4 protons and 5 neutrons. Which nucleus has a radius that is approximately 3 times that of this isotope?
   (a) $^{27}_{13}$Al
   (b) $^{24}_{12}$Mg
   (c) $^{243}_{94}$Pu
   (d) $^{135}_{56}$Ba
   (e) $^{81}_{36}$Kr

10. Which of the following is *not* an assumption involved in the expression: $r = (1.2 \times 10^{-15} \text{ m})A^{1/3}$?
    (a) Nuclei are incompressible.
    (b) The nucleus is spherical in shape.
    (c) All nucleons have roughly the same mass.
    (d) Nuclear densities are proportional to the mass numbers.
    (e) The number of nucleons is proportional to the nuclear mass.

11. What is the approximate radius of a carbon nucleus that has six protons and six neutrons?
    (a) $1.2 \times 10^{-15}$ m
    (b) $2.2 \times 10^{-15}$ m
    (c) $2.7 \times 10^{-15}$ m
    (d) $2.9 \times 10^{-15}$ m
    (e) $1.2 \times 10^{-15}$ m

12. Assuming the radius of a hydrogen atom is given by the Bohr radius, $r_{Bohr} = 5.29 \times 10^{-11}$ m, what is the ratio of the nuclear density of a hydrogen atom to its atomic density?
    **Note:** *Assume for this calculation that the mass of the atom is equal to the mass of the proton.*
    (a) $1.2 \times 10^{-14}$
    (b) $4.4 \times 10^{4}$
    (c) $8.6 \times 10^{13}$
    (d) $3.9 \times 10^{17}$
    (e) $2.3 \times 10^{-5}$

13. $^{207}_{82}$Pb has a mass of $3.4368 \times 10^{-25}$ kg. What is the approximate density of this lead nucleus?
    (a) $2.3 \times 10^{17}$ kg/m$^3$
    (b) $3.5 \times 10^{18}$ kg/m$^3$
    (c) $4.8 \times 10^{19}$ kg/m$^3$
    (d) $5.2 \times 10^{20}$ kg/m$^3$
    (e) $6.1 \times 10^{21}$ kg/m$^3$

14. Which one of the following statements concerning stable nuclei is true?
    (a) Stable nuclei have nucleon numbers less than 83.
    (b) Stable nuclei generally have odd atomic numbers.
    (c) Stable nuclei have atomic numbers greater than 83.
    (d) Stable nuclei generally have an odd number of neutrons.
    (e) Stable nuclei generally have more neutrons than protons.

15. This question refers to the figure shown. Which one of the following concepts explains why heavy nuclei do not follow the $N = Z$ line (or trend) in the figure?
    (a) transmutation
    (b) Coulomb repulsion
    (c) particle-wave duality
    (d) Pauli exclusion principle
    (e) Heisenberg uncertainty principle

16. Which one of the following terms *does not* apply to nuclear forces?
    (a) strong
    (b) charge-independent
    (c) weak
    (d) short-range
    (e) long-range

## Section 31.3 The Mass Defect of the Nucleus and Nuclear Binding Energy

17. Which one of the following expressions relates the terms *binding energy* and *mass defect*?
    (a) $\Delta E_0 = (\Delta m)c^2$
    (b) $\Delta(mc) = h/\Delta\lambda$
    (c) $\Delta E = h\Delta f$
    (d) $\lambda = hc/\Delta E$
    (e) $E_0 = (1/2)mv^2$

18. The binding energy of an isotope of chlorine is 298 MeV. What is the mass defect of this chlorine nucleus in atomic mass units?
    (a) 3.13 u
    (b) 2.30 u
    (c) 0.882 u
    (d) 0.320 u
    (e) 0.034 u

19. How much energy is required to remove a neutron ($m_n$ = 1.008 665 u) from $^{15}_{7}$N that has an atomic mass of 15.000 108 u to make $^{14}_{7}$N that has an atomic mass of 14.003 074 u?
    Note: *The energy equivalent of the atomic mass unit is 931.5 MeV.*
    (a) 1.163 MeV
    (b) 6.423 MeV
    (c) 10.83 MeV
    (d) 928.7 MeV
    (e) 939.6 MeV

20. How much energy is required to split a $^{52}_{24}$Cr atom of mass 51.940 509 u into two identical $^{26}_{12}$Mg atoms? The mass of this isotope of magnesium is 25.982 59 u.
    (a) 22.98 MeV
    (b) 14.20 MeV
    (c) 0.0247 MeV
    (d) 0.6936 MeV
    (e) 0.1562 MeV

21. The proton has a mass of 1.007 28 u; and the neutron has a mass of 1.008 67. Use this information to determine the binding energy per nucleon of $^{232}_{90}$Th which has an atomic mass of 232.038 054 u.
    (a) 6.5 MeV
    (b) 7.4 MeV
    (c) 8.7 MeV
    (d) 9.8 MeV
    (e) 10.2 MeV

22. What is the binding energy per nucleon of $^{202}_{80}$Hg that has an atomic mass of 201.970 617 u?
    Note: *Use the following atomic masses in your calculation:* $^{1}_{1}$H=1.007 825 u and $^{1}_{0}$n=1.008 665 u.
    (a) 8.647 MeV
    (b) 11.47 MeV
    (c) 9.151 MeV
    (d) 7.897 MeV
    (e) 8.361 MeV

## Section 31.4 Radioactivity

23. Which one of the following processes is illustrated by the reaction: $^{238}_{90}$Th → $^{234}_{88}$Ra + $^{4}_{2}$He ?
    (a) beta decay
    (b) alpha decay
    (c) gamma decay
    (d) neutrino emission
    (e) positron emission

24. Which one of the following thicknesses of lead would be least effective in stopping β rays?
    (a) 0.04 mm
    (b) 0.25 mm
    (c) 0.30 mm
    (d) 0.40 mm
    (e) 0.50 mm

25. Which one of the following types of nuclear radiation is not affected by a magnetic field?
    (a) alpha particles
    (b) β⁻ rays
    (c) gamma rays
    (d) β⁺ rays
    (e) helium nuclei

26. Which particle(s) is(are) emitted when $^{40}_{19}$K decays into $^{40}_{20}$Ca ?
    (a) a photon
    (b) a proton
    (c) an alpha particle
    (d) a positron and a neutrino
    (e) an electron and an antineutrino

27. Consider the nuclear decay process: $^{90}_{39}$Y → $^{90}_{38}$Sr + ? What is(are) the missing product(s)?
    (a) a photon
    (b) a proton
    (c) an alpha particle
    (d) a positron and a neutrino
    (e) an electron and an alpha particle

28. Which process is used in the operation of the smoke detector discussed in *Cutnell and Johnson's* text?
    (a) alpha decay
    (b) beta decay
    (c) gamma decay
    (d) X-ray absorption
    (e) proton absorption

29. Which one of the following isotopes is produced when $^{145}_{61}$Pm decays by emitting an α particle?
    (a) $^{143}_{57}$La
    (b) $^{141}_{59}$Pr
    (c) $^{145}_{60}$Nd
    (d) $^{145}_{61}$Pm
    (e) $^{145}_{62}$Sm

30. Which entry in the table below describes the daughter nucleus when $^{31}_{14}$Si decays by β⁻ emission?

    |     | number of protons | number of neutrons |
    | --- | --- | --- |
    | (a) | 15 | 16 |
    | (b) | 15 | 31 |
    | (c) | 13 | 17 |
    | (d) | 13 | 18 |
    | (e) | 14 | 30 |

31. Which one of the following quantities is not a conserved quantity according to the laws of physics?
    (a) electric charge
    (b) nucleon number
    (c) angular momentum
    (d) linear momentum
    (e) kinetic energy

32. Complete the following sentence: In a β⁺ decay process, the emitted particle is
    (a) an electron.
    (b) a neutron.
    (c) a positron.
    (d) a proton.
    (e) a photon.

33. What is the wavelength of the 0.059-MeV γ ray photon emitted by $^{60}_{27}$Co ?
    (a) $6.7 \times 10^{-12}$ m
    (b) $1.4 \times 10^{-11}$ m
    (c) $2.1 \times 10^{-11}$ m
    (d) $2.8 \times 10^{-11}$ m
    (e) $3.4 \times 10^{-11}$ m

# 310 NUCLEAR PHYSICS AND RADIOACTIVITY

34. Radiation or(and) particles emerge(s) from a radioactive sample. These products from the sample are allowed to pass through a narrow slit and may be considered a beam. The beam is passed between two plates that carry opposite electrical charge. The experimental region contains no magnetic fields. It is observed that the beam is deflected toward the negatively charged plate.

Which one of the following statements is the best conclusion for this situation?
(a) The beam is only α rays.
(b) The beam is only β⁻ rays.
(c) The beam is only γ rays.
(d) The beam could be either α rays or β⁺ rays.
(e) The beam could be α rays, β⁺ rays, or γ rays.

## Section 31.5 The Neutrino
## Section 31.6 Radioactive Decay and Activity

35. Complete the following nuclear reaction: $^{12}_{7}N \rightarrow ^{12}_{6}C + \underline{?} + \nu$.
(a) α
(b) p
(c) β⁺
(d) β⁻
(e) γ

36. What is the SI unit for activity?
(a) Ci
(b) counts/min
(c) Hz
(d) Gy
(e) Bq

37. In a beta decay process, not all of the released energy is carried by the beta particle. Who proposed the existence of the neutrino in 1930 to account for the missing energy?
(a) Niels Bohr
(b) Erwin Schrödinger
(c) Werner Heisenberg
(d) Wolfgang Pauli
(e) Enrico Fermi

38. An isotope of krypton has a half-life of 3 minutes. A sample of this isotope produces 1000 counts per minute in a Geiger counter. Determine the number of counts per minute produced after 15 minutes.
(a) zero
(b) 15
(c) 30
(d) 60
(e) 1000

39. Which one of the following isotopes is produced when $^{214}_{83}Bi$ decays by emitting an alpha particle?
(a) $^{210}_{81}Tl$
(b) $^{212}_{81}Tl$
(c) $^{210}_{79}Au$
(d) $^{212}_{79}Au$
(e) $^{210}_{83}Bi$

40. Which one of the following particles is emitted when $^{22}_{11}Na$ decays into $^{20}_{10}Ne$?
(a) α
(b) p
(c) β⁺
(d) β⁻
(e) n

41. The half-life of a particular isotope of iodine is 8.0 days. How much of a 10.0-g sample of this isotope will remain after 30 days?
    (a) 0.37 g
    (b) 0.45 g
    (c) 0.60 g
    (d) 0.74 g
    (e) 1.25 g

42. The half-life a particular isotope of barium is 12 s. What is the activity of a $10^{-6}$ kg sample of this isotope?
    (a) $1.2 \times 10^{15}$ Bq
    (b) $1.8 \times 10^{16}$ Bq
    (c) $2.4 \times 10^{17}$ Bq
    (d) $3.6 \times 10^{18}$ Bq
    (e) $6.0 \times 10^{23}$ Bq

43. A sample contains 1000 nuclei of a radioactive isotope of barium. It is found that 60 s later, 970 nuclei in the sample have decayed. Determine the half-life of this isotope.
    (a) 10 s
    (b) 12 s
    (c) 14 s
    (d) 16 s
    (e) 18 s

44. Determine the activity of $6 \times 10^{12}$ atoms of Rn-220 that has a half-life of 56 s.
    (a) 2.0 Ci
    (b) 2.5 Ci
    (c) 3.0 Ci
    (d) 3.5 Ci
    (e) 4.0 Ci

45. The half-life of $^{200}_{79}$Au is $2.88 \times 10^3$ s. What is the mass of a sample of $^{200}_{79}$Au that has an activity of $1.42 \times 10^{12}$ Bq?
    (a) $9.80 \times 10^{-9}$ g
    (b) $1.96 \times 10^{-6}$ g
    (c) $5.89 \times 10^{-12}$ g
    (d) $2.41 \times 10^{-3}$ g
    (e) $2.78 \times 10^{-15}$ g

## Section 31.7 Radioactive Dating
## Section 31.8 Radioactive Decay Series

46. Which process is involved in determining the age of a prehistoric object?
    (a) alpha decay
    (b) beta decay
    (c) gamma decay
    (d) X-ray absorption
    (e) proton absorption

47. Which one of the following statements is true concerning the radioisotope carbon-14 that is used in carbon dating?
    (a) It is produced by living cells.
    (b) It is produced during $\beta^-$ decay.
    (c) It is produced by the decay of carbon-12.
    (d) It is produced by cells after they have died.
    (e) It is produced by cosmic rays striking the atmosphere.

48. The activity of carbon-14 in a sample of charcoal from an archaeological site is 0.04 Bq. Determine the age of the sample. The half-life of carbon-14 is 5730 years.
    (a) 10 500 yr
    (b) 12 500 yr
    (c) 14 500 yr
    (d) 16 500 yr
    (e) 18 500 yr

49. The ratio of the abundance of carbon-14 to carbon-12 in a sample of dead wood is one quarter the ratio for living wood. If the half-life of carbon-14 is 5730 years, which one of the following expressions determines how many years ago the wood died?
    (a) $2 \times 5730$
    (b) $4 \times 5730$
    (c) $0.75 \times 5730$
    (d) $0.50 \times 5730$
    (e) $0.25 \times 5730$

# NUCLEAR PHYSICS AND RADIOACTIVITY

50. The activity of a carbon-14 sample is 0.1 Ci. If this sample is burned, what is the activity of the resulting $CO_2$?
    (a) zero
    (b) 0.1 Ci
    (c) 0.2 Ci
    (d) 0.3 Ci
    (e) 2.00 Ci

51. Tritium is an isotope of hydrogen which has two neutrons in addition to its proton. Tritium undergoes $\beta^-$ decay with a half-life of 12.3 years. What percentage of an initially pure sample of tritium will remain undecayed after 35 years?
    (a) 2.9 %
    (b) 6.0 %
    (c) 7.0 %
    (d) 14 %
    (e) 19 %

## Additional Problems

52. $^{14}_{6}C$ (14.003 241) undergoes $\beta^-$ decay into $^{14}_{7}N$ (14.003 074 u). What is the maximum kinetic energy of the beta rays emitted in this decay process?
    (a) 0.156 MeV
    (b) 0.342 MeV
    (c) 3.09 MeV
    (d) 17.7 MeV
    (e) 28.0 MeV

53. In a certain $\gamma$ decay process, an excited neon atom emits a $\gamma$-ray with an energy 1.630 MeV. The neon atom in the ground state has an mass of 19.992 435 u. What is the mass of the excited neon atom?
    (a) 17.498 658 u
    (b) 18.746 422 u
    (c) 19.994 185 u
    (d) 19.999 685 u
    (e) 20.003 185 u

*Questions 54 through 56 pertain to the reaction shown below:*

Consider the following nuclear decay: $^{236}_{92}U \rightarrow {}^{232}_{90}Th + X$

54. What is $X$?
    (a) $\alpha$
    (b) p
    (c) $\beta^+$
    (d) $\beta^-$
    (e) n

55. Determine the amount of energy released in this decay. Use the following atomic masses:
    $^{236}_{92}U = 236.045\ 562$ u;  $^{232}_{90}Th = 232.038\ 054$ u;  $^{4}_{2}He = 4.002\ 603$ u;
    $^{1}_{0}n = 1.008\ 665$ u; and $^{1}_{1}p = 1.007\ 277$ u
    Conversion factors: 1 u = 931.5 MeV; 1 eV = $1.602 \times 10^{-19}$ J
    (a) $3.5 \times 10^{-8}$ J
    (b) $6.0 \times 10^{-10}$ J
    (c) $4.6 \times 10^{-12}$ J
    (d) $7.3 \times 10^{-13}$ J
    (e) $2.9 \times 10^{25}$ J

56. If the uranium nucleus is at rest before its decay, which one of the following statements is true concerning the final nuclei?
    (a) They have equal kinetic energies, but the thorium nucleus has much more momentum.
    (b) They have momenta of equal magnitudes, but the thorium nucleus has much more kinetic energy.
    (c) They have equal kinetic energies and momenta of equal magnitudes.
    (d) They have equal kinetic energies, but $X$ has much more momentum.
    (e) They have momenta of equal magnitudes, but $X$ has much more kinetic energy.

# CHAPTER 32

# Ionizing Radiation, Nuclear Energy, and Elementary Particles

## Section 32.1 Biological Effects of Ionizing Radiation

1. Complete the following statement: The term *ionizing radiation* does not apply to
   (a) alpha particles.
   (b) electrons
   (c) X-ray photons.
   (d) positrons.
   (e) radio photons.

2. What absorbed dose of protons with an RBE of 10 will cause the same damage to biological tissue as a 200 rd dose of neutrons that have an RBE of 2.6?
   (a) 3.8 rd
   (b) 20 rd
   (c) 26 rd
   (d) 52 rd
   (e) 520 rd

3. A radiologist absorbs $3.6 \times 10^{-5}$ J of radiation. Determine the absorbed dose if his mass is 70.0 kg.
   (a) $1.9 \times 10^{-7}$ Gy
   (b) $2.3 \times 10^{-7}$ Gy
   (c) $2.6 \times 10^{-7}$ Gy
   (d) $3.6 \times 10^{-7}$ Gy
   (e) $5.1 \times 10^{-7}$ Gy

4. A physicist wishes to measure the exposure of a beam of gamma rays. The beam is passed through $2.00 \times 10^{-2}$ kg of dry air at STP. The beam produces positive ions in the air which have a total charge of $3.87 \times 10^{-6}$ C. What is the exposure (in roentgens) of the beam?
   (a) $7.74 \times 10^{-8}$ R
   (b) $1.94 \times 10^{-4}$ R
   (c) $3.25 \times 10^{-2}$ R
   (d) 0.750 R
   (e) 1.25 R

5. A medical researcher wishes to compare the tissue damage produced by slow neutrons, which have a relative biological effectiveness (RBE) of 2.2, to that produced by fast neutrons with an RBE of 8.8. For the slow neutrons, the absorbed dose is 560 rd. What absorbed dose (in rd) of fast neutrons will produce the same biologically equivalent dose (in rem) as that for the slow neutrons?
   (a) 140 rd
   (b) 29 rd
   (c) 560 rd
   (d) 310 rd
   (e) 2240 rd

6. A beam of 4.5-MeV neutrons is directed at a 0.030-kg tissue sample. Each second, $1.5 \times 10^6$ neutrons strike the sample. If the relative biological effectiveness of these neutrons is 7.0, what biologically equivalent dose (in rem) is received by the sample in 65 seconds?
   (a) 0.23 rem
   (b) 0.55 rem
   (c) 1.6 rem
   (d) 19 rem
   (e) 33 rem

7. A biological tissue is irradiated with neutrons. The biologically equivalent dose of the neutrons is $5.0 \times 10^2$ rem. Determine the RBE of the neutrons if the absorbed dose is 250 rd.
   (a) 0.5
   (b) 2.0
   (c) 4.0
   (d) 5.0
   (e) 25

8. A single, whole-body dose of 450 rem is considered a lethal dose for approximately fifty percent of all individuals receiving such a dose. If a 62-kg person were exposed to such a dose of radiation that has an RBE of 0.845, how much energy has the person absorbed?
   (a) 480 J
   (b) 5.3 J
   (c) 75 J
   (d) 120 J
   (e) 330 J

314  IONIZING RADIATION, NUCLEAR ENERGY AND ELEMENTARY PARTICLES

9. What radiation absorbed dose of slow neutrons (RBE = 2.5) is equivalent to a dose of 35.0 rad of fast neutrons (RBE = 9.0)?
   (a) 9.7 rad
   (b) 130 rad
   (c) 160 rad
   (d) 260 rad
   (e) 320 rad

10. Complete the following statement: The average biologically equivalent dose of radiation from consumer products received by a resident of the United States is about
    (a) 10 mrem/yr.
    (b) 15 mrem/yr.
    (c) 20 mrem/yr.
    (d) 50 mrem/yr.
    (e) 200 mrem/yr.

11. Which source of radiation contributes most to the average biological equivalent dose received by a United States resident?
    (a) radon gas
    (b) cosmic rays
    (c) consumer products
    (d) internal radioactive nuclei
    (e) medical diagnostics

## Section 32.2  Induced Nuclear Reactions

12. Consider the following nuclear reaction: $^{4}_{2}He + ^{10}_{5}B \rightarrow ^{13}_{6}C + X$. Which one of the following correctly identifies $X$ and represents the reaction?
    (a) $^{10}_{5}B(\alpha, \beta^{+})^{13}_{6}C$
    (b) $^{13}_{6}C(p, \alpha)^{10}_{5}B$
    (c) $^{13}_{6}C(\alpha, p)^{10}_{5}B$
    (d) $^{10}_{5}B(\alpha, p)^{13}_{6}C$
    (e) $^{10}_{5}B(\alpha, n)^{13}_{6}C$

13. Complete the following nuclear reaction: $^{4}_{2}He + ^{12}_{6}C \rightarrow n + p + \underline{?}$
    (a) $^{16}_{8}O$
    (b) $^{15}_{8}O$
    (c) $^{14}_{7}N$
    (d) $^{14}_{9}F$
    (e) $^{15}_{7}N$

14. Consider the nuclear reaction: $^{122}_{52}Te(X, d)^{124}_{53}I$. The symbol "d" indicates a deuterium nucleus $^{2}_{1}H$. Which one of the following statements is true concerning X?
    (a) X could be a proton.
    (b) X could be a gamma ray.
    (c) X must contain 2 neutrons.
    (d) X must have mass number 2.
    (e) X must have atomic number 4.

15. Consider the following reaction: $^{14}_{7}N(d, \alpha)^{12}_{6}C$ where $^{14}_{7}N$ has a mass of 14.003 074 u; $^{12}_{6}C$ has a mass of 12.000 000 u; d has a mass of 2.014 102 u; and α has a mass of 4.002 603 u. How much energy is released in this reaction (**Note**: 1 u = 931.5 MeV)?
    (a) 1.2 MeV
    (b) 8.3 MeV
    (c) 13.6 MeV
    (d) 27.4 MeV
    (e) 58.3 MeV

16. Which one of the following quantities is not necessarily conserved in nuclear reactions?
    (a) electric charge
    (b) number of protons
    (c) linear momentum
    (d) angular momentum
    (e) number of protons and neutrons

17. Consider the reaction: $_1^1H + {}_{62}^{150}Sm \rightarrow {}_{61}^{147}Pm + {}_2^4He$ where the masses are $_1^1H = 1.007\,825$ u; $_2^4He = 4.002\,603$ u; $_{62}^{150}Sm = 149.917\,276$ u; $_{61}^{147}Pm = 146.915\,108$ u. How much energy is released in the reaction?
    (a) 3.14 MeV
    (b) 6.88 MeV
    (c) 12.6 MeV
    (d) 19.8 MeV
    (e) 36.2 MeV

18. The first induced nuclear reaction, $_7^{14}N(\alpha,p)_8^{17}O$, in a laboratory was studied by Rutherford in 1919. How much energy is absorbed in this reaction if the atomic masses are: $_7^{14}N = 14.003\,074$ u, $_8^{17}O = 16.999\,133$ u, $\alpha = 4.002\,603$ u, and $p = 1.007\,825$ u? **Note:** 1 u = 931.5 MeV.
    (a) 1.193 MeV
    (b) 3.338 MeV
    (c) 6.603 MeV
    (d) 15.08 MeV
    (e) 27.91 MeV

*Section 32.3 Nuclear Fission*
*Section 32.4 Nuclear Reactors*

19. What is the importance of *thermal neutrons* in nuclear processes?
    (a) Thermal neutron capture results in uranium fission.
    (b) Thermal neutrons are released in radioactive decay.
    (c) Thermal neutrons are necessary in the fusion of deuterium.
    (d) Thermal neutrons are commonly released in fusion reactions.
    (e) Thermal neutrons are sources of gamma rays.

20. Which one of the following energy values would be characteristic of a *thermal neutron*?
    (a) 0.03 eV
    (b) 0.4 eV
    (c) 3 eV
    (d) 100 eV
    (e) 0.04 MeV

21. A particular nuclear fission reaction produces 150 MeV per fission. How many fissions per second are required to generate $3.00 \times 10^8$ W of power?
    (a) $2.00 \times 10^{16}$
    (b) $5.25 \times 10^{17}$
    (c) $3.20 \times 10^{18}$
    (d) $1.25 \times 10^{19}$
    (e) $6.02 \times 10^{23}$

22. A nuclear reaction that uses one nucleus of $_{92}^{236}U$ generates 170 MeV. How much energy is released when 5.0 kg of this isotope are used?
    (a) $1.4 \times 10^{10}$ J
    (b) $6.9 \times 10^{13}$ J
    (c) $3.5 \times 10^{14}$ J
    (d) $8.9 \times 10^{14}$ J
    (e) $8.2 \times 10^{16}$

23. How many neutrons are produced in the reaction: $n + {}_{92}^{235}U \rightarrow {}_{56}^{144}Ba + {}_{36}^{89}Kr + ?\,n$ ?
    (a) 1
    (b) 2
    (c) 3
    (d) 4
    (e) 5

24. Identify X in the following reaction: $_{92}^{235}U + \gamma \rightarrow {}_{56}^{142}Ba + {}_{36}^{90}Kr + X$.
    (a) one alpha particle
    (b) two alpha particles
    (c) three protons
    (d) three neutrons
    (e) six neutrons

# 316  IONIZING RADIATION, NUCLEAR ENERGY AND ELEMENTARY PARTICLES

25. Which one of the following processes causes the explosion of a nuclear bomb?
    (a) beta decay
    (b) alpha decay
    (c) moderation
    (d) photon absorption
    (e) chain reaction

26. How many neutrons are produced in the following reaction: $n + {}^{235}_{92}U \rightarrow {}^{88}_{38}Sr + {}^{136}_{54}Xe + ?$
    (a) 11
    (b) 12
    (c) 22
    (d) 24
    (e) 48

27. What is the function of the moderator in a fission reactor?
    (a) The moderator absorbs gamma rays.
    (b) The moderator absorbs slow neutrons.
    (c) The moderator decreases the speeds of fast neutrons.
    (d) The moderator prevents heat loss from the reactor core.
    (e) The moderator prevents the reactor from reaching a critical state.

28. A nuclear reactor continuously generates 150 MW of power through the fissioning of uranium. Suppose that each fission releases 190 MeV. If one mole of uranium ($6.023 \times 10^{23}$ nuclei) has a mass of 0.235 kg, what mass of uranium has undergone fission in a 4.0 day period?
    (a) 0.33 kg
    (b) 0.67 kg
    (c) 1.3 kg
    (d) 2.6 kg
    (e) 5.2 kg

## Section 32.5  Nuclear Fusion

29. Determine the amount of energy released in the following reaction: ${}^{2}_{1}H + {}^{2}_{1}H \rightarrow {}^{4}_{2}He$. Use the following information for your calculation: ${}^{2}_{1}H$ has a mass of 2.014 102 u, ${}^{4}_{2}He$ has a mass of 4.002 603 u, and 1 u = 931.5 MeV.
    (a) 0.20 MeV
    (b) 11.9 MeV
    (c) 23.8 MeV
    (d) 257 MeV
    (e) 7480 MeV

30. Which one of the following statements is true concerning the reaction ${}^{2}_{1}H + {}^{6}_{3}Li \rightarrow {}^{4}_{2}He + {}^{4}_{2}He$ where ${}^{2}_{1}H$ has a mass of 2.014 u; ${}^{4}_{2}He$ has a mass of 4.003 u; ${}^{6}_{3}Li$ has a mass of 6.015 u; and 1 u = 931.5 MeV?
    (a) The reaction releases 14 MeV.
    (b) The reaction releases 21 MeV.
    (c) The reaction releases 36 MeV.
    (d) The reaction requires 14 MeV to occur.
    (e) The reaction requires 21 MeV to occur.

31. Determine the amount of energy released in the following reaction: ${}^{4}_{2}He + {}^{4}_{2}He + {}^{4}_{2}He \rightarrow {}^{12}_{6}C$ where ${}^{4}_{2}He$ = 4.002 603 u and ${}^{12}_{6}C$ = 12.000 000 u.
    (a) 2.27 MeV
    (b) 3.01 MeV
    (c) 3.73 MeV
    (d) 4.37 MeV
    (e) 7.27 MeV

32. Which one of the following statements is the best explanation as to why *nuclear fusion* is not at present used to generate electric power?
    (a) Fusion produces too much radiation.
    (b) Fusion requires isotopes that are scarce.
    (c) Fusion processes can result in nuclear explosions.
    (d) Fusion results in large amounts of radioactive waste.
    (e) Fusion requires very high temperatures which are difficult to contain.

33. Determine the amount of energy released in the following reaction:
$$^2_1H + ^2_1H \rightarrow ^1_1H + ^3_1H$$
where the masses are
$$^2_1H = 2.014\ 102\ u; \qquad ^1_1H = 1.007\ 825\ u; \qquad ^3_1H = 3.016\ 049\ u$$
(a) 2.02 eV  
(b) 4.03 eV  
(c) 2.02 MeV  
(d) 4.03 MeV  
(e) 8.00 MeV

34. How many kilowatt • hours of energy are released from 25 g of deuterium $^2_1H$ fuel in the fusion reaction: $^2_1H + ^2_1H \rightarrow ^4_2He + \gamma$ where the masses are $^2_1H = 2.014\ 102\ u$ and $^4_2He = 4.002\ 603\ u$.
**Notes:** Ignore the energy carried off by the gamma ray. Conversion factors: 1 kWh = $3.600 \times 10^6$ J; 1 eV = $1.602 \times 10^{-19}$ J.
(a) $1 \times 10^6$ kWh  
(b) $2 \times 10^6$ kWh  
(c) $3 \times 10^6$ kWh  
(d) $4 \times 10^6$ kWh  
(e) $5 \times 10^6$ kWh

35. One of the nuclear fusion reactions which occurs in stars is: $^{21}_{10}Ne + ^4_2He \rightarrow ^{24}_{12}Mg + ^1_0n$ where the masses are $^{21}_{10}Ne = 20.993\ 849\ u$; $^4_2He = 4.002\ 603\ u$; $^{24}_{12}Mg = 23.985\ 042\ u$; and $^1_0n = 1.008\ 665\ u$. How much energy is released in this reaction?
(a) 2.557 MeV  
(b) 4.572 MeV  
(c) 6.452 MeV  
(d) 8.493 MeV  
(e) 9.370 MeV

## Section 32.6 Elementary Particles

36. Note the forces:
    (1) weak nuclear force
    (2) strong nuclear force
    (3) gravitational force
    (4) electromagnetic force
Through which force(s) can leptons interact?
(a) only 1  
(b) only 2  
(c) only 1 and 2  
(d) only 2, 3 and 4  
(e) only 1, 3, and 4

37. In the medical diagnostic technique known as positron emission tomography (PET), a positron and an electron annihilate each other and two γ–ray photons are emitted. What is the angle between the momentum vectors of the two photons?
(a) zero  
(b) 45°  
(c) 90°  
(d) 180°  
(e) Any angle is possible.

38. Of the reactions listed below, which *will not* proceed via the strong interaction?
(a) $n + p \rightarrow e^- + \bar{\nu}_e$
(b) $p + p \rightarrow \Sigma^+ + K^+$
(c) $\pi^- + p \rightarrow \Sigma^0 + K^0$
(d) $\pi^- + p \rightarrow n + \gamma$
(e) $\bar{K}^- + p \rightarrow \bar{K}^+ + \Xi^-$

39. A sigma particle, initially at rest, decays into a lambda particle and a photon: $\Sigma^0 \rightarrow \Lambda^0 + \gamma$. Determine the kinetic energy of the lambda particle if the energy of the photon is 74 MeV. The rest energies are 1192 MeV and 1116 MeV for the sigma and lambda particles, respectively.
(a) 2 MeV  
(b) 9 MeV  
(c) 74 MeV  
(d) 248 MeV  
(e) 1118 MeV

# 318   IONIZING RADIATION, NUCLEAR ENERGY AND ELEMENTARY PARTICLES

40. In the reaction $\pi^- + p \rightarrow n + \pi^0$, the proton was initially at rest. The final kinetic energies of the neutron and the pion are 0.4 MeV and 2.9 MeV, respectively. Determine the initial kinetic energy of the $\pi^-$. The rest energies are: $\pi^- = 139.6$ MeV; $\pi^0 = 135.0$ MeV; p = 938.3 MeV; and n = 939.6 MeV.
    (a) 2.5 MeV
    (b) 3.4 MeV
    (c) 139.6 MeV
    (d) 279.2 MeV
    (e) zero

41. Which one of the following particles completes the reaction: $K^+ \rightarrow \pi^+ + \pi^+ + \underline{?}$
    (a) $\pi^+$
    (b) $\pi^-$
    (c) $\pi^0$
    (d) $K^0$
    (e) $\eta^0$

42. Which one of the following statements is true concerning the proton?
    (a) The proton cannot be further subdivided.
    (b) The proton is composed of two up quarks and a down quark.
    (c) The proton is composed of two down quarks and an up quark.
    (d) The proton is composed of a down quark and an up antiquark.
    (e) The proton is composed of an up quark and a down antiquark.

43. Which one of the following particles is not a baryon?
    (a) proton
    (b) neutron
    (c) pion
    (d) sigma particle
    (e) lambda particle

44. Which one of the following particles is not a member of the *hadron* family?
    (a) pion
    (b) neutron
    (c) muon
    (d) kaon
    (e) proton

45. Which one of the following particles is not composed of quarks?
    (a) neutron
    (b) muon
    (c) pion
    (d) kaon
    (e) proton

46. How many members are in the *photon* family?
    (a) 1
    (b) 2
    (c) 3
    (d) 4
    (e) 5

47. Which one of the following statements concerning *pions* is true?
    (a) They are stable particles.
    (b) They belong to the lepton family.
    (c) They are composed of three quarks.
    (d) They only exist in two charge states.
    (e) They interact with protons via the strong interaction.

48. What is the antiparticle of a electron?
    (a) $\pi^+$
    (b) $\nu^+$
    (c) electron (self)
    (d) photon
    (e) $\beta^+$

49. Which one of the following names *is not* one that is used to name quarks?
    (a) charm
    (b) top
    (c) strange
    (d) exotic
    (e) down

50. Which one of the following statements about the *standard model* is false?
    (a) The weak nuclear force and the electromagnetic force are manifestations of a more fundamental interaction called the electroweak interaction.
    (b) The strong nuclear force between quarks is described in terms of the concept of color.
    (c) The standard model provides an explanation for the strong nuclear and weak nuclear forces.
    (d) The gravitational force and the strong nuclear force are manifestations of a more fundamental interaction called the quark interaction.
    (e) Nucleons are composed of quarks.

## Section 32.7 Cosmology

51. Astronomers studying the light from calcium atoms located in a galaxy in the constellation Boötes find that the spectral lines are shifted toward the red end of the spectrum. The redshift indicates that the galaxy is moving away at a speed of $3.9 \times 10^6$ m/s. What is the distance (in light • years) to the galaxy?
    (a) $4.7 \times 10^9$ light • years
    (b) $2.3 \times 10^8$ light • years
    (c) $1.1 \times 10^7$ light • years
    (d) $8.4 \times 10^6$ light • years
    (e) $6.6 \times 10^4$ light • years

52. Which one of the following statements concerning the standard cosmological model is false?
    (a) Shortly after the Big Bang, all of the fundamental forces behaved as a single force.
    (b) About 0.5 million years after the Big Bang, hydrogen and helium atoms began to form.
    (c) The first distinguishable particles in existence after the Big Bang were quarks and leptons.
    (d) The Grand Unified Theory describes the universe immediately before and shortly after the Big Bang.
    (e) At $10^{-43}$ s after the Big Bang, the gravitational force was distinguishable from the other fundamental forces.

## Additional Problems

53. Nucleus **A** has $Z$ protons and $N$ neutrons. Nucleus **B** has $2Z$ protons and $2N$ neutrons. Nucleus **A** has a smaller binding energy per nucleon than **B**. Which entry in the table below is correct?

| | The fusion of two A's to form B | The fission of B to form two A's |
|---|---|---|
| (a) | process releases energy | process absorbs energy |
| (b) | process absorbs energy | process absorbs energy |
| (c) | process releases energy | process releases energy |
| (d) | process absorbs energy | process releases energy |
| (e) | process is not possible | process releases energy |

54. Consider the nuclear reaction $^{235}_{92}\text{U} \rightarrow \text{X} + ^{94}_{40}\text{Zr} + \text{n}$ and the masses: $^{235}_{92}\text{U} = 235.0439$ u; $^{94}_{40}\text{Zr} = 93.9063$ u; n = 1.008 67 u. If 208.66 MeV of energy is released in this reaction, determine the mass of X.
    (a) 38.970 u
    (b) 39.962 u
    (c) 40.962 u
    (d) 84.589 u
    (e) 139.905 u

# ANSWERS — Test Bank Questions

## Chapter 1
### Introduction and Mathematical Concepts

| | | | | | | | | | | |
|---|---|---|---|---|---|---|---|---|---|---|
| 1. e | 12. a | 23. b | 34. e | 45. d | 56. c |
| 2. e | 13. d | 24. a | 35. e | 46. b | 57. b |
| 3. b | 14. d | 25. d | 36. d | 47. b | 58. b |
| 4. a | 15. d | 26. c | 37. a | 48. e | 59. a |
| 5. d | 16. a | 27. e | 38. b | 49. e | 60. d |
| 6. c | 17. c | 28. d | 39. c | 50. a | 61. c |
| 7. c | 18. e | 29. a | 40. a | 51. c | 62. a |
| 8. e | 19. b | 30. b | 41. d | 52. d | 63. b |
| 9. c | 20. c | 31. c | 42. a | 53. e | |
| 10. b | 21. a | 32. b | 43. c | 54. a | |
| 11. e | 22. d | 33. a | 44. d | 55. e | |

## Chapter 2
### Kinematics in One Dimension

| | | | | | |
|---|---|---|---|---|---|
| 1. c | 17. d | 33. c | 49. c | 65. c | 81. b |
| 2. b | 18. a | 34. a | 50. d | 66. d | 82. d |
| 3. a | 19. c | 35. b | 51. e | 67. e | 83. b |
| 4. e | 20. c | 36. c | 52. d | 68. b | 84. b |
| 5. a | 21. e | 37. d | 53. b | 69. d | 85. a |
| 6. d | 22. b | 38. b | 54. c | 70. b | 86. a |
| 7. e | 23. a | 39. c | 55. a | 71. e | 87. c |
| 8. e | 24. d | 40. e | 56. b | 72. d | 88. e |
| 9. e | 25. c | 41. b | 57. b | 73. e | 89. b |
| 10. e | 26. b | 42. c | 58. a | 74. b | 90. c |
| 11. c | 27. b | 43. c | 59. b | 75. d | 91. a |
| 12. a | 28. d | 44. d | 60. d | 76. a | 92. c |
| 13. b | 29. e | 45. c | 61. a | 77. c | 93. b |
| 14. d | 30. d | 46. b | 62. d | 78. a | 94. d |
| 15. e | 31. a | 47. b | 63. a | 79. c | 95. d |
| 16. c | 32. e | 48. c | 64. e | 80. d | 96. e |

## Chapter 3
### Kinematics in Two Dimensions

| | | | | | |
|---|---|---|---|---|---|
| 1. e | 12. e | 23. c | 34. c | 45. d | 56. a |
| 2. a | 13. a | 24. b | 35. d | 46. c | 57. d |
| 3. a | 14. b | 25. e | 36. b | 47. a | 58. e |
| 4. c | 15. d | 26. c | 37. e | 48. d | 59. e |
| 5. b | 16. c | 27. c | 38. b | 49. a | 60. d |
| 6. a | 17. d | 28. d | 39. e | 50. c | 61. c |
| 7. b | 18. a | 29. e | 40. a | 51. e | 62. a |
| 8. c | 19. b | 30. e | 41. d | 52. a | |
| 9. b | 20. b | 31. b | 42. b | 53. c | |
| 10. e | 21. d | 32. e | 43. c | 54. b | |
| 11. c | 22. c | 33. a | 44. c | 55. c | |

# Chapter 4
## Forces and Newton's Laws of Motion

| | | | | | | | | | | | |
|---|---|---|---|---|---|---|---|---|---|---|---|
| 1. | a | 18. | c | 35. | b | 52. | b | 69. | d | 86. | c |
| 2. | e | 19. | d | 36. | b | 53. | e | 70. | e | 87. | c |
| 3. | e | 20. | b | 37. | c | 54. | e | 71. | b | 88. | d |
| 4. | a | 21. | b | 38. | a | 55. | b | 72. | a | 89. | b |
| 5. | c | 22. | d | 39. | e | 56. | e | 73. | e | 90. | e |
| 6. | a | 23. | c | 40. | d | 57. | c | 74. | d | 91. | b |
| 7. | e | 24. | a | 41. | a | 58. | d | 75. | d | 92. | c |
| 8. | d | 25. | e | 42. | d | 59. | b | 76. | b | 93. | b |
| 9. | e | 26. | c | 43. | c | 60. | d | 77. | c | 94. | a |
| 10. | b | 27. | d | 44. | b | 61. | e | 78. | c | 95. | c |
| 11. | c | 28. | a | 45. | b | 62. | a | 79. | b | 96. | e |
| 12. | c | 29. | b | 46. | c | 63. | e | 80. | b | 97. | d |
| 13. | a | 30. | d | 47. | a | 64. | a | 81. | e | | |
| 14. | e | 31. | c | 48. | a | 65. | a | 82. | b | | |
| 15. | b | 32. | d | 49. | d | 66. | c | 83. | e | | |
| 16. | e | 33. | d | 50. | d | 67. | b | 84. | a | | |
| 17. | d | 34. | a | 51. | b | 68. | d | 85. | e | | |

# Chapter 5
## Dynamics of Uniform Circular Motion

| | | | | | | | | | | | |
|---|---|---|---|---|---|---|---|---|---|---|---|
| 1. | b | 10. | e | 19. | e | 28. | c | 37. | b | 46. | c |
| 2. | d | 11. | b | 20. | a | 29. | b | 38. | e | 47. | c |
| 3. | c | 12. | a | 21. | d | 30. | a | 39. | d | 48. | a |
| 4. | e | 13. | e | 22. | e | 31. | a | 40. | a | 49. | b |
| 5. | d | 14. | b | 23. | c | 32. | d | 41. | d | 50. | b |
| 6. | c | 15. | d | 24. | d | 33. | a | 42. | d | 51. | c |
| 7. | e | 16. | a | 25. | c | 34. | e | 43. | a | 52. | d |
| 8. | a | 17. | c | 26. | e | 35. | a | 44. | a | 53. | e |
| 9. | d | 18. | b | 27. | b | 36. | c | 45. | e | | |

# Chapter 6
## Work and Energy

| | | | | | | | | | | | |
|---|---|---|---|---|---|---|---|---|---|---|---|
| 1. | d | 13. | b | 25. | c | 37. | e | 49. | d | 61. | c |
| 2. | e | 14. | d | 26. | e | 38. | d | 50. | d | 62. | b |
| 3. | a | 15. | b | 27. | b | 39. | c | 51. | c | 63. | d |
| 4. | b | 16. | a | 28. | d | 40. | a | 52. | e | 64. | b |
| 5. | e | 17. | d | 29. | d | 41. | e | 53. | c | 65. | c |
| 6. | e | 18. | a | 30. | e | 42. | e | 54. | a | 66. | e |
| 7. | a | 19. | b | 31. | b | 43. | a | 55. | e | 67. | a |
| 8. | a | 20. | e | 32. | b | 44. | d | 56. | b | 68. | e |
| 9. | c | 21. | c | 33. | c | 45. | a | 57. | d | 69. | b |
| 10. | c | 22. | b | 34. | d | 46. | c | 58. | a | 70. | d |
| 11. | d | 23. | d | 35. | a | 47. | b | 59. | e | 71. | d |
| 12. | c | 24. | a | 36. | b | 48. | c | 60. | c | | |

## Chapter 7
### Impulse and Momentum

1. e
2. a
3. e
4. e
5. c
6. c
7. d
8. b
9. e
10. c
11. a
12. a
13. b
14. b
15. d
16. d
17. e
18. d
19. c
20. b
21. c
22. c
23. b
24. a
25. a
26. e
27. e
28. d
29. d
30. b
31. a
32. a
33. c
34. d
35. b
36. d
37. a
38. d
39. b
40. e
41. e
42. e
43. d
44. d
45. b
46. c
47. e
48. b
49. a
50. d
51. a
52. d
53. c
54. b
55. b
56. e
57. a
58. a
59. e
60. c
61. e

## Chapter 8
### Rotational Kinematics

1. b
2. e
3. b
4. a
5. c
6. d
7. c
8. b
9. a
10. d
11. a
12. d
13. e
14. c
15. b
16. a
17. e
18. b
19. e
20. b
21. d
22. e
23. d
24. c
25. b
26. d
27. c
28. e
29. b
30. a
31. d
32. b
33. c
34. e
35. c
36. a
37. b
38. e
39. c
40. d
41. b
42. d
43. e
44. a
45. c
46. c
47. a
48. c
49. d
50. c

## Chapter 9
### Rotational Dynamics

1. e
2. b
3. a
4. d
5. c
6. d
7. d
8. d
9. b
10. e
11. c
12. a
13. e
14. a
15. d
16. c
17. e
18. d
19. c
20. b
21. a
22. e
23. b
24. e
25. a
26. b
27. a
28. a
29. d
30. c
31. a
32. b
33. e
34. c
35. d
36. c
37. d
38. c
39. e
40. d
41. d
42. e
43. e
44. c
45. b
46. b
47. e
48. a
49. c
50. e
51. a
52. c

## Chapter 10
### Elasticity and Sinple Harmonic Motion

1. a
2. b
3. d
4. a
5. e
6. d
7. b
8. c
9. a
10. e
11. e
12. d
13. c
14. a
15. b
16. c
17. b
18. a
19. e
20. e
21. d
22. c
23. d
24. b
25. b
26. e
27. c
28. a
29. b
30. c
31. a
32. d
33. c
34. a
35. c
36. a

## Chapter 10 (continued)

| | | | | | |
|---|---|---|---|---|---|
| 37. b | 41. e | 45. c | 49. c | 53. e | 57. b |
| 38. a | 42. a | 46. d | 50. b | 54. c | |
| 39. d | 43. e | 47. d | 51. e | 55. a | |
| 40. d | 44. d | 48. b | 52. a | 56. e | |

## Chapter 11
**Fluids**

| | | | | | |
|---|---|---|---|---|---|
| 1. d | 11. b | 21. a | 31. e | 41. d | 51. a |
| 2. a | 12. e | 22. d | 32. b | 42. a | 52. d |
| 3. d | 13. c | 23. b | 33. a | 43. d | 53. b |
| 4. c | 14. b | 24. c | 34. b | 44. b | 54. e |
| 5. c | 15. e | 25. d | 35. c | 45. e | 55. d |
| 6. e | 16. e | 26. c | 36. d | 46. a | 56. b |
| 7. b | 17. d | 27. e | 37. a | 47. c | 57. a |
| 8. d | 18. c | 28. b | 38. e | 48. b | 58. c |
| 9. e | 19. a | 29. e | 39. c | 49. c | |
| 10. a | 20. b | 30. b | 40. e | 50. a | |

## Chapter 12
**Temperature and Heat**

| | | | | | |
|---|---|---|---|---|---|
| 1. e | 11. a | 21. e | 31. d | 41. e | 51. d |
| 2. b | 12. b | 22. d | 32. a | 42. b | 52. e |
| 3. e | 13. c | 23. b | 33. a | 43. d | 53. c |
| 4. d | 14. d | 24. a | 34. b | 44. c | 54. b |
| 5. a | 15. a | 25. c | 35. b | 45. a | 55. a |
| 6. c | 16. c | 26. b | 36. c | 46. a | 56. d |
| 7. a | 17. e | 27. c | 37. c | 47. c | 57. b |
| 8. d | 18. d | 28. b | 38. e | 48. b | 58. c |
| 9. c | 19. e | 29. e | 39. a | 49. b | 59. a |
| 10. c | 20. d | 30. e | 40. d | 50. e | 60. b |

## Chapter 13
**The Transfer of Heat**

| | | | | | |
|---|---|---|---|---|---|
| 1. e | 8. a | 15. b | 22. b | 29. e | 36. c |
| 2. b | 9. e | 16. b | 23. c | 30. c | 37. e |
| 3. c | 10. c | 17. c | 24. d | 31. b | 38. a |
| 4. a | 11. d | 18. a | 25. a | 32. d | 39. d |
| 5. b | 12. a | 19. e | 26. b | 33. d | 40. e |
| 6. d | 13. e | 20. d | 27. d | 34. c | |
| 7. c | 14. a | 21. a | 28. e | 35. e | |

## Chapter 14
### The Ideal Gas Law and Kinetic Theory

| | | | | | |
|---|---|---|---|---|---|
| 1. e | 10. d | 19. c | 28. b | 37. e | 46. c |
| 2. d | 11. b | 20. a | 29. d | 38. a | 47. c |
| 3. d | 12. d | 21. c | 30. d | 39. d | 48. b |
| 4. b | 13. b | 22. a | 31. c | 40. c | 49. e |
| 5. a | 14. e | 23. e | 32. b | 41. b | 50. d |
| 6. e | 15. c | 24. b | 33. a | 42. a | |
| 7. a | 16. a | 25. a | 34. b | 43. a | |
| 8. e | 17. c | 26. e | 35. b | 44. d | |
| 9. c | 18. b | 27. d | 36. c | 45. d | |

## Chapter 15
### Thermodynamics

| | | | | | |
|---|---|---|---|---|---|
| 1. a | 13. e | 25. a | 37. a | 49. b | 61. e |
| 2. e | 14. e | 26. b | 38. d | 50. e | 62. b |
| 3. d | 15. d | 27. c | 39. c | 51. a | 63. e |
| 4. a | 16. c | 28. c | 40. b | 52. c | 64. d |
| 5. a | 17. b | 29. d | 41. a | 53. d | 65. a |
| 6. c | 18. c | 30. e | 42. e | 54. b | 66. a |
| 7. c | 19. e | 31. e | 43. b | 55. a | 67. c |
| 8. d | 20. a | 32. b | 44. a | 56. b | 68. a |
| 9. b | 21. b | 33. d | 45. a | 57. b | 69. a |
| 10. e | 22. b | 34. d | 46. a | 58. d | 70. d |
| 11. b | 23. d | 35. b | 47. c | 59. c | 71. e |
| 12. d | 24. c | 36. c | 48. c | 60. e | 72. d |

## Chapter 16
### Wave Motion

| | | | | | |
|---|---|---|---|---|---|
| 1. a | 11. c | 21. e | 31. e | 41. c | 51. c |
| 2. b | 12. d | 22. b | 32. e | 42. c | 52. a |
| 3. d | 13. d | 23. a | 33. b | 43. e | 53. b |
| 4. e | 14. c | 24. c | 34. c | 44. d | 54. d |
| 5. b | 15. a | 25. a | 35. e | 45. a | 55. d |
| 6. e | 16. c | 26. d | 36. d | 46. b | 56. d |
| 7. c | 17. e | 27. c | 37. c | 47. b | 57. a |
| 8. b | 18. b | 28. d | 38. e | 48. a | 58. e |
| 9. b | 19. a | 29. b | 39. c | 49. e | 59. a |
| 10. e | 20. d | 30. a | 40. b | 50. c | |

## Chapter 17
### The Principle of Linear Superposition and Interference Phenomena

| | | | | | |
|---|---|---|---|---|---|
| 1. d | 5. b | 9. c | 13. e | 17. b | 21. c |
| 2. a | 6. a | 10. c | 14. a | 18. a | 22. d |
| 3. b | 7. e | 11. a | 15. e | 19. d | 23. e |
| 4. c | 8. d | 12. e | 16. c | 20. b | 24. d |

**Chapter 17** *(continued)*

| | | | | | |
|---|---|---|---|---|---|
| 25. c | 29. c | 33. b | 37. d | 41. d | 45. e |
| 26. b | 30. b | 34. c | 38. d | 42. e | 46. b |
| 27. c | 31. c | 35. e | 39. a | 43. a | |
| 28. a | 32. d | 36. d | 40. b | 44. b | |

**Chapter 18**
**Electric Forces and Electric Fields**

| | | | | | |
|---|---|---|---|---|---|
| 1. c | 11. d | 21. b | 31. e | 41. d | 51. e |
| 2. c | 12. c | 22. e | 32. d | 42. b | 52. c |
| 3. e | 13. b | 23. e | 33. c | 43. a | 53. b |
| 4. d | 14. a | 24. b | 34. e | 44. b | 54. a |
| 5. c | 15. a | 25. d | 35. b | 45. d | 55. e |
| 6. d | 16. c | 26. e | 36. c | 46. d | 56. b |
| 7. b | 17. d | 27. a | 37. e | 47. c | 57. d |
| 8. b | 18. e | 28. c | 38. a | 48. e | 58. e |
| 9. a | 19. d | 29. d | 39. a | 49. a | |
| 10. c | 20. a | 30. e | 40. b | 50. b | |

**Chapter 19**
**Electric Potential Energy and the Electric Potential**

| | | | | | |
|---|---|---|---|---|---|
| 1. e | 12. e | 23. e | 34. e | 45. b | 56. c |
| 2. d | 13. a | 24. a | 35. b | 46. e | 57. c |
| 3. a | 14. a | 25. e | 36. d | 47. e | 58. d |
| 4. d | 15. b | 26. d | 37. a | 48. b | 59. b |
| 5. e | 16. e | 27. b | 38. a | 49. b | 60. b |
| 6. a | 17. c | 28. d | 39. c | 50. c | 61. c |
| 7. c | 18. d | 29. d | 40. c | 51. a | 62. e |
| 8. c | 19. b | 30. b | 41. a | 52. d | 63. d |
| 9. d | 20. c | 31. c | 42. e | 53. b | 64. a |
| 10. b | 21. a | 32. e | 43. c | 54. d | 65. b |
| 11. a | 22. b | 33. d | 44. d | 55. a | 66. c |

**Chapter 20**
**Electric Circuits**

| | | | | | |
|---|---|---|---|---|---|
| 1. d | 12. e | 23. a | 34. c | 45. b | 56. c |
| 2. a | 13. d | 24. b | 35. b | 46. c | 57. b |
| 3. e | 14. a | 25. d | 36. c | 47. a | 58. a |
| 4. d | 15. b | 26. b | 37. b | 48. b | 59. c |
| 5. d | 16. c | 27. c | 38. d | 49. c | 60. a |
| 6. c | 17. c | 28. a | 39. a | 50. c | 61. c |
| 7. e | 18. e | 29. c | 40. b | 51. a | 62. b |
| 8. e | 19. b | 30. b | 41. c | 52. e | 63. e |
| 9. c | 20. b | 31. e | 42. b | 53. d | 64. d |
| 10. c | 21. e | 32. a | 43. e | 54. e | 65. c |
| 11. d | 22. a | 33. a | 44. d | 55. a | 66. c |

**Chapter 20** *(continued)*

| | | | | | | | | | |
|---|---|---|---|---|---|---|---|---|---|
| 67. | b | 72. | a | 77. | c | 82. | b | 87. | e |
| 68. | d | 73. | d | 78. | c | 83. | b | 88. | d |
| 69. | c | 74. | b | 79. | a | 84. | e | 89. | b |
| 70. | a | 75. | e | 80. | e | 85. | d | 90. | d |
| 71. | d | 76. | e | 81. | e | 86. | a | 91. | d |

92. a

**Chapter 21**
**Electric Circuits**

| | | | | | | | | | | | |
|---|---|---|---|---|---|---|---|---|---|---|---|
| 1. | e | 12. | c | 23. | e | 34. | c | 45. | e | 56. | a |
| 2. | d | 13. | c | 24. | e | 35. | c | 46. | c | 57. | d |
| 3. | e | 14. | b | 25. | d | 36. | a | 47. | e | 58. | d |
| 4. | a | 15. | d | 26. | c | 37. | a | 48. | b | 59. | c |
| 5. | b | 16. | a | 27. | b | 38. | c | 49. | d | 60. | d |
| 6. | a | 17. | b | 28. | a | 39. | a | 50. | c | 61. | b |
| 7. | d | 18. | b | 29. | d | 40. | e | 51. | b | | |
| 8. | c | 19. | c | 30. | e | 41. | d | 52. | e | | |
| 9. | a | 20. | b | 31. | c | 42. | a | 53. | e | | |
| 10. | e | 21. | c | 32. | b | 43. | d | 54. | b | | |
| 11. | a | 22. | d | 33. | b | 44. | e | 55. | d | | |

**Chapter 22**
**Electromagnetic Induction**

| | | | | | | | | | | | |
|---|---|---|---|---|---|---|---|---|---|---|---|
| 1. | a | 12. | d | 23. | d | 34. | b | 45. | d | 56. | b |
| 2. | d | 13. | c | 24. | d | 35. | c | 46. | e | 57. | d |
| 3. | e | 14. | d | 25. | c | 36. | e | 47. | a | 58. | c |
| 4. | e | 15. | c | 26. | e | 37. | b | 48. | c | 59. | d |
| 5. | e | 16. | e | 27. | d | 38. | b | 49. | b | 60. | c |
| 6. | b | 17. | b | 28. | e | 39. | e | 50. | e | 61. | e |
| 7. | a | 18. | d | 29. | a | 40. | c | 51. | a | 62. | b |
| 8. | b | 19. | a | 30. | c | 41. | a | 52. | e | 63. | a |
| 9. | e | 20. | b | 31. | c | 42. | e | 53. | e | 64. | d |
| 10. | a | 21. | d | 32. | a | 43. | d | 54. | d | 65. | b |
| 11. | a | 22. | b | 33. | e | 44. | b | 55. | a | 66. | c |

**Chapter 23**
**Alternating Current Circuits**

| | | | | | | | | | | | |
|---|---|---|---|---|---|---|---|---|---|---|---|
| 1. | e | 9. | d | 17. | e | 25. | c | 33. | a | 41. | a |
| 2. | c | 10. | c | 18. | e | 26. | e | 34. | e | 42. | a |
| 3. | a | 11. | b | 19. | a | 27. | d | 35. | c | 43. | e |
| 4. | c | 12. | b | 20. | c | 28. | a | 36. | a | 44. | d |
| 5. | d | 13. | d | 21. | b | 29. | e | 37. | c | 45. | d |
| 6. | e | 14. | d | 22. | c | 30. | d | 38. | d | 46. | b |
| 7. | a | 15. | b | 23. | e | 31. | d | 39. | b | 47. | b |
| 8. | a | 16. | c | 24. | b | 32. | b | 40. | b | 48. | c |

# Chapter 23 *(continued)*

| | | | | | |
|---|---|---|---|---|---|
| 49. b | 55. d | 61. e | 67. c | 73. d | 79. a |
| 50. b | 56. c | 62. b | 68. d | 74. a | |
| 51. a | 57. d | 63. e | 69. a | 75. b | |
| 52. b | 58. c | 64. a | 70. d | 76. c | |
| 53. e | 59. a | 65. e | 71. a | 77. e | |
| 54. b | 60. c | 66. c | 72. b | 78. d | |

# Chapter 24
# Electromagnetic Waves

| | | | | | |
|---|---|---|---|---|---|
| 1. a | 11. d | 21. b | 31. b | 41. e | 51. b |
| 2. a | 12. e | 22. b | 32. a | 42. a | 52. c |
| 3. c | 13. a | 23. a | 33. c | 43. d | 53. d |
| 4. d | 14. b | 24. d | 34. d | 44. b | 54. e |
| 5. b | 15. c | 25. c | 35. e | 45. c | 55. a |
| 6. b | 16. c | 26. a | 36. d | 46. b | 56. b |
| 7. e | 17. c | 27. c | 37. a | 47. e | 57. c |
| 8. d | 18. d | 28. b | 38. e | 48. d | 58. e |
| 9. c | 19. c | 29. b | 39. a | 49. a | 59. e |
| 10. e | 20. e | 30. a | 40. d | 50. c | |

# Chapter 25
# The Reflection of Light: Mirrors

| | | | | | |
|---|---|---|---|---|---|
| 1. d | 8. a | 15. e | 22. b | 29. b | 36. c |
| 2. c | 9. b | 16. b | 23. a | 30. d | 37. e |
| 3. a | 10. e | 17. e | 24. a | 31. b | 38. c |
| 4. d | 11. a | 18. a | 25. b | 32. a | 39. e |
| 5. c | 12. e | 19. e | 26. e | 33. b | |
| 6. c | 13. d | 20. a | 27. d | 34. d | |
| 7. c | 14. b | 21. c | 28. c | 35. e | |

# Chapter 26
# The Refraction of Light: Lenses and Optical Instruments

| | | | | | |
|---|---|---|---|---|---|
| 1. c | 13. a | 25. c | 37. a | 49. b | 61. b |
| 2. e | 14. b | 26. b | 38. b | 50. e | 62. e |
| 3. d | 15. d | 27. c | 39. a | 51. d | 63. a |
| 4. c | 16. a | 28. b | 40. b | 52. c | 64. c |
| 5. a | 17. c | 29. a | 41. d | 53. a | 65. c |
| 6. d | 18. d | 30. e | 42. e | 54. b | 66. a |
| 7. b | 19. e | 31. d | 43. d | 55. d | 67. c |
| 8. e | 20. a | 32. b | 44. a | 56. c | 68. b |
| 9. b | 21. b | 33. b | 45. d | 57. e | 69. b |
| 10. a | 22. c | 34. c | 46. c | 58. d | 70. e |
| 11. d | 23. d | 35. e | 47. e | 59. d | 71. e |
| 12. c | 24. e | 36. d | 48. c | 60. e | 72. b |

**Chapter 26** *(continued)*

| | | | | | | | | | | |
|---|---|---|---|---|---|---|---|---|---|---|
| 73. | a | 77. | d | 81. | c | 85. | d | 89. | b | 93. d |
| 74. | b | 78. | c | 82. | c | 86. | a | 90. | d | |
| 75. | a | 79. | a | 83. | a | 87. | b | 91. | e | |
| 76. | e | 80. | e | 84. | a | 88. | b | 92. | d | |

## Chapter 27
### Interference and the Wave Nature of Light

| | | | | | | | | | | | |
|---|---|---|---|---|---|---|---|---|---|---|---|
| 1. | a | 10. | d | 19. | e | 28. | a | 37. | a | 46. | c |
| 2. | b | 11. | d | 20. | c | 29. | c | 38. | b | 47. | b |
| 3. | d | 12. | d | 21. | b | 30. | e | 39. | b | 48. | c |
| 4. | c | 13. | c | 22. | e | 31. | d | 40. | a | 49. | e |
| 5. | c | 14. | b | 23. | a | 32. | c | 41. | e | 50. | e |
| 6. | b | 15. | e | 24. | a | 33. | b | 42. | d | 51. | e |
| 7. | e | 16. | a | 25. | d | 34. | b | 43. | a | 52. | a |
| 8. | a | 17. | d | 26. | d | 35. | d | 44. | b | 53. | c |
| 9. | e | 18. | e | 27. | c | 36. | b | 45. | c | 54. | d |

## Chapter 28
### Special Relativity

| | | | | | | | | | | | |
|---|---|---|---|---|---|---|---|---|---|---|---|
| 1. | e | 12. | c | 23. | e | 34. | a | 45. | a | 56. | e |
| 2. | a | 13. | e | 24. | d | 35. | d | 46. | b | 57. | b |
| 3. | d | 14. | b | 25. | c | 36. | a | 47. | b | 58. | c |
| 4. | e | 15. | a | 26. | c | 37. | c | 48. | a | 59. | e |
| 5. | e | 16. | d | 27. | d | 38. | c | 49. | d | 60. | b |
| 6. | b | 17. | c | 28. | c | 39. | e | 50. | e | 61. | d |
| 7. | c | 18. | a | 29. | d | 40. | b | 51. | a | 62. | e |
| 8. | a | 19. | c | 30. | a | 41. | c | 52. | d | | |
| 9. | b | 20. | b | 31. | d | 42. | b | 53. | a | | |
| 10. | b | 21. | d | 32. | d | 43. | d | 54. | c | | |
| 11. | a | 22. | c | 33. | b | 44. | e | 55. | b | | |

## Chapter 29
### Particles and Waves

| | | | | | | | | | | | |
|---|---|---|---|---|---|---|---|---|---|---|---|
| 1. | e | 10. | b | 19. | a | 28. | d | 37. | a | 46. | d |
| 2. | d | 11. | c | 20. | b | 29. | d | 38. | c | 47. | d |
| 3. | c | 12. | a | 21. | d | 30. | c | 39. | e | 48. | b |
| 4. | c | 13. | e | 22. | b | 31. | b | 40. | e | 49. | e |
| 5. | a | 14. | e | 23. | d | 32. | a | 41. | b | | |
| 6. | c | 15. | d | 24. | e | 33. | b | 42. | c | | |
| 7. | b | 16. | b | 25. | d | 34. | c | 43. | d | | |
| 8. | a | 17. | a | 26. | e | 35. | a | 44. | c | | |
| 9. | e | 18. | d | 27. | a | 36. | c | 45. | b | | |

# Chapter 30
## The Nature of the Atom

| | | | | | | | | | | | |
|---|---|---|---|---|---|---|---|---|---|---|---|
| 1. | b | 13. | c | 25. | e | 37. | a | 49. | b | 61. | d |
| 2. | b | 14. | e | 26. | d | 38. | d | 50. | c | 62. | b |
| 3. | d | 15. | c | 27. | a | 39. | e | 51. | a | 63. | c |
| 4. | c | 16. | a | 28. | a | 40. | a | 52. | e | 64. | d |
| 5. | d | 17. | c | 29. | a | 41. | b | 53. | b | 65. | b |
| 6. | b | 18. | b | 30. | d | 42. | d | 54. | e | 66. | e |
| 7. | b | 19. | d | 31. | e | 43. | d | 55. | e | 67. | c |
| 8. | e | 20. | c | 32. | e | 44. | e | 56. | c | 68. | b |
| 9. | a | 21. | a | 33. | b | 45. | c | 57. | b | 69. | c |
| 10. | d | 22. | a | 34. | c | 46. | c | 58. | d | 70. | d |
| 11. | d | 23. | e | 35. | a | 47. | b | 59. | a | 71. | c |
| 12. | e | 24. | c | 36. | e | 48. | c | 60. | b | 72. | e |

# Chapter 31
## Nuclear Physics and Radioactivity

| | | | | | | | | | | | |
|---|---|---|---|---|---|---|---|---|---|---|---|
| 1. | b | 11. | c | 21. | b | 31. | e | 41. | d | 51. | d |
| 2. | b | 12. | c | 22. | d | 32. | c | 42. | c | 52. | a |
| 3. | e | 13. | a | 23. | b | 33. | c | 43. | b | 53. | c |
| 4. | d | 14. | e | 24. | a | 34. | d | 44. | a | 54. | a |
| 5. | d | 15. | b | 25. | c | 35. | c | 45. | b | 55. | d |
| 6. | b | 16. | e | 26. | e | 36. | e | 46. | b | 56. | e |
| 7. | e | 17. | a | 27. | d | 37. | d | 47. | e | | |
| 8. | a | 18. | d | 28. | a | 38. | c | 48. | c | | |
| 9. | c | 19. | c | 29. | b | 39. | a | 49. | a | | |
| 10. | d | 20. | a | 30. | a | 40. | c | 50. | b | | |

# Chapter 32
## Ionizing Radiation, Nuclear Energy, and Elementary Particles

| | | | | | | | | | | | |
|---|---|---|---|---|---|---|---|---|---|---|---|
| 1. | e | 10. | a | 19. | a | 28. | b | 37. | d | 46. | a |
| 2. | d | 11. | a | 20. | a | 29. | c | 38. | a | 47. | e |
| 3. | e | 12. | d | 21. | d | 30. | b | 39. | a | 48. | e |
| 4. | d | 13. | c | 22. | c | 31. | e | 40. | e | 49. | d |
| 5. | a | 14. | c | 23. | c | 32. | e | 41. | b | 50. | d |
| 6. | c | 15. | c | 24. | d | 33. | d | 42. | b | 51. | b |
| 7. | b | 16. | b | 25. | e | 34. | d | 43. | c | 52. | d |
| 8. | e | 17. | b | 26. | b | 35. | a | 44. | c | 53. | a |
| 9. | b | 18. | a | 27. | c | 36. | e | 45. | b | 54. | e |

# Homework Choices
## CHAPTER 1 — Introduction and Mathematical Concepts

### Section 1.3 The Role of Units in Problem Solving

2. Chapter 1, Problem 2
   - (A) 138 000 m
   - (B) 42 200 m
   - (C) 2760 m
   - (D) 1320 m
   - (E) 1160 m

4. Chapter 1, Problem 4
   - (A) 161 km/h
   - (B) 275 km/h
   - (C) 420 km/h
   - (D) 840 km/h
   - (E) 2760 km/h

6. Chapter 1, Problem 6
   - (A) 1.4 magnums
   - (B) 1.7 magnums
   - (C) 2.0 magnums
   - (D) 2.5 magnums
   - (E) 2.9 magnums

8. Chapter 1, Problem 8
   - (A) 1
   - (B) 2
   - (C) 3
   - (D) 4
   - (E) 5

10. Chapter 1, Problem 10
    - (A) $1.78 \times 10^{-2}$ P
    - (B) $1.78 \times 10^{-1}$ P
    - (C) $1.78 \times 10^{-4}$ P
    - (D) $1.78 \times 10^{-3}$ P
    - (E) $1.78 \times 10^{1}$ P

### Section 1.4 Trigonometry

12. Chapter 1, Problem 12
    - (A) 6.84°
    - (B) 13.6°
    - (C) 20.9°
    - (D) 72.4°
    - (E) 83.1°

14. Chapter 1, Problem 14
    - (A) 1.2 km
    - (B) 1.9 km
    - (C) 3.8 km
    - (D) 4.7 km
    - (E) 5.5 km

16. Chapter 1, Problem 16
    - (A) 13.5 m
    - (B) 1.21 m
    - (C) 11.6 m
    - (D) 30.1 m
    - (E) 0.364 m

18. Chapter 1, Problem 18
    - (A) 0.158 m
    - (B) 0.397 m
    - (C) 0.487 m
    - (D) 0.237 m
    - (E) 0.846 m

20. Chapter 1, Problem 20
    *The solution is a proof.*

## Section 1.6 Vector Addition and Subtraction

22a. Chapter 1, Problem 22a
- (A) 72 m
- (B) 58 m
- (C) 45 m
- (D) 64 m
- (E) 90 m

22b. Chapter 1, Problem 22b
- (A) 53° south of east
- (B) 53° north of east
- (C) 45° south of east
- (D) 37° north of east
- (E) 37° south of east

24. Chapter 1, Problem 24
- (A) 18 cm
- (B) 49 cm
- (C) 25 cm
- (D) 12 cm
- (E) 15 cm

26a. Chapter 1, Problem 26a
- (A) 2.63 km
- (B) 12.1 km
- (C) 8.76 km
- (D) 4.25 km
- (E) 5.75 km

26b. Chapter 1, Problem 26b
- (A) 57.7° west of south
- (B) 57.7° south of west
- (C) 50.8° west of south
- (D) 50.8° south of west
- (E) 9.2° west of south

28. Chapter 1, Problem 28
- (A) 19 m, 71° angle with respect to the side of the court
- (B) 20 m, 15° angle with respect to the side of the court
- (C) 21 m, 45° angle with respect to the side of the court
- (D) 22 m, 67° angle with respect to the side of the court
- (E) 23 m, 19° angle with respect to the side of the court

30a. Chapter 1, Problem 30a
- (A) 2201 m
- (B) 2628 m
- (C) 3088 m
- (D) 4746 m
- (E) 6176 m

30b. Chapter 1, Problem 30b
- (A) 66.2°, south of east
- (B) 66.2°, north of east
- (C) due east
- (D) 21.9°, north of east
- (E) 21.9°, south of east

## Section 1.7 The Components of a Vector

32a. Chapter 1, Problem 32a
- (A) 104 m
- (B) 143 m
- (C) 177 m
- (D) 63.7 m
- (E) 129 m

32b. Chapter 1, Problem 32b
- (A) 104 m
- (B) 143 m
- (C) 177 m
- (D) 63.7 m
- (E) 129 m

# Homework: Introduction and Mathematical Concepts

34. Chapter 1, Problem 34
    - (A) 2.05 m/s
    - (B) 2.87 m/s
    - (C) 3.50 m/s
    - (D) 4.10 m/s
    - (E) 6.10 m/s

36a. Chapter 1, Problem 36a
    - (A) $1.6 \times 10^2$ N
    - (B) $1.8 \times 10^2$ N
    - (C) $2.0 \times 10^2$ N
    - (D) $2.4 \times 10^2$ N
    - (E) $2.8 \times 10^2$ N

36b. Chapter 1, Problem 36b
    - (A) 33°
    - (B) 37°
    - (C) 41°
    - (D) 49°
    - (E) 53°

38. Chapter 1, Problem 38
    - (A) 75 m/s
    - (B) 100 m/s
    - (C) 125 m/s
    - (D) 150 m/s
    - (E) 180 m/s

40a. Chapter 1, Problem 40a
    - (A) $A_x$ = 650 units, $A_y$ = 380 units
    - (B) $A_x$ = 570 units, $A_y$ = −480 units
    - (C) $A_x$ = 380 units, $A_y$ = 570 units
    - (D) $A_x$ = 650 units, $A_y$ = −480 units
    - (E) $A_x$ = 380 units, $A_y$ = 650 units

40b. Chapter 1, Problem 40b
    - (A) $A_x$ = 650 units, $A_y$ = 380 units
    - (B) $A_x$ = 570 units, $A_y$ = −480 units
    - (C) $A_x$ = 380 units, $A_y$ = 570 units
    - (D) $A_x$ = 650 units, $A_y$ = −480 units
    - (E) $A_x$ = 380 units, $A_y$ = 650 units

## Section 1.8 Addition of Vectors by Means of Components

42a. Chapter 1, Problem 42a
    - (A) 21 paces
    - (B) 36 paces
    - (C) 42 paces
    - (D) 46 paces
    - (E) 88 paces

42b. Chapter 1, Problem 42b
    - (A) 21 paces
    - (B) 36 paces
    - (C) 42 paces
    - (D) 46 paces
    - (E) 88 paces

44. Chapter 1, Problem 44
    - (A) 105 N, 68.2° north of east
    - (B) 105 N, 21.8° north of east
    - (C) 137 N, 21.8° north of east
    - (D) 137 N, 68.2° north of east
    - (E) 120 N, 30.0° north of east

46a. Chapter 1, Problem 46a
    - (A) 1.2 km
    - (B) 2.4 km
    - (C) 2.7 km
    - (D) 3.3 km
    - (E) 4.8 km

# HOMEWORK CHOICES

**46b.** Chapter 1, Problem 46b
- (A) $3.0 \times 10^1$ degrees, south of east
- (B) $3.0 \times 10^1$ degrees, north of east
- (C) due east
- (D) $6.0 \times 10^1$ degrees, south of east
- (E) $6.0 \times 10^1$ degrees, north of east

**48.** Chapter 1, Problem 48
- (A) 0.292
- (B) 0.532
- (C) 0.684
- (D) 1.06
- (E) 1.88

**50.** Chapter 1, Problem 50
- (A) 450 N, 53.9° below the $-x$ axis
- (B) 450 N, 53.9° above the $-x$ axis
- (C) 436 N, 41.0° below the $-x$ axis
- (D) 436 N, 41.0° above the $-x$ axis
- (E) 166 N, along the $-x$ axis

**52.** Chapter 1, Problem 52
- (A) 4.94 cm, 16.7° north of west
- (B) 9.42 cm, 18.4° south of west
- (C) 13.9 cm, due west
- (D) 14.7 cm, 19.6° south of west
- (E) 18.8 cm, 21.7° south of west

## Additional Problems

**54.** Chapter 1, Problem 54
- (A) 0.43 m
- (B) 0.47 m
- (C) 0.50 m
- (D) 0.54 m
- (E) 0.59 m

**56.** Chapter 1, Problem 56
- (A) $[L]/[T]^2$
- (B) $[T]/[L]^3$
- (C) $1/[T]$
- (D) $[T]^2/[L]$
- (E) $1/[T]^3$

**58.** Chapter 1, Problem 58
- (A) 1076 m$^2$
- (B) 3281 m$^2$
- (C) 4048 m$^2$
- (D) 5280 m$^2$
- (E) 7610 m$^2$

**60.** Chapter 1, Problem 60
- (A) 110 cm
- (B) 89 cm
- (C) 79 cm
- (D) 66 cm
- (E) 46 cm

**62a.** Chapter 1, Problem 62a
- (A) 974 m
- (B) 818 m
- (C) 636 m
- (D) 487 m
- (E) 408 m

**62b.** Chapter 1, Problem 62b
- (A) −408 m
- (B) −487 m
- (C) −636 m
- (D) −818 m
- (E) −974 m

# Homework: Introduction and Mathematical Concepts

64. Chapter 1, Problem 64
    - (A) 160 m
    - (B) 120 m
    - (C) 76 m
    - (D) 59 m
    - (E) 43 m

66a. Chapter 1, Problem 66a
    - (A) 114 N, 33° south of east
    - (B) 126 N, 54° north of east
    - (C) 126 N, 54° south of east
    - (D) 142 N, 67° north of east
    - (E) 142 N, 67° south of east

66b. Chapter 1, Problem 66b
    - (A) 114 N, 33° south of west
    - (B) 126 N, 54° north of west
    - (C) 126 N, 54° south of west
    - (D) 142 N, 67° north of west
    - (E) 142 N, 67° south of west

# Homework Choices
# CHAPTER 2
# Kinematics in One Dimension

### Section 2.2 Speed and Velocity

2a. Chapter 2, Problem 2a
   (A) 7.07 km
   (B) 6.36 km
   (C) 3.00 km
   (D) 2.12 km
   (E) 1.50 km

2b. Chapter 2, Problem 2b
   (A) 7.07 km, 45° north of east
   (B) 6.36 km, 45° south of east
   (C) 3.00 km, 45° north of east
   (D) 2.12 km, 45° north of east
   (E) 1.50 km, 45° south of east

4. Chapter 2, Problem 4
   (A) 0.213 s
   (B) 0.1.33 s
   (C) 2.45 s
   (D) 3.57 s
   (E) 4.69 s

6. Chapter 2, Problem 6
   (A) 55 m
   (B) 12 m
   (C) 35 m
   (D) 19 m
   (E) 28 m

8. Chapter 2, Problem 8
   (A) 13 m
   (B) 52 m
   (C) 4.3 m
   (D) 8.6 m
   (E) 26 m

10. Chapter 2, Problem 10
   (A) 2.27 m/s, downward
   (B) 3.93 m/s, upward
   (C) 5.09 m/s, upward
   (D) 6.25 m/s, downward
   (E) 7.41 m/s, downward

12. Chapter 2, Problem 12
   (A) 0.49 s
   (B) 1.3 s
   (C) 2.1 s
   (D) 4.2 s
   (E) 6.2 s

### Section 2.3 Acceleration

14. Chapter 2, Problem 14
   (A) 8.18 m/s$^2$
   (B) 6.14 m/s$^2$
   (C) 4.09 m/s$^2$
   (D) 2.04 m/s$^2$
   (E) 1.02 m/s$^2$

16. Chapter 2, Problem 16
   (A) 6.50 s
   (B) 11.1 s
   (C) 12.0 s
   (D) 14.8 s
   (E) 17.6 s

18. Chapter 2, Problem 18
   (A) 3.2 m/s
   (B) 4.8 m/s
   (C) 6.2 m/s
   (D) 8.0 m/s
   (E) 9.6 m/s

20. Chapter 2, Problem 20
    - (A) −2.0 m/s
    - (B) +8.0 m/s
    - (C) −6.0 m/s
    - (D) +1.0 m/s
    - (E) −4.0 m/s

## Section 2.4 Equations of Kinematics for Constant Acceleration
## Section 2.5 Applications of the Equations of Kinematics

22a. Chapter 2, Problem 22a
   - (A) −11 m/s$^2$
   - (B) +0.40 m/s$^2$
   - (C) −5.8 m/s$^2$
   - (D) +11 m/s$^2$
   - (E) −0.40 m/s$^2$

22b. Chapter 2, Problem 22b
   - (A) −22 m/s$^2$
   - (B) +0.80 m/s$^2$
   - (C) −12 m/s$^2$
   - (D) +22 m/s$^2$
   - (E) −0.80 m/s$^2$

24. Chapter 2, Problem 24
   - (A) 3.0 m
   - (B) 3.7 m
   - (C) 4.5 m
   - (D) 6.8 m
   - (E) 9.0 m

26. Chapter 2, Problem 26
   - (A) $1.5 \times 10^0$ m
   - (B) $2.5 \times 10^1$ m
   - (C) $5.0 \times 10^1$ m
   - (D) $7.5 \times 10^1$ m
   - (E) $1.0 \times 10^2$ m

28. Chapter 2, Problem 28
   - (A) 0.52 s
   - (B) 0.26 s
   - (C) 0.17 s
   - (D) 0.085 s
   - (E) 0.34 s

30. Chapter 2, Problem 30
   - (A) 96.9 m/s
   - (B) 82.9 m/s
   - (C) 65.3 m/s
   - (D) 20.1 m/s
   - (E) zero

32a. Chapter 2, Problem 32a
   - (A) 10.9 m/s
   - (B) 8.33 m/s
   - (C) 7.40 m/s
   - (D) 6.46 m/s
   - (E) 5.26 m/s

32b. Chapter 2, Problem 32b
   - (A) 281 m
   - (B) 233 m
   - (C) 189 m
   - (D) 93.7 m
   - (E) 49.2 m

34. Chapter 2, Problem 34

| | acceleration phase | deceleration phase |
|---|---|---|
| (A) | 0.25 km | 1.75 km |
| (B) | 0.50 km | 1.50 km |
| (C) | 1.00 km | 1.00 km |
| (D) | 1.50 km | 0.50 km |
| (E) | 1.75 km | 0.25 km |

## HOMEWORK CHOICES — Chapter 2

36a. Chapter 2, Problem 36a
- (A) 8.5 s
- (B) 11 s
- (C) 14 s
- (D) 17 s
- (E) 22 s

36b. Chapter 2, Problem 36b
- (A) $3.0 \times 10^1$ m
- (B) $2.5 \times 10^1$ m
- (C) $5.0 \times 10^1$ m
- (D) $7.5 \times 10^1$ m
- (E) $1.0 \times 10^2$ m

38. Chapter 2, Problem 38
- (A) 1.7 s
- (B) 2.2 s
- (C) 3.4 s
- (D) 4.4 s
- (E) 7.9 s

## Section 2.6 Freely Falling Bodies

40a. Chapter 2, Problem 40a
- (A) zero
- (B) 9.8 m/s², upward
- (C) 9.8 m/s², downward
- (D) 1.1 m/s², upward
- (E) 1.1 m/s², downward

40b. Chapter 2, Problem 40b
- (A) 4.0 m
- (B) 4.5 m
- (C) 4.9 m
- (D) 5.7 m
- (E) 6.9 m

42. Chapter 2, Problem 42

| | $d_1$ | $d_2$ | $d_3$ |
|---|---|---|---|
| (A) | 0.29 m | 0.59 m | 0.88 m |
| (B) | 0.036 m | 0.14 m | 0.32 m |
| (C) | 0.15 m | 0.30 m | 0.44 m |
| (D) | 0.018 m | 0.071 m | 0.16 m |
| (E) | 0.012 m | 0.047 m | 0.11 m |

44. Chapter 2, Problem 44
- (A) 0.26 m
- (B) 1.3 m
- (C) 2.2 m
- (D) 3.6 m
- (E) 4.9 m

46. Chapter 2, Problem 46
- (A) 19 m/s, upward
- (B) 26 m/s, upward
- (C) 39 m/s, upward
- (D) 52 m/s, upward
- (E) 78 m/s, upward

48. Chapter 2, Problem 48
- (A) 3.06 s
- (B) 3.82 s
- (C) 4.59 s
- (D) 5.36 s
- (E) 6.12 s

50. Chapter 2, Problem 50
- (A) 4.5 m
- (B) 15 m
- (C) 22 m
- (D) 29 m
- (E) 37 m

# Homework: Kinematics in One Dimension

52. **Chapter 2, Problem 52**
    - (A) 1
    - (B) $1/\sqrt{2}$
    - (C) $\sqrt{2}$
    - (D) $2\sqrt{2}$
    - (E) 2

54. **Chapter 2, Problem 54**
    - (A) 9.8 m/s
    - (B) 10.4 m/s
    - (C) 11.8 m/s
    - (D) 13.2 m/s
    - (E) 25.0 m/s

56a. **Chapter 2, Problem 56a**
    - (A) $H$
    - (B) $1.5H$
    - (C) $2.0H$
    - (D) $2.5H$
    - (E) $3.0H$

56b. **Chapter 2, Problem 56b**
    - (A) If the acceleration due to gravity is *less* than 9.80 m/s$^2$, the additional distance would be *less* than that found for part (a).
    - (B) If the acceleration due to gravity is *greater* than 9.80 m/s$^2$, the additional distance would be *less* than that found for part (a).
    - (C) If the acceleration due to gravity is less than 9.80 m/s$^2$, the additional distance would be *more* than that found for part (a).
    - (D) If the acceleration due to gravity is *greater* than 9.80 m/s$^2$, the additional distance would be *more* than that found for part (a).
    - (E) The additional distance would be the same as that found for part (a).

58. **Chapter 2, Problem 58**
    - (A) 8.18 m/s
    - (B) 7.20 m/s
    - (C) 0.122 m/s
    - (D) 0.139 m/s
    - (E) 25.0 m/s

60. **Chapter 2, Problem 60**
    - (A) 14 m and 29 m
    - (B) 16 m and 41 m
    - (C) 15 m and 29 m
    - (D) 16 m and 48 m
    - (E) 14 m and 48 m

## Section 2.7 Graphical Analysis of Velocity and Acceleration for Linear Motion

62. **Chapter 2, Problem 62**

    |     | $v_A$ | $v_B$ | $v_C$ |
    | --- | --- | --- | --- |
    | (A) | $-2.0 \times 10^1$ km/h | $1.0 \times 10^1$ km/h | $2.0 \times 10^1$ km/h |
    | (B) | $2.0 \times 10^1$ km/h | $-1.0 \times 10^1$ km/h | $-2.0 \times 10^1$ km/h |
    | (C) | $-3.0 \times 10^1$ km/h | $1.0 \times 10^1$ km/h | $2.0 \times 10^1$ km/h |
    | (D) | $3.0 \times 10^1$ km/h | $-1.0 \times 10^1$ km/h | $-2.0 \times 10^1$ km/h |
    | (E) | $-1.0 \times 10^1$ km/h | $1.0 \times 10^1$ km/h | $2.0 \times 10^1$ km/h |

64a. **Chapter 2, Problem 64a**

    |     | A | B | C | D |
    | --- | --- | --- | --- | --- |
    | (a) | − | 0 | + | − |
    | (b) | + | − | + | 0 |
    | (c) | + | 0 | − | + |
    | (d) | + | − | 0 | + |
    | (e) | − | 0 | − | + |

## HOMEWORK CHOICES

64b. Chapter 2, Problem 64b

|   | $v_A$ | $v_B$ | $v_C$ | $v_D$ |
|---|---|---|---|---|
| (A) | 2.5 km/h | −1.9 km/h | 0.0 km/h | 4.0 km/h |
| (B) | −5.0 km/h | 0.0 km/h | 3.8 km/h | −2.5 km/h |
| (C) | 3.8 k m/h | −2.5 km/h | 4.0 km/h | 0.0 km/h |
| (D) | −2.5 km/h | 0.0 km/h | −1.9 km/h | 5.0 km/h |
| (E) | 5.0 km/h | 0.0 km/h | −3.8 km/h | 2.5 km/h |

66. Chapter 2, Problem 66
    *Solutions are graphs.*

## Additional Problems

68. Chapter 2, Problem 68
    - (A) 16 m/s
    - (B) 19 m/s
    - (C) 21 m/s
    - (D) 25 m/s
    - (E) 28 m/s

70. Chapter 2, Problem 70
    - (A) 11.2 m/s$^2$, due west
    - (B) 11.2 m/s$^2$, due east
    - (C) 5.60 m/s$^2$, due west
    - (D) 5.60 m/s$^2$, due east
    - (E) 0.179 m/s$^2$, due east

72. Chapter 2, Problem 72
    - (A) 153 m/s, upward
    - (B) 72.2 m/s, upward
    - (C) 187 m/s, upward
    - (D) 113 m/s, upward
    - (E) 221 m/s, upward

74. Chapter 2, Problem 74
    - (A) 6.143 m/s
    - (B) 6.740 m/s
    - (C) 7.143 m/s
    - (D) 7.343 m/s
    - (E) 7.749 m/s

76. Chapter 2, Problem 76
    - (A) 3.85 m/s, downward
    - (B) 7.62 m/s, upward
    - (C) 21.8 m/s, upward
    - (D) 43.7 m/s, downward
    - (E) 54.6 m/s, downward

78. Chapter 2, Problem 78
    - (A) 1.2
    - (B) 1.4
    - (C) 1.6
    - (D) 1.8
    - (E) 2.0

80. Chapter 2, Problem 80
    - (A) 3.7 s
    - (B) 7.7 s
    - (C) 11 s
    - (D) 15 s
    - (E) 19 s

82. Chapter 2, Problem 82
    - (A) 47.2 m
    - (B) 52.8 m
    - (C) 56.9 m
    - (D) 60.0 m
    - (E) 62.1 m

84. Chapter 2, Problem 84
    **Note**: the following distances are measured below the top of the cliff.
    (A) 18 m
    (B) 15 m
    (C) 12 m
    (D) 9.0 m
    (E) 6.0 m

86. Chapter 2, Problem 86
    (A) 5.6 m/s$^2$
    (B) 4.2 m/s$^2$
    (C) 3.7 m/s$^2$
    (D) 2.0 m/s$^2$
    (E) 1.4 m/s$^2$

# Homework Choices
## CHAPTER 3 — Kinematics in Two Dimensions

*Section 3.1 Displacement, Velocity, and Acceleration*

2. Chapter 3, Problem 2
   (A) $x = 61.7$ km, $y = 115$ km
   (B) $x = 143$ km, $y = 75.3$ km
   (C) $x = 85.1$ km, $y = 85.1$ km
   (D) $x = 115$ km, $y = 61.7$ km
   (E) $x = 75.3$ km, $y = 143$ km

4. Chapter 3, Problem 4
   (A) 4.72 m/s
   (B) 9.44 m/s
   (C) 14.1 m/s
   (D) 16.9 m/s
   (E) 23.6 m/s

6. Chapter 3, Problem 6
   (A) 3.6 m/s
   (B) 3.9 m/s
   (C) 4.7 m/s
   (D) 5.2 m/s
   (E) 6.4 m/s

8. Chapter 3, Problem 8
   (A) $6.0 \times 10^1$ m/s
   (B) $6.5 \times 10^1$ m/s
   (C) $7.0 \times 10^1$ m/s
   (D) $8.5 \times 10^1$ m/s
   (E) $1.0 \times 10^2$ m/s

10a. Chapter 3, Problem 10a
   (A) 0.540 km, 17° south of west
   (B) 1.08 km, 23° south of west
   (C) 1.43 km, 18° north of west
   (D) 1.35 km, 21° north of west
   (E) 1.22 km, 25° north of west

10b. Chapter 3, Problem 10b
   (A) 0.540 km/h, 21° north of west
   (B) 1.22 km/h, 18° north of west
   (C) 1.08 km/h, 25° north of west
   (D) 1.43 km/h, 17° south of west
   (E) 1.35 km/h, 23° south of west

*Section 3.2 Equations of Kinematics in Two Dimensions*
*Section 3.3 Projectile Motion*

12. Chapter 3, Problem 12
   (A) $5.28 \times 10^3$ m/s
   (B) $5.92 \times 10^3$ m/s
   (C) $2.29 \times 10^3$ m/s
   (D) $2.68 \times 10^3$ m/s
   (E) $3.52 \times 10^3$ m/s

14. Chapter 3, Problem 14
   (A) 4.33 m/s
   (B) 6.89 m/s
   (C) 13.6 m/s
   (D) 15.0 m/s
   (E) zero

16. Chapter 3, Problem 16
   (A) 2.0 m
   (B) 5.1 m
   (C) 11 m
   (D) 14 m
   (E) 19 m

# Homework: Kinematics in Two Dimensions

18a. Chapter 3, Problem 18a
    (A) $1.5 \times 10^1$ m      (C) $4.0 \times 10^1$ m      (E) $1.2 \times 10^2$ m
    (B) $3.0 \times 10^1$ m      (D) $6.0 \times 10^1$ m

18b. Chapter 3, Problem 18b
    (A) 152 m      (C) 221 m      (E) 290 m
    (B) 188 m      (D) 255 m

20. Chapter 3, Problem 20
    (A) the ball, 0.70 s      (C) the runner, 0.70 s      (E) the runner, 1.0 s
    (B) the ball, 1.4 s      (D) the runner, 1.4 s

22. Chapter 3, Problem 22
    (A) 1.44 m      (C) 5.59 m      (E) 10.3 m
    (B) 3.52 m      (D) 9.11 m

24. Chapter 3, Problem 24
    (A) 11 m/s      (C) 18 m/s      (E) 35 m/s
    (B) 13 m/s      (D) 25 m/s

26. Chapter 3, Problem 26
    (A) 2.4 m      (C) 1.2 m      (E) 0.60 m
    (B) 1.8 m      (D) 0.90 m

28. Chapter 3, Problem 28
    (A) 34 m      (C) 54 m      (E) 71 m
    (B) 48 m      (D) 62 m

30. Chapter 3, Problem 30
    (A) 2.4 m/s      (C) 5.3 m/s      (E) 4.9 m/s
    (B) 4.1 m/s      (D) 6.2 m/s

32. Chapter 3, Problem 32
    (A) 14°      (C) 7.0°      (E) 9.6°
    (B) 16°      (D) 8.0°

34. Chapter 3, Problem 34
    (A) 13 m/s      (C) 9.5 m/s      (E) 10 m/s
    (B) 6.3 m/s      (D) 8.8 m/s

36. Chapter 3, Problem 36
    (A) 66 m      (C) 44 m      (E) 22 m
    (B) 56 m      (D) 28 m

38. Chapter 3, Problem 38
    (A) 4.52 m/s, 59.4°      (C) 4.21 m/s, 36.7°      (E) 2.33 m/s, 48.1°
    (B) 2.49 m/s, 59.4°      (D) 1.99 m/s, 36.7°

# HOMEWORK CHOICES  Chapter 3

40. Chapter 3, Problem 40
    (A) 1.5
    (B) 2.5
    (C) 2
    (D) 3
    (E) 4

42. Chapter 3, Problem 42
    (A) 4.98 m/s
    (B) 5.14 m/s
    (C) 5.79 m/s
    (D) 9.64 m/s
    (E) 11.5 m/s

44. Chapter 3, Problem 44
    (A) 4.2 m/s
    (B) 5.0 m/s
    (C) 5.8 m/s
    (D) 7.0 m/s
    (E) 9.7 m/s

46. Chapter 3, Problem 46
    (A) $v_{0B} = 9.37$ m/s, $\theta_B = 68.0°$
    (B) $v_{0B} = 8.44$ m/s, $\theta_B = 81.5°$
    (C) $v_{0B} = 9.12$ m/s, $\theta_B = 74.2°$
    (D) $v_{0B} = 8.79$ m/s, $\theta_B = 81.5°$
    (E) $v_{0B} = 8.69$ m/s, $\theta_B = 70.7°$

## Section 3.4 Relative Velocity

48. Chapter 3, Problem 48
    (A) 24 m/s, due north
    (B) 24 m/s, due south
    (C) 26 m/s, due north
    (D) 26 m/s, due south
    (E) 25 m/s, due north

50. Chapter 3, Problem 50
    (A) 4.0 s
    (B) 6.3 s
    (C) 31 s
    (D) 43 s
    (E) 75 s

52. Chapter 3, Problem 52
    (A) 2.3 m/s
    (B) 2.7 m/s
    (C) 3.6 m/s
    (D) 4.5 m/s
    (E) 5.4 m/s

54. Chapter 3, Problem 54
    (A) 5.59 m/s, 67.3° south of east
    (B) 5.59 m/s, 67.3° north of east
    (C) 24.2 m/s, 21.1° north of east
    (D) 24.2 m/s, 21.1° south of east
    (E) 22.6 m/s, due east

56. Chapter 3, Problem 56
    (A) 1.6 m/s
    (B) 2.1 m/s
    (C) 2.3 m/s
    (D) 3.7 m/s
    (E) 4.5 m/s

58. Chapter 3, Problem 58
    (A) 16.0 m/s, 10.6°, north of east
    (B) 8.81 m/s, 66.9° south of east
    (C) 5.75 m/s, 37.0° north of east
    (D) 6.25 m/s, 79.4°, south of east
    (E) 7.50 m/s, 23.1°, north of east

60. Chapter 3, Problem 60
    (A) 0.57
    (B) 0.82
    (C) 1.4
    (D) 1.2
    (E) 0.69

## Additional Problems

62. Chapter 3, Problem 62
    - (A) 8600 m
    - (B) 9100 m
    - (C) 9800 m
    - (D) 10 100 m
    - (E) 11 700 m

64a. Chapter 3, Problem 64a
    - (A) 0.24 s
    - (B) 0.32 s
    - (C) 0.57 s
    - (D) 0.63 s
    - (E) 0.40 s

64b. Chapter 3, Problem 64b
    - (A) 693 m
    - (B) 264 m
    - (C) 440 m
    - (D) 352 m
    - (E) 630 m

66. Chapter 3, Problem 66
    - (A) 6.0 m/s, 36° west of north
    - (B) 26 m/s, 36° west of north
    - (C) 6.0 m/s, 43° east of north
    - (D) 26 m/s, 47° east of north
    - (E) 23 m/s, 52° east of north

68a. Chapter 3, Problem 68a
    - (A) 11.8 m/s
    - (B) 16.5 m/s
    - (C) 19.8 m/s
    - (D) 23.5 m/s
    - (E) 32.9 m/s

68b. Chapter 3, Problem 68b
    - (A) 53.5°, above the horizontal
    - (B) 37.0°, above the horizontal
    - (C) 41.7°, above the horizontal
    - (D) 49.5°, above the horizontal
    - (E) 60.0°, above the horizontal

70a. Chapter 3, Problem 70a
    - (A) 8.2 m
    - (B) 14 m
    - (C) 7.8 m
    - (D) 12 m
    - (E) 9.0 m

70b. Chapter 3, Problem 70b
    - (A) 0.55 s
    - (B) 1.1 s
    - (C) 0.70 s
    - (D) 1.4 s
    - (E) 0.91 s

72. Chapter 3, Problem 72
    - (A) 0.05 m/s
    - (B) 0.10 m/s
    - (C) 0.15 m/s
    - (D) 0.20 m/s
    - (E) 0.30 m/s

74. Chapter 3, Problem 74
    - (A) 0.96 s
    - (B) 1.04 s
    - (C) 1.12 s
    - (D) 1.16 s
    - (E) 1.24 s

76. Chapter 3, Problem 76
    - (A) 1870 km
    - (B) 1950 km
    - (C) 2160 km
    - (D) 2280 km
    - (E) 2440 km

78. Chapter 3, Problem 78
    *The solution is a proof.*

# Homework Choices
# CHAPTER 4
# Forces and Newton's Laws of Motion

*Section 4.3 Newton's Second Law of Motion*

2. Chapter 4, Problem 2
   (A) 0.114 m/s$^2$
   (B) 0.746 m/s$^2$
   (C) 0.120 m/s$^2$
   (D) 0.867 m/s$^2$
   (E) 0.103 m/s$^2$

4. Chapter 4, Problem 4
   (A) $8.2 \times 10^{-3}$ s
   (B) $4.1 \times 10^{-2}$ s
   (C) $9.7 \times 10^{-2}$ s
   (D) $2.2 \times 10^{-1}$ s
   (E) $1.0 \times 10^{1}$ s

6. Chapter 4, Problem 6
   (A) 1450 N
   (B) 2750 N
   (C) 2900 N
   (D) 1300 N
   (E) 2100 N

8. Chapter 4, Problem 8
   (A) 35.4 m/s
   (B) 39.0 m/s
   (C) 44.1 m/s
   (D) 50.0 m/s
   (E) 62.5 m/s

*Section 4.4 The Vector Nature of Newton's Second Law of Motion*
*Section 4.5 Newton's Third Law of Motion*

10. Chapter 4, Problem 10
    (A) $F_E$ = 360 N, $F_N$ = 360 N
    (B) $F_E$ = 440 N, $F_N$ = 570 N
    (C) $F_E$ = 570 N, $F_N$ = 440 N
    (D) $F_E$ = 220 N, $F_N$ = 285 N
    (E) $F_E$ = 285 N, $F_N$ = 220 N

12. Chapter 4, Problem 12
    (A) 18.0 m/s$^2$, 56.3° above the *x* axis
    (B) 20.0 m/s$^2$, 57.0° above the *x* axis
    (C) 22.0 m/s$^2$, 54.9° above the *x* axis
    (D) 24.0 m/s$^2$, 49.1° above the *x* axis
    (E) 26.0 m/s$^2$, 60.0° above the *x* axis

14. Chapter 4, Problem 14

    |     | man | woman |
    | --- | --- | --- |
    | (A) | 4.59 m/s$^2$, east | 4.59 m/s$^2$, west |
    | (B) | 4.59 m/s$^2$, west | 4.59 m/s$^2$, east |
    | (C) | 2.30 m/s$^2$, east | 1.15 m/s$^2$, west |
    | (D) | 1.15 m/s$^2$, east | 2.30 m/s$^2$, west |
    | (E) | 0.55 m/s$^2$, east | 0.94 m/s$^2$, west |

16. Chapter 4, Problem 16
    (A) 28 s
    (B) 35 s
    (C) 41 s
    (D) 33 s
    (E) 44 s

18. Chapter 4, Problem 18
    - (A) 28 s
    - (B) 36 s
    - (C) 51 s
    - (D) 64 s
    - (E) 77 s

## Section 4.7 The Gravitational Force

20. Chapter 4, Problem 20
    - (A) 15 m/s$^2$
    - (B) 12 m/s$^2$
    - (C) 9.8 m/s$^2$
    - (D) 5.5 m/s$^2$
    - (E) 4.2 m/s$^2$

22. Chapter 4, Problem 22
    - (A) $4.0 \times 10^{-6}$ N
    - (B) $1.8 \times 10^{-7}$ N
    - (C) $9.4 \times 10^{-8}$ N
    - (D) $3.0 \times 10^{-8}$ N
    - (E) $1.0 \times 10^{-6}$ N

24. Chapter 4, Problem 24
    - (A) $2.06 \times 10^{19}$ N
    - (B) $7.62 \times 10^{19}$ N
    - (C) $1.01 \times 10^{20}$ N
    - (D) $4.77 \times 10^{20}$ N
    - (E) $5.35 \times 10^{20}$ N

26. Chapter 4, Problem 26
    - (A) 0.157
    - (B) 0.169
    - (C) 0.163
    - (D) 0.159
    - (E) 0.165

28. Chapter 4, Problem 28
    - (A) 130 N
    - (B) 260 N
    - (C) 390 N
    - (D) 420 N
    - (E) 870 N

30. Chapter 4, Problem 30
    - (A) $2.59 \times 10^{-11}$ m/s$^2$
    - (B) $1.13 \times 10^{-10}$ m/s$^2$
    - (C) $2.25 \times 10^{-10}$ m/s$^2$
    - (D) $1.46 \times 10^{-9}$ m/s$^2$
    - (E) $2.70 \times 10^{-9}$ m/s$^2$

32. Chapter 4, Problem 32
    - (A) $2.16 \times 10^{8}$ m
    - (B) $3.46 \times 10^{8}$ m
    - (C) $3.85 \times 10^{8}$ m
    - (D) $4.29 \times 10^{8}$ m
    - (E) $4.63 \times 10^{8}$ m

34. Chapter 4, Problem 34
    - (A) 0.394
    - (B) 0.628
    - (C) 0.433
    - (D) 0.188
    - (E) 0.0561

## Section 4.8 The Normal Force
## Section 4.9 Static and Kinetic Frictional Forces

36a. Chapter 4, Problem 36a
    - (A) $9.31 \times 10^{2}$ N
    - (B) $8.08 \times 10^{2}$ N
    - (C) $1.25 \times 10^{3}$ N
    - (D) $1.10 \times 10^{3}$ N
    - (E) zero

36b. Chapter 4, Problem 36b
  (A) $9.31 \times 10^2$ N
  (B) $8.08 \times 10^2$ N
  (C) $1.25 \times 10^3$ N
  (D) $1.10 \times 10^3$ N
  (E) zero

36c. Chapter 4, Problem 36c
  (A) $9.31 \times 10^2$ N
  (B) $8.08 \times 10^2$ N
  (C) $1.25 \times 10^3$ N
  (D) $1.10 \times 10^3$ N
  (E) zero

38. Chapter 4, Problem 38
  (A) 11 m/s$^2$
  (B) 8.6 m/s$^2$
  (C) 5.8 m/s$^2$
  (D) 1.5 m/s$^2$
  (E) 0.91 m/s$^2$

40. Chapter 4, Problem 40
  (A) The block will move. $a = 3.72$ m/s$^2$
  (B) The block will move. $a = 3.44$ m/s$^2$
  (C) The block will move. $a = 2.89$ m/s$^2$
  (D) The block will move. $a = 2.11$ m/s$^2$
  (E) The block will not move.

42. Chapter 4, Problem 42
  (A) 0.78 m/s$^2$
  (B) 1.4 m/s$^2$
  (C) 2.9 m/s$^2$
  (D) 3.3 m/s$^2$
  (E) 3.8 m/s$^2$

44a. Chapter 4, Problem 44a
  (A) 23.8 N
  (B) 18.6 N
  (C) 22.4 N
  (D) 24.9 N
  (E) 21.2 N

44b. Chapter 4, Problem 44b
  (A) 23.8 N
  (B) 18.6 N
  (C) 22.4 N
  (D) 24.9 N
  (E) 21.2 N

44c. Chapter 4, Problem 44c
  (A) 23.8 N
  (B) 18.6 N
  (C) 22.4 N
  (D) 24.9 N
  (E) 21.2 N

46. Chapter 4, Problem 46
  (A) 0.694 s
  (B) 0.836 s
  (C) 1.44 s
  (D) 2.39 s
  (E) 8.24 s

48. Chapter 4, Problem 48
  (A) 45°
  (B) 53°
  (C) 68°
  (D) 72°
  (E) 77°

*Section 4.11 Equilibrium Applications of Newton's Laws of Motion*

50a. Chapter 4, Problem 50a
- (A) $1.33 \times 10^7$ N
- (B) $8.02 \times 10^6$ N
- (C) $4.98 \times 10^6$ N
- (D) $7.40 \times 10^5$ N
- (E) $5.17 \times 10^7$ N

50b. Chapter 4, Problem 50b
- (A) $1.67 \times 10^9$ N
- (B) $7.98 \times 10^8$ N
- (C) $5.88 \times 10^8$ N
- (D) $9.02 \times 10^7$ N
- (E) $4.33 \times 10^7$ N

52. Chapter 4, Problem 52
- (A) 4440 N
- (B) 309 N
- (C) 3390 N
- (D) 2180 N
- (E) 618 N

54. Chapter 4, Problem 54
- (A) 16 N
- (B) 27 N
- (C) 39 N
- (D) 43 N
- (E) 54 N

56. Chapter 4, Problem 56
- (A) 4300 N
- (B) 4500 N
- (C) 8100 N
- (D) 8700 N
- (E) 16 000 N

58. Chapter 4, Problem 58
- (A) 220 N
- (B) 310 N
- (C) 440 N
- (D) 680 N
- (E) 890 N

60. Chapter 4, Problem 60

|  | $T_1$ | $T_2$ |
|---|---|---|
| (A) | 317 N | 177 N |
| (B) | 249 N | 317 N |
| (C) | 177 N | 340 N |
| (D) | 249 N | 177 N |
| (E) | 340 N | 249 N |

62. Chapter 4, Problem 62
- (A) 99.3 N
- (B) 118 N
- (C) 155 N
- (D) 261 N
- (E) 286 N

64a. Chapter 4, Problem 64a
- (A) 44.2 N
- (B) 79.0 N
- (C) 88.9 N
- (D) 158 N
- (E) 219 N

64b. Chapter 4, Problem 64b
- (A) 44.2 N
- (B) 79.0 N
- (C) 88.9 N
- (D) 158 N
- (E) 219 N

## HOMEWORK CHOICES

66. Chapter 4, Problem 66
    (A) 0.20
    (B) 0.25
    (C) 0.29
    (D) 0.36
    (E) 0.48

## Section 4.12 Nonequilibrium Applications of Newton's Laws of Motion

68a. Chapter 4, Problem 68a
    (A) $3.8 \times 10^1$ N
    (B) $6.5 \times 10^1$ N
    (C) $1.0 \times 10^2$ N
    (D) $1.4 \times 10^2$ N
    (E) $2.8 \times 10^2$ N

68b. Chapter 4, Problem 68b
    (A) $3.8 \times 10^1$ N
    (B) $6.5 \times 10^1$ N
    (C) $1.0 \times 10^2$ N
    (D) $1.4 \times 10^2$ N
    (E) $2.8 \times 10^2$ N

70. Chapter 4, Problem 70
    (A) 706 N
    (B) 93.6 N
    (C) 543 N
    (D) 187 N
    (E) 612 N

72a. Chapter 4, Problem 72a
    (A) 6.81 m/s$^2$
    (B) 3.40 m/s$^2$
    (C) 8.51 m/s$^2$
    (D) 2.99 m/s$^2$
    (E) 9.80 m/s$^2$

72b. Chapter 4, Problem 72b
    (A) 237 N
    (B) 157 N
    (C) 111 N
    (D) 129 N
    (E) 258 N

74. Chapter 4, Problem 74
    (A) 1050 N
    (B) 1560 N
    (C) 450 N
    (D) 270 N
    (E) 160 N

76. Chapter 4, Problem 76
    (A) 3.82 m/s$^2$
    (B) 2.52 m/s$^2$
    (C) 1.30 m/s$^2$
    (D) 6.34 m/s$^2$
    (E) 2.67 m/s$^2$

78. Chapter 4, Problem 78
    (A) 29 400 N
    (B) 36 200 N
    (C) 95 700 N
    (D) 18 900 N
    (E) 134 000 N

80. Chapter 4, Problem 80
    (A) 2730 N
    (B) 1810 N
    (C) 904 N
    (D) 356 N
    (E) 232 N

82a. Chapter 4, Problem 82a
    (A) No change in **A** or **B**. Increase of 4.7 N in **C**.
    (B) No change in **A** or **C**. Decrease of 4.7 N in **B**.
    (C) No change in **A**, **B**, or **C**.
    (D) No change in **B** or **C**. Decrease of 4.7 N in **A**.
    (E) No change in **C**. Decrease of 4.7 N in **A** and an increase of 4.7 N in **B**.

82b. Chapter 4, Problem 82b
 (A) No change in **A** or **B**. Increase of 4.7 N in **C**.
 (B) No change in **A** or **C**. Decrease of 4.7 N in **B**.
 (C) No change in **A**, **B**, or **C**.
 (D) No change in **B** or **C**. Decrease of 4.7 N in **A**.
 (E) No change in **C**. Decrease of 4.7 N in **A** and an increase of 4.7 N in **B**.

84. Chapter 4, Problem 84
 (A) 5.77 s
 (B) 5.24 s
 (C) 6.67 s
 (D) 2.74 s
 (E) 8.17 s

86a. Chapter 4, Problem 86a
 (A) 2.6 m/s$^2$
 (B) 3.3 m/s$^2$
 (C) 5.2 m/s$^2$
 (D) 6.6 m/s$^2$
 (E) 8.6 m/s$^2$

86b. Chapter 4, Problem 86b
 (A) 18 m/s$^2$
 (B) 11 m/s$^2$
 (C) 9.8 m/s$^2$
 (D) 8.6 m/s$^2$
 (E) 8.2 m/s$^2$

88a. Chapter 4, Problem 88a
 (A) 4.44 m/s$^2$
 (B) 6.34 m/s$^2$
 (C) 2.66 m/s$^2$
 (D) 3.68 m/s$^2$
 (E) 9.80 m/s$^2$

88b. Chapter 4, Problem 88b
 (A) 11.8 m/s$^2$
 (B) 21.6 m/s$^2$
 (C) 9.80 m/s$^2$
 (D) 6.34 m/s$^2$
 (E) 3.68 m/s$^2$

88c. Chapter 4, Problem 88c
 (A) The tension in the rope in the first case is greater than it is the second case.
 (B) The inertia of the 412-N block need only be considered in the first case, and it is not important in the second case.
 (C) The acceleration due to gravity is larger in the first case than it is in the second case.
 (D) In the first case, the inertia of both blocks affects the acceleration whereas, in the second case, only the lighter block's inertia remains.
 (E) The answers in parts a and b are identical because the 908-N downward force is the same in both cases.

90a. Chapter 4, Problem 90a
 (A) 0.81 m/s$^2$
 (B) 1.6 m/s$^2$
 (C) 3.2 m/s$^2$
 (D) 9.8 m/s$^2$
 (E) 0.60 m/s$^2$

90b. Chapter 4, Problem 90b
 (A) $T_1 = 104$ N, $T_3 = 230$ N
 (B) $T_1 = 98$ N, $T_3 = 245$ N
 (C) $T_1 = 104$ N, $T_3 = 275$ N
 (D) $T_1 = 245$ N, $T_3 = 104$ N
 (E) $T_1 = 245$ N, $T_3 = 98$ N

*HOMEWORK CHOICES*  Chapter 4   *351*

## Additional Problems

92a. Chapter 4, Problem 92a
- (A) 120 N
- (B) 175 N
- (C) 350 N
- (D) 610 N
- (E) 870 N

92b. Chapter 4, Problem 92b
- (A) 120 N
- (B) 175 N
- (C) 350 N
- (D) 610 N
- (E) 870 N

94a. Chapter 4, Problem 94a
- (A) 161 N
- (B) 1230 N
- (C) 1580 N
- (D) 1710 N
- (E) 1850 N

94b. Chapter 4, Problem 94b
- (A) zero
- (B) 161 N
- (C) 427 N
- (D) 925 N
- (E) 1230 N

96. Chapter 4, Problem 96
- (A) $8 \times 10^{11}$ N
- (B) $2 \times 10^{18}$ N
- (C) $4 \times 10^{21}$ N
- (D) $7 \times 10^{28}$ N
- (E) $1 \times 10^{39}$ N

98. Chapter 4, Problem 98
- (A) 1.09 m/s$^2$, upward
- (B) 1.20 m/s$^2$, upward
- (C) 2.18 m/s$^2$, upward
- (D) 2.40 m/s$^2$, upward
- (E) 11.4 m/s$^2$, upward

100. Chapter 4, Problem 100
- (A) 588 N
- (B) 606 N
- (C) 645 N
- (D) 720 N
- (E) 960 N

102. Chapter 4, Problem 102
- (A) 170 N
- (B) 130 N
- (C) 94 N
- (D) 38 N
- (E) zero

104. Chapter 4, Problem 104
- (A) 1030 N
- (B) 2490 N
- (C) 3560 N
- (D) 4010 N
- (E) 4750 N

106. Chapter 4, Problem 106
- (A) 27.9 m
- (B) 21.8 m
- (C) 13.9 m
- (D) 18.4 m
- (E) 36.8 m

108a. Chapter 4, Problem 108a
- (A) 302 N
- (B) 484 N
- (C) 498 N
- (D) 576 N
- (E) 591 N

108b. Chapter 4, Problem 108b
- (A) 302 N
- (B) 484 N
- (C) 498 N
- (D) 576 N
- (E) 591 N

110. Chapter 4, Problem 110
- (A) +1240 N
- (B) −640 N
- (C) −470 N
- (D) −230 N
- (E) +110 N

112. Chapter 4, Problem 112
- (A) $1.2 \times 10^4$ m/s
- (B) $4.4 \times 10^4$ m/s
- (C) $8.1 \times 10^4$ m/s
- (D) $1.8 \times 10^5$ m/s
- (E) $3.3 \times 10^5$ m/s

114a. Chapter 4, Problem 114a
- (A) $a = g \sin \theta$
- (B) $a = g \tan \theta$
- (C) $a = g \cos \theta$
- (D) $a = (\sin \theta)/g$
- (E) $a = (\tan \theta)/g$

114b. Chapter 4, Problem 114b
- (A) 5.55 m/s$^2$
- (B) 2.82 m/s$^2$
- (C) 9.65 m/s$^2$
- (D) 3.15 m/s$^2$
- (E) 1.73 m/s$^2$

114c. Chapter 4, Problem 114c
- (A) 60°
- (B) 45°
- (C) 30°
- (D) 0°
- (E) 90°

116a. Chapter 4, Problem 116a
- (A) 7.39 N
- (B) 13.7 N
- (C) 27.4 N
- (D) 48.6 N
- (E) 54.8 N

116b. Chapter 4, Problem 116b
- (A) 0.739 m/s$^2$
- (B) 1.37 m/s$^2$
- (C) 2.74 m/s$^2$
- (D) 4.86 m/s$^2$
- (E) 5.48 m/s$^2$

# Homework Choices
## CHAPTER 5 — Dynamics of Uniform Circular Motion

### Section 5.1 Uniform Circular Motion
### Section 5.2 Centripetal Acceleration

2. Chapter 5, Problem 2
   - (A) 0.016 s
   - (B) 0.008 s
   - (C) 0.032 s
   - (D) 0.064 s
   - (E) 0.005 s

4. Chapter 5, Problem 4
   - (A) 1.2 m
   - (B) 1.3 m
   - (C) 1.5 m
   - (D) 1.6 m
   - (E) 3.2 m

6. Chapter 5, Problem 6
   - (A) 84°
   - (B) 61°
   - (C) 48°
   - (D) 35°
   - (E) 24°

8a. Chapter 5, Problem 8a
   - (A) $v = 420$ m/s, $a_c = 3.37 \times 10^{-2}$ m/s$^2$
   - (B) $v = 464$ m/s, $a_c = 2.39 \times 10^{-2}$ m/s$^2$
   - (C) $v = 328$ m/s, $a_c = 2.39 \times 10^{-2}$ m/s$^2$
   - (D) $v = 464$ m/s, $a_c = 3.37 \times 10^{-2}$ m/s$^2$
   - (E) $v = 328$ m/s, $a_c = 3.37 \times 10^{-2}$ m/s$^2$

8b. Chapter 5, Problem 8b
   - (A) $v = 464$ m/s, $a_c = 3.89 \times 10^{-2}$ m/s$^2$
   - (B) $v = 232$ m/s, $a_c = 1.69 \times 10^{-2}$ m/s$^2$
   - (C) $v = 232$ m/s, $a_c = 9.74 \times 10^{-3}$ m/s$^2$
   - (D) $v = 328$ m/s, $a_c = 1.95 \times 10^{-2}$ m/s$^2$
   - (E) $v = 420$ m/s, $a_c = 3.19 \times 10^{-2}$ m/s$^2$

10. Chapter 5, Problem 10
   - (A) 10 600 rev/min
   - (B) 7910 rev/min
   - (C) 3160 rev/min
   - (D) 2260 rev/min
   - (E) 176 rev/min

### Section 5.3 Centripetal Force

12. Chapter 5, Problem 12
   - (A) 7.0 m/s
   - (B) 36 m/s
   - (C) 21 m/s
   - (D) 16 m/s
   - (E) 12 m/s

14a. Chapter 5, Problem 14a
   - (A) 0.189 N
   - (B) 0.153 N
   - (C) 0.117 N
   - (D) 0.0945 N
   - (E) 0.0601 N

# Homework: Dynamics of Uniform Circular Motion

14b. Chapter 5, Problem 14b
- (A) 1.5
- (B) 2.0
- (C) 2.5
- (D) 3.0
- (E) 4.0

16. Chapter 5, Problem 16
- (A) 0.24 m
- (B) 0.27 m
- (C) 0.31 m
- (D) 0.38 m
- (E) 0.46 m

18. Chapter 5, Problem 18
- (A) 0.356
- (B) 0.186
- (C) 0.242
- (D) 0.491
- (E) 0.158

20. Chapter 5, Problem 20
- (A) 1/2
- (B) 2/1
- (C) 1/4
- (D) 4/1
- (E) 1/16

## Section 5.4 Banked Curves

22. Chapter 5, Problem 22
- (A) 42°
- (B) 39°
- (C) 32°
- (D) 26°
- (E) 23°

24a. Chapter 5, Problem 24a
- (A) $\tan \theta = rv^2/g$
- (B) $\tan \theta = v^2/(rg)$
- (C) $\sin \theta = rv^2/g$
- (D) $\cos \theta = v^2/(rg)$
- (E) $\sin \theta = rg/v^2$

24b. Chapter 5, Problem 24b
- (A) 28.1°
- (B) 61.9°
- (C) 64.8°
- (D) 34.9°
- (E) 25.2°

26. Chapter 5, Problem 26
- (A) 32 s
- (B) 38 s
- (C) 45 s
- (D) 53 s
- (E) 61 s

## Section 5.5 Satellites in Circular Orbits
## Section 5.6 Apparent Weightlessness and Artificial Gravity

28. Chapter 5, Problem 28
- (A) 131 m
- (B) 174 m
- (C) 218 m
- (D) 262 m
- (E) 305 m

30. Chapter 5, Problem 30
- (A) $6.97 \times 10^8$ m
- (B) $7.86 \times 10^8$ m
- (C) $1.33 \times 10^9$ m
- (D) $1.54 \times 10^9$ m
- (E) $6.02 \times 10^{10}$ m

# HOMEWORK CHOICES

32a. Chapter 5, Problem 32a
(A) $2.99 \times 10^4$ m/s
(B) $1.98 \times 10^4$ m/s
(C) $3.44 \times 10^4$ m/s
(D) $3.13 \times 10^4$ m/s
(E) $3.71 \times 10^4$ m/s

32b. Chapter 5, Problem 32b
(A) $1.99 \times 10^{30}$ kg
(B) $1.97 \times 10^{30}$ kg
(C) $2.03 \times 10^{30}$ kg
(D) $2.05 \times 10^{30}$ kg
(E) $2.01 \times 10^{30}$ kg

34. Chapter 5, Problem 34
(A) 411 days
(B) 174 days
(C) 597 days
(D) 223 days
(E) 262 days

36a. Chapter 5, Problem 36a
(A) 912 m
(B) 1370 m
(C) 456 m
(D) 1140 m
(E) 228 m

36b. Chapter 5, Problem 36b
(A) 456 m
(B) 228 m
(C) 912 m
(D) 1140 m
(E) 1370 m

36c. Chapter 5, Problem 36c
(A) 2.25 m/s$^2$
(B) 2.50 m/s$^2$
(C) 1.25 m/s$^2$
(D) 1.50 m/s$^2$
(E) 1.00 m/s$^2$

## Section 5.7 Vertical Circular Motion

38. Chapter 5, Problem 38
(A) $1.3 \times 10^3$ m
(B) $2.1 \times 10^3$ m
(C) $2.7 \times 10^3$ m
(D) $3.9 \times 10^3$ m
(E) $5.4 \times 10^3$ m

40. Chapter 5, Problem 40
(A) 327 N
(B) 492 N
(C) 815 N
(D) 1310 N
(E) 1830 N

42. Chapter 5, Problem 42
(A) 4.47 m/s
(B) 6.62 m/s
(C) 10.0 m/s
(D) 11.7 m/s
(E) 14.0 m/s

44. Chapter 5, Problem 44
(A) 0.34 rev/s
(B) 0.56 rev/s
(C) 0.85 rev/s
(D) 1.0 rev/s
(E) 1.2 rev/s

## Additional Problems

46a. Chapter 5, Problem 46a
- (A) zero
- (B) $2.0 \times 10^1$ m/s$^2$
- (C) $4.0 \times 10^1$ m/s$^2$
- (D) $5.0 \times 10^1$ m/s$^2$
- (E) $7.0 \times 10^1$ m/s$^2$

46b. Chapter 5, Problem 46b
- (A) zero
- (B) $2.0 \times 10^1$ m/s$^2$
- (C) $4.0 \times 10^1$ m/s$^2$
- (D) $5.0 \times 10^1$ m/s$^2$
- (E) $7.0 \times 10^1$ m/s$^2$

46c. Chapter 5, Problem 46c
- (A) zero
- (B) $2.0 \times 10^1$ m/s$^2$
- (C) $4.0 \times 10^1$ m/s$^2$
- (D) $5.0 \times 10^1$ m/s$^2$
- (E) $7.0 \times 10^1$ m/s$^2$

48. Chapter 5, Problem 48
- (A) 1180 m/s
- (B) 1960 m/s
- (C) 2550 m/s
- (D) 2790 m/s
- (E) 3070 m/s

50. Chapter 5, Problem 50
- (A) $2.94 \times 10^3$ s
- (B) $7.15 \times 10^3$ s
- (C) $1.43 \times 10^4$ s
- (D) $4.32 \times 10^4$ s
- (E) $8.64 \times 10^4$ s

52. Chapter 5, Problem 52
- (A) 37 m
- (B) 51 m
- (C) 66 m
- (D) 82 m
- (E) 93 m

54. Chapter 5, Problem 54
- (A) $2.0 \times 10^2$ m/s$^2$
- (B) $1.2 \times 10^2$ m/s$^2$
- (C) $3.0 \times 10^2$ m/s$^2$
- (D) $8.4 \times 10^1$ m/s$^2$
- (E) $4.0 \times 10^2$ m/s$^2$

56a. Chapter 5, Problem 56a
- (A) $1.5 \times 10^5$ m/s$^2$
- (B) $3.0 \times 10^5$ m/s$^2$
- (C) $4.5 \times 10^5$ m/s$^2$
- (D) $6.0 \times 10^5$ m/s$^2$
- (E) $7.5 \times 10^5$ m/s$^2$

56b. Chapter 5, Problem 56b
- (A) $7.7 \times 10^4$ g
- (B) $6.1 \times 10^4$ g
- (C) $4.6 \times 10^4$ g
- (D) $3.1 \times 10^4$ g
- (E) $1.5 \times 10^4$ g

58a. Chapter 5, Problem 58a
- (a) the frictional force exerted on the rider
- (b) the gravitational force exerted on the rider by the earth
- (c) the normal force exerted on the rider by the wall
- (d) the centrifugal force exerted on the rider by the wall
- (e) the gravitational force exerted on the rider by the wall

*HOMEWORK CHOICES*

58b. Chapter 5, Problem 58b
- (A) 1670 N
- (B) 1420 N
- (C) 539 N
- (D) 167 N
- (E) 142 N

58c. Chapter 5, Problem 58c
- (A) 1.00
- (B) 0.380
- (C) 0.619
- (D) 0.263
- (E) 0.323

**Homework Choices**
# CHAPTER 6     Work and Energy

## Section 6.1 Work

2. Chapter 6, Problem 2
   - (A) $1.9 \times 10^2$ N
   - (B) $2.2 \times 10^2$ N
   - (C) $2.5 \times 10^2$ N
   - (D) $2.8 \times 10^2$ N
   - (E) $3.1 \times 10^2$ N

4a. Chapter 6, Problem 4a
   - (A) 1390 J
   - (B) 2190 J
   - (C) 2980 J
   - (D) 3290 J
   - (E) 3770 J

4b. Chapter 6, Problem 4b
   - (A) 1390 J
   - (B) 2190 J
   - (C) 2980 J
   - (D) 3290 J
   - (E) 3770 J

6. Chapter 6, Problem 6
   - (A) $9.40 \times 10^6$ J
   - (B) $1.41 \times 10^7$ J
   - (C) $1.88 \times 10^7$ J
   - (D) $2.35 \times 10^7$ J
   - (E) $2.82 \times 10^7$ J

8a. Chapter 6, Problem 8a
   - (A) More net work is done during the dive because the net force is greater during the dive than it is during the climb.
   - (B) More net work is done during the climb because the net force is greater during the climb than it is during the dive.
   - (C) More net work is done during the climb because the net force is greater during the dive than it is during the climb.
   - (D) More net work is done during the dive because the net force is greater during the climb than it is during the dive.

8b. Chapter 6, Problem 8b
   - (A) $6.8 \times 10^7$ J
   - (B) $5.2 \times 10^7$ J
   - (C) $4.4 \times 10^7$ J
   - (D) $3.8 \times 10^7$ J
   - (E) $2.1 \times 10^7$ J

10a. Chapter 6, Problem 10a
   - (A) $-1.80 \times 10^3$ J
   - (B) $+1.80 \times 10^3$ J
   - (C) $-1.20 \times 10^3$ J
   - (D) $+1.20 \times 10^3$ J
   - (E) $+1.92 \times 10^3$ J

10b. Chapter 6, Problem 10b
   - (A) $-1.80 \times 10^3$ J
   - (B) $+1.80 \times 10^3$ J
   - (C) $-1.20 \times 10^3$ J
   - (D) $+1.20 \times 10^3$ J
   - (E) $+1.92 \times 10^3$ J

12. Chapter 6, Problem 12
   - (A) $5.0 \times 10^2$ N
   - (B) $2.0 \times 10^3$ N
   - (C) $3.0 \times 10^4$ N
   - (D) $1.5 \times 10^3$ N
   - (E) $1.7 \times 10^4$ N

*HOMEWORK CHOICES*     **Chapter 6**   359

## *Section 6.2  The Work-Energy Theorem and Kinetic Energy*

14. Chapter 6, Problem 14
    - (A) 39 m/s
    - (B) 82 m/s
    - (C) 28 m/s
    - (D) 66 m/s
    - (E) 150 m/s

16a. Chapter 6, Problem 16a
    - (A) zero
    - (B) $-5.70 \times 10^{11}$ J
    - (C) $5.70 \times 10^{11}$ J
    - (D) $-2.35 \times 10^{11}$ J
    - (E) $2.35 \times 10^{11}$ J

16b. Chapter 6, Problem 16b
    - (A) zero
    - (B) $-5.70 \times 10^{11}$ J
    - (C) $5.70 \times 10^{11}$ J
    - (D) $-2.35 \times 10^{11}$ J
    - (E) $2.35 \times 10^{11}$ J

18. Chapter 6, Problem 18
    - (A) 6.8
    - (B) 0.51
    - (C) 3.4
    - (D) 0.15
    - (E) 2.6

20a. Chapter 6, Problem 20a
    - (A) 76 J
    - (B) 91 J
    - (C) 0.92 J
    - (D) 1.8 J
    - (E) 38 J

20b. Chapter 6, Problem 20b
    - (A) $7.6 \times 10^3$ N
    - (B) $9.1 \times 10^3$ N
    - (C) $9.2 \times 10^1$ N
    - (D) $1.8 \times 10^2$ N
    - (E) $3.8 \times 10^3$ N

22. Chapter 6, Problem 22
    - (A) 0.048
    - (B) 0.094
    - (C) 0.020
    - (D) 0.062
    - (E) 0.036

24. Chapter 6, Problem 24
    - (A) 9.8 m/s
    - (B) $2.1 \times 10^1$ m/s
    - (C) $3.0 \times 10^1$ m/s
    - (D) $6.0 \times 10^1$ m/s
    - (E) $9.0 \times 10^1$ m/s

26. Chapter 6, Problem 26
    - (A) $-4.7 \times 10^1$ J
    - (B) $+6.8 \times 10^1$ J
    - (C) $-2.5 \times 10^2$ J
    - (D) $+5.4 \times 10^2$ J
    - (E) $-1.9 \times 10^3$ J

## *Section 6.3  Gravitational Potential Energy*
## *Section 6.4  Conservative Forces and Nonconservative Forces*

28a. Chapter 6, Problem 28a
    - (A) −66 J
    - (B) +66 J
    - (C) −13 J
    - (D) +13 J
    - (E) +26 J

# Homework: Work and Energy

28b. Chapter 6, Problem 28b
- (A) −66 J
- (B) +66 J
- (C) −13 J
- (D) +13 J
- (E) +26 J

30a. Chapter 6, Problem 30a
- (A) −43 J
- (B) +43 J
- (C) −260 J
- (D) +260 J
- (E) zero

30b. Chapter 6, Problem 30b
- (A) −43 J
- (B) +43 J
- (C) −260 J
- (D) +260 J
- (E) zero

32. Chapter 6, Problem 32
- (A) $2.18 \times 10^5$ J
- (B) $3.34 \times 10^5$ J
- (C) $5.24 \times 10^5$ J
- (D) $6.68 \times 10^5$ J
- (E) $2.03 \times 10^6$ J

## Section 6.5 The Conservation of Mechanical Energy

34. Chapter 6, Problem 34
- (A) 6.6 m/s
- (B) 4.7 m/s
- (C) 3.3 m/s
- (D) 2.3 m/s
- (E) 1.9 m/s

36. Chapter 6, Problem 36
- (A) 2.0 m
- (B) 1.8 m
- (C) 1.6 m
- (D) 1.4 m
- (E) 1.2 m

38. Chapter 6, Problem 38
- (A) 4.09 m/s
- (B) 5.78 m/s
- (C) 7.16 m/s
- (D) 8.19 m/s
- (E) 11.6 m/s

40. Chapter 6, Problem 40
- (A) 5.9 m/s
- (B) 7.0 m/s
- (C) 4.2 m/s
- (D) 4.9 m/s
- (E) 6.3 m/s

42. Chapter 6, Problem 42
- (A) 3.0 m
- (B) 11 m
- (C) 27 m
- (D) 35 m
- (E) 42 m

44. Chapter 6, Problem 44
- (A) 9.0 m
- (B) 12 m
- (C) 18 m
- (D) 27 m
- (E) 36 m

46. Chapter 6, Problem 46
- (A) 20.4 kg
- (B) 40.8 kg
- (C) 51.0 kg
- (D) 61.2 kg
- (E) 81.6 kg

# HOMEWORK CHOICES

## Section 6.6 Nonconservative Forces and the Work-Energy Theorem

48. Chapter 6, Problem 48
    - (A) zero
    - (B) −4.4 J
    - (C) −5.3 J
    - (D) −7.1 J
    - (E) −8.5 J

50. Chapter 6, Problem 50
    - (A) 26 m/s
    - (B) 18 m/s
    - (C) 12 m/s
    - (D) 9.4 m/s
    - (E) 8.6 m/s

52. Chapter 6, Problem 52
    - (A) 45.9 m/s
    - (B) 64.9 m/s
    - (C) 35.6 m/s
    - (D) 50.2 m/s
    - (E) 32.4 m/s

54. Chapter 6, Problem 54
    - (A) 1240 N
    - (B) 2070 N
    - (C) 2480 N
    - (D) 4130 N
    - (E) 4960 N

56. Chapter 6, Problem 56
    - (A) 30.7°
    - (B) 33.8°
    - (C) 21.5°
    - (D) 24.6°
    - (E) 27.7°

58. Chapter 6, Problem 58
    - (A) 111 m
    - (B) 127 m
    - (C) 135 m
    - (D) 142 m
    - (E) 159 m

## Section 6.7 Power

60. Chapter 6, Problem 60
    - (A) 54.6 W
    - (B) 220 W
    - (C) 1300 W
    - (D) 1800 W
    - (E) 2100 W

62. Chapter 6, Problem 62
    - (A) 73.5 s
    - (B) 69.5 s
    - (C) 64.7 s
    - (D) 60.0 s
    - (E) 55.1 s

64. Chapter 6, Problem 64
    - (A) 3.53
    - (B) 4.00
    - (C) 11.0
    - (D) 12.5
    - (E) 14.1

66a. Chapter 6, Problem 66a
    - (A) $2.00 \times 10^3$ W
    - (B) $3.62 \times 10^3$ W
    - (C) $4.00 \times 10^3$ W
    - (D) $5.75 \times 10^3$ W
    - (E) $4.90 \times 10^4$ W

66b. Chapter 6, Problem 66b
    - (A) $5.75 \times 10^3$ W
    - (B) $1.95 \times 10^4$ W
    - (C) $2.17 \times 10^4$ W
    - (D) $4.90 \times 10^4$ W
    - (E) $3.35 \times 10^4$ W

## Section 6.9 Work Done by a Variable Force

68a. Chapter 6, Problem 68a
    (A) zero
    (B) 3.0 J
    (C) −6.0 J
    (D) 6.0 J
    (E) −9.0 J

68b. Chapter 6, Problem 68b
    (A) zero
    (B) 3.0 J
    (C) −6.0 J
    (D) 6.0 J
    (E) −9.0 J

68c. Chapter 6, Problem 68c
    (A) zero
    (B) 3.0 J
    (C) −6.0 J
    (D) 6.0 J
    (E) −9.0 J

70a. Chapter 6, Problem 70a
    (A) 46 %
    (B) 60 %
    (C) 40 %
    (D) 54 %
    (E) 66 %

70b. Chapter 6, Problem 70b
    (A) 46 %
    (B) 60 %
    (C) 40 %
    (D) 54 %
    (E) 66 %

72a. Chapter 6, Problem 72a
    (A) zero
    (B) $5.00 \times 10^1$ J
    (C) $7.50 \times 10^1$ J
    (D) $1.00 \times 10^2$ J
    (E) $1.50 \times 10^2$ J

72b. Chapter 6, Problem 72b
    (A) 4.78 m/s
    (B) 5.00 m/s
    (C) 5.77 m/s
    (D) 7.07 m/s
    (E) 9.99 m/s

## Additional Problems

74a. Chapter 6, Problem 74a
    (A) $3.9 \times 10^4$ J
    (B) $2.2 \times 10^5$ J
    (C) $1.3 \times 10^4$ J
    (D) $3.4 \times 10^5$ J
    (E) $1.7 \times 10^5$ J

74b. Chapter 6, Problem 74b
    (A) The work is *negative* since the force must be greater than the weight of the ball.
    (B) The work is *positive* since the force is perpendicular to the displacement.
    (C) The work is *negative* since the force and displacement are in opposite directions.
    (D) The work is *positive* since the force and displacement are in the same direction.
    (E) The work is *negative* since the force must do work against gravity.

76. Chapter 6, Problem 76
    (A) 1590 J
    (B) 3180 J
    (C) 4490 J
    (D) 5230 J
    (E) 6260 J

## HOMEWORK CHOICES

78. Chapter 6, Problem 78
    - (A) 16.5 m
    - (B) 12.8 m
    - (C) 11.4 m
    - (D) 9.23 m
    - (E) 7.62 m

80. Chapter 6, Problem 80
    **Note**: in the following table the forces are designated as follows: $W_P$ = work done by the applied force, $W_f$ = work done by the frictional force, $W_N$ = work done by the normal force, and $W_g$ = work done by the gravitational force.

    |     | $W_P$ | $W_f$ | $W_N$ | $W_g$ |
    | --- | --- | --- | --- | --- |
    | (A) | $1.0 \times 10^3$ J | $-1.0 \times 10^3$ J | zero | zero |
    | (B) | $1.6 \times 10^3$ J | $-4.7 \times 10^2$ J | $+3.8 \times 10^3$ J | $-3.8 \times 10^3$ J |
    | (C) | $1.0 \times 10^3$ J | $-9.4 \times 10^2$ J | zero | zero |
    | (D) | $1.0 \times 10^3$ J | $-1.0 \times 10^3$ J | $-3.8 \times 10^3$ J | $+3.8 \times 10^3$ J |
    | (E) | $9.4 \times 10^2$ J | $-9.4 \times 10^2$ J | zero | zero |

82. Chapter 6, Problem 82
    - (A) 2.98 m/s
    - (B) 3.13 m/s
    - (C) 4.43 m/s
    - (D) 8.28 m/s
    - (E) 11.7 m/s

84. Chapter 6, Problem 84
    - (A) 38°
    - (B) 41°
    - (C) 45°
    - (D) 48°
    - (E) 56°

Homework Choices
# CHAPTER 7  Impulse and Momentum

## Section 7.1 The Impulse-Momentum Theorem

2. Chapter 7, Problem 2
   (A) $1.8 \times 10^3$ m/s
   (B) $1.0 \times 10^3$ m/s
   (C) $6.0 \times 10^2$ m/s
   (D) $3.5 \times 10^2$ m/s
   (E) $1.8 \times 10^2$ m/s

4a. Chapter 7, Problem 4a
   (A) $1.79 \times 10^{29}$ kg · m/s
   (B) $2.00 \times 10^{20}$ kg · m/s
   (C) $4.00 \times 10^{29}$ kg · m/s
   (D) $8.94 \times 10^{19}$ kg · m/s
   (E) $1.11 \times 10^{26}$ kg · m/s

4b. Chapter 7, Problem 4b
   (A) Yes, the direction of the earth's linear momentum is constant.
   (B) No, the earth's velocity changes direction as it travels in its nearly circular orbit around the sun; and the earth's gravitational force on the sun causes the earth's linear momentum to change.
   (C) No, the earth's velocity changes direction as it travels in its nearly circular orbit around the sun; and the sun's gravitational force causes the earth's linear momentum to change.
   (D) No, the earth's velocity changes direction as it travels in its nearly circular orbit around the sun, but no force is required to change the earth's linear momentum.
   (E) No, the direction of the earth's velocity is constant, but the sun's gravitational force causes the earth's linear momentum to change.

6a. Chapter 7, Problem 6a
   (A) 478 kg · m/s, parallel to the ball's velocity
   (B) 2.6 kg · m/s, perpendicular to the ball's velocity
   (C) 1.3 kg · m/s, parallel to the ball's velocity
   (D) 55 kg · m/s, perpendicular to the ball's velocity
   (E) 220 kg · m/s, parallel to the ball's velocity

6b. Chapter 7, Problem 6b
   (A) 1.3 N, parallel to the ball's velocity
   (B) 55 N, perpendicular to the ball's velocity
   (C) 478 N, parallel to the ball's velocity
   (D) 2.6 N, perpendicular to the ball's velocity
   (E) 220 N, parallel to the ball's velocity

8. Chapter 7, Problem 8
   (A) 3.6 N, upward
   (B) 0.90 N, upward
   (C) 0.90 N, downward
   (D) 1.8 N, upward
   (E) 1.8 N, downward

10. Chapter 7, Problem 10
    (A) 0.75 m
    (B) 1.4 m
    (C) 1.9 m
    (D) 2.4 m
    (E) 2.8 m

12. Chapter 7, Problem 12
    (A) 4.28 N·s, upward
    (B) 3.02 N·s, upward
    (C) 3.02 N·s, downward
    (D) 6.05 N·s, upward
    (E) 6.05 N·s, downward

# HOMEWORK CHOICES

14. Chapter 7, Problem 14
    - (A) 626 N
    - (B) 214 N
    - (C) 539 N
    - (D) 344 N
    - (E) 2160 N

## Section 7.2 The Principle of Conservation of Linear Momentum

16. Chapter 7, Problem 16
    - (A) $-1.0 \times 10^{-4}$ m/s
    - (B) $-1.5 \times 10^{-4}$ m/s
    - (C) $-3.0 \times 10^{-4}$ m/s
    - (D) $-4.5 \times 10^{-4}$ m/s
    - (E) $-6.0 \times 10^{-4}$ m/s

18. Chapter 7, Problem 18
    - (A) 5.0 kg
    - (B) 3.0 kg
    - (C) 1.5 kg
    - (D) 2.3 kg
    - (E) 4.5 kg

20a. Chapter 7, Problem 20a
    - (A) −0.14 m/s
    - (B) −0.22 m/s
    - (C) −0.082 m/s
    - (D) −0.31 m/s
    - (E) −0.047 m/s

20b. Chapter 7, Problem 20b
    - (A) $-4.3 \times 10^{-2}$ m/s
    - (B) $-2.2 \times 10^{-3}$ m/s
    - (C) $-8.2 \times 10^{-3}$ m/s
    - (D) $-7.1 \times 10^{-3}$ m/s
    - (E) $-7.4 \times 10^{-3}$ m/s

22. Chapter 7, Problem 22
    - (A) 0.94 m
    - (B) 1.2 m
    - (C) 1.5 m
    - (D) 1.8 m
    - (E) 2.1 m

24. Chapter 7, Problem 24
    - (A) 0.49
    - (B) 0.58
    - (C) 1.5
    - (D) 1.8
    - (E) 2.0

## Section 7.3 and Section 7.4 Collisions in One and Two Dimensions

26. Chapter 7, Problem 26
    - (A) 0.25 m/s
    - (B) 0.50 m/s
    - (C) 2.0 m/s
    - (D) 4.0 m/s
    - (E) 6.0 m/s

28. Chapter 7, Problem 28
    - (A) 0.072 m
    - (B) 0.104 m
    - (C) 0.149 m
    - (D) 0.208 m
    - (E) 0.360 m

30. Chapter 7, Problem 30
    - (A) +24.8 m/s
    - (B) +4.54 m/s
    - (C) +12.4 m/s
    - (D) +6.42 m/s
    - (E) +9.09 m/s

32a. Chapter 7, Problem 32a
    - (A) 1.7 m/s
    - (B) 2.6 m/s
    - (C) 3.4 m/s
    - (D) 4.1 m/s
    - (E) 7.3 m/s

# Homework: Impulse and Momentum

32b. Chapter 7, Problem 32b
(A) 1.7 m/s
(B) 2.6 m/s
(C) 3.4 m/s
(D) 4.1 m/s
(E) 7.3 m/s

34a. Chapter 7, Problem 34a
The direction of the velocity of the person on the sled is in the same direction as the person was moving before jumping onto the sled. Choose the correct speed:
(A) 1.20 m/s
(B) 2.96 m/s
(C) 1.56 m/s
(D) 3.46 m/s
(E) 3.17 m/s

34b. Chapter 7, Problem 34b
(A) 0.005 39
(B) 0.0120
(C) 0.0148
(D) 0.0171
(E) 0.0207

36. Chapter 7, Problem 36
(A) $2.54 \times 10^{-3}$ kg
(B) $3.06 \times 10^{-3}$ kg
(C) $3.58 \times 10^{-3}$ kg
(D) $4.50 \times 10^{-3}$ kg
(E) $5.75 \times 10^{-3}$ kg

38. Chapter 7, Problem 38
(A) 0.56 m/s, to the left
(B) 0.56 m/s, to the right
(C) 0.28 m/s, to the left
(D) 0.28 m/s, to the right
(E) 0.76 m/s, to the right

40. Chapter 7, Problem 40
Note: The ball with an initial speed of 7.0 m/s is labeled "ball **1**;" and the other ball with an initial speed of −4.0 m/s is labeled "ball **2**."

|     | ball **1** | ball **2** |
| --- | --- | --- |
| (a) | −4.0 m/s | +7.0 m/s |
| (b) | −7.0 m/s | +4.0 m/s |
| (c) | zero | zero |
| (d) | −3.0 m/s | +3.0 m/s |
| (e) | −5.5 m/s | +5.5 m/s |

## Section 7.5  *Center of Mass*

42a. Chapter 7, Problem 42a
(A) −1.0 m/s
(B) +1.0 m/s
(C) −2.0 m/s
(D) +2.0 m/s
(E) −0.40 m/s

42b. Chapter 7, Problem 42b
(A) −1.0 m/s
(B) +1.0 m/s
(C) −2.0 m/s
(D) +2.0 m/s
(E) −0.40 m/s

42c. Chapter 7, Problem 42c
(A) The velocity of the center of mass should be *less than* the common velocity after the collision because the combined masses after the collision is greater than the individual masses before the collision.
(B) The velocity of the center of mass should be *the same* as the common velocity after the collision because all points on the two coupled cars move at the same speed.
(C) The velocity of the center of mass should be *greater than* the common velocity after the collision because the initial speed of the second car was greater than the common velocity after the collision.

# HOMEWORK CHOICES

44. Chapter 7, Problem 44
    - (A) 0.25
    - (B) 0.50
    - (C) 1.0
    - (D) 2.0
    - (E) 4.0

## Additional Problems

46. Chapter 7, Problem 46
    - (A) $7.2 \times 10^{-2}$
    - (B) $3.6 \times 10^{-2}$
    - (C) $1.4 \times 10^{-3}$
    - (D) $9.1 \times 10^{-3}$
    - (E) $2.8 \times 10^{-4}$

48a. Chapter 7, Problem 48a

| | total linear momentum | total kinetic energy |
|---|---|---|
| (A) | $1.40 \times 10^2$ kg·m/s | $1.16 \times 10^2$ J |
| (B) | $2.20 \times 10^2$ kg·m/s | $4.40 \times 10^2$ J |
| (C) | $3.60 \times 10^2$ kg·m/s | $7.20 \times 10^2$ J |
| (D) | $5.80 \times 10^2$ kg·m/s | $1.16 \times 10^3$ J |
| (E) | $7.20 \times 10^3$ kg·m/s | $2.20 \times 10^3$ J |

48b. Chapter 7, Problem 48b

| | total linear momentum | total kinetic energy |
|---|---|---|
| (A) | $1.40 \times 10^2$ kg·m/s | $1.16 \times 10^2$ J |
| (B) | $2.20 \times 10^2$ kg·m/s | $4.40 \times 10^2$ J |
| (C) | $3.60 \times 10^2$ kg·m/s | $7.20 \times 10^2$ J |
| (D) | $5.80 \times 10^2$ kg·m/s | $1.16 \times 10^3$ J |
| (E) | $7.20 \times 10^3$ kg·m/s | $2.20 \times 10^3$ J |

50a. Chapter 7, Problem 50a
   - (A) $-4.4 \times 10^3$ N
   - (B) $-2.2 \times 10^3$ N
   - (C) $-1.1 \times 10^3$ N
   - (D) $+2.2 \times 10^3$ N
   - (E) $+4.4 \times 10^3$ N

50b. Chapter 7, Problem 50b
   - (A) $-4.4 \times 10^3$ N
   - (B) $-2.2 \times 10^3$ N
   - (C) $-1.1 \times 10^3$ N
   - (D) $+2.2 \times 10^3$ N
   - (E) $+4.4 \times 10^3$ N

52a. Chapter 7, Problem 52a
   - (A) −1.05 m/s
   - (B) −2.10 m/s
   - (C) −0.83 m/s
   - (D) +1.27 m/s
   - (E) +2.53 m/s

52b. Chapter 7, Problem 52b
   - (A) −1.05 m/s
   - (B) −2.10 m/s
   - (C) −0.83 m/s
   - (D) +1.27 m/s
   - (E) +2.53 m/s

54. Chapter 7, Problem 54
   - (A) $2.0 \times 10^3$ N, due west
   - (B) $5.5 \times 10^2$ N, due west
   - (C) $5.5 \times 10^2$ N, due east
   - (D) $2.6 \times 10^3$ N, due west
   - (E) $2.6 \times 10^3$ N, due east

# Homework: Impulse and Momentum

56a. Chapter 7, Problem 56a
- (A) 1.91 m/s, upward
- (B) 2.02 m/s, upward
- (C) 2.13 m/s, upward
- (D) 2.29 m/s, upward
- (E) zero

56b. Chapter 7, Problem 56b
- (A) 0.119 m
- (B) 0.220 m
- (C) 0.231 m
- (D) 0.292 m
- (E) zero

58. Chapter 7, Problem 58

| | $x_{cm}$ | $y_{cm}$ |
|---|---|---|
| (A) | zero | zero |
| (B) | $2.53 \times 10^{-10}$ m | $5.16 \times 10^{-11}$ m |
| (C) | $2.28 \times 10^{-10}$ m | $1.11 \times 10^{-10}$ m |
| (D) | $2.53 \times 10^{-10}$ m | zero |
| (E) | $2.28 \times 10^{-10}$ m | zero |

60. Chapter 7, Problem 60
- (A) 8.19 m/s
- (B) 9.28 m/s
- (C) 9.66 m/s
- (D) 18.6 m/s
- (E) 19.4 m/s

# Homework Choices
## CHAPTER 8 — Rotational Kinematics

*Section 8.1 Rotational Motion and Angular Displacement*
*Section 8.2 Angular Velocity and Angular Momentum*

2a. Chapter 8, Problem 2a
    (A) 1.3 rad      (C) 4.7 rad      (E) 12 rad
    (B) 2.2 rad      (D) 9.4 rad

2b. Chapter 8, Problem 2b
    (A) 1.3 rad      (C) 4.7 rad      (E) 12 rad
    (B) 2.2 rad      (D) 9.4 rad

4. Chapter 8, Problem 4
    (A) 61 rad/s      (C) 190 rad/s      (E) 310 rad/s
    (B) 95 rad/s      (D) 230 rad/s

6. Chapter 8, Problem 6
    (A) zero      (C) $-3.2 \times 10^{-3}$ rad/s$^2$      (E) $-9.0 \times 10^{-4}$ rad/s$^2$
    (B) $-6.4 \times 10^{-3}$ rad/s$^2$      (D) $-1.8 \times 10^{-3}$ rad/s$^2$

8. Chapter 8, Problem 8
    (A) 1.3 s      (C) 0.91 s      (E) 0.55 s
    (B) 1.1 s      (D) 0.77 s

10. Chapter 8, Problem 10
    (A) 825 m      (C) 977 m      (E) 1470 m
    (B) 850 m      (D) 1120 m

12. Chapter 8, Problem 12
    (A) 600 s      (C) 2400 s      (E) 1200 s
    (B) 3600 s      (D) 1800 s

14. Chapter 8, Problem 14
    (A) $1.00 \times 10^{-2}$ s      (C) $2.00 \times 10^{-2}$ s      (E) $4.50 \times 10^{-2}$ s
    (B) $1.50 \times 10^{-2}$ s      (D) $3.00 \times 10^{-2}$ s

16a. Chapter 8, Problem 16a
    (A) $\theta_{moon} = 5.07 \times 10^{-3}$ rad, $\theta_{sun} = 5.19 \times 10^{-3}$ rad
    (B) $\theta_{moon} = 1.43 \times 10^{-3}$ rad, $\theta_{sun} = 2.11 \times 10^{-3}$ rad
    (C) $\theta_{moon} = 8.83 \times 10^{-3}$ rad, $\theta_{sun} = 1.03 \times 10^{-2}$ rad
    (D) $\theta_{moon} = 6.72 \times 10^{-3}$ rad, $\theta_{sun} = 7.51 \times 10^{-3}$ rad
    (E) $\theta_{moon} = 9.04 \times 10^{-3}$ rad, $\theta_{sun} = 9.27 \times 10^{-3}$ rad

# Homework: Rotational Kinematics

16b. Chapter 8, Problem 16b
   (A) A total eclipse is possible since the moon subtends a slightly larger angle than the sun, as measured by a person standing on the earth.
   (B) A total eclipse is possible since an observer can look along the line connecting the center of the moon and the center of the sun.
   (C) A total eclipse is not possible since the path of the moon never crosses exactly in front of the sun for any observer on earth.
   (D) A total eclipse is not possible since the sun subtends a slightly larger angle than the moon, as measured by a person standing on the earth.
   (E) A total eclipse is not possible since no observer can look along the line connecting the center of the moon and the center of the sun.

16c. Chapter 8, Problem 16c
   (A) 100.0 %
   (B) 95.1 %
   (C) 92.7 %
   (D) 90.1%
   (E) 88.7 %

## Section 8.3 The Equations of Rotational Kinematics

18a. Chapter 8, Problem 18a
   (A) −44 rad/s$^2$
   (B) −22 rad/s$^2$
   (C) −17 rad/s$^2$
   (D) −5.9 rad/s$^2$
   (E) −2.9 rad/s$^2$

18b. Chapter 8, Problem 18b
   (A) 5.1 s
   (B) 2.5 s
   (C) 0.88 s
   (D) 0.68 s
   (E) 0.34 s

20. Chapter 8, Problem 20
   (A) 2.30 s
   (B) 3.11 s
   (C) 4.50 s
   (D) 4.76 s
   (E) 5.22 s

22. Chapter 8, Problem 22
   (A) 11 rad/s$^2$
   (B) 22 rad/s$^2$
   (C) 44 rad/s$^2$
   (D) 66 rad/s$^2$
   (E) 88 rad/s$^2$

24. Chapter 8, Problem 24
   (A) +0.0803 rad/s$^2$
   (B) −0.272 rad/s$^2$
   (C) −0.504 rad/s$^2$
   (D) +0.544 rad/s$^2$
   (E) +0.880 rad/s$^2$

26. Chapter 8, Problem 26
   (A) 5.62
   (B) 1.79
   (C) 4.52 s
   (D) 2.83 s
   (E) 3.37 s

28. Chapter 8, Problem 28
   (A) 0.74 rev
   (B) 0.81 rev
   (C) 1.6 rev
   (D) 2.1 rev
   (E) 2.4 rev

## Section 8.4 Angular Variables and Tangential Variables

30. Chapter 8, Problem 30
    - (A) 0.013 rad/s
    - (B) 0.042 rad/s
    - (C) 4.7 rad/s
    - (D) 9.4 rad/s
    - (E) 49 rad/s

32a. Chapter 8, Problem 32a
    - (A) $2.6 \times 10^5$ m/s
    - (B) $1.7 \times 10^5$ m/s
    - (C) $5.5 \times 10^4$ m/s
    - (D) $4.3 \times 10^4$ m/s
    - (E) $3.6 \times 10^3$ m/s

32b. Chapter 8, Problem 32b
    - (A) $2.6 \times 10^8$ y
    - (B) $1.7 \times 10^8$ y
    - (C) $5.5 \times 10^7$ y
    - (D) $4.3 \times 10^7$ y
    - (E) $3.6 \times 10^6$ y

34a. Chapter 8, Problem 34a
    - (A) $1.20 \times 10^2$ m/s
    - (B) $4.66 \times 10^2$ m/s
    - (C) $6.38 \times 10^2$ m/s
    - (D) $2.39 \times 10^3$ m/s
    - (E) $3.19 \times 10^3$ m/s

34b. Chapter 8, Problem 34b
    - (A) 62.4°
    - (B) 65.9°
    - (C) 68.1°
    - (D) 70.6°
    - (E) 74.0°

36. Chapter 8, Problem 36
    - (A) 0.817 m
    - (B) 0.693 m
    - (C) 0.500 m
    - (D) 0.125 m
    - (E) 0.118 m

38a. Chapter 8, Problem 38a
    - (A) 5.42 rad/s
    - (B) 2.56 rad/s
    - (C) 4.43 rad/s
    - (D) 3.61 rad/s
    - (E) 13.1 rad/s

38b. Chapter 8, Problem 38b
    - (A) 1.31 rad/s$^2$
    - (B) 3.26 rad/s$^2$
    - (C) 5.42 rad/s$^2$
    - (D) 6.53 rad/s$^2$
    - (E) 4.17 rad/s$^2$

40. Chapter 8, Problem 40
    - (A) 0.393 s
    - (B) 0.202 s
    - (C) 0.143 s
    - (D) 0.314 s
    - (E) 0.101 s

## Section 8.5 Centripetal Acceleration and Tangential Acceleration

42. Chapter 8, Problem 42
    - (A) 1.8
    - (B) 0.77
    - (C) 1.6
    - (D) 0.56
    - (E) 1.3

# Homework: Rotational Kinematics

44a. Chapter 8, Problem 44a
- (A) 1.29 m/s
- (B) 1.35 m/s
- (C) 1.41 m/s
- (D) 1.47 m/s
- (E) 1.53 m/s

44b. Chapter 8, Problem 44b
- (A) 40.4 rad/s$^2$
- (B) 35.9 rad/s$^2$
- (C) 31.4 rad/s$^2$
- (D) 46.2 rad/s$^2$
- (E) 52.1 rad/s$^2$

46a. Chapter 8, Problem 46a
- (A) 94.1 cm/s
- (B) 84.6 cm/s
- (C) 78.0 cm/s
- (D) 53.2 cm/s
- (E) 42.0 cm/s

46b. Chapter 8, Problem 46b
- (A) $1.65 \times 10^2$ cm/s$^2$
- (B) $2.60 \times 10^2$ cm/s$^2$
- (C) $1.40 \times 10^3$ cm/s$^2$
- (D) $2.45 \times 10^3$ cm/s$^2$
- (E) $1.95 \times 10^3$ cm/s$^2$

48. Chapter 8, Problem 48
- (A) 0.213 s
- (B) 0.422 s
- (C) 0.111 s
- (D) 0.045 s
- (E) zero

## Section 8.6  Rolling Motion

50. Chapter 8, Problem 50
- (A) 3900 m
- (B) 4300 m
- (C) 7800 m
- (D) 8600 m
- (E) 19 100 m

52a. Chapter 8, Problem 52a
- (A) 15.0 rad/s
- (B) 32.6 rad/s
- (C) 45.5 rad/s
- (D) 62.7 rad/s
- (E) 79.6 rad/s

52b. Chapter 8, Problem 52b
- (A) 1.04 m/s
- (B) 1.50 m/s
- (C) 4.39 m/s
- (D) 7.96 m/s
- (E) 9.45 m/s

54a. Chapter 8, Problem 54a
- (A) 3.97 rad/s
- (B) 7.50 rad/s
- (C) 7.93 rad/s
- (D) 1.30 rad/s
- (E) 1.73 rad/s

54b. Chapter 8, Problem 54b
- (A) $-3.97 \times 10^{-3}$ rad/s$^2$
- (B) $-7.50 \times 10^{-2}$ rad/s$^2$
- (C) $-7.93 \times 10^{-3}$ rad/s$^2$
- (D) $-1.30 \times 10^{-3}$ rad/s$^2$
- (E) $-1.73 \times 10^{-3}$ rad/s$^2$

56. Chapter 8, Problem 56
- (A) 8.34 rad
- (B) 11.8 rad
- (C) 14.3 rad
- (D) 16.7 rad
- (E) 20.1 rad

# HOMEWORK CHOICES    Chapter 8

58a. Chapter 8, Problem 58a
- (A) 9.45 rad/s, clockwise
- (B) 8.33 rad/s, counterclockwise
- (C) 18.5 rad/s, counterclockwise
- (D) 14.7 rad/s, clockwise
- (E) 16.7 rad/s, clockwise

58b. Chapter 8, Problem 58b
- (A) 14.7 rad/s, clockwise
- (B) 18.5 rad/s, counterclockwise
- (C) 16.7 rad/s clockwise
- (D) 8.33 rad/s, counterclockwise
- (E) 9.45 rad/s, clockwise

60. Chapter 8, Problem 60
- (A) 8.26 rad
- (B) 17.2 rad
- (C) 20.6 rad
- (D) 28.0 rad
- (E) 31.1 rad

## *Additional Problems*

62a. Chapter 8, Problem 62a
- (A) 0.105 rad/s, clockwise
- (B) 0.130 rad/s, counterclockwise
- (C) 0.0127 rad/s, clockwise
- (D) 0.0167 rad/s, counterclockwise
- (E) $1.45 \times 10^{-4}$ rad/s, clockwise

62b. Chapter 8, Problem 62b
- (A) 0.105 rad/s, clockwise
- (B) 0.130 rad/s, counterclockwise
- (C) 0.0127 rad/s, clockwise
- (D) 0.0167 rad/s, counterclockwise
- (E) $1.45 \times 10^{-4}$ rad/s, clockwise

64a. Chapter 8, Problem 64a
- (A) 98.2 rad/s
- (B) 72.0 rad/s
- (C) 54.0 rad/s
- (D) 27.0 rad/s
- (E) 48.6 rad/s

64b. Chapter 8, Problem 64b
- (A) 884 rad
- (B) 648 rad
- (C) 486 rad
- (D) 243 rad
- (E) 437 rad

66a. Chapter 8, Problem 66a
- (A) 0.339 m/s$^2$
- (B) 0.360 m/s$^2$
- (C) 0.583 m/s$^2$
- (D) 1.72 m/s$^2$
- (E) 6.46 m/s$^2$

66b. Chapter 8, Problem 66b
- (A) 18.7°
- (B) 19.8°
- (C) 26.3°
- (D) 31.0°
- (E) 38.8°

68. Chapter 8, Problem 68
- (A) 10.7 m
- (B) 9.15 m
- (C) 8.00 m
- (D) 7.40 m
- (E) 6.05 m

70. Chapter 8, Problem 70
- (A) 4.49 s
- (B) 3.28 s
- (C) 7.45 s
- (D) 6.49 s
- (E) 2.36 s

# Homework: Rotational Kinematics

72. Chapter 8, Problem 72
    - (A) 9.80 rad/s
    - (B) 14.8 rad/s
    - (C) 19.6 rad/s
    - (D) 22.6 rad/s
    - (E) 29.7 rad/s

74. Chapter 8, Problem 74
    - (A) 3/4
    - (B) 2/5
    - (C) 2/3
    - (D) 1/4
    - (E) 1/2

# Homework Choices
## CHAPTER 9 — Rotational Dynamics

### Section 9.1 The Effects of Forces and Torques on the Motion of Rigid Objects

2. Chapter 9, Problem 2
   - (A) 58.7 N
   - (B) 160 N
   - (C) 120 N
   - (D) 210 N
   - (E) 250 N

4. Chapter 9, Problem 4
   - (A) $8.50 \times 10^2$ N·m
   - (B) $4.87 \times 10^3$ N·m
   - (C) $9.73 \times 10^3$ N·m
   - (D) $1.70 \times 10^3$ N·m
   - (E) $1.35 \times 10^3$ N·m

6. Chapter 9, Problem 6
   - (A) 23.3 N·m, CW
   - (B) 23.3 N·m, CCW
   - (C) 14.3 N·m, CW
   - (D) 14.3 N·m, CCW
   - (E) 48.6 N·m, CW

8a. Chapter 9, Problem 8a
   - (A) $\tau = FL/4$
   - (B) $\tau = FL/2$
   - (C) $\tau = FL$
   - (D) $\tau = 2FL$
   - (E) $\tau = 4FL$

8b. Chapter 9, Problem 8b
   - (A) $\tau = FL/4$
   - (B) $\tau = FL/2$
   - (C) $\tau = FL$
   - (D) $\tau = 2FL$
   - (E) $\tau = 4FL$

8c. Chapter 9, Problem 8c
   - (A) $\tau = FL/4$
   - (B) $\tau = FL/2$
   - (C) $\tau = FL$
   - (D) $\tau = 2FL$
   - (E) $\tau = 4FL$

10. Chapter 9, Problem 10
    - (A) $a = 0.100$ m, $b = 0.300$ m
    - (B) $a = 0.150$ m, $b = 0.200$ m
    - (C) $a = 0.200$ m, $b = 0.200$ m
    - (D) $a = 0.200$ m, $b = 0.300$ m
    - (E) $a = 0.300$ m, $b = 0.100$ m

### Section 9.2 Rigid Objects in Equilibrium
### Section 9.3 Center of Gravity

12. Chapter 9, Problem 12
    - (A) 5.1 kg
    - (B) 7.2 kg
    - (C) 21 kg
    - (D) 14 kg
    - (E) 16 kg

14. Chapter 9, Problem 14
    - (A) 0.60 m
    - (B) 0.70 m
    - (C) 0.80 m
    - (D) 0.90 m
    - (E) 1.0 m

## Homework: Rotational Dynamics

16. Chapter 9, Problem 16
    - (A) 1200 N
    - (B) 880 N
    - (C) 640 N
    - (D) 510 N
    - (E) 440 N

18a. Chapter 9, Problem 18a
    - (A) 0.574 m
    - (B) 0.761 m
    - (C) 0.776 m
    - (D) 0.488 m
    - (E) 0.529 m

18b. Chapter 9, Problem 18b
    - (A) 0.574 m
    - (B) 0.761 m
    - (C) 0.776 m
    - (D) 0.488 m
    - (E) 0.529 m

20a. Chapter 9, Problem 20a
    - (A) $1.60 \times 10^5$ N
    - (B) $2.40 \times 10^5$ N
    - (C) $3.10 \times 10^5$ N
    - (D) $4.20 \times 10^5$ N
    - (E) $5.30 \times 10^5$ N

20b. Chapter 9, Problem 20b
    - (A) $1.60 \times 10^5$ N
    - (B) $2.40 \times 10^5$ N
    - (C) $3.10 \times 10^5$ N
    - (D) $4.20 \times 10^5$ N
    - (E) $5.30 \times 10^5$ N

22a. Chapter 9, Problem 22a
    - (A) $4.33 \times 10^2$ N
    - (B) $5.00 \times 10^2$ N
    - (C) $2.12 \times 10^2$ N
    - (D) $6.35 \times 10^2$ N
    - (E) $2.50 \times 10^2$ N

22b. Chapter 9, Problem 22b
    - (A) $2.12 \times 10^2$ N
    - (B) $2.50 \times 10^2$ N
    - (C) $4.33 \times 10^2$ N
    - (D) $5.00 \times 10^2$ N
    - (E) $6.35 \times 10^2$ N

22c. Chapter 9, Problem 22c
    - (A) $4.33 \times 10^2$ N
    - (B) $6.35 \times 10^2$ N
    - (C) $5.00 \times 10^2$ N
    - (D) $2.50 \times 10^2$ N
    - (E) $2.12 \times 10^2$ N

24a. Chapter 9, Problem 24a
    - (A) $1.3 \times 10^4$ N
    - (B) $1.6 \times 10^4$ N
    - (C) $2.2 \times 10^4$ N
    - (D) $4.2 \times 10^4$ N
    - (E) $8.2 \times 10^3$ N

24b. Chapter 9, Problem 24b
    - (A) $1.3 \times 10^4$ N
    - (B) $1.6 \times 10^4$ N
    - (C) $2.2 \times 10^4$ N
    - (D) $4.2 \times 10^4$ N
    - (E) $8.2 \times 10^3$ N

26. Chapter 9, Problem 26
    - (A) 21 N
    - (B) 37 N
    - (C) 52 N
    - (D) 61 N
    - (E) 69 N

HOMEWORK CHOICES    Chapter 9    377

28. Chapter 9, Problem 28
    (A) 2.3 m
    (B) 1.9 m
    (C) 2.6 m
    (D) 1.7 m
    (E) 3.4 m

## Section 9.4 Newton's Second Law for Rotational Motion about a Fixed Axis

30a. Chapter 9, Problem 30a

|     | $I_1$ | $I_2$ | $I_3$ |
|-----|-------|-------|-------|
| (A) | 1.20 kg·m² | 7.00 kg·m² | 1.44 kg·m² |
| (B) | 0.360 kg·m² | 4.90 kg·m² | 1.38 kg·m² |
| (C) | 0.414 kg·m² | 3.81 kg·m² | 0.904 kg·m² |
| (D) | 1.44 kg·m² | 6.64 kg·m² | 1.20 kg·m² |
| (E) | 0.360 kg·m² | 6.64 kg·m² | 1.44 kg·m² |

30b. Chapter 9, Problem 30b
    (A) 9.64 kg·m²
    (B) 4.90 kg·m²
    (C) 5.12 kg·m²
    (D) 9.28 kg·m²
    (E) 6.64 kg·m²

30c. Chapter 9, Problem 30c
    (A) The smallest mass does not necessarily contribute the smallest amount to the moment of inertia because the moment of inertia also depends on the position squared.
    (B) The smallest mass does not necessarily contribute the smallest amount to the moment of inertia because the moment of inertia also depends linearly on the position.
    (C) The smallest mass does necessarily contribute the smallest amount to the moment of inertia because the moment of inertia depends linearly on the mass.
    (D) The smallest mass does necessarily contribute the smallest amount to the moment of inertia because the moment of inertia also on the mass squared.
    (E) The smallest mass does not necessarily contribute the smallest amount to the moment of inertia because the moment of inertia also depends on the inverse of the position squared.

32a. Chapter 9, Problem 32a
    (A) 22 N·m
    (B) −22 N·m
    (C) 11 N·m
    (D) −11 N·m
    (E) −35 N·m

32b. Chapter 9, Problem 32b
    (A) −11 rad/s²
    (B) −9.2 rad/s²
    (C) −7.4 rad/s²
    (D) −5.8 rad/s²
    (E) −3.6 rad/s²

34. Chapter 9, Problem 34
    (A) 1.00 rad/s²
    (B) 0.88 rad/s²
    (C) 0.60 rad/s²
    (D) 0.50 rad/s²
    (E) 0.16 rad/s²

36. Chapter 9, Problem 36
    (A) 1.28 kg
    (B) 6.49 kg
    (C) 2.02 kg
    (D) 4.66 kg
    (E) 2.37 kg

## Homework: Rotational Dynamics

38. Chapter 9, Problem 38
    - (A) 9.7 rev
    - (B) 6.6 rev
    - (C) 8.1 rev
    - (D) 5.5 rev
    - (E) 7.3 rev

40. Chapter 9, Problem 40
    - (A) 460 N
    - (B) 510 N
    - (C) 370 N
    - (D) 420 N
    - (E) 530 N

42. Chapter 9, Problem 42
    - (A) $I = (3/2)MR^2$
    - (B) $I = (1/2)MR^2$
    - (C) $I = (7/2)MR^2$
    - (D) $I = (5/2)MR^2$
    - (E) $I = 2MR^2$

44. Chapter 9, Problem 44
    - (A) 550 N·m
    - (B) 1100 N·m
    - (C) 1400 N·m
    - (D) 1700 N·m
    - (E) 1900 N·m

## Section 9.5 Rotational Work and Energy

46. Chapter 9, Problem 46
    - (A) 0.011
    - (B) 0.018
    - (C) 0.032
    - (D) 0.14
    - (E) 0.65

48a. Chapter 9, Problem 48a
    - (A) $7.92 \times 10^5$ J
    - (B) $9.02 \times 10^5$ J
    - (C) $4.37 \times 10^5$ J
    - (D) $4.51 \times 10^5$ J
    - (E) $3.55 \times 10^5$ J

48b. Chapter 9, Problem 48b
    - (A) $7.92 \times 10^4$ J
    - (B) $1.38 \times 10^4$ J
    - (C) $5.23 \times 10^5$ J
    - (D) $4.65 \times 10^5$ J
    - (E) $9.60 \times 10^4$ J

48c. Chapter 9, Problem 48c
    - (A) $9.74 \times 10^5$ J
    - (B) $9.02 \times 10^5$ J
    - (C) $4.37 \times 10^5$ J
    - (D) $4.51 \times 10^5$ J
    - (E) $8.88 \times 10^5$ J

50. Chapter 9, Problem 50
    - (A) $\sqrt{\dfrac{2}{5}}$
    - (B) $\sqrt{\dfrac{10}{7}}$
    - (C) $\sqrt{\dfrac{7}{5}}$
    - (D) $\sqrt{\dfrac{5}{7}}$
    - (E) $\sqrt{\dfrac{5}{2}}$

52. Chapter 9, Problem 52
    - (A) 3.59 m/s
    - (B) 7.67 m/s
    - (C) 4.26 m/s
    - (D) 6.83 m/s
    - (E) 5.89 m/s

54. Chapter 9, Problem 54
    (A) 2.4 m        (C) 1.6 m        (E) 2.0 m
    (B) 3.5 m        (D) 3.0 m

## Section 9.6 Angular Momentum

56. Chapter 9, Problem 56
    (A) 3.52 rad/s   (C) 4.10 rad/s   (E) 10.1 rad/s
    (B) 7.11 rad/s   (D) 1.22 rad/s

58. Chapter 9, Problem 58
    (A) 1.21 rad/s   (C) 1.83 rad/s   (E) 5.74 rad/s
    (B) 1.43 rad/s   (D) 2.87 rad/s

60a. Chapter 9, Problem 60a
    (A) 0.500 rad/s  (C) 2.00 rad/s   (E) 1.25 rad/s
    (B) 0.250 rad/s  (D) 1.00 rad/s

60b. Chapter 9, Problem 60b
    (A) 1.00 s       (C) 1.50 s       (E) 3.31 s
    (B) 1.33 s       (D) 2.99 s

62. Chapter 9, Problem 62
    (A) 0.249        (C) 0.188        (E) 0.573
    (B) 0.831        (D) 0.328

## Additional Problems

64. Chapter 9, Problem 64
    (A) $-3.32 \times 10^{-2}$ N·m   (C) $-4.92 \times 10^{-2}$ N·m   (E) $-5.18 \times 10^{-2}$ N·m
    (B) $-2.55 \times 10^{-1}$ N·m   (D) $-6.40 \times 10^{-1}$ N·m

66a. Chapter 9, Problem 66a
    (A) $1.16 \times 10^{4}$ N   (C) $1.92 \times 10^{3}$ N   (E) $2.28 \times 10^{3}$ N
    (B) $1.73 \times 10^{4}$ N   (D) $3.40 \times 10^{3}$ N

66b. Chapter 9, Problem 66b
    (A) $1.16 \times 10^{4}$ N   (C) $1.92 \times 10^{3}$ N   (E) $2.28 \times 10^{3}$ N
    (B) $1.73 \times 10^{4}$ N   (D) $3.40 \times 10^{3}$ N

68. Chapter 9, Problem 68

|     | T           | F           |
| --- | ----------- | ----------- |
| (a) | 61.3 N, down | 14.2 N, up  |
| (b) | 56.4 N, down | 70.6 N, up  |
| (c) | 14.2 N, down | 61.3 N, up  |
| (d) | 70.6 N, up  | 56.4 N, down |
| (e) | 56.4 N, up  | 70.6 N, down |

## Homework: Rotational Dynamics

70. Chapter 9, Problem 70
    - (A) 1/2
    - (B) 1/3
    - (C) 2
    - (D) 3
    - (E) 4

72a. Chapter 9, Problem 72a
    - (A) $2.12 \times 10^4$ N
    - (B) $2.35 \times 10^4$ N
    - (C) $2.52 \times 10^4$ N
    - (D) $1.06 \times 10^4$ N
    - (E) $5.85 \times 10^3$ N

72b. Chapter 9, Problem 72b
    - (A) $2.12 \times 10^4$ N
    - (B) $2.35 \times 10^4$ N
    - (C) $2.52 \times 10^4$ N
    - (D) $1.06 \times 10^4$ N
    - (E) $5.85 \times 10^3$ N

74. Chapter 9, Problem 74
    - (A) 3.5 rad/s
    - (B) 7.0 rad/s
    - (C) 14 rad/s
    - (D) 28 rad/s
    - (E) 56 rad/s

76. Chapter 9, Problem 76
    - (A) 6.00 s
    - (B) 2.81 s
    - (C) 1.50 s
    - (D) 4.24 s
    - (E) 2.12 s

# Homework Choices
## CHAPTER 10
# Elasticity and Simple Harmonic Motion

*Section 10.1 Elastic Deformation*
*Section 10.2 Stress, Strain, and Hooke's Law*

2. Chapter 10, Problem 2
   - (A) 35 m
   - (B) 31 m
   - (C) 27 m
   - (D) 16 m
   - (E) 11 m

4a. Chapter 10, Problem 4a
   - (A) $4.8 \times 10^7$ N/m$^2$
   - (B) $3.1 \times 10^7$ N/m$^2$
   - (C) $6.0 \times 10^6$ N/m$^2$
   - (D) $4.9 \times 10^6$ N/m$^2$
   - (E) $3.0 \times 10^5$ N/m$^2$

4b. Chapter 10, Problem 4b
   - (A) $3.0 \times 10^{-6}$ m
   - (B) $3.5 \times 10^{-6}$ m
   - (C) $4.0 \times 10^{-6}$ m
   - (D) $5.3 \times 10^{-6}$ m
   - (E) $6.0 \times 10^{-6}$ m

6. Chapter 10, Problem 6
   - (A) 36 m
   - (B) 75 m
   - (C) 190 m
   - (D) 260 m
   - (E) 380 m

8. Chapter 10, Problem 8
   - (A) $1.8 \times 10^{-4}$ m
   - (B) $2.7 \times 10^{-4}$ m
   - (C) $3.2 \times 10^{-4}$ m
   - (D) $4.1 \times 10^{-5}$ m
   - (E) $9.0 \times 10^{-5}$ m

10. Chapter 10, Problem 10
   - (A) $1.9 \times 10^{-4}$
   - (B) $2.8 \times 10^{-4}$
   - (C) $3.1 \times 10^{-4}$
   - (D) $3.6 \times 10^{-4}$
   - (E) $4.2 \times 10^{-4}$

12. Chapter 10, Problem 12
   - (A) $4.9 \times 10^4$ N
   - (B) $1.7 \times 10^4$ N
   - (C) $3.8 \times 10^4$ N
   - (D) $6.6 \times 10^4$ N
   - (E) $5.4 \times 10^4$ N

14a. Chapter 10, Problem 14a
   - (A) $2.5 \times 10^{-4}$
   - (B) $3.5 \times 10^{-4}$
   - (C) $4.9 \times 10^{-4}$
   - (D) $6.4 \times 10^{-4}$
   - (E) $8.9 \times 10^{-4}$

14b. Chapter 10, Problem 14b
   - (A) $2.1 \times 10^{-5}$ m
   - (B) $4.3 \times 10^{-5}$ m
   - (C) $7.5 \times 10^{-5}$ m
   - (D) $8.9 \times 10^{-5}$ m
   - (E) $9.7 \times 10^{-5}$ m

# Homework: Elasticity and Simple Harmonic Motion

16. Chapter 10, Problem 16
    - (A) $8.8 \times 10^{-7}$ m
    - (B) $3.4 \times 10^{-6}$ m
    - (C) $1.3 \times 10^{-6}$ m
    - (D) $1.7 \times 10^{-5}$ m
    - (E) $2.5 \times 10^{-5}$ m

18. Chapter 10, Problem 18
    - (A) $5.0 \times 10^{-4}$ m
    - (B) $2.5 \times 10^{-3}$ m
    - (C) $7.5 \times 10^{-4}$ m
    - (D) $2.0 \times 10^{-3}$ m
    - (E) $1.0 \times 10^{-3}$ m

20. Chapter 10, Problem 20
    - (A) 48
    - (B) 28
    - (C) 36
    - (D) 21
    - (E) 12

22. Chapter 10, Problem 22
    - (A) $4.3 \times 10^{-5}$ m
    - (B) $3.2 \times 10^{-5}$ m
    - (C) $2.1 \times 10^{-5}$ m
    - (D) $1.4 \times 10^{-5}$ m
    - (E) $2.5 \times 10^{-6}$ m

## *Section 10.3 The Ideal Spring and Simple Harmonic Motion*

24. Chapter 10, Problem 24
    - (A) $6.7 \times 10^{2}$ N/m
    - (B) $3.7 \times 10^{2}$ N/m
    - (C) $6.5 \times 10^{1}$ N/m
    - (D) $3.8 \times 10^{1}$ N/m
    - (E) $1.6 \times 10^{1}$ N/m

26. Chapter 10, Problem 26
    - (A) $1.2 \times 10^{-2}$ m
    - (B) $4.8 \times 10^{-2}$ m
    - (C) $6.9 \times 10^{-2}$ m
    - (D) $8.3 \times 10^{-2}$ m
    - (E) $7.5 \times 10^{-2}$ m

28. Chapter 10, Problem 28
    - (A) $1.0 \times 10^{-2}$ m
    - (B) $4.6 \times 10^{-2}$ m
    - (C) $6.3 \times 10^{-2}$ m
    - (D) $8.1 \times 10^{-2}$ m
    - (E) $5.5 \times 10^{-2}$ m

30. Chapter 10, Problem 30
    - (A) 0.311
    - (B) 0.236
    - (C) 0.437
    - (D) 0.367
    - (E) 0.218

32. Chapter 10, Problem 32
    - (A) 0.040 m
    - (B) 0.080 m
    - (C) 0.100 m
    - (D) 0.120 m
    - (E) 0.020 m

34a. Chapter 10, Problem 34a
    - (A) $3.7 \times 10^{3}$ N/m
    - (B) $1.0 \times 10^{3}$ N/m
    - (C) $3.0 \times 10^{3}$ N/m
    - (D) $1.7 \times 10^{3}$ N/m
    - (E) $4.8 \times 10^{3}$ N/m

HOMEWORK CHOICES        Chapter 10    383

34b. Chapter 10, Problem 34b
  (A) 0.510
  (B) 0.446
  (C) 0.402
  (D) 0.371
  (E) 0.340

## Section 10.4 Simple Harmonic Motion and the Reference Circle

36a. Chapter 10, Problem 36a
  (A) 0.152 m/s
  (B) 0.229 m/s
  (C) 0.435 m/s
  (D) 0.530 m/s
  (E) 0.337 m/s

36b. Chapter 10, Problem 36b
  (A) 0.283 m/s$^2$
  (B) 0.530 m/s$^2$
  (C) 1.06 m/s$^2$
  (D) 1.85 m/s$^2$
  (E) 3.70 m/s$^2$

38a. Chapter 10, Problem 38a
  (A) 0.080 m
  (B) 0.040 m
  (C) 0.020 m
  (D) 0.160 m
  (E) 0.320 m

38b. Chapter 10, Problem 38b
  (A) 1.6 rad/s
  (B) 0.80 rad/s
  (C) 0.40 rad/s
  (D) 0.20 rad/s
  (E) 3.2 rad/s

38c. Chapter 10, Problem 38c
  (A) 0.512 N/m
  (B) 0.750 N/m
  (C) 2.0 N/m
  (D) 5.0 N/m
  (E) 8.2 N/m

38d. Chapter 10, Problem 38d
  (A) −0.13 m/s
  (B) +0.13 m/s
  (C) −0.26 m/s
  (D) +0.26 m/s
  (E) zero

38e. Chapter 10, Problem 38e
  (A) 0.10 m/s$^2$
  (B) 0.20 m/s$^2$
  (C) 0.40 m/s$^2$
  (D) 0.80 m/s$^2$
  (E) 1.6 m/s$^2$

40. Chapter 10, Problem 40
  (A) $8.9 \times 10^3$ m/s
  (B) $5.6 \times 10^3$ m/s
  (C) $3.4 \times 10^3$ m/s
  (D) $2.8 \times 10^3$ m/s
  (E) $7.7 \times 10^3$ m/s

42a. Chapter 10, Problem 42a
  (A) 0.206 m
  (B) 0.362 m
  (C) 0.405 m
  (D) 0.552 m
  (E) 0.810 m

42b. Chapter 10, Problem 42b
  (A) 3.69 m/s
  (B) 2.26 m/s
  (C) 5.62 m/s
  (D) 4.59 m/s
  (E) 6.62 m/s

**Homework: Elasticity and Simple Harmonic Motion**

42c. Chapter 10, Problem 42c
(A) 36.2 m/s$^2$
(B) 56.7 m/s$^2$
(C) 63.9 m/s$^2$
(D) 24.4 m/s$^2$
(E) 42.1 m/s$^2$

44. Chapter 10, Problem 44
(A) 1.00
(B) 3.00
(C) 5.00
(D) 8.00
(E) 9.00

46. Chapter 10, Problem 46
(A) 26 Hz
(B) 31 Hz
(C) 42 Hz
(D) 66 Hz
(E) 120 Hz

## Section 10.5 *Energy and Simple Harmonic Motion*

48. Chapter 10, Problem 48
(A) 1.5 m/s
(B) 0.50 m/s
(C) 0.44 m/s
(D) 0.32 m/s
(E) 0.19 m/s

50. Chapter 10, Problem 50
(A) 66 m/s
(B) 32 m/s
(C) 23 m/s
(D) 18 m/s
(E) 7.8 m/s

52. Chapter 10, Problem 52
(A) 27.6 N/m
(B) 49.7 N/m
(C) 99.3 N/m
(D) 151 N/m
(E) 303 N/m

54. Chapter 10, Problem 54
(A) 2.3 m/s
(B) 3.4 m/s
(C) 4.0 m/s
(D) 8.0 m/s
(E) 16 m/s

56a. Chapter 10, Problem 56a

| | amplitude | frequency |
|---|---|---|
| (A) | $5.08 \times 10^{-2}$ m | 4.24 Hz |
| (B) | $3.59 \times 10^{-2}$ m | 4.24 Hz |
| (C) | $2.54 \times 10^{-2}$ m | 3.00 Hz |
| (D) | $3.59 \times 10^{-2}$ m | 6.00 Hz |
| (E) | $2.54 \times 10^{-2}$ m | 6.00 Hz |

56b. Chapter 10, Problem 56b

| | amplitude | frequency |
|---|---|---|
| (A) | $5.08 \times 10^{-2}$ m | 4.24 Hz |
| (B) | $3.59 \times 10^{-2}$ m | 4.24 Hz |
| (C) | $2.54 \times 10^{-2}$ m | 3.00 Hz |
| (D) | $3.59 \times 10^{-2}$ m | 6.00 Hz |
| (E) | $2.54 \times 10^{-2}$ m | 6.00 Hz |

58a. Chapter 10, Problem 58a
- (A) 24 m/s
- (B) 19 m/s
- (C) 4.9 m/s
- (D) 1.3 m/s
- (E) zero

58b. Chapter 10, Problem 58b
- (A) 24 rad/s
- (B) 19 rad/s
- (C) 4.9 rad/s
- (D) 2.2 rad/s
- (E) 1.3 rad/s

60. Chapter 10, Problem 60
- (A) 6.40 m/s
- (B) 5.06 m/s
- (C) 4.32 m/s
- (D) 2.53 m/s
- (E) 2.08 m/s

62. Chapter 10, Problem 62
- (A) 0.267 m
- (B) 0.138 m
- (C) 0.462 m
- (D) 0.365 m
- (E) 0.582 m

## *Section 10.6 The Pendulum*

64. Chapter 10, Problem 64
- (A) 36 m
- (B) 14 m
- (C) 21 m
- (D) 18 m
- (E) 45 m

66. Chapter 10, Problem 66
- (A) 0.44
- (B) 0.51
- (C) 0.72
- (D) 0.85
- (E) 0.89

68. Chapter 10, Problem 68
- (A) $\omega = \sqrt{\dfrac{g}{R}}$
- (B) $\omega = \sqrt{gR}$
- (C) $\omega = gR$
- (D) $\omega = \sqrt{\dfrac{mg}{R}}$
- (E) $\omega = \dfrac{mg}{R}$

70. Chapter 10, Problem 70
- (A) $L = (1/2)R$
- (B) $L = (4/3)R$
- (C) $L = (5/3)R$
- (D) $L = (7/5)R$
- (E) $L = (9/2)R$

## *Additional Problems*

72. Chapter 10, Problem 72
- (A) 0.982 m
- (B) 1.01 m
- (C) 0.983 m
- (D) 0.978 m
- (E) 0.995 m

74. Chapter 10, Problem 74
- (A) $6.8 \times 10^{-2}$ m
- (B) $2.4 \times 10^{-2}$ m
- (C) $8.9 \times 10^{-2}$ m
- (D) $3.8 \times 10^{-2}$ m
- (E) $1.3 \times 10^{-2}$ m

# Homework: Elasticity and Simple Harmonic Motion

76. Chapter 10, Problem 76
    - (A) 0.22 m
    - (B) 0.11 m
    - (C) 0.31 m
    - (D) 0.16 m
    - (E) 0.08 m

78. Chapter 10, Problem 78
    - (A) 0.039°
    - (B) 0.091°
    - (C) 0.056°
    - (D) 0.082°
    - (E) 0.043°

80. Chapter 10, Problem 80
    - (A) 8400
    - (B) 1200
    - (C) 4600
    - (D) 3800
    - (E) 9700

82. Chapter 10, Problem 82
    - (A) $F/2$
    - (B) $2F$
    - (C) $4F$
    - (D) $F/4$
    - (E) $F$

84. Chapter 10, Problem 84
    - (A) 206 m/s
    - (B) 618 m/s
    - (C) 848 m/s
    - (D) 921 m/s
    - (E) 1380 m/s

86. Chapter 10, Problem 86
    - (A) 0.206
    - (B) 0.508
    - (C) 0.612
    - (D) 0.806
    - (E) 0.962

88a. Chapter 10, Problem 88a
    - (A) $3.5 \times 10^3$ N/m
    - (B) $1.0 \times 10^3$ N/m
    - (C) $8.4 \times 10^2$ N/m
    - (D) $6.4 \times 10^2$ N/m
    - (E) $3.2 \times 10^2$ N/m

88b. Chapter 10, Problem 88b
    - (A) $1.1 \times 10^{-5}$ J
    - (B) $5.6 \times 10^{-6}$ J
    - (C) $2.8 \times 10^{-6}$ J
    - (D) $1.4 \times 10^{-6}$ J
    - (E) $7.0 \times 10^{-7}$ J

# Homework Choices
# CHAPTER 11    Fluids

## Section 11.1 Mass Density

2a. Chapter 11, Problem 2a
    (A) $1.2 \times 10^{18}$ kg/m$^2$
    (B) $3.7 \times 10^{18}$ kg/m$^2$
    (C) $4.9 \times 10^{18}$ kg/m$^2$
    (C) $7.4 \times 10^{18}$ kg/m$^2$
    (E) $1.5 \times 10^{19}$ kg/m$^2$

2b. Chapter 11, Problem 2b
    (A) $1.6 \times 10^{12}$ lb
    (B) $3.4 \times 10^{13}$ lb
    (C) $2.2 \times 10^{12}$ lb
    (D) $7.3 \times 10^{12}$ lb
    (E) $7.4 \times 10^{11}$ lb

4. Chapter 11, Problem 4
    (A) $7.7 \times 10^{-2}$ m
    (B) $6.2 \times 10^{-2}$ m
    (C) $4.1 \times 10^{-2}$ m
    (D) $1.9 \times 10^{-2}$ m
    (E) $7.0 \times 10^{-2}$ m

6. Chapter 11, Problem 6
    (A) $6.2 \times 10^{-5}$ m$^3$
    (B) $2.3 \times 10^{-5}$ m$^3$
    (C) $4.2 \times 10^{-5}$ m$^3$
    (D) $1.8 \times 10^{-5}$ m$^3$
    (E) $2.9 \times 10^{-4}$ m$^3$

8. Chapter 11, Problem 8
    (A) $1.70 \times 10^{-5}$
    (B) $2.42 \times 10^{-5}$
    (C) $7.12 \times 10^{-6}$
    (D) $5.68 \times 10^{-6}$
    (E) $1.80 \times 10^{-6}$

## Section 11.2 Pressure

10. Chapter 11, Problem 10
    (A) 6500 N
    (B) 3300 N
    (C) 1100 N
    (D) 980 N
    (E) 140 N

12a. Chapter 11, Problem 12a
    (A) $6.60 \times 10^2$ N
    (B) $5.10 \times 10^2$ N
    (C) $4.50 \times 10^2$ N
    (D) $2.97 \times 10^2$ N
    (E) $1.43 \times 10^2$ N

12b. Chapter 11, Problem 12b
    (A) 90.0 N
    (B) 76.7 N
    (C) 63.0 N
    (D) 53.2 N
    (E) 38.0 N

14. Chapter 11, Problem 14
    (A) 630 000 N
    (B) 60 000 N
    (C) 7700 N
    (D) 11 000 N
    (E) 33 000 N

# Homework: Fluids

16a. Chapter 11, Problem 16a
- (A) 0.055 m
- (B) 0.11 m
- (C) 0.036 m
- (D) 0.25 m
- (E) 0.018 m

16b. Chapter 11, Problem 16b
- (A) 5.4 J
- (B) 11 J
- (C) 3.6 J
- (D) 6.3 J
- (E) 98 J

18. Chapter 11, Problem 18
- (A) $6.2 \times 10^3$ Pa
- (B) $5.7 \times 10^3$ Pa
- (C) $2.2 \times 10^3$ Pa
- (D) $1.8 \times 10^3$ Pa
- (E) $9.0 \times 10^2$ Pa

## Section 11.3 Pressure and Depth in a Static Fluid
## Section 11.4 Pressure Gauges

20. Chapter 11, Problem 20
- (A) $4.30 \times 10^4$ Pa
- (B) $1.30 \times 10^5$ Pa
- (C) $1.01 \times 10^5$ Pa
- (D) $1.19 \times 10^5$ Pa
- (E) $2.90 \times 10^4$ Pa

22. Chapter 11, Problem 22
- (A) 0.15 m
- (B) 0.071 m
- (C) 0.17 m
- (D) 0.086 m
- (E) 0.12 m

24. Chapter 11, Problem 24
- (A) 6.53 m
- (B) 8.87 m
- (C) 10.3 m
- (D) 16.4 m
- (E) 101 m

26. Chapter 11, Problem 26
- (A) $3.5 \times 10^5$ Pa
- (B) $7.0 \times 10^5$ Pa
- (C) $4.9 \times 10^5$ Pa
- (D) $9.8 \times 10^5$ Pa
- (E) $2.5 \times 10^5$ Pa

28. Chapter 11, Problem 28
- (A) 62.0 m
- (B) 83.4 m
- (C) 105 m
- (D) 137 m
- (E) 168 m

30. Chapter 11, Problem 30
- (A) 0.46 m
- (B) 1.73 m
- (C) 0.92 m
- (D) 1.00 m
- (E) 1.46 m

32a. Chapter 11, Problem 32a
- (A) 4.85 m
- (B) 2.89 m
- (C) 16.1 m
- (D) 17.1 m
- (E) 18.1 m

32b. Chapter 11, Problem 32b
- (A) 1.59 m
- (B) 11.4 m
- (C) 13.2 m
- (D) 12.3 m
- (E) 1.30 m

## Section 11.5 Pascal's Principle

34. Chapter 11, Problem 34
    (A) $9.0 \times 10^0$
    (B) $6.0 \times 10^1$
    (C) $7.5 \times 10^1$
    (D) $8.0 \times 10^1$
    (E) $1.2 \times 10^2$

36. Chapter 11, Problem 36
    (A) $5.7 \times 10^{-2}$ m
    (B) $3.3 \times 10^{-2}$ m
    (C) $7.4 \times 10^{-2}$ m
    (D) $1.1 \times 10^{-1}$ m
    (E) $2.5 \times 10^{-1}$ m

38. Chapter 11, Problem 38
    (A) 12 N
    (B) 54 N
    (C) 72 N
    (D) 88 N
    (E) 108 N

## Section 11.6 Archimedes' Principle

40. Chapter 11, Problem 40
    (A) 250 N
    (B) 280 N
    (C) 290 N
    (D) 220 N
    (E) 270 N

42. Chapter 11, Problem 42
    (A) ice
    (B) silver
    (C) quartz
    (D) yellow pine
    (E) diamond

44. Chapter 11, Problem 44
    (A) $6.0 \times 10^2$ kg/m$^3$
    (B) $5.7 \times 10^2$ kg/m$^3$
    (C) $4.0 \times 10^2$ kg/m$^3$
    (D) $3.6 \times 10^2$ kg/m$^3$
    (E) $2.0 \times 10^2$ kg/m$^3$

46. Chapter 11, Problem 46
    (A) 8
    (B) 12
    (C) 16
    (D) 20
    (E) 24

48. Chapter 11, Problem 48
    (A) 0.10
    (B) 0.20
    (C) 0.25
    (D) 0.30
    (E) 0.40

50. Chapter 11, Problem 50
    (A) $R_1 = 5.28 \times 10^{-2}$ m, $R_2 = 6.20 \times 10^{-2}$ m
    (B) $R_1 = 8.76 \times 10^{-3}$ m, $R_2 = 1.21 \times 10^{-2}$ m
    (C) $R_1 = 3.05 \times 10^{-2}$ m, $R_2 = 3.58 \times 10^{-2}$ m
    (D) $R_1 = 1.05 \times 10^{-2}$ m, $R_2 = 1.29 \times 10^{-2}$ m
    (E) $R_1 = 4.54 \times 10^{-2}$ m, $R_2 = 6.75 \times 10^{-2}$ m

52. Chapter 11, Problem 52
    (A) 66.7 %
    (B) 60.3 %
    (C) 56.2 %
    (D) 51.4 %
    (E) 50.0 %

# Homework: Fluids

## Section 11.8 The Equation of Continuity

54. Chapter 11, Problem 54
    - (A) $9.55 \times 10^6$ gal
    - (B) $8.12 \times 10^6$ gal
    - (C) $7.27 \times 10^6$ gal
    - (D) $6.96 \times 10^6$ gal
    - (E) $2.58 \times 10^6$ gal

56a. Chapter 11, Problem 56a
    - (A) 0.55 cm
    - (B) 0.60 cm
    - (C) 0.63 cm
    - (D) 0.75 cm
    - (E) 1.0 cm

56b. Chapter 11, Problem 56b
    - (A) 62 s
    - (B) 88 s
    - (C) 95 s
    - (D) 110 s
    - (E) 130 s

58. Chapter 11, Problem 58
    - (A) 0.577
    - (B) 0.333
    - (C) 0.816
    - (D) 0.667
    - (E) 0.707

## Section 11.9 Bernoulli's Equation
## Section 11.10 Applications of Bernoulli's Equation

60. Chapter 11, Problem 60
    - (A) 85 Pa
    - (B) 290 Pa
    - (C) 330 Pa
    - (D) 470 Pa
    - (E) 600 Pa

62a. Chapter 11, Problem 62a
    - (A) $5.3 \times 10^4$ Pa
    - (B) $7.1 \times 10^4$ Pa
    - (C) $9.2 \times 10^4$ Pa
    - (D) $1.4 \times 10^5$ Pa
    - (E) $4.9 \times 10^5$ Pa

62b. Chapter 11, Problem 62b
    - (A) $5.3 \times 10^4$ Pa
    - (B) $7.1 \times 10^4$ Pa
    - (C) $9.2 \times 10^4$ Pa
    - (D) $1.4 \times 10^5$ Pa
    - (E) $4.9 \times 10^5$ Pa

64a. Chapter 11, Problem 64a
    - (A) $1.01 \times 10^5$ Pa
    - (B) $3.76 \times 10^5$ Pa
    - (C) $4.92 \times 10^5$ Pa
    - (D) $1.99 \times 10^5$ Pa
    - (E) $2.48 \times 10^5$ Pa

64b. Chapter 11, Problem 64b
    - (A) $4.92 \times 10^5$ Pa
    - (B) $1.01 \times 10^5$ Pa
    - (C) $3.76 \times 10^5$ Pa
    - (D) $2.48 \times 10^5$ Pa
    - (E) $1.99 \times 10^5$ Pa

64c. Chapter 11, Problem 64c
    - (A) 0.0772 m³/s
    - (B) 0.106 m³/s
    - (C) 0.518 m³/s
    - (D) 0.342 m³/s
    - (E) 0.189 m³/s

# HOMEWORK CHOICES

66a. Chapter 11, Problem 66a
    (A) 21.2 m/s      (C) 30.0 m/s      (E) 42.4 m/s
    (B) 2.56 m/s      (D) 36.2 m/s

66b. Chapter 11, Problem 66b
    (A) $2.38 \times 10^5$ Pa      (C) $1.37 \times 10^5$ Pa      (E) $1.68 \times 10^5$ Pa
    (B) $3.00 \times 10^5$ Pa      (D) $2.74 \times 10^5$ Pa

68. Chapter 11, Problem 68
    (A) 0.14 m      (C) 0.28 m      (E) 0.23 m
    (B) 0.19 m      (D) 0.38 m

70a. Chapter 11, Problem 70a
    (A) 13.3 m/s      (C) 21.6 m/s      (E) 32.8 m/s
    (B) 15.7 m/s      (D) 26.7 m/s

70b. Chapter 11, Problem 70b
    (A) 36.4 m      (C) 54.9 m      (E) 84.0 m
    (B) 47.6 m      (D) 69.4 m

72a. Chapter 11, Problem 72a
    (A) The upper hole must have the smaller area.
    (B) The upper hole must have the larger area.

72b. Chapter 11, Problem 72b
    (A) 0.840      (C) 1.54      (E) 2.00
    (B) 0.707      (D) 1.19

74a. Chapter 11, Problem 74a
    *The solution is a proof.*

74b. Chapter 11, Problem 74b
    *The solution is a proof.*

### Section 11.11 Viscous Flow

76. Chapter 11, Problem 76
    (A) 4 Pa      (C) 12 Pa      (E) 20 Pa
    (B) 8 Pa      (D) 16 Pa

78. Chapter 11, Problem 78
    (A) 4.4 Pa      (C) 14 Pa      (E) 28 Pa
    (B) 6.2 Pa      (D) 22 Pa

## Homework: Fluids

80. Chapter 11, Problem 80
    - (A) $9.6 \times 10^{-3}$ m
    - (B) $1.5 \times 10^{-2}$ m
    - (C) $2.9 \times 10^{-2}$ m
    - (D) $4.7 \times 10^{-2}$ m
    - (E) $5.1 \times 10^{-2}$ m

82. Chapter 11, Problem 82
    - (A) 2.25
    - (B) 1.75
    - (C) 1.50
    - (D) 1.25
    - (E) 0.75

## Additional Problems

84a. Chapter 11, Problem 84a
    - (A) 11.2 m/s
    - (B) 18.5 m/s
    - (C) 22.2 m/s
    - (D) 36.2 m/s
    - (E) 44.4 m/s

84b. Chapter 11, Problem 84b
    - (A) $2.58 \times 10^5$ Pa
    - (B) $3.06 \times 10^5$ Pa
    - (C) $1.41 \times 10^5$ Pa
    - (D) $3.47 \times 10^5$ Pa
    - (E) $1.68 \times 10^5$ Pa

86a. Chapter 11, Problem 86a
    - (A) $1.28 \times 10^5$ Pa
    - (B) $1.18 \times 10^5$ Pa
    - (C) $1.08 \times 10^5$ Pa
    - (D) $9.98 \times 10^4$ Pa
    - (E) $7.68 \times 10^4$ Pa

86b. Chapter 11, Problem 86b
    - (A) The meat baster will work better at the top of a mountain because the pressure of the atmosphere increases with increasing altitude.
    - (B) The meat baster will work better at the top of a mountain because the pressure of the atmosphere decreases with increasing altitude.
    - (C) The meat baster will work worse at the top of a mountain because the pressure of the atmosphere increases with increasing altitude.
    - (D) The meat baster will work worse at the top of a mountain because the pressure of the atmosphere decreases with increasing altitude.
    - (E) The meat baster will work the same at the top of a mountain because altitude does not affect the operation of this device.

88. Chapter 11, Problem 88
    - (A) $5.04 \times 10^{-6}$ m$^3$
    - (B) $4.74 \times 10^{-6}$ m$^3$
    - (C) $6.71 \times 10^{-6}$ m$^3$
    - (D) $3.08 \times 10^{-6}$ m$^3$
    - (E) $3.91 \times 10^{-6}$ m$^3$

90. Chapter 11, Problem 90
    - (A) 2.5 m/s
    - (B) 2.0 m/s
    - (C) 1.5 m/s
    - (D) 1.0 m/s
    - (E) 0.5 m/s

92. Chapter 11, Problem 92
    - (A) $2.8 \times 10^4$ Pa
    - (B) $5.6 \times 10^4$ Pa
    - (C) $1.9 \times 10^4$ Pa
    - (D) $4.7 \times 10^4$ Pa
    - (E) $9.4 \times 10^4$ Pa

*HOMEWORK CHOICES*

94. Chapter 11, Problem 94
    - (A) 98.3 %
    - (B) 89.5 %
    - (C) 72.8 %
    - (D) 38.6 %
    - (E) 23.6 %

96. Chapter 11, Problem 96
    - (A) 0.13 m
    - (B) 0.25 m
    - (C) 0.20 m
    - (D) 0.10 m
    - (E) 0.16 m

98. Chapter 11, Problem 98
    - (A) 0.57 m
    - (B) 0.63 m
    - (C) 0.74 m
    - (D) 0.82 m
    - (E) 0.89 m

100. Chapter 11, Problem 100
    - (A) $1.81 \times 10^{-2}$ m$^3$/s
    - (B) $2.63 \times 10^{-2}$ m$^3$/s
    - (C) $3.57 \times 10^{-2}$ m$^3$/s
    - (D) $4.25 \times 10^{-2}$ m$^3$/s
    - (E) $4.98 \times 10^{-2}$ m$^3$/s

102. Chapter 11, Problem 102
    - (A) $1.83 \times 10^4$ Pa
    - (B) $2.88 \times 10^4$ Pa
    - (C) $3.50 \times 10^4$ Pa
    - (D) $4.07 \times 10^4$ Pa
    - (E) $5.75 \times 10^4$ Pa

# Homework Choices
## CHAPTER 12 — Temperature and Heat

*Section 12.1 Common Temperature Scales*
*Section 12.2 The Kelvin Temperature Scale*
*Section 12.3 Thermometers*

2a. Chapter 12, Problem 2a
(A) from 18.0 °C to 32.4 °C
(B) from 10.0 °C to 32.4 °C
(C) from 10.0 °C to 40.0 °C
(D) from 18.0 °C to 40.0 °C
(E) from 32.4 °C to 40.0 °C

2b. Chapter 12, Problem 2b
(A) from 291.15 K to 305.55 K
(B) from 283.15 K to 313.15 K
(C) from 305.55 K to 313.15 K
(D) from 283.15 K to 305.55 K
(E) from 291.15 K to 313.15 K

4a. Chapter 12, Problem 4a
(A) −196 °C
(B) −164 °C
(C) −321 °C
(D) −242 °C
(E) −289 °C

4b. Chapter 12, Problem 4b
(A) −164 °F
(B) −196 °F
(C) −242 °F
(D) −289 °F
(E) −321 °F

6a. Chapter 12, Problem 6a
(A) 111 °C
(B) 104 °C
(C) 81.4 °C
(D) 67.7 °C
(E) 36.7 °C

6b. Chapter 12, Problem 6b
(A) 36.7 K
(B) 67.7 K
(C) 81.4 K
(D) 104 K
(E) 111 K

8. Chapter 12, Problem 8
(A) 19 °C
(B) 23 °C
(C) 25 °C
(D) 29 °C
(E) 31 °C

*Section 12.4 Linear Thermal Expansion*

10a. Chapter 12, Problem 10a
(A) $8.4 \times 10^{-4}$ m
(B) $7.3 \times 10^{-4}$ m
(C) $3.1 \times 10^{-4}$ m
(D) $1.9 \times 10^{-4}$ m
(E) $8.5 \times 10^{-5}$ m

10b. Chapter 12, Problem 10b
(A) $8.4 \times 10^{-5}$ m
(B) $7.3 \times 10^{-5}$ m
(C) $3.1 \times 10^{-5}$ m
(D) $1.9 \times 10^{-5}$ m
(E) $8.5 \times 10^{-6}$ m

*HOMEWORK CHOICES*    **Chapter 12**    395

12a. Chapter 12, Problem 12a
   (A) The radius of the hole will remain the same size because only the copper around it expands.
   (B) The radius of the hole will be smaller when the plate is heated because the copper expands inward as well as outward.
   (C) The radius of the hole will be larger when the plate is heated because the hole expands as if it were made of copper.

12b. Chapter 12, Problem 12b
   (A) 0.0017
   (B) 0.0028
   (C) 0.0039
   (D) 0.0056
   (E) 0.011

14. Chapter 12, Problem 14
   (A) 17 °C
   (B) 21 °C
   (C) 25 °C
   (D) 29 °C
   (E) 33 °C

16. Chapter 12, Problem 16
   (A) 0.0017 m
   (B) 1.7 m
   (C) 2.9 m
   (D) 5.8 m
   (E) 12 m

18. Chapter 12, Problem 18
   (A) 6.0 C°
   (B) 12 C°
   (C) 3.0 C°
   (D) 15 C°
   (E) 9.0 C°

20. Chapter 12, Problem 20
   (A) 49 °C
   (B) 20 °C
   (C) 40 °C
   (D) 22 °C
   (E) 33 °C

22. Chapter 12, Problem 22
   (A) $3.1 \times 10^{-4}$ m
   (B) $7.3 \times 10^{-4}$ m
   (C) $8.4 \times 10^{-4}$ m
   (D) $1.2 \times 10^{-5}$ m
   (E) $2.5 \times 10^{-5}$ m

24. Chapter 12, Problem 24
   (A) 20 °C
   (B) 26 °C
   (C) 31 °C
   (D) 35 °C
   (E) 38 °C

26a. Chapter 12, Problem 26a
   (A) The angular speed of the wheel will increase because as the wheel is heated, its radius increases, thereby decreasing its moment of inertia.
   (B) The angular speed of the wheel will increase because as the wheel is heated, its radius decreases, thereby decreasing its moment of inertia.
   (C) The angular speed of the wheel will decrease because as the wheel is heated, its radius increases, thereby increasing its moment of inertia.
   (D) The angular speed of the wheel will decrease because as the wheel is heated, its radius decreases, thereby increasing its moment of inertia.
   (E) The angular speed will remain the same as before because the change in temperature has no effect on the moment of inertia.

26b. Chapter 12, Problem 26b
   (A) 17.91 rad/s
   (B) 17.62 rad/s
   (C) 17.50 rad/s
   (D) 17.74 rad/s
   (E) 17.83 rad/s

## Section 12.5 Volume Thermal Expansion

28. Chapter 12, Problem 28
    - (A) $3.1 \times 10^{-3}$ m$^3$
    - (B) $9.8 \times 10^{-3}$ m$^3$
    - (C) $4.9 \times 10^{-3}$ m$^3$
    - (D) $7.4 \times 10^{-3}$ m$^3$
    - (E) $1.1 \times 10^{-3}$ m$^3$

30. Chapter 12, Problem 30
    - (A) gasoline
    - (B) water
    - (C) mercury
    - (D) carbon tetrachloride
    - (E) methyl alcohol

32. Chapter 12, Problem 32
    - (A) $3.1 \times 10^{-6}$ m$^3$
    - (B) $9.8 \times 10^{-6}$ m$^3$
    - (C) $4.9 \times 10^{-6}$ m$^3$
    - (D) $7.3 \times 10^{-6}$ m$^3$
    - (E) $1.1 \times 10^{-6}$ m$^3$

34. Chapter 12, Problem 34
    - (A) $1.41 \times 10^{-4}$/C°
    - (B) $2.10 \times 10^{-4}$/C°
    - (C) $2.79 \times 10^{-4}$/C°
    - (D) $3.40 \times 10^{-4}$/C°
    - (E) $4.17 \times 10^{-4}$/C°

36. Chapter 12, Problem 36
    - (A) 12 800 kg/m$^3$
    - (B) 13 000 kg/m$^3$
    - (C) 13 200 kg/m$^3$
    - (D) 13 400 kg/m$^3$
    - (E) 13 600 kg/m$^3$

38a. Chapter 12, Problem 38a
    - (A) The apparent weight will be the same before and after heating since a change in volume of the sphere with temperature does not change its apparent weight.
    - (B) The apparent weight will be smaller after it cools since the sphere expands while cooling, displacing more water, and decreasing the buoyant force acting on it.
    - (C) The apparent weight will be larger after it cools since the sphere expands while cooling, displacing more water, and increasing the buoyant force acting on it.
    - (D) The apparent weight will be smaller after it cools since the sphere shrinks while cooling, displacing less water, and increasing the buoyant force acting on it.
    - (E) The apparent weight will be larger after it cools since the sphere shrinks while cooling, displacing less water, and decreasing the buoyant force acting on it.

38b. Chapter 12, Problem 38b
    - (A) zero
    - (B) 9.0 N
    - (C) 15 N
    - (D) 18 N
    - (E) 21 N

40. Chapter 12, Problem 40
    - (A) 0.284 m
    - (B) 0.483 m
    - (C) 0.591 m
    - (D) 0.765 m
    - (E) 0.927 m

## Section 12.6 Heat and Internal Energy
## Section 12.7 Heat and Temperature Change: Specific Heat Capacity

42. Chapter 12, Problem 42
    - (A) 13 °C
    - (B) 22 °C
    - (C) 15 °C
    - (D) 7.5 °C
    - (E) 19 °C

*HOMEWORK CHOICES*

44. Chapter 12, Problem 44
    (A) steel
    (B) aluminum
    (C) silver
    (D) gold
    (E) lead

46a. Chapter 12, Problem 46a
    (A) 120 s
    (B) 180 s
    (C) 270 s
    (D) 305 s
    (E) 425 s

46b. Chapter 12, Problem 46b
    (A) 120 s
    (B) 180 s
    (C) 270 s
    (D) 305 s
    (E) 425 s

48. Chapter 12, Problem 48
    (A) 121 kg
    (B) 127 kg
    (C) 134 kg
    (D) 139 kg
    (E) 143 kg

50. Chapter 12, Problem 50
    (A) 210 J
    (B) 780 J
    (C) 1400 J
    (D) 1900 J
    (E) 2200 J

52. Chapter 12, Problem 52
    (A) $1.25 \times 10^{-2}$ C°
    (B) $1.48 \times 10^{-3}$ C°
    (C) $1.67 \times 10^{-3}$ C°
    (D) $1.17 \times 10^{-2}$ C°
    (E) $1.36 \times 10^{-2}$ C°

54. Chapter 12, Problem 54
    (A) $3.1 \times 10^{3}$ N
    (B) $6.2 \times 10^{3}$ N
    (C) $1.1 \times 10^{3}$ N
    (D) $4.8 \times 10^{3}$ N
    (E) $2.2 \times 10^{3}$ N

## *Section 12.8 Heat and Phase Change: Latent Heat*

56. Chapter 12, Problem 56
    (A) copper
    (B) gold
    (C) lead
    (D) ammonia
    (E) water

58. Chapter 12, Problem 58
    (A) 0.58 °C
    (B) 4.6 °C
    (C) 9.2 °C
    (D) 13 °C
    (E) 17 °C

60. Chapter 12, Problem 60
    (A) $1.85 \times 10^{5}$ J
    (B) $1.97 \times 10^{5}$ J
    (C) $2.09 \times 10^{5}$ J
    (D) $2.18 \times 10^{5}$ J
    (E) $1.38 \times 10^{5}$ J

62. Chapter 12, Problem 62
    (A) 0.084 kg
    (B) 0.098 kg
    (C) 0.11 kg
    (D) 0.14 kg
    (E) 0.16 kg

64. Chapter 12, Problem 64
    (A) 0.0028 kg
    (B) 0.0044 kg
    (C) 0.0050 kg
    (D) 0.0056 kg
    (E) 0.0061 kg

66. Chapter 12, Problem 66
    (A) 44 hours
    (B) 36 hours
    (C) 21 hours
    (D) 15 hours
    (E) 11 hours

68. Chapter 12, Problem 68
    (A) 1.0 ton
    (B) 1.5 tons
    (C) 2.0 tons
    (D) 2.5 tons
    (E) 3.0 tons

70. Chapter 12, Problem 70
    (A) $1.74 \times 10^3$ m/s
    (B) $2.31 \times 10^3$ m/s
    (C) $1.16 \times 10^3$ m/s
    (D) $2.46 \times 10^3$ m/s
    (E) $1.92 \times 10^3$ m/s

## Section 12.9  Equilibrium Between Phases of Matter

72. Chapter 12, Problem 72
    (A) $2.0 \times 10^6$ Pa
    (B) $3.8 \times 10^6$ Pa
    (C) $4.5 \times 10^6$ Pa
    (D) $6.1 \times 10^6$ Pa
    (E) $7.4 \times 10^6$ Pa

74. Chapter 12, Problem 74
    (A) 40 °C
    (B) 83 °C
    (C) 100 °C
    (D) 120 °C
    (E) 140 °C

## Section 12.10  Humidity

76. Chapter 12, Problem 76
    (A) 1800 Pa
    (B) 2100 Pa
    (C) 2400 Pa
    (D) 3300 Pa
    (E) 4200 Pa

78. Chapter 12, Problem 78
    (A) 16 %
    (B) 25 %
    (C) 34 %
    (D) 47 %
    (E) 53 %

80. Chapter 12, Problem 80
    (A) 18 %
    (B) 39 %
    (C) 56 %
    (D) 68 %
    (E) 80 %

## Additional Problems

82. Chapter 12, Problem 82
    (A) 0.042 m
    (B) 0.061 m
    (C) 0.012 m
    (D) 0.10 m
    (E) 0.084 m

*HOMEWORK CHOICES*  **Chapter 12**

84. Chapter 12, Problem 84
    - (A) 110 C°
    - (B) 33 C°
    - (C) 220 C°
    - (D) 17 C°
    - (E) 84 C°

86. Chapter 12, Problem 86
    - (A) 56.7 °C
    - (B) 81.1 °C
    - (C) 61.9 °C
    - (D) 70.4 °C
    - (E) 78.3 °C

88. Chapter 12, Problem 88
    - (A) 190 J/(kg · C°)
    - (B) 320 J/(kg · C°)
    - (C) 690 J/(kg · C°)
    - (D) 750 J/(kg · C°)
    - (E) 810 J/(kg · C°)

90. Chapter 12, Problem 90
    - (A) 10 °C
    - (B) 15 °C
    - (C) 20 °C
    - (D) 5 °C
    - (E) 0 °C

92. Chapter 12, Problem 92
    - (A) 0.28 ohms
    - (B) 0.33 ohms
    - (C) 0.47 ohms
    - (D) 0.54 ohms
    - (E) 0.69 ohms

94. Chapter 12, Problem 94
    - (A) $3.4 \times 10^{-4}/\text{C}°$
    - (B) $4.0 \times 10^{-4}/\text{C}°$
    - (C) $4.8 \times 10^{-4}/\text{C}°$
    - (D) $6.0 \times 10^{-4}/\text{C}°$
    - (E) $7.2 \times 10^{-4}/\text{C}°$

96. Chapter 12, Problem 96
    - (A) $8.3 \times 10^{-5}$ m$^2$
    - (B) $5.6 \times 10^{-5}$ m$^2$
    - (C) $2.8 \times 10^{-5}$ m$^2$
    - (D) $1.4 \times 10^{-5}$ m$^2$
    - (E) $1.0 \times 10^{-5}$ m$^2$

98. Chapter 12, Problem 98
    - (A) 18 N
    - (B) 32 N
    - (C) 50 N
    - (D) 66 N
    - (E) 82 N

100. Chapter 12, Problem 100
    - (A) $5.0 \times 10^4$ J
    - (B) $7.0 \times 10^4$ J
    - (C) $9.0 \times 10^4$ J
    - (D) $1.1 \times 10^5$ J
    - (E) $1.3 \times 10^5$ J

# Homework Choices
## CHAPTER 13 — The Transfer of Heat

### Section 13.2 Conduction

2. Chapter 13, Problem 2
   - (A) 8.1 J/s
   - (B) 18 J/s
   - (C) 32 J/s
   - (D) 42 J/s
   - (E) 49 J/s

4a. Chapter 13, Problem 4a
   - (A) $8.6 \times 10^6$ J
   - (B) $7.2 \times 10^6$ J
   - (C) $4.9 \times 10^6$ J
   - (D) $3.8 \times 10^6$ J
   - (E) $2.7 \times 10^6$ J

4b. Chapter 13, Problem 4b
   - (A) $ 0.13
   - (B) $ 0.39
   - (C) $ 0.24
   - (D) $ 0.42
   - (E) $ 0.48

6. Chapter 13, Problem 6
   - (A) 2.7 J
   - (B) 6.0 J
   - (C) 8.5 J
   - (D) 12 J
   - (E) 16 J

8. Chapter 13, Problem 8
   - (A) 85 %
   - (B) 79 %
   - (C) 69 %
   - (D) 21 %
   - (E) 15 %

10a. Chapter 13, Problem 10a
   - (A) $1.54 \times 10^4$ J/(s · m$^2$)
   - (B) $1.10 \times 10^4$ J/(s · m$^2$)
   - (C) $6.58 \times 10^3$ J/(s · m$^2$)
   - (D) $4.40 \times 10^3$ J/(s · m$^2$)
   - (E) $2.99 \times 10^3$ J/(s · m$^2$)

10b. Chapter 13, Problem 10b
   - (A) $1.60 \times 10^6$ J/(s · m$^2$)
   - (B) $2.40 \times 10^6$ J/(s · m$^2$)
   - (C) $4.00 \times 10^6$ J/(s · m$^2$)
   - (D) $5.60 \times 10^6$ J/(s · m$^2$)
   - (E) $6.40 \times 10^6$ J/(s · m$^2$)

12. Chapter 13, Problem 12
   - (A) 44 min
   - (B) 33 min
   - (C) 22 min
   - (D) 16 min
   - (E) 11 min

14a. Chapter 13, Problem 14a
   - (A) 26 °C
   - (B) 24 °C
   - (C) 22 °C
   - (D) 21 °C
   - (E) 20 °C

14b. Chapter 13, Problem 14b
   - (A) 23 °C
   - (B) 21 °C
   - (C) 18 °C
   - (D) 15 °C
   - (E) 12 °C

# HOMEWORK CHOICES  Chapter 13

16. Chapter 13, Problem 16
    - (A) 0.11 mm
    - (B) 0.22 mm
    - (C) 0.33 mm
    - (D) 0.45 mm
    - (E) 0.67 mm

## Section 13.3 Radiation

18. Chapter 13, Problem 18
    - (A) $9.4 \times 10^{-6}$ m$^2$
    - (B) $3.7 \times 10^{-5}$ m$^2$
    - (C) $1.5 \times 10^{-5}$ m$^2$
    - (D) $4.2 \times 10^{-5}$ m$^2$
    - (E) $2.6 \times 10^{-5}$ m$^2$

20. Chapter 13, Problem 20
    - (A) $1.45 \times 10^2$ J/(s · m$^2$)
    - (B) $2.04 \times 10^2$ J/(s · m$^2$)
    - (C) $3.34 \times 10^2$ J/(s · m$^2$)
    - (D) $4.29 \times 10^2$ J/(s · m$^2$)
    - (E) $5.18 \times 10^2$ J/(s · m$^2$)

22a. Chapter 13, Problem 22a
    - (A) 207 K
    - (B) 390 K
    - (C) 1700 K
    - (D) 64 000 K
    - (E) 150 000 K

22b. Chapter 13, Problem 22b
    - (A) 1700 K
    - (B) 780 K
    - (C) 390 K
    - (D) 207 K
    - (E) 63 K

24. Chapter 13, Problem 24
    - (A) 1.93
    - (B) 3.72
    - (C) 5.26
    - (D) 7.44
    - (E) 9.59

26a. Chapter 13, Problem 26a
    - (A) 42 W
    - (B) 540 W
    - (C) 67 W
    - (D) 610 W
    - (E) 380 W

26b. Chapter 13, Problem 26b
    - (A) 12 Cal
    - (B) 58 Cal
    - (C) 24 Cal
    - (D) 46 Cal
    - (E) 39 Cal

28. Chapter 13, Problem 28
    - (A) 1.1 kg
    - (B) 0.81 kg
    - (C) 0.57 kg
    - (D) 0.39 kg
    - (E) 0.19 kg

30. Chapter 13, Problem 30
    - (A) 642 °C
    - (B) 558 °C
    - (C) 689 °C
    - (D) 519 °C
    - (E) 482 °C

## Additional Problems

32. Chapter 13, Problem 32
    - (A) 1.3
    - (B) 1.6
    - (C) 1.8
    - (D) 2.1
    - (E) 2.7

## Homework: The Transfer of Heat

34. Chapter 13, Problem 34
    - (A) 300 K
    - (B) 380 K
    - (C) 320 K
    - (D) 360 K
    - (E) 340 K

36. Chapter 13, Problem 36
    - (A) 0.70
    - (B) 0.61
    - (C) 0.50
    - (D) 0.33
    - (E) 0.10

38. Chapter 13, Problem 38
    - (A) 0.0086 kg
    - (B) 0.017 kg
    - (C) 0.0043 kg
    - (D) 0.030 kg
    - (E) 0.055 kg

40. Chapter 13, Problem 40
    - (A) 623 K
    - (B) 674 K
    - (C) 711 K
    - (D) 732 K
    - (E) 754 K

42. Chapter 13, Problem 42
    *The solution is a proof.*

# Homework Choices
## CHAPTER 14
# The Ideal Gas Law and Kinetic Theory

### Section 14.1 Molecular Mass, the Mole, and Avogadro's Number

2a. Chapter 14, Problem 2a
- (A) 43.025 u
- (B) 86.050 u
- (C) 147.153 u
- (D) 208.106 u
- (E) 294.307 u

2b. Chapter 14, Problem 2b
- (A) $1.152 \times 10^{-25}$ kg
- (B) $2.444 \times 10^{-25}$ kg
- (C) $2.693 \times 10^{-25}$ kg
- (D) $3.456 \times 10^{-25}$ kg
- (E) $4.887 \times 10^{-25}$ kg

4. Chapter 14, Problem 4
- (A) 38 mol
- (B) 33 mol
- (C) 28 mol
- (D) 23 mol
- (E) 18 mol

6a. Chapter 14, Problem 6a
- (A) $5.84 \times 10^{20}$
- (B) $6.02 \times 10^{20}$
- (C) $1.29 \times 10^{21}$
- (D) $2.58 \times 10^{21}$
- (E) $6.65 \times 10^{21}$

6b. Chapter 14, Problem 6b
- (A) $5.84 \times 10^{20}$
- (B) $6.02 \times 10^{20}$
- (C) $1.29 \times 10^{21}$
- (D) $2.58 \times 10^{21}$
- (E) $6.65 \times 10^{21}$

8a. Chapter 14, Problem 8a
- (A) $7.48 \times 10^{-26}$ kg
- (B) $3.98 \times 10^{-26}$ kg
- (C) $7.65 \times 10^{-26}$ kg
- (D) $4.82 \times 10^{-26}$ kg
- (E) $6.81 \times 10^{-26}$ kg

8b. Chapter 14, Problem 8b
- (A) $4.02 \times 10^{25}$
- (B) $5.27 \times 10^{25}$
- (C) $1.04 \times 10^{25}$
- (D) $2.11 \times 10^{25}$
- (E) $3.34 \times 10^{25}$

### Section 14.2 The Ideal Gas Law

10. Chapter 14, Problem 10
- (A) $7.6 \times 10^{13}$
- (B) $5.8 \times 10^{13}$
- (C) $8.5 \times 10^{13}$
- (D) $3.6 \times 10^{13}$
- (E) $2.8 \times 10^{13}$

12. Chapter 14, Problem 12
- (A) 22.4 liters
- (B) 15.4 liters
- (C) 11.2 liters
- (D) 9.80 liters
- (E) 1.00 liter

14. Chapter 14, Problem 14
    (A) $6.0 \times 10^3$ Pa
    (B) $1.1 \times 10^4$ Pa
    (C) $2.4 \times 10^4$ Pa
    (D) $3.6 \times 10^4$ Pa
    (E) $4.0 \times 10^4$ Pa

16. Chapter 14, Problem 16
    (A) 3.29 kg
    (B) 0.796 kg
    (C) 0.550 kg
    (D) 4.60 kg
    (E) 1.81 kg

18. Chapter 14, Problem 18
    (A) 0.140 m
    (B) 0.130 m
    (C) 0.150 m
    (D) 0.125 m
    (E) 0.160 m

20. Chapter 14, Problem 20
    (A) 175 K
    (B) 392 K
    (C) 623 K
    (D) 882 K
    (E) 1250 K

22. Chapter 14, Problem 22
    (A) 1.00
    (B) 1.02
    (C) 1.04
    (D) 1.06
    (E) 1.08

24. Chapter 14, Problem 24
    (A) 1.7 kg
    (B) 3.0 kg
    (C) 2.4 kg
    (D) 5.6 kg
    (E) 6.3 kg

26a. Chapter 14, Problem 26a
    (A) 540 K
    (B) 180 K
    (C) 240 K
    (D) 330 K
    (E) 480 K

26b. Chapter 14, Problem 26b
    (A) $2.8 \times 10^5$ Pa
    (B) $8.4 \times 10^4$ Pa
    (C) $1.1 \times 10^5$ Pa
    (D) $3.3 \times 10^5$ Pa
    (E) $6.7 \times 10^5$ Pa

28. Chapter 14, Problem 28
    (A) $6.30 \times 10^2$ N/m
    (B) $1.00 \times 10^3$ N/m
    (C) $1.41 \times 10^3$ N/m
    (D) $1.77 \times 10^3$ N/m
    (E) $1.98 \times 10^3$ N/m

## *Section 14.3 Kinetic Theory of Gases*

30. Chapter 14, Problem 30
    (A) $2.4 \times 10^4$ m/s
    (B) $6.9 \times 10^3$ m/s
    (C) $3.6 \times 10^4$ m/s
    (D) $4.8 \times 10^4$ m/s
    (E) $1.2 \times 10^4$ m/s

32. Chapter 14, Problem 32
    (A) $1.52 \times 10^4$ K
    (B) $2.02 \times 10^4$ K
    (C) $2.27 \times 10^4$ K
    (D) $1.01 \times 10^4$ K
    (E) $3.03 \times 10^4$ K

# HOMEWORK CHOICES

34. Chapter 14, Problem 34
    - (A) 0.53
    - (B) 1.1
    - (C) 1.9
    - (D) 2.7
    - (E) 3.1

36. Chapter 14, Problem 36
    - (A) $4.5 \times 10^2$ K
    - (B) $3.0 \times 10^2$ K
    - (C) 57 K
    - (D) 31 K
    - (E) 19 K

38. Chapter 14, Problem 38
    - (A) $1.0 \times 10^1$ s
    - (B) $2.5 \times 10^1$ s
    - (C) $3.0 \times 10^1$ s
    - (D) $4.3 \times 10^1$ s
    - (E) $5.0 \times 10^1$ s

40. Chapter 14, Problem 40
    - (A) 594 m/s
    - (B) 343 m/s
    - (C) 297 m/s
    - (D) 198 m/s
    - (E) 114 m/s

## *Section 14.4 Diffusion*

42. Chapter 14, Problem 42
    - (A) 11 s
    - (B) 9.7 s
    - (C) 8.4 s
    - (D) 6.9 s
    - (E) 2.3 s

44. Chapter 14, Problem 44
    - (A) $2.0 \times 10^{-3}$ m/s
    - (B) $1.6 \times 10^{-3}$ m/s
    - (C) $8.3 \times 10^{-3}$ m/s
    - (D) $7.0 \times 10^{-3}$ m/s
    - (E) $5.9 \times 10^{-3}$ m/s

46a. Chapter 14, Problem 46a
    - (A) 1.1 s
    - (B) 1.6 s
    - (C) 2.1 s
    - (D) 2.6 s
    - (E) 3.1 s

46b. Chapter 14, Problem 46b
    - (A) $1.1 \times 10^{-5}$ s
    - (B) $1.6 \times 10^{-5}$ s
    - (C) $2.1 \times 10^{-5}$ s
    - (D) $2.6 \times 10^{-5}$ s
    - (E) $3.1 \times 10^{-5}$ s

46c. Chapter 14, Problem 46c
    - (A) The time calculated in part (b) is much shorter because water vapor does not really behave as an ideal gas does.
    - (B) The time calculated in part (b) is much shorter because it does not take into account any collisions that water molecules make in traveling the 0.010 m.
    - (C) The time calculated in part (b) is much shorter because Fick's law does not take into account the *rms* speed of the water molecules.

48. Chapter 14, Problem 48
    - (A) $1.1 \times 10^3$ s
    - (B) $1.6 \times 10^4$ s
    - (C) $2.1 \times 10^5$ s
    - (D) $2.3 \times 10^6$ s
    - (E) $1.5 \times 10^7$ s

## Additional Problems

50. Chapter 14, Problem 50
    - (A) $1.0 \times 10^3$ kg
    - (B) $1.6 \times 10^4$ kg
    - (C) $2.1 \times 10^4$ kg
    - (D) $1.5 \times 10^5$ kg
    - (E) $2.6 \times 10^5$ kg

52. Chapter 14, Problem 52
    - (A) $1.1$ kg/m$^3$
    - (B) $1.4$ kg/m$^3$
    - (C) $2.2$ kg/m$^3$
    - (D) $2.8$ kg/m$^3$
    - (E) $3.3$ kg/m$^3$

54. Chapter 14, Problem 54
    - (A) 512 m/s
    - (B) 206 m/s
    - (C) 396 m/s
    - (D) 189 m/s
    - (E) 414 m/s

56. Chapter 14, Problem 56
    - (A) 0.27 m
    - (B) 0.34 m
    - (C) 0.42 m
    - (D) 1.3 m
    - (E) 2.4 m

58. Chapter 14, Problem 58
    - (A) $2.0 \times 10^4$ homes
    - (B) $6.1 \times 10^4$ homes
    - (C) $8.2 \times 10^4$ homes
    - (D) $1.0 \times 10^5$ homes
    - (E) $2.5 \times 10^5$ homes

60a. Chapter 14, Problem 60a
    - (A) −12 N
    - (B) −18 N
    - (C) −120 N
    - (D) −180 N
    - (E) −210 N

60b. Chapter 14, Problem 60b
    - (A) 12 N
    - (B) 18 N
    - (C) 120 N
    - (D) 180 N
    - (E) 210 N

60c. Chapter 14, Problem 60c
    - (A) $4.0 \times 10^4$ Pa
    - (B) $7.0 \times 10^5$ Pa
    - (C) $6.0 \times 10^5$ Pa
    - (D) $4.0 \times 10^5$ Pa
    - (E) $6.0 \times 10^4$ Pa

62. Chapter 14, Problem 62
    - (A) 390 K
    - (B) 440 K
    - (C) 520 K
    - (D) 670 K
    - (E) 710 K

# Homework Choices
# CHAPTER 15   Thermodynamics

## Section 15.3  The First Law of Thermodynamics

2a. Chapter 15, Problem 2a
- (A)  $-4.7 \times 10^5$ J
- (B)  $-6.4 \times 10^5$ J
- (C)  $-7.5 \times 10^5$ J
- (D)  $+2.8 \times 10^5$ J
- (E)  $+3.2 \times 10^5$ J

2b. Chapter 15, Problem 2b
- (A)  $-4.7 \times 10^5$ J
- (B)  $-6.4 \times 10^5$ J
- (C)  $-7.5 \times 10^5$ J
- (D)  $+2.8 \times 10^5$ J
- (E)  $+3.2 \times 10^5$ J

2c. Chapter 15, Problem 2c
- (A)  $-4.7 \times 10^5$ J
- (B)  $-6.4 \times 10^5$ J
- (C)  $-7.5 \times 10^5$ J
- (D)  $+2.8 \times 10^5$ J
- (E)  $+3.2 \times 10^5$ J

4a. Chapter 15, Problem 4a
- (A)  $-5.0 \times 10^1$ J, decrease
- (B)  $-2.4 \times 10^1$ J, decrease
- (C)  $3.0 \times 10^1$ J, increase
- (D)  $1.0 \times 10^1$ J, increase
- (E)  $-4.2 \times 10^2$ J, decrease

4b. Chapter 15, Problem 4b
- (A)  $3.2 \times 10^2$ J, increase
- (B)  $-1.3 \times 10^2$ J, decrease
- (C)  $-4.2 \times 10^2$ J, increase
- (D)  $5.0 \times 10^2$ J, increase
- (E)  $-2.4 \times 10^2$ J, decrease

4c. Chapter 15, Problem 4c
- (A)  $-3.0 \times 10^2$ J, decrease
- (B)  $-1.0 \times 10^2$ J, decrease
- (C)  $5.0 \times 10^2$ J, increase
- (D)  $4.2 \times 10^2$ J, increase
- (E)  $-2.4 \times 10^2$ J, decrease

6a. Chapter 15, Problem 6a
- (A)  $-1.10 \times 10^6$ J
- (B)  $-8.50 \times 10^5$ J
- (C)  $-7.10 \times 10^5$ J
- (D)  $-6.40 \times 10^5$ J
- (E)  $-5.03 \times 10^5$ J

6b. Chapter 15, Problem 6b
- (A)  $7.10 \times 10^3$
- (B)  $2.62 \times 10^3$
- (C)  $1.69 \times 10^3$
- (D)  $1.52 \times 10^3$
- (E)  $1.20 \times 10^2$

## Section 15.4  Thermal Processes

8a. Chapter 15, Problem 8a
- (A) Work is done on the system.
- (B) Work is done by the system.

## Homework: Thermodynamics

8b. Chapter 15, Problem 8b
   - (A) −420 J
   - (B) −105 J
   - (C) −210 J
   - (D) +210 J
   - (E) +420 J

10a. Chapter 15, Problem 10a
   - (A) $1.2 \times 10^3$ J
   - (B) $1.8 \times 10^3$ J
   - (C) $2.0 \times 10^3$ J
   - (D) $2.4 \times 10^3$ J
   - (E) $3.0 \times 10^3$ J

10b. Chapter 15, Problem 10b
   - (A) The work is done by the system and is positive.
   - (B) The work is done by the system and is negative.
   - (C) The work is done on the system and is positive.
   - (D) The work is done on the system and is negative.

12. Chapter 15, Problem 12
   - (A) 0.11 m
   - (B) 0.16 m
   - (C) 0.24 m
   - (D) 0.37 m
   - (E) 0.42 m

14a. Chapter 15, Problem 14a
   - (A) $-1.5 \times 10^3$ J
   - (B) $-3.0 \times 10^3$ J
   - (C) $4.0 \times 10^2$ J
   - (D) $9.0 \times 10^2$ J
   - (E) zero

14b. Chapter 15, Problem 14b
   - (A) $-1.5 \times 10^3$ J
   - (B) $-3.0 \times 10^3$ J
   - (C) $4.0 \times 10^2$ J
   - (D) $9.0 \times 10^2$ J
   - (E) zero

14c. Chapter 15, Problem 14c
   - (A) $-1.5 \times 10^3$ J
   - (B) $-3.0 \times 10^3$ J
   - (C) $4.0 \times 10^2$ J
   - (D) $9.0 \times 10^2$ J
   - (E) zero

16a. Chapter 15, Problem 16a
   - (A) $-2.1 \times 10^3$ J
   - (B) $+4.0 \times 10^3$ J
   - (C) $-2.5 \times 10^3$ J
   - (D) $+3.2 \times 10^3$ J
   - (E) $-1.2 \times 10^3$ J

16b. Chapter 15, Problem 16b
   - (A) Heat flows into the gas.
   - (B) Heat flows out of the gas.

18. Chapter 15, Problem 18
   - (A) 99.1 %
   - (B) 96.2 %
   - (C) 84.3 %
   - (D) 72.4 %
   - (E) 68.5 %

## Section 15.5 Thermal Processes that Utilize an Ideal Gas

20. Chapter 15, Problem 20
    - (A) $2.08 \times 10^3$ J
    - (B) $3.23 \times 10^3$ J
    - (C) $9.69 \times 10^3$ J
    - (D) $1.29 \times 10^4$ J
    - (E) $3.49 \times 10^4$ J

22. Chapter 15, Problem 22
    - (A) 366 K
    - (B) 397 K
    - (C) 434 K
    - (D) 452 K
    - (E) 474 K

24. Chapter 15, Problem 24
    - (A) 1.66
    - (B) 3.17
    - (C) 3.33
    - (D) 2.83
    - (E) 1.20

26. Chapter 15, Problem 26
    - (A) +1.36 J
    - (B) +1.18 J
    - (C) +0.81 J
    - (D) +0.68 J
    - (E) +0.59 J

28. Chapter 15, Problem 28

|     | path   | $\Delta U$ | W      | Q       |
|-----|--------|------------|--------|---------|
| (A) | A to B | 4990 J     | 3320 J | 8310 J  |
|     | B to C | −4990 J    | zero   | −4990 J |
|     | C to D | −2490 J    | −1660 J| −4150 J |
|     | D to A | 2490 J     | zero   | 2490 J  |
| (B) | A to B | 4450 J     | 6650 J | 11 000J |
|     | B to C | −4450 J    | zero   | −4450 J |
|     | C to D | −2220 J    | zero   | −2220 J |
|     | D to A | 2220 J     | zero   | 2220 J  |
| (C) | A to B | 1660 J     | 1330 J | 2990 J  |
|     | B to C | −1330 J    | zero   | −1330 J |
|     | C to D | −664 J     | −666 J | −1330 J |
|     | D to A | 334 J      | zero   | 334 J   |
| (D) | A to B | 3320 J     | 2220 J | 5540 J  |
|     | B to C | −3320 J    | zero   | −3320 J |
|     | C to D | −1660 J    | −1110 J| −2770 J |
|     | D to A | 1660 J     | zero   | 1660 J  |
| (E) | A to B | 5890 J     | 9860 J | 15750 J |
|     | B to C | −5890 J    | zero   | −5890 J |
|     | C to D | −2940 J    | −1960 J| −4900 J |
|     | D to A | 2940 J     | zero   | 2940 J  |

30a. Chapter 15, Problem 30a
   - (A) $-2.40 \times 10^5$ J
   - (B) $+2.40 \times 10^5$ J
   - (C) $-8.00 \times 10^4$ J
   - (D) $+8.00 \times 10^4$ J
   - (E) $-1.60 \times 10^4$ J

30b. Chapter 15, Problem 30b
   - (A) Heat flows into the gas.
   - (B) Heat flows out of the gas.

32. Chapter 15, Problem 32
    (A) 3640 J, absorbed
    (B) −1820 J, given off
    (C) 1570 J, absorbed
    (D) −1110 J, given off
    (E) −3990 J, given off

## Section 15.6 Specific Heat Capacities and the First Law of Thermodynamics

34a. Chapter 15, Problem 34a
   (A) −7.2 K
   (B) −9.5 K
   (C) −12 K
   (D) −15 K
   (E) −22 K

34b. Chapter 15, Problem 34b
   (A) −7.2 K
   (B) −9.5 K
   (C) −12 K
   (D) −15 K
   (E) −22 K

36a. Chapter 15, Problem 36a
   (A) 85 K
   (B) 98 K
   (C) 110 K
   (D) 140 K
   (E) 170 K

36b. Chapter 15, Problem 36b
   (A) 2450 J
   (B) 3120 J
   (C) 3390 J
   (D) 4780 J
   (E) 5240 J

36c. Chapter 15, Problem 36c
   (A) 5170 Pa
   (B) 4240 Pa
   (C) 3450 Pa
   (D) 2330 Pa
   (E) 2170 Pa

38. Chapter 15, Problem 38
    (A) 1
    (B) 3/2
    (C) 7/2
    (D) 9/2
    (E) 5/2

40a. Chapter 15, Problem 40a
   (A) 12.6 J
   (B) 24.4 J
   (C) 37.3 J
   (D) 50.2 J
   (E) 71.0 J

40b. Chapter 15, Problem 40b
   (A) 71.0 J/(mol · K)
   (B) 50.2 J/(mol · K)
   (C) 12.6 J/(mol · K)
   (D) 37.3 J/(mol · K)
   (E) 24.4 J/(mol · K)

42. Chapter 15, Problem 42
    (A) $1.70 \times 10^5$ J
    (B) $2.61 \times 10^4$ J
    (C) $3.41 \times 10^5$ J
    (D) $2.38 \times 10^4$ J
    (E) $2.07 \times 10^4$ J

## Section 15.8 Heat Engines

44. Chapter 15, Problem 44
    (A) 1.2
    (B) 1.4
    (C) 1.8
    (D) 2.4
    (E) 3.3

*HOMEWORK CHOICES*  **Chapter 15**  *411*

46a. Chapter 15, Problem 46a
    (A) 3100 J      (C) 8600 J      (E) 9000 J
    (B) 3500 J      (D) 7500 J

46b. Chapter 15, Problem 46b
    (A) 3100 J      (C) 8600 J      (E) 9000 J
    (B) 3500 J      (D) 7500 J

48. Chapter 15, Problem 48
    (A) 88 %      (C) 79 %      (E) 44 %
    (B) 24 %      (D) 67 %

## *Section 15.9 Carnot's Principle and the Carnot Engine*

50. Chapter 15, Problem 50
    (A) 0.14      (C) 0.35      (E) 0.43
    (B) 0.21      (D) 0.40

52. Chapter 15, Problem 52
    (A) 350 K      (C) 400 K      (E) 750 K
    (B) 970 K      (D) 500 K

54a. Chapter 15, Problem 54a
    (A) 0.66      (C) 0.25      (E) 0.33
    (B) 0.50      (D) 0.75

54b. Chapter 15, Problem 54b
    (A) 0.66      (C) 1.5      (E) 0.33
    (B) 3.0      (D) 2.5

56. Chapter 15, Problem 56
    (A) 0.0410 kg      (C) 0.0202 kg      (E) 0.0054 kg
    (B) 0.0269 kg      (D) 0.0149 kg

58a. Chapter 15, Problem 58a
    (A) 271 K      (C) 735 K      (E) 1590 K
    (B) 344 K      (D) 1170 K

58b. Chapter 15, Problem 58b
    (A) 271 K      (C) 735 K      (E) 1590 K
    (B) 344 K      (D) 1170 K

60. Chapter 15, Problem 60
    (A) $2.26 \times 10^3$ J      (C) $3.89 \times 10^3$ J      (E) $8.63 \times 10^3$ J
    (B) $2.40 \times 10^3$ J      (D) $4.98 \times 10^3$ J

## Section 15.10 Refrigerators, Air Conditioners, and Heat Pumps

62. Chapter 15, Problem 62
    - (A) 1.8
    - (B) 5.4
    - (C) 2.7
    - (D) 4.9
    - (E) 3.2

64. Chapter 15, Problem 64
    - (A) 284 K
    - (B) 647 K
    - (C) 236 K
    - (D) 491 K
    - (E) 127 K

66. Chapter 15, Problem 66
    - (A) 270 J
    - (B) 510 J
    - (C) 890 J
    - (D) 750 J
    - (E) 930 J

68. Chapter 15, Problem 68
    - (A) 0.30
    - (B) 0.49
    - (C) 0.70
    - (D) 1.4
    - (E) 3.3

70a. Chapter 15, Problem 70a
    - (A) 0.50
    - (B) 0.67
    - (C) 1.0
    - (D) 1.5
    - (E) 2.0

70b. Chapter 15, Problem 70b
    - (A) 0.50
    - (B) 0.67
    - (C) 1.0
    - (D) 1.5
    - (E) 2.0

70c. Chapter 15, Problem 70c
    - (A) 0.50
    - (B) 0.67
    - (C) 1.0
    - (D) 1.5
    - (E) 2.0

72. Chapter 15, Problem 72
    - (A) 8.8 K
    - (B) 9.5 K
    - (C) 1.4 K
    - (D) 1.9 K
    - (E) 3.6 K

## Section 15.11 Entropy and the Second Law of Thermodynamics

74. Chapter 15, Problem 74
    - (A) $3.5 \times 10^1$ J/K
    - (B) $6.0 \times 10^1$ J/K
    - (C) $2.6 \times 10^2$ J/K
    - (D) $1.2 \times 10^3$ J/K
    - (E) $3.7 \times 10^3$ J/K

76. Chapter 15, Problem 76
    - (A) 267 K
    - (B) 276 K
    - (C) 288 K
    - (D) 310 K
    - (E) 375 K

78a. Chapter 15, Problem 78a
    - (A) 119 K
    - (B) 267 K
    - (C) 313 K
    - (D) 446 K
    - (E) 514 K

78b. Chapter 15, Problem 78b
- (A) $2.37 \times 10^2$ J/K
- (B) $1.10 \times 10^2$ J/K
- (C) $4.29 \times 10^2$ J/K
- (D) $5.60 \times 10^2$ J/K
- (E) $3.19 \times 10^2$ J/K

78c. Chapter 15, Problem 78c
- (A) $2.00 \times 10^4$ J
- (B) $3.00 \times 10^4$ J
- (C) $4.62 \times 10^4$ J
- (D) $5.40 \times 10^4$ J
- (E) $6.80 \times 10^4$ J

## Additional Problems

80a. Chapter 15, Problem 80a
- (A) $-5.0 \times 10^3$ J
- (B) $-2.5 \times 10^3$ J
- (C) zero
- (D) $+2.5 \times 10^3$ J
- (E) $+5.0 \times 10^3$ J

80b. Chapter 15, Problem 80b
- (A) $-5.0 \times 10^3$ J
- (B) $-2.5 \times 10^3$ J
- (C) zero
- (D) $+2.5 \times 10^3$ J
- (E) $+5.0 \times 10^3$ J

82. Chapter 15, Problem 82
- (A) 4.0
- (B) 2.0
- (C) 0.75
- (D) 0.50
- (E) 0.25

84. Chapter 15, Problem 84
- (A) $3.24 \times 10^4$ J
- (B) $2.94 \times 10^4$ J
- (C) $7.62 \times 10^4$ J
- (D) $6.92 \times 10^4$ J
- (E) $8.48 \times 10^4$ J

86. Chapter 15, Problem 86
- (A) 820 J
- (B) 220 J
- (C) 690 J
- (D) 1100 J
- (E) 490 J

88. Chapter 15, Problem 88
- (A) Increasing $T_H$ by 40 K gives an efficiency of 0.493. Decreasing $T_C$ by 40 K gives an efficiency of 0.523. The greatest improvement is made by decreasing $T_C$.
- (B) Increasing $T_H$ by 40 K gives an efficiency of 0.507. Decreasing $T_C$ by 40 K gives an efficiency of 0.477. The greatest improvement is made by increasing $T_H$.
- (C) Increasing $T_H$ by 40 K gives an efficiency of 0.819. Decreasing $T_C$ by 40 K gives an efficiency of 0.903. The greatest improvement is made by decreasing $T_C$.
- (D) Increasing $T_H$ by 40 K gives an efficiency of 0.971. Decreasing $T_C$ by 40 K gives an efficiency of 0.911. The greatest improvement is made by increasing $T_H$.
- (E) Increasing $T_H$ by 40 K gives an efficiency of 0.660. Decreasing $T_C$ by 40 K gives an efficiency of 0.749. The greatest improvement is made by decreasing $T_C$.

90a. Chapter 15, Problem 90a
- (A) −1100 J
- (B) −1900 J
- (C) −2800 J
- (D) +3900 J
- (E) +4800 J

# Homework: Thermodynamics

90b. Chapter 15, Problem 90b
 - (A) Heat flows into the gas.
 - (B) Heat flows out of the gas.

92. Chapter 15, Problem 92
 - (A) 2700 J
 - (B) zero
 - (C) 2200 J
 - (D) 1200 J
 - (E) 1500 J

94. Chapter 15, Problem 94
 - (A) 0.185 m
 - (B) 0.539 m
 - (C) 0.389 m
 - (D) 0.264 m
 - (E) 0.689 m

96. Chapter 15, Problem 96
 - (A) 1500 J
 - (B) 2000 J
 - (C) 2500 J
 - (D) 3000 J
 - (E) 3500 J

98a. Chapter 15, Problem 98a
 - (A) 206 K
 - (B) 254 K
 - (C) 524 K
 - (D) 477 K
 - (E) 419 K

98b. Chapter 15, Problem 98b
 - (A) 381 K
 - (B) 418 K
 - (C) 319 K
 - (D) 276 K
 - (E) 323 K

# Homework Choices
# CHAPTER 16    Waves and Sound

### Section 16.1 The Nature of Waves
### Section 16.2 Periodic Waves

2.   Chapter 16, Problem 2
     (A)   $1.7 \times 10^{-3}$ s      (C)   $1.7 \times 10^{-5}$ s      (E)   $1.7 \times 10^{-7}$ s
     (B)   $3.4 \times 10^{-3}$ s      (D)   $3.4 \times 10^{-5}$ s

4.   Chapter 16, Problem 4
     (A)   $2.0 \times 10^{14}$ Hz      (C)   $4.5 \times 10^{14}$ Hz      (E)   $6.4 \times 10^{14}$ Hz
     (B)   $3.2 \times 10^{14}$ Hz      (D)   $5.7 \times 10^{14}$ Hz

6.   Chapter 16, Problem 6
     (A)   0.23 m      (C)   0.56 m      (E)   0.83 m
     (B)   0.49 m      (D)   0.71 m

8a.   Chapter 16, Problem 8a
     (A)   0.100 Hz      (C)   0.200 Hz      (E)   0.500 Hz
     (B)   0.150 Hz      (D)   0.350 Hz

8b.   Chapter 16, Problem 8b
     (A)   15.0 m/s      (C)   2.00 m/s      (E)   6.00 m/s
     (B)   8.00 m/s      (D)   4.00 m/s

10.   Chapter 16, Problem 10
     (A)   $5.0 \times 10^{1}$ s      (C)   $1.4 \times 10^{1}$ s      (E)   $7.0 \times 10^{1}$ s
     (B)   $3.7 \times 10^{1}$ s      (D)   $2.2 \times 10^{0}$ s

12a.   Chapter 16, Problem 12a
     (A)   6.5 m/s      (C)   3.5 m/s      (E)   1.1 m/s
     (B)   5.3 m/s      (D)   2.7 m/s

12b.   Chapter 16, Problem 12b
     (A)   1.14 m      (C)   3.49 m      (E)   6.55 m
     (B)   2.72 m      (D)   5.28 m

### Section 16.3 The Speed of a Wave on a String

14.   Chapter 16, Problem 14
     (A)   450 m/s      (C)   750 m/s      (E)   1200 m/s
     (B)   600 m/s      (D)   900 m/s

16.   Chapter 16, Problem 16
     (A)   2.00      (C)   1.00      (E)   1.41
     (B)   0.500      (D)   0.707

18. Chapter 16, Problem 18
    (A) 0.17 s
    (B) 0.25 s
    (C) 2.8 s
    (D) 4.0 s
    (E) 5.9 s

20a. Chapter 16, Problem 20a
    (A) $4.84 \times 10^4$ m/s
    (B) $1.81 \times 10^4$ m/s
    (C) $1.58 \times 10^3$ m/s
    (D) $4.54 \times 10^2$ m/s
    (E) $2.20 \times 10^2$ m/s

20b. Chapter 16, Problem 20b
    (A) 2.93 m/s
    (B) 3.77 m/s
    (C) 4.60 m/s
    (D) 7.12 m/s
    (E) 9.19 m/s

22. Chapter 16, Problem 22
    (A) $1.06 \times 10^{-3}$ s
    (B) $2.88 \times 10^{-3}$ s
    (C) $3.26 \times 10^{-3}$ s
    (D) $5.19 \times 10^{-3}$ s
    (E) $7.45 \times 10^{-3}$ s

## Section 16.4 The Mathematical Description of a Wave

24a. Chapter 16, Problem 24a

| | $A$ | $f$ | $\lambda$ | $v$ |
|---|---|---|---|---|
| (A) | 0.45 m | 2.0 Hz | 0.90 m | 1.8 m/s |
| (B) | 0.45 m | 4.0 Hz | 2.0 m | 8.0 m/s |
| (C) | 0.45 m | 4.0 Hz | 0.50 m | 2.0 m/s |
| (D) | 0.23 m | 4.4 Hz | 2.0 m | 8.8 m/s |
| (E) | 0.23 m | 3.6 Hz | 1.0 m | 3.6 m/s |

24b. Chapter 16, Problem 24b
    (A) $+x$ direction
    (B) $-x$ direction

26. Chapter 16, Problem 26
    (A) $y = (0.010 \text{ m}) \sin(0.40\pi t - 130x)$
    (B) $y = (0.010 \text{ m}) \sin(5.0\pi t - 100\pi x)$
    (C) $y = (0.010 \text{ m}) \sin(0.40 t + 160\pi x)$
    (D) $y = (0.010 \text{ m}) \sin(10\pi - 50\pi x)$
    (E) $y = (0.010 \text{ m}) \sin(2.5\pi t - 130x)$

28. Chapter 16, Problem 28
    (A) 0.20 m/s
    (B) 0.72 m/s
    (C) 0.55 m/s
    (D) 0.40 m/s
    (E) 0.12 m/s

30. Chapter 16, Problem 30
    (A) $1.9 \times 10^{-3}$ s
    (B) $2.9 \times 10^{-3}$ s
    (C) $3.8 \times 10^{-3}$ s
    (D) $5.7 \times 10^{-3}$ s
    (E) $7.4 \times 10^{-3}$ s

## HOMEWORK CHOICES

### Section 16.5 The Nature of Sound
### Section 16.6 The Speed of Sound

32. Chapter 16, Problem 32
    - (A) 9.60 m
    - (B) 11.8 m
    - (C) 18.4 m
    - (D) 32.5 m
    - (E) 45.2 m

34. Chapter 16, Problem 34
    - (A) 9
    - (B) 11
    - (C) 13
    - (D) 15
    - (E) 17

36. Chapter 16, Problem 36
    - (A) $2.2 \times 10^{-4}$ s
    - (B) $3.4 \times 10^{-4}$ s
    - (C) $4.4 \times 10^{-4}$ s
    - (D) $6.7 \times 10^{-4}$ s
    - (E) $1.3 \times 10^{-3}$ s

38a. Chapter 16, Problem 38a

|     | first | second | third |
|-----|-------|--------|-------|
| (A) | air   | water  | steel |
| (B) | water | steel  | air   |
| (C) | steel | air    | water |
| (D) | water | air    | steel |
| (E) | steel | water  | air   |

38b. Chapter 16, Problem 38b

|     | second  | third   |
|-----|---------|---------|
| (A) | 0.059 s | 0.339 s |
| (B) | 0.019 s | 0.304 s |
| (C) | 0.045 s | 0.296 s |
| (D) | 0.029 s | 0.352 s |
| (E) | 0.033 s | 0.310 s |

40. Chapter 16, Problem 40
    - (A) 110 rad/s
    - (B) 690 rad/s
    - (C) 220 rad/s
    - (D) 570 rad/s
    - (E) 470 rad/s

42. Chapter 16, Problem 42
    - (A) 3
    - (B) 5
    - (C) 7
    - (D) 9
    - (E) 11

44. Chapter 16, Problem 44
    - (A) 45 m
    - (B) 52 m
    - (C) 61 m
    - (D) 76 m
    - (E) 110 m

46. Chapter 16, Problem 46
    - (A) $8.0 \times 10^2$ m/s
    - (B) $5.1 \times 10^2$ m/s
    - (C) $6.2 \times 10^2$ m/s
    - (D) $4.8 \times 10^2$ m/s
    - (E) $7.0 \times 10^2$ m/s

418     **Homework: Waves and Sound**

48. Chapter 16, Problem 48
    - (A) 192 m/s
    - (B) 204 m/s
    - (C) 217 m/s
    - (D) 239 m/s
    - (E) 246 m/s

50a. Chapter 16, Problem 50a
    - (A) 49.9 m
    - (B) 61.0 m
    - (C) 99.5 m
    - (D) 125 m
    - (E) 75.0 m

50b. Chapter 16, Problem 50b
    - (A) 0.3 %
    - (B) 0.6 %
    - (C) 0.9 %
    - (D) 1.3 %
    - (E) 1.7 %

## Section 16.7  Sound Intensity

52. Chapter 16, Problem 52
    - (A) $1.2 \times 10^{-3}$ J
    - (B) $6.0 \times 10^{-4}$ J
    - (C) $2.4 \times 10^{-4}$ J
    - (D) $8.0 \times 10^{-5}$ J
    - (E) $1.8 \times 10^{-5}$ J

54. Chapter 16, Problem 54
    - (A) $8.5 \times 10^{-5}$ W/m$^2$
    - (B) $1.2 \times 10^{-4}$ W/m$^2$
    - (C) $1.5 \times 10^{-5}$ W/m$^2$
    - (D) $1.8 \times 10^{-4}$ W/m$^2$
    - (E) $2.4 \times 10^{-5}$ W/m$^2$

56. Chapter 16, Problem 56
    - (A) $6.5 \times 10^{-3}$ W/m$^2$
    - (B) $5.1 \times 10^{-3}$ W/m$^2$
    - (C) $4.6 \times 10^{-3}$ W/m$^2$
    - (D) $3.7 \times 10^{-3}$ W/m$^2$
    - (E) $5.7 \times 10^{-2}$ W/m$^2$

58. Chapter 16, Problem 58
    - (A) 2050 m
    - (B) 2510 m
    - (C) 2900 m
    - (D) 3350 m
    - (E) 3550 m

60. Chapter 16, Problem 60
    - (A) 30.8 m and 92.4 m
    - (B) 41.0 m and 164 m
    - (C) 61.6 m and 246 m
    - (D) 82.0 m and 246 m
    - (E) 82.0 m and 164 m

## Section 16.8  Decibels

62. Chapter 16, Problem 62
    - (A) 0.580 W/m$^2$
    - (B) 0.150 W/m$^2$
    - (C) 0.223 W/m$^2$
    - (D) 0.440 W/m$^2$
    - (E) 0.316 W/m$^2$

64. Chapter 16, Problem 64
    - (A) 66 dB
    - (B) 58 dB
    - (C) 38 dB
    - (D) 15 dB
    - (E) 10 dB

# HOMEWORK CHOICES

66. Chapter 16, Problem 66
    (A) $4.0 \times 10^2$
    (B) $2.0 \times 10^2$
    (C) $2.3 \times 10^1$
    (D) $3.0 \times 10^0$
    (E) $1.2 \times 10^0$

68. Chapter 16, Problem 68
    (A) 17.5 W
    (B) 28.1 W
    (C) 35.0 W
    (D) 56.2 W
    (E) 67.1 W

70. Chapter 16, Problem 70
    (A) 32 dB
    (B) 43 dB
    (C) 56 dB
    (D) 66 dB
    (E) 77 dB

72. Chapter 16, Problem 72
    (A) 1
    (B) 3
    (C) 4
    (D) 6
    (E) 8

74. Chapter 16, Problem 74
    (A) $r_1 = 1.6$ m, $r_2 = 2.6$ m
    (B) $r_1 = 3.9$ m, $r_2 = 4.9$ m
    (C) $r_1 = 4.5$ m, $r_2 = 5.5$ m
    (D) $r_1 = 2.2$ m, $r_2 = 3.2$ m
    (E) $r_1 = 5.4$ m, $r_2 = 6.4$ m

## *Section 16.10 The Doppler Effect*

76. Chapter 16, Problem 76
    (A) 811 Hz
    (B) 824 Hz
    (C) 838 Hz
    (D) 852 Hz
    (E) 892 Hz

78. Chapter 16, Problem 78
    (A) 0.188 m
    (B) 0.159 m
    (C) 0.395 m
    (D) 0.340 m
    (E) 0.271 m

80. Chapter 16, Problem 80
    (A) 8.4 m/s
    (B) 13 m/s
    (C) 9.7 m/s
    (D) 6.3 m/s
    (E) 11 m/s

82. Chapter 16, Problem 82
    (A) 1240 m/s
    (B) 1090 m/s
    (C) 776 m/s
    (D) 548 m/s
    (E) 403 m/s

84. Chapter 16, Problem 84
    (A) 600 Hz
    (B) 615 Hz
    (C) 622 Hz
    (D) 652 Hz
    (E) 660 Hz

86. Chapter 16, Problem 86
    (A) 0.52 m
    (B) 0.33 m
    (C) 0.26 m
    (D) 0.17 m
    (E) 0.11 m

## Additional Problems

88. Chapter 16, Problem 88
    (A) $1.3 \times 10^{-4}$ m
    (B) $8.1 \times 10^{-5}$ m
    (C) $1.0 \times 10^{-6}$ m
    (D) $1.6 \times 10^{-6}$ m
    (E) $3.2 \times 10^{-6}$ m

90. Chapter 16, Problem 90
    (A) 230 N
    (B) 162 N
    (C) 116 N
    (D) 82 N
    (E) 76 N

92. Chapter 16, Problem 92
    (A) 0.99 m/s
    (B) 3.6 m/s
    (C) 5.8 m/s
    (D) 1.7 m/s
    (E) 9.2 m/s

94. Chapter 16, Problem 94
    (A) 48 m
    (B) 72 m
    (C) 96 m
    (D) 140 m
    (E) 190 m

96. Chapter 16, Problem 96
    (A) $y = (0.15 \text{ m}) \sin(130\pi t + 92\pi x)$
    (B) $y = (0.15 \text{ m}) \sin(160\pi t - 22\pi x)$
    (C) $y = (0.15 \text{ m}) \sin(130t + 92x)$
    (D) $y = (0.15 \text{ m}) \sin(160\pi t + 22\pi x)$
    (E) $y = (0.15 \text{ m}) \sin(120\pi t + 6.9\pi x)$

98. Chapter 16, Problem 98
    (A) $1.43 \times 10^{-2}$ m
    (B) $8.19 \times 10^{-2}$ m
    (C) $3.84 \times 10^{-2}$ m
    (D) $7.68 \times 10^{-2}$ m
    (E) $5.27 \times 10^{-2}$ m

100. Chapter 16, Problem 100
    (A) 3.8 J
    (B) 8.6 J
    (C) 9.5 J
    (D) 12 J
    (E) 2.9 J

102. Chapter 16, Problem 102
    (A) 3.4 m/s$^2$
    (B) 2.6 m/s$^2$
    (C) 0.38 m/s$^2$
    (D) 0.14 m/s$^2$
    (E) 0.022 m/s$^2$

104. Chapter 16, Problem 104
    (A) 0.18 m/s
    (B) 0.29 m/s
    (C) 0.37 m/s
    (D) 0.52 m/s
    (E) 0.62 m/s

106. Chapter 16, Problem 106
    (A) 20
    (B) 40
    (C) 30
    (D) 26
    (E) 13

108. Chapter 16, Problem 108
    (A) 9.2 m
    (B) 8.5 m
    (C) 9.7 m
    (D) 8.9 m
    (E) 10 m

# Homework Choices
# CHAPTER 17

# The Principle of Linear Superposition and Interference Phenomena

## Section 17.1 The Principle of Linear Superposition
## Section 17.2 Constructive and Destructive Interference of Sound Waves

2. Chapter 17, Problem 2
   *The solution is a graph.*

4. Chapter 17, Problem 4
   (A) Constructive interference occurs because the path difference is one wavelength.
   (B) Constructive interference occurs because the path difference is one-half of a wavelength.
   (C) Destructive interference occurs because the path difference is one-half of a wavelength.
   (D) Destructive interference occurs because the path difference is one wavelength.

6. Chapter 17, Problem 6
   (A) 380 m/s      (C) 540 m/s      (E) 1100 m/s
   (B) 960 m/s      (D) 790 m/s

8. Chapter 17, Problem 8
   (A) 428 Hz       (C) 303 Hz       (E) 161 Hz
   (B) 321 Hz       (D) 214 Hz

10. Chapter 17, Problem 10
    **Note**: *All distances below are measured from speaker* **A**.
    (A) 1.55 m, 3.90 m, and 6.25 m       (D) 1.75 m, 3.25 m, and 5.65 m
    (B) 1.95 m, 3.90 m, and 5.85 m       (E) 1.65 m, 3.25 m, and 6.15 m
    (C) 1.30 m, 2.60 m, and 6.50 m

## Section 17.3 Diffraction

12. Chapter 17, Problem 12
    (A) 1.22         (C) 1.00         (E) 1.33
    (B) 1.44         (D) 1.11

14. Chapter 17, Problem 14
    (A) $8.1 \times 10^3$ Hz     (C) $1.2 \times 10^4$ Hz     (E) $5.7 \times 10^3$ Hz
    (B) $9.4 \times 10^4$ Hz     (D) $1.5 \times 10^4$ Hz

16. Chapter 17, Problem 16
    (A) 1.4°         (C) 2.5°         (E) 3.7°
    (B) 1.8°         (D) 3.1°

## Section 17.4 Beats

18. Chapter 17, Problem 18
    - (A) 0.1 s
    - (B) 0.2 s
    - (C) 0.5 s
    - (D) 2 s
    - (E) 5 s

20. Chapter 17, Problem 20
    - (A) 427 Hz
    - (B) 435 Hz
    - (C) 445 Hz
    - (D) 449 Hz
    - (E) 451 Hz

22. Chapter 17, Problem 22
    - (A) 0 Hz
    - (B) 2 Hz
    - (C) 4 Hz
    - (D) 6 Hz
    - (E) 8 Hz

24. Chapter 17, Problem 24
    - (A) 1.7 Hz
    - (B) 3.5 Hz
    - (C) 2.1 Hz
    - (D) 4.4 Hz
    - (E) 6.8 Hz

## Section 17.5 Transverse Standing Waves

26. Chapter 17, Problem 26
    - (A) 1.0 m/s
    - (B) 1.2 m/s
    - (C) 2.4 m/s
    - (D) 3.0 m/s
    - (E) 4.8 m/s

28. Chapter 17, Problem 28
    - (A) 3.00
    - (B) 5.30
    - (C) 10.6
    - (D) 15.1
    - (E) 17.0

30a. Chapter 17, Problem 30a
    - (A) 1.00 m
    - (B) 2.00 m
    - (C) 3.00 m
    - (D) 4.00 m
    - (E) 5.00 m

30b. Chapter 17, Problem 30b
    - (A) 15.0 m/s
    - (B) 25.0 m/s
    - (C) 50.0 m/s
    - (D) 70.0 m/s
    - (E) 85.0 m/s

30c. Chapter 17, Problem 30c
    - (A) 12.1 Hz
    - (B) 14.2 Hz
    - (C) 17.0 Hz
    - (D) 21.3 Hz
    - (E) 28.3 Hz

32. Chapter 17, Problem 32
    - (A) 2
    - (B) 3
    - (C) 4
    - (D) 5
    - (E) 6

34. Chapter 17, Problem 34
    - (A) 27.0 Hz
    - (B) 45.0 Hz
    - (C) 54.0 Hz
    - (D) 60.5 Hz
    - (E) 72.1 Hz

*HOMEWORK CHOICES*       **Chapter 17**   *423*

36a. Chapter 17, Problem 36a
    (A)  0.0559 m     (C)  0.0347 m     (E)  0.0394 m
    (B)  0.0249 m     (D)  0.0352 m

36b. Chapter 17, Problem 36b
    (A)  0.0559 m     (C)  0.0347 m     (E)  0.0394 m
    (B)  0.0249 m     (D)  0.0352 m

## *Section 17.6 Longitudinal Standing Waves*
## *Section 17.7 Complex Sound Waves*

38a. Chapter 17, Problem 38a
    (A)  50 Hz     (C)  200 Hz     (E)  150 Hz
    (B)  25 Hz     (D)  100 Hz

38b. Chapter 17, Problem 38b
    (A)  The tube is open at both ends.
    (B)  The tube is open at one end only.

40. Chapter 17, Problem 40
    (A)  19 m and 0.0019 m     (C)  5.5 m and 0.0055 m     (E)  8.6 m and 0.0086 m
    (B)  11 m and 0.0011 m     (D)  7.2 m and 0.0072 m

42. Chapter 17, Problem 42
    (A)  256 m/s     (C)  512 m/s     (E)  908 m/s
    (B)  362 m/s     (D)  724 m/s

44. Chapter 17, Problem 44
    (A)  2.99     (C)  1.98     (E)  1.47
    (B)  2.49     (D)  1.73

46. Chapter 17, Problem 46
    (A)  $2.66 \times 10^{-25}$ kg     (C)  $6.37 \times 10^{-25}$ kg     (E)  $8.82 \times 10^{-25}$ kg
    (B)  $1.84 \times 10^{-25}$ kg     (D)  $4.78 \times 10^{-25}$ kg

48. Chapter 17, Problem 48
    (A)  25 m     (C)  7.9 m     (E)  6.1 m
    (B)  5.3 m     (D)  11 m

50. Chapter 17, Problem 50
    (A)  127 Hz     (C)  162 Hz     (E)  209 Hz
    (B)  148 Hz     (D)  183 Hz

## *Additional Problems*

52a. Chapter 17, Problem 52a
    (A)  600, 1000, and 1400 Hz     (C)  700, 1500, and 2300 Hz     (E)  1200, 2000, and 2800 Hz
    (B)  800, 1200, and 1600 Hz     (D)  1000, 1500, and 2000 Hz

52b. Chapter 17, Problem 52b
 (A) 600, 1000, and 1400 Hz
 (B) 800, 1200, and 1600 Hz
 (C) 700, 1500, and 2300 Hz
 (D) 1000, 1500, and 2000 Hz
 (E) 1200, 2000, and 2800 Hz

52c. Chapter 17, Problem 52c
 (A) 600, 1000, and 1400 Hz
 (B) 800, 1200, and 1600 Hz
 (C) 700, 1500, and 2300 Hz
 (D) 1000, 1500, and 2000 Hz
 (E) 1200, 2000, and 2800 Hz

54. Chapter 17, Problem 54
 (A) 488 Hz
 (B) 490 Hz
 (C) 492 Hz
 (D) 494 Hz
 (E) 496 Hz

56a. Chapter 17, Problem 56a
 (A) 65.40 Hz
 (B) 92.49 Hz
 (C) 261.6 Hz
 (D) 277.5 Hz
 (E) 370.0 Hz

56b. Chapter 17, Problem 56b
 (A) 0.233 m
 (B) 0.329 m
 (C) 0.464 m
 (D) 0.656 m
 (E) 0.928 m

58a. Chapter 17, Problem 58a
 (A) 8.0 cm
 (B) 6.0 cm
 (C) 4.0 cm
 (D) 2.0 cm
 (E) 1.0 cm

58b. Chapter 17, Problem 58b
 (A) +12.0 cm/s
 (B) +6.0 cm/s
 (C) −3.0 cm/s
 (D) −6.0 cm/s
 (E) −12.0 cm/s

58c. Chapter 17, Problem 58c
 (A) 0.5 Hz
 (B) 1.0 Hz
 (C) 1.5 Hz
 (D) 2.5 Hz
 (E) 3.0 Hz

58d. Chapter 17, Problem 58d
 (A) 28.3 cm/s
 (B) 24.0 cm/s
 (C) 16.8 cm/s
 (D) 14.2 cm/s
 (E) 12.0 cm/s

60a. Chapter 17, Problem 60a
 (A) $f_x = 397$ Hz, $f_y = 395$ Hz
 (B) $f_x = 387$ Hz, $f_y = 395$ Hz
 (C) $f_x = 389$ Hz, $f_y = 397$ Hz
 (D) $f_x = 397$ Hz, $f_y = 389$ Hz
 (E) $f_x = 395$ Hz, $f_y = 387$ Hz

60b. Chapter 17, Problem 60b
 (A) $f_x = 397$ Hz, $f_y = 395$ Hz
 (B) $f_x = 387$ Hz, $f_y = 395$ Hz
 (C) $f_x = 389$ Hz, $f_y = 397$ Hz
 (D) $f_x = 397$ Hz, $f_y = 389$ Hz
 (E) $f_x = 395$ Hz, $f_y = 387$ Hz

# Homework Choices
## CHAPTER 18
# Electric Forces and Electric Fields

*Section 18.1 The Origin of Electricity*
*Section 18.2 Charged Objects and the Electric Force*
*Section 18.3 Conductors and Insulators*
*Section 18.4 Charging by Contact and by Induction*

2. Chapter 18, Problem 2
   (A) +9.6 µC
   (B) +6.6 µC
   (C) −2.4 µC
   (D) −1.6 µC
   (E) −18 µC

4. Chapter 18, Problem 4
   (A) −1.6 µC
   (B) −6.4 µC
   (C) −9.0 µC
   (D) +1.6 µC
   (E) +5.5 µC

6. Chapter 18, Problem 6
   (A) $4.8 \times 10^{-17}$ kg, object A has more mass
   (B) $5.2 \times 10^{-17}$ kg, object B has more mass
   (C) $2.9 \times 10^{-17}$ kg, object A has more mass
   (D) $1.7 \times 10^{-17}$ kg, object B has more mass
   (E) $3.4 \times 10^{-17}$ kg, object A has more mass

*Section 18.5 Coulomb's Law*

8. Chapter 18, Problem 8
   (A) $2.27 \times 10^{39}$
   (B) $4.41 \times 10^{39}$
   (C) $4.54 \times 10^{38}$
   (D) $1.61 \times 10^{38}$
   (E) $6.84 \times 10^{39}$

10. Chapter 18, Problem 10
    (A) $8.6 \times 10^{-3}$ m
    (B) $7.8 \times 10^{-3}$ m
    (C) $3.8 \times 10^{-3}$ m
    (D) $3.2 \times 10^{-4}$ m
    (E) $6.1 \times 10^{-5}$ m

12. Chapter 18, Problem 12
    (A) $2.06 \times 10^3$ C
    (B) $2.18 \times 10^3$ C
    (C) $3.09 \times 10^3$ C
    (D) $4.12 \times 10^3$ C
    (E) $6.18 \times 10^3$ C

14. Chapter 18, Problem 14
    (A) $x = -0.42$ m
    (B) $x = -0.21$ m
    (C) $x = +0.25$ m
    (D) $x = +0.64$ m
    (E) $x = +0.71$ m

16. Chapter 18, Problem 16
    (A) 3.0 N
    (B) 5.1 N
    (C) 6.8 N
    (D) 7.2 N
    (E) 9.8 N

# Homework: Electric Forces and Electric Fields

18. Chapter 18, Problem 18
    - (A) +0.70 μC
    - (B) +2.0 μC
    - (C) +1.4 μC
    - (D) −0.70 μC
    - (E) −1.4 μC

20a. Chapter 18, Problem 20a
    - (A) $2.10 \times 10^7$ m/s$^2$
    - (B) $4.20 \times 10^7$ m/s$^2$
    - (C) $5.88 \times 10^7$ m/s$^2$
    - (D) $6.42 \times 10^7$ m/s$^2$
    - (E) $6.80 \times 10^7$ m/s$^2$

20b. Chapter 18, Problem 20b
    Note: all angles are measured relative to the +x axis.
    - (A) 67.9°
    - (B) 70.9°
    - (C) 41.8°
    - (D) 58.3°
    - (E) 20.3°

22. Chapter 18, Problem 22
    - (A) $q_1 = 7.00$ μC, $q_2 = 2.00$ μC
    - (B) $q_1 = 4.50$ μC, $q_2 = 4.50$ μC
    - (C) $q_1 = 8.00$ μC, $q_2 = 1.00$ μC
    - (D) $q_1 = 5.00$ μC, $q_2 = 4.00$ μC
    - (E) $q_1 = 6.00$ μC, $q_2 = 3.00$ μC

24. Chapter 18, Problem 24
    - (A) 0.77 N
    - (B) 0.63 N
    - (C) 0.57 N
    - (D) 0.43 N
    - (E) 0.37 N

26. Chapter 18, Problem 26
    - (A) $q_1 = -1.91$ μC and $q_2 = +0.479$ μC
      or $q_1 = -0.479$ μC and $q_2 = +1.91$ μC
    - (B) $q_1 = -2.79$ μC and $q_2 = +0.613$ μC
      or $q_1 = -0.613$ μC and $q_2 = +2.79$ μC
    - (C) $q_1 = -3.82$ μC and $q_2 = +0.722$ μC
      or $q_1 = -0.722$ μC and $q_2 = +3.82$ μC
    - (D) $q_1 = -4.64$ μC and $q_2 = +0.851$ μC
      or $q_1 = -0.851$ μC and $q_2 = +4.64$ μC
    - (E) $q_1 = -5.58$ μC and $q_2 = +0.957$ μC
      or $q_1 = -0.957$ μC and $q_2 = +5.58$ μC

## Section 18.6 The Electric Field
## Section 18.7 Electric Field Lines
## Section 18.8 The Electric Field Inside a Conductor: Shielding

28. Chapter 18, Problem 28
    *The solution is a drawing.*

30. Chapter 18, Problem 30
    - (A) 0.16 N · m
    - (B) 0.40 N · m
    - (C) 1.8 N · m
    - (D) 2.0 N · m
    - (E) zero

32. Chapter 18, Problem 32
    - (A) $+3q$ on the exterior surface and $-q$ on the interior surface
    - (B) $+2q$ on the exterior surface and $+q$ on the interior surface
    - (C) $+q$ on the exterior surface and $-2q$ on the interior surface
    - (D) $+q/2$ on the exterior surface and $-q$ on the interior surface
    - (E) $+2q$ on the exterior surface and $-2q$ on the interior surface

## HOMEWORK CHOICES

**34a.** Chapter 18, Problem 34a
- (A) $2.70 \times 10^{-2}$ N
- (B) $6.00 \times 10^{-2}$ N
- (C) $7.50 \times 10^{-2}$ N
- (D) $4.50 \times 10^{-2}$ N
- (E) $8.25 \times 10^{-2}$ N

**34b.** Chapter 18, Problem 34b
Note: all angles are measured relative to the $+x$ axis.
- (A) 36.9°
- (B) 41.4°
- (C) 48.5°
- (D) 53.1°
- (E) 64.6°

**36a.** Chapter 18, Problem 36a
- (A) $5.3 \times 10^5$ N/C, $+x$ direction
- (B) $4.2 \times 10^5$ N/C, $-x$ direction
- (C) $2.9 \times 10^5$ N/C, $+x$ direction
- (D) $2.3 \times 10^5$ N/C, $-x$ direction
- (E) $1.5 \times 10^5$ N/C, $+x$ direction

**36b.** Chapter 18, Problem 36b
- (A) 2.3 N, $-x$ direction
- (B) 1.5 N, $+x$ direction
- (C) 4.2 N, $-x$ direction
- (D) 5.3 N, $+x$ direction
- (E) 2.9 N, $-x$ direction

**38.** Chapter 18, Problem 38

|     | Figure 18.22$b$ | Figure 18.22$c$ |
|-----|-----------------|-----------------|
| (A) | 181 N/C         | 362 N/C         |
| (B) | 219 N/C         | 311 N/C         |
| (C) | 266 N/C         | 362 N/C         |
| (D) | 181 N/C         | 311 N/C         |
| (E) | 219 N/C         | 266 N/C         |

**40.** Chapter 18, Problem 40
- (A) $3.3 \times 10^{-3}$ s
- (B) $1.6 \times 10^{-2}$ s
- (C) $2.8 \times 10^{-2}$ s
- (D) $5.2 \times 10^{-2}$ s
- (E) $7.6 \times 10^{-2}$ s

**42.** Chapter 18, Problem 42
- (A) $3.3 \times 10^7$ m/s
- (B) $1.0 \times 10^7$ m/s
- (C) $7.5 \times 10^6$ m/s
- (D) $5.0 \times 10^6$ m/s
- (E) $2.0 \times 10^6$ m/s

**44a.** Chapter 18, Problem 44a
- (A) $5.8 \times 10^{-9}$ C/m$^2$
- (B) $3.3 \times 10^{-9}$ C/m$^2$
- (C) $4.2 \times 10^{-9}$ C/m$^2$
- (D) $2.7 \times 10^{-9}$ C/m$^2$
- (E) $1.4 \times 10^{-9}$ C/m$^2$

**44b.** Chapter 18, Problem 44b
- (A) $2.7 \times 10^{-12}$ C
- (B) $4.8 \times 10^{-12}$ C
- (C) $5.4 \times 10^{-12}$ C
- (D) $6.3 \times 10^{-12}$ C
- (E) $7.6 \times 10^{-12}$ C

**46.** Chapter 18, Problem 46
- (A) $1.04 \times 10^3$ N/C
- (B) $4.18 \times 10^3$ N/C
- (C) $3.12 \times 10^3$ N/C
- (D) $5.23 \times 10^3$ N/C
- (E) $2.09 \times 10^3$ N/C

## Section 18.9 Gauss' Law

48. Chapter 18, Problem 48
    - (A) 35 N · m$^2$/C
    - (B) 92 N · m$^2$/C
    - (C) 220 N · m$^2$/C
    - (D) 46 N · m$^2$/C
    - (E) 84 N · m$^2$/C

50a. Chapter 18, Problem 50a
    - (A) zero
    - (B) $1.6 \times 10^5$ N · m$^2$/C
    - (C) $4.0 \times 10^5$ N · m$^2$/C
    - (D) $3.1 \times 10^5$ N · m$^2$/C
    - (E) $2.0 \times 10^5$ N · m$^2$/C

50b. Chapter 18, Problem 50b
    - (A) $-2.6 \times 10^5$ N · m$^2$/C
    - (B) $-4.1 \times 10^5$ N · m$^2$/C
    - (C) $-3.0 \times 10^5$ N · m$^2$/C
    - (D) $-1.0 \times 10^5$ N · m$^2$/C
    - (E) zero

50c. Chapter 18, Problem 50c
    - (A) zero
    - (B) $6.6 \times 10^5$ N · m$^2$/C
    - (C) $3.3 \times 10^5$ N · m$^2$/C
    - (D) $4.8 \times 10^5$ N · m$^2$/C
    - (E) $1.4 \times 10^5$ N · m$^2$/C

52a. Chapter 18, Problem 52a
    - (A) $1.2 \times 10^5$ N · m$^2$/C
    - (B) $3.4 \times 10^5$ N · m$^2$/C
    - (C) $2.3 \times 10^5$ N · m$^2$/C
    - (D) $4.6 \times 10^5$ N · m$^2$/C
    - (E) zero

52b. Chapter 18, Problem 52b
    - (A) $3.4 \times 10^5$ N · m$^2$/C
    - (B) $4.6 \times 10^5$ N · m$^2$/C
    - (C) $1.2 \times 10^5$ N · m$^2$/C
    - (D) $2.3 \times 10^5$ N · m$^2$/C
    - (E) zero

52c. Chapter 18, Problem 52c
    - (A) zero
    - (B) $2.3 \times 10^5$ N · m$^2$/C
    - (C) $3.4 \times 10^5$ N · m$^2$/C
    - (D) $4.6 \times 10^5$ N · m$^2$/C
    - (E) $1.2 \times 10^5$ N · m$^2$/C

54a. Chapter 18, Problem 54a
    - (A) $7.9 \times 10^5$ N/C, radially outward
    - (B) $3.2 \times 10^5$ N/C, radially outward
    - (C) $1.4 \times 10^6$ N/C, radially inward
    - (D) $5.9 \times 10^5$ N/C, radially inward
    - (E) zero

54b. Chapter 18, Problem 54b
    - (A) $7.9 \times 10^5$ N/C, radially outward
    - (B) $3.2 \times 10^5$ N/C, radially outward
    - (C) $1.4 \times 10^6$ N/C, radially inward
    - (D) $5.9 \times 10^5$ N/C, radially inward
    - (E) zero

54c. Chapter 18, Problem 54c
    - (A) $7.9 \times 10^5$ N/C, radially outward
    - (B) $3.2 \times 10^5$ N/C, radially outward
    - (C) $1.4 \times 10^6$ N/C, radially inward
    - (D) $5.9 \times 10^5$ N/C, radially inward
    - (E) zero

*HOMEWORK CHOICES*  Chapter 18  429

56. Chapter 18, Problem 56
    *The solution is a proof.*


## Additional Problems

58. Chapter 18, Problem 58
    *The solution is a drawing.*

60. Chapter 18, Problem 60
    (A) 3.60 N, 51.3° north of east
    (B) 34.6 N, 36.3° north of east
    (C) 14.4 N, 44.2° north of east
    (D) 17.3 N, 38.7° south of east
    (E) 14.4 N, 36.3° south of east

62. Chapter 18, Problem 62
    (A) $+5.0 \times 10^5$ C
    (B) $-1.6 \times 10^5$ C
    (C) $-2.3 \times 10^5$ C
    (D) $-5.0 \times 10^5$ C
    (E) $+1.6 \times 10^5$ C

64a. Chapter 18, Problem 64a
    (A) 3.8 m on the side opposite from that of the negative charge
    (B) 6.2 m on the same side as the negative charge
    (C) 3.0 m on the side opposite from that of the negative charge
    (D) 4.9 m on the same side as the negative charge
    (E) 5.7 m on the side opposite from that of the negative charge

64b. Chapter 18, Problem 64b
    (A) zero
    (B) 1.4 N
    (C) 2.8 N
    (D) 3.6 N
    (E) 4.7 N

66. Chapter 18, Problem 66
    (A) 65 N/C
    (B) 15 N/C
    (C) 41 N/C
    (D) 35 N/C
    (E) 49 N/C

68a. Chapter 18, Problem 68a
    (A) one positive and one negative
    (B) either both positive or both negative

68b. Chapter 18, Problem 68b
    (A) 6.4 µC
    (B) 3.3 µC
    (C) 4.0 µC
    (D) 8.4 µC
    (E) 7.1 µC

70. Chapter 18, Problem 70
    (A) $1.0 \times 10^4$ m/s
    (B) $1.7 \times 10^4$ m/s
    (C) $2.6 \times 10^4$ m/s
    (D) $3.9 \times 10^4$ m/s
    (E) $4.0 \times 10^4$ m/s

# Homework Choices
## CHAPTER 19
# Electric Potential Energy and the Electric Potential

### Section 19.1 Potential Energy
### Section 19.2 The Electric Potential Difference

2a. Chapter 19, Problem 2a
- (A) $-3.2 \times 10^{-15}$ J
- (B) $+3.2 \times 10^{-15}$ J
- (C) $-4.8 \times 10^{-16}$ J
- (D) $+4.8 \times 10^{-16}$ J
- (E) $+7.7 \times 10^{-16}$ J

2b. Chapter 19, Problem 2b
- (A) $-3.2 \times 10^{-15}$ J
- (B) $+3.2 \times 10^{-15}$ J
- (C) $-4.8 \times 10^{-16}$ J
- (D) $+4.8 \times 10^{-16}$ J
- (E) $+7.7 \times 10^{-16}$ J

4a. Chapter 19, Problem 4a
- (A) $4.75 \times 10^{-14}$ J
- (B) $1.05 \times 10^{-14}$ J
- (C) $3.50 \times 10^{-14}$ J
- (D) $2.00 \times 10^{-14}$ J
- (E) $5.60 \times 10^{-14}$ J

4b. Chapter 19, Problem 4b
- (A) $2.00 \times 10^{-14}$ J
- (B) $3.50 \times 10^{-14}$ J
- (C) $4.75 \times 10^{-14}$ J
- (D) $5.60 \times 10^{-14}$ J
- (E) $1.05 \times 10^{-14}$ J

6. Chapter 19, Problem 6
- (A) $4.8 \times 10^2$ eV
- (B) $2.1 \times 10^2$ eV
- (C) $8.0 \times 10^2$ eV
- (D) $6.2 \times 10^2$ eV
- (E) $9.4 \times 10^2$ eV

8. Chapter 19, Problem 8
- (A) 5100 kg
- (B) 4400 kg
- (C) 3700 kg
- (D) 2200 kg
- (E) 1800 kg

10. Chapter 19, Problem 10
- (A) 234
- (B) 5.45
- (C) 29.7
- (D) 42.8
- (E) 1830

### Section 19.3 The Electric Potential Difference Created by Point Charges

12. Chapter 19, Problem 12
- (A) 13 V
- (B) 21 V
- (C) 26 V
- (D) 33 V
- (E) 41 V

14. Chapter 19, Problem 14
- (A) $+2.18 \times 10^{-18}$ J
- (B) $-4.35 \times 10^{-18}$ J
- (C) $+3.02 \times 10^{-18}$ J
- (D) $-6.15 \times 10^{-18}$ J
- (E) $+4.00 \times 10^{-18}$ J

## HOMEWORK CHOICES

16. Chapter 19, Problem 16
    (A) 0.37 m
    (B) 0.43 m
    (C) 0.26 m
    (D) 0.30 m
    (E) 0.48 m

18a. Chapter 19, Problem 18a
    (A) $+1.1 \times 10^6$ V
    (B) $-1.1 \times 10^6$ V
    (C) $+2.2 \times 10^6$ V
    (D) $-2.2 \times 10^6$ V
    (E) $-2.0 \times 10^6$ V

18b. Chapter 19, Problem 18b
    (A) +3.6 J
    (B) −3.8 J
    (C) +4.8 J
    (D) −4.4 J
    (E) +4.0 J

20. Chapter 19, Problem 20
    (A) $d/3$ and $2d/3$
    (B) $2d/3$ and $2d$
    (C) $d/3$ and $d$
    (D) $d$ and $2d$
    (E) $d/3$ and $2d$

22. Chapter 19, Problem 22
    (A) 8.0 J
    (B) 6.2 J
    (C) 4.0 J
    (D) 3.1 J
    (E) 2.4 J

24. Chapter 19, Problem 24
    (A) $-4.8$ μC
    (B) $+5.9$ μC
    (C) $-3.4$ μC
    (D) $+1.7$ μC
    (E) $-5.1$ μC

26. Chapter 19, Problem 26
    **Note**: *the following locations are relative to the negative charge.*
    (A) 0.176 m to the left, 0.200 m to the right
    (B) 0.208 m to the left, 0.294 m to the right
    (C) 0.388 m to the left, 0.276 m to the right
    (D) 0.222 m to the left, 0.169 m to the right
    (E) 0.164 m to the left, 0.371 m to the right

28. Chapter 19, Problem 28
    (A) $6.63 \times 10^{-2}$ m
    (B) $2.06 \times 10^{-2}$ m
    (C) $8.83 \times 10^{-2}$ m
    (D) $1.41 \times 10^{-2}$ m
    (E) $3.96 \times 10^{-2}$ m

## *Section 19.4  Equipotential Surfaces and Their Relation to the Electric Field*

30. Chapter 19, Problem 30
    (A) 2.6 m
    (B) 1.8 m
    (C) 0.75 m
    (D) 0.38 m
    (E) 0.15 m

32. Chapter 19, Problem 32
    (A) $7.2 \times 10^6$ V/m
    (B) $1.1 \times 10^6$ V/m
    (C) $8.8 \times 10^6$ V/m
    (D) $5.6 \times 10^6$ V/m
    (E) $3.4 \times 10^6$ V/m

34. Chapter 19, Problem 34
    (A) 1.41
    (B) 0.707
    (C) 1.00
    (D) 0.500
    (E) 0.841

# Homework: Electric Potential Energy and the Electric Potential

36a. Chapter 19, Problem 36a
(A) 5.0 V/m
(B) $2.0 \times 10^1$ V/m
(C) $3.5 \times 10^1$ V/m
(D) $1.0 \times 10^1$ V/m
(E) zero

36b. Chapter 19, Problem 36b
(A) $6.0 \times 10^1$ V/m
(B) $1.0 \times 10^1$ V/m
(C) 5.0 V/m
(D) $2.0 \times 10^1$ V/m
(E) zero

36c. Chapter 19, Problem 36c
(A) $1.0 \times 10^1$ V/m
(B) $2.0 \times 10^1$ V/m
(C) $6.0 \times 10^1$ V/m
(D) 5.0 V/m
(E) zero

38. Chapter 19, Problem 38
(A) 1
(B) 2
(C) 3
(D) 4
(E) 5

## Section 19.5 Capacitors and Dielectrics

40. Chapter 19, Problem 40
(A) $4.36 \times 10^4$ mi$^2$
(B) $3.92 \times 10^4$ mi$^2$
(C) $2.06 \times 10^4$ mi$^2$
(D) $1.79 \times 10^4$ mi$^2$
(E) $9.81 \times 10^3$ mi$^2$

42. Chapter 19, Problem 42
(A) 0.008 µF
(B) 0.06 µF
(C) 0.004 µF
(D) 0.1 µF
(E) 0.02 µF

44. Chapter 19, Problem 44
(A) $1.0 \times 10^{-4}$ m
(B) $1.5 \times 10^{-4}$ m
(C) $2.0 \times 10^{-4}$ m
(D) $3.0 \times 10^{-4}$ m
(E) $1.9 \times 10^{-5}$ m

46. Chapter 19, Problem 46
(A) 1.0 µF
(B) 1.4 µF
(C) 2.7 µF
(D) 7.3 µF
(E) 13 µF

48. Chapter 19, Problem 48
(A) $1.4 \times 10^{-7}$ J
(B) $9.1 \times 10^{-8}$ J
(C) $5.9 \times 10^{-8}$ J
(D) $3.4 \times 10^{-8}$ J
(E) $1.2 \times 10^{-8}$ J

50. Chapter 19, Problem 50
(A) $3.5 \times 10^{-2}$ m
(B) $2.7 \times 10^{-2}$ m
(C) $1.1 \times 10^{-2}$ m
(D) $2.0 \times 10^{-2}$ m
(E) $1.6 \times 10^{-2}$ m

52. Chapter 19, Problem 52
(A) $2.18 \times 10^{-5}$ m
(B) $7.33 \times 10^{-5}$ m
(C) $5.45 \times 10^{-5}$ m
(D) $1.83 \times 10^{-5}$ m
(E) $3.64 \times 10^{-5}$ m

## Additional Problems

54. **Chapter 19, Problem 54**
    - (A) $2.3 \times 10^6$ V
    - (B) $3.6 \times 10^6$ V
    - (C) $5.6 \times 10^6$ V
    - (D) $7.2 \times 10^6$ V
    - (E) $9.0 \times 10^6$ V

56a. **Chapter 19, Problem 56a**
    - (A) 32.2 J
    - (B) $1.92 \times 10^{-2}$ J
    - (C) $3.10 \times 10^{-2}$ J
    - (D) $5.80 \times 10^{-3}$ J
    - (E) $1.04 \times 10^{-6}$ J

56b. **Chapter 19, Problem 56b**
    - (A) $3.10 \times 10^{-2}$ V
    - (B) 1.61 V
    - (C) 18.9 V
    - (D) 32.2 V
    - (E) 173 V

56c. **Chapter 19, Problem 56c**
    - (A) Point **A** is at the higher potential.
    - (B) Point **B** is at the higher potential.

58a. **Chapter 19, Problem 58a**
    - (A) −1500 V
    - (B) +1500 V
    - (C) +1140 V
    - (D) −1140 V
    - (E) +800 V

58b. **Chapter 19, Problem 58b**
    - (A) Point **B** has the higher potential since the change in potential is positive.
    - (B) Point **A** has the higher potential since the change in potential is positive.
    - (C) Point **A** has the higher potential since the change in potential is negative.
    - (D) Point **B** has the higher potential since the change in potential is negative.

60. **Chapter 19, Problem 60**
    Note: assume the dashed line is the $y$ axis and the negative charge is located at $y = 0$.
    The following distances are measured relative to the negative charge.
    - (A) −2.4 m, +2.4 m
    - (B) −0.80 m, +0.80 m
    - (C) −3.6 m, +3.6 m
    - (D) −0.40 m, +0.40 m
    - (E) −1.2 m, +1.2 m

62. **Chapter 19, Problem 62**
    - (A) 9.2 V, increase
    - (B) 7.7 V, decrease
    - (C) 7.7 V, increase
    - (D) 9.2 V, decrease
    - (E) 3.3 V, increase

64. **Chapter 19, Problem 64**
    - (A) 7.3 m/s
    - (B) 3.9 m/s
    - (C) 6.8 m/s
    - (D) 5.4 m/s
    - (E) 6.1 m/s

66a. **Chapter 19, Problem 66a**
    - (A) 0.637 m
    - (B) 0.442 m
    - (C) 0.773 m
    - (D) 0.187 m
    - (E) 0.839 m

66b. **Chapter 19, Problem 66b**
    - (A) 6.83
    - (B) 3.67
    - (C) 5.03
    - (D) 2.66
    - (E) 1.95

# Homework Choices
# CHAPTER 20    Electric Circuits

## Section 20.1 Electromotive Force and Current
## Section 20.2 Ohm's Law

2. Chapter 20, Problem 2
   - (A) $6.2 \times 10^{23}$
   - (B) $2.0 \times 10^{23}$
   - (C) $4.5 \times 10^{22}$
   - (D) $5.0 \times 10^{21}$
   - (E) $1.0 \times 10^{20}$

4. Chapter 20, Problem 4
   - (A) 0.21 A
   - (B) 0.46 A
   - (C) 2.2 A
   - (D) 4.8 A
   - (E) 7.0 A

6. Chapter 20, Problem 6
   - (A) $2.1 \times 10^{6}$ J
   - (B) $1.3 \times 10^{6}$ J
   - (C) $1.1 \times 10^{5}$ J
   - (D) $4.3 \times 10^{2}$ J
   - (E) $3.6 \times 10^{2}$ J

8. Chapter 20, Problem 8
   - (A) $1.0 \times 10^{5}$ J
   - (B) $6.2 \times 10^{4}$ J
   - (C) $2.9 \times 10^{3}$ J
   - (D) $5.1 \times 10^{2}$ J
   - (E) $2.0 \times 10^{5}$ J

## Section 20.3 Resistance and Resistivity

10. Chapter 20, Problem 10
    - (A) 0.58 Ω
    - (B) 0.28 Ω
    - (C) 0.67 Ω
    - (D) 0.34 Ω
    - (E) 0.49 Ω

12. Chapter 20, Problem 12
    - (A) 4.5 m
    - (B) 9.0 m
    - (C) 11 m
    - (D) 14 m
    - (E) 18 m

14. Chapter 20, Problem 14
    - (A) $1.2 \times 10^{-2}$ Ω/m
    - (B) $1.6 \times 10^{-2}$ Ω/m
    - (C) $2.2 \times 10^{-2}$ Ω/m
    - (D) $3.6 \times 10^{-2}$ Ω/m
    - (E) $4.2 \times 10^{-2}$ Ω/m

16. Chapter 20, Problem 16
    - (A) −28.6 °C
    - (B) −14.6 °C
    - (C) −48.6 °C
    - (D) −34.6 °C
    - (E) −68.6 °C

18. Chapter 20, Problem 18
    - (A) 1.65
    - (B) 1.53
    - (C) 1.37
    - (D) 1.22
    - (E) 1.11

# HOMEWORK CHOICES     Chapter 20

20. Chapter 20, Problem 20
   - (A) 57 °C
   - (B) 71 °C
   - (C) 84 °C
   - (D) 110 °C
   - (E) 140 °C

## Section 20.4 Electric Power

22. Chapter 20, Problem 22
   - (A) 144 W
   - (B) 228 W
   - (C) 523 W
   - (D) 2740 W
   - (E) 4330 W

24. Chapter 20, Problem 24
   - (A) 0.83 A
   - (B) 0.77 A
   - (C) 1.2 A
   - (D) 1.0 A
   - (E) 1.4 A

26. Chapter 20, Problem 26
   - (A) $ 1.2
   - (B) $ 4.4
   - (C) $ 3.6
   - (D) $ 5.8
   - (E) $ 7.2

28. Chapter 20, Problem 28
   - (A) 50 m
   - (B) 25 m
   - (C) 90 m
   - (D) 45 m
   - (E) 70 m

30. Chapter 20, Problem 30
   - (A) 33 °C
   - (B) 26 °C
   - (C) 48 °C
   - (D) 18 °C
   - (E) 12 °C

## Section 20.5 Alternating Current

32. Chapter 20, Problem 32
   - (A) $4.63 \times 10^7$
   - (B) $3.82 \times 10^7$
   - (C) $1.04 \times 10^7$
   - (D) $2.28 \times 10^7$
   - (E) $5.73 \times 10^7$

34. Chapter 20, Problem 34
   - (A) 7.5 V
   - (B) 14 V
   - (C) 18 V
   - (D) 21 V
   - (E) 3.4 V

36. Chapter 20, Problem 36
   - (A) $2.3 \times 10^{-3}$ m
   - (B) $3.2 \times 10^{-3}$ m
   - (C) $4.2 \times 10^{-3}$ m
   - (D) $1.3 \times 10^{-3}$ m
   - (E) $5.6 \times 10^{-3}$ m

38a. Chapter 20, Problem 38a
   - (A) 83 W
   - (B) $7.0 \times 10^3$ W
   - (C) $3.5 \times 10^2$ W
   - (D) $9.5 \times 10^2$ W
   - (E) $2.0 \times 10^3$ W

436     Homework: Electric Circuits

38b. Chapter 20, Problem 38b
    (A) 450 s      (C) 340 s      (E) 180 s
    (B) 250 s      (D) 150 s

## *Section 20.6 Series Wiring*

40. Chapter 20, Problem 40
    (A) 1.0 V      (C) 9.0 V      (E) 24 V
    (B) 5.0 V      (D) 12 V

42a. Chapter 20, Problem 42a
    (A) 5.00 V      (C) 9.00 V      (E) 12.0 V
    (B) 10.0 V      (D) 24.0 V

42b. Chapter 20, Problem 42b
    (A) 9.00 V      (C) 5.00 V      (E) 12.0 V
    (B) 24.0 V      (D) 10.0 V

44a. Chapter 20, Problem 44a
    (A) 2.4 A      (C) 1.6 A      (E) 3.8 A
    (B) 6.2 A      (D) 4.4 A

44b. Chapter 20, Problem 44b
    (A) $V_{1.0\Omega} = 2.6$ V, $V_{5.0\Omega} = 4.0$ V, $V_{9.0\Omega} = 14$ V
    (B) $V_{1.0\Omega} = 1.8$ V, $V_{5.0\Omega} = 4.0$ V, $V_{9.0\Omega} = 12$ V
    (C) $V_{1.0\Omega} = 2.9$ V, $V_{5.0\Omega} = 6.0$ V, $V_{9.0\Omega} = 10$ V
    (D) $V_{1.0\Omega} = 1.6$ V, $V_{5.0\Omega} = 5.0$ V, $V_{9.0\Omega} = 12$ V
    (E) $V_{1.0\Omega} = 1.6$ V, $V_{5.0\Omega} = 8.0$ V, $V_{9.0\Omega} = 14$ V

44c. Chapter 20, Problem 44c
    (A) $P_{1.0\Omega} = 2.6$ W, $P_{5.0\Omega} = 13$ W, $P_{9.0\Omega} = 23$ W
    (B) $P_{1.0\Omega} = 2.3$ W, $P_{5.0\Omega} = 18$ W, $P_{9.0\Omega} = 26$ W
    (C) $P_{1.0\Omega} = 2.8$ W, $P_{5.0\Omega} = 19$ W, $P_{9.0\Omega} = 24$ W
    (D) $P_{1.0\Omega} = 1.3$ W, $P_{5.0\Omega} = 26$ W, $P_{9.0\Omega} = 23$ W
    (E) $P_{1.0\Omega} = 1.3$ W, $P_{5.0\Omega} = 13$ W, $P_{9.0\Omega} = 26$ W

46. Chapter 20, Problem 46
    (A) 0.80 V      (C) 1.8 V      (E) 2.6 V
    (B) 1.2 V      (D) 2.2 V

48a. Chapter 20, Problem 48a
    (A) $1.0 \times 10^2$ $\Omega$      (C) $7.0 \times 10^1$ $\Omega$      (E) $5.0 \times 10^1$ $\Omega$
    (B) $1.5 \times 10^2$ $\Omega$      (D) $3.5 \times 10^1$ $\Omega$

48b. Chapter 20, Problem 48b
    (A) $3.5 \times 10^1$ $\Omega$      (C) $1.0 \times 10^2$ $\Omega$      (E) $7.0 \times 10^1$ $\Omega$
    (B) $5.0 \times 10^1$ $\Omega$      (D) $1.5 \times 10^2$ $\Omega$

*Section 20.7 Parallel Wiring*

50. Chapter 20, Problem 50
    - (A) 601 Ω
    - (B) 446 Ω
    - (C) 272 Ω
    - (D) 119 Ω
    - (E) 66.0 Ω

52. Chapter 20, Problem 52
    - (A) $3R$
    - (B) $R$
    - (C) $R/3$
    - (D) $R/6$
    - (E) $R/9$

54. Chapter 20, Problem 54
    - (A) $R_{50}$ = 209 Ω, $R_{100}$ = 105 Ω
    - (B) $R_{50}$ = 288 Ω, $R_{100}$ = 144 Ω
    - (C) $R_{50}$ = 244 Ω, $R_{100}$ = 122 Ω
    - (D) $R_{50}$ = 228 Ω, $R_{100}$ = 114 Ω
    - (E) $R_{50}$ = 196 Ω, $R_{100}$ = 98.0 Ω

56a. Chapter 20, Problem 56a
    - (A) 8.0 Ω
    - (B) 16 Ω
    - (C) 24 Ω
    - (D) 28 Ω
    - (E) 32 Ω

56b. Chapter 20, Problem 56b
    - (A) 8.0 Ω
    - (B) 16 Ω
    - (C) 24 Ω
    - (D) 28 Ω
    - (E) 32 Ω

58. Chapter 20, Problem 58
    - (A) $3.62 \times 10^{-3}$ Ω
    - (B) $4.56 \times 10^{-3}$ Ω
    - (C) $1.16 \times 10^{-3}$ Ω
    - (D) $2.27 \times 10^{-3}$ Ω
    - (E) $5.67 \times 10^{-3}$ Ω

*Section 20.8 Circuits Wired Partially in Series and Partially in Parallel*

60. Chapter 20, Problem 60
    - (A) 2.2 A
    - (B) 4.3 A
    - (C) 7.0 A
    - (D) 9.2 A
    - (E) 21 A

62. Chapter 20, Problem 62
    - (A) 670 Ω
    - (B) 410 Ω
    - (C) 290 Ω
    - (D) 930 Ω
    - (E) 120 Ω

64a. Chapter 20, Problem 64a
    - (A) $6.04 \times 10^{-2}$ A
    - (B) $8.33 \times 10^{-2}$ A
    - (C) $1.18 \times 10^{-1}$ A
    - (D) $3.12 \times 10^{-1}$ A
    - (E) $3.65 \times 10^{-1}$ A

64b. Chapter 20, Problem 64b
    - (A) 0.833 W
    - (B) 0.972 W
    - (C) 1.88 W
    - (D) 2.81 W
    - (E) 5.63 W

# 438  Homework: Electric Circuits

66a. Chapter 20, Problem 66a

66b. Chapter 20, Problem 66b

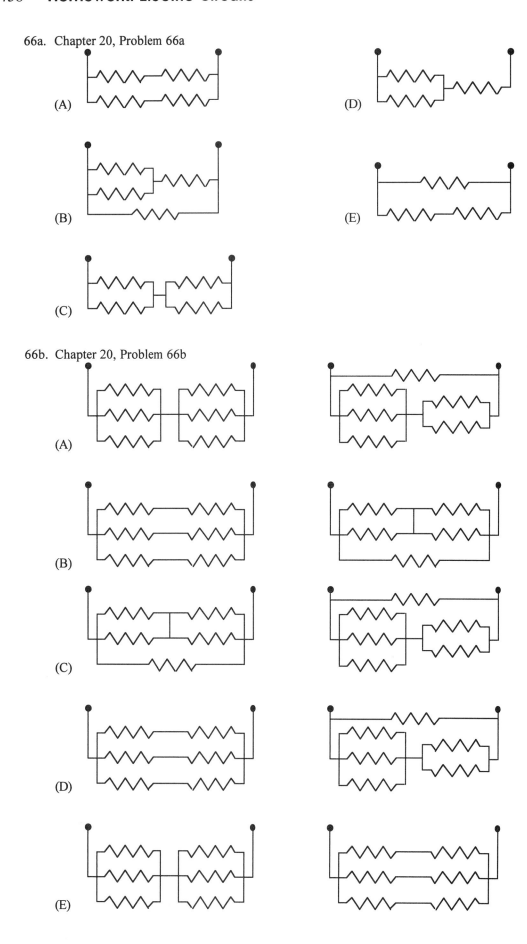

68. Chapter 20, Problem 68
    (A) 400 Ω
    (B) 500 Ω
    (C) 600 Ω
    (D) 700 Ω
    (E) 800 Ω

## Section 20.9 Internal Resistance

70. Chapter 20, Problem 70
    (A) 7.27 V
    (B) 9.00 V
    (C) 9.60 V
    (D) 10.9 V
    (E) 12.0 V

72. Chapter 20, Problem 72
    (A) 0.12 Ω
    (B) 0.15 Ω
    (C) 0.18 Ω
    (D) 0.10 Ω
    (E) 0.060 Ω

74. Chapter 20, Problem 74
    (A) 0.942
    (B) 0.934
    (C) 0.925
    (D) 0.916
    (E) 0.907

## Section 20.10 Kirchhoff's Rules

76a. Chapter 20, Problem 76a
    (A) 12 V
    (B) 18 V
    (C) 48 V
    (D) 36 V
    (E) 24 V

76b. Chapter 20, Problem 76b
    (A) 12 V
    (B) 18 V
    (C) 48 V
    (D) 36 V
    (E) 24 V

78a. Chapter 20, Problem 78a
    (A) 0.50 A
    (B) 3.0 A
    (C) 1.0 A
    (D) 2.0 A
    (E) 4.0 A

78b. Chapter 20, Problem 78b
    *The solution is a verification of the answer found in part (a).*

80. Chapter 20, Problem 80
    (A) 2.0 A, left to right
    (B) 1.6 A, right to left
    (C) 1.9 A, right to left
    (D) 1.7 A, left to right
    (E) 1.8 A, left to right

82. Chapter 20, Problem 82
    (A) 1.71 A, from the bottom to the top
    (B) 1.71 A, from the top to the bottom
    (C) 7.21 A, from the top to the bottom
    (D) 0.712 A, from the top to the bottom
    (E) 0.712 A, from the bottom to the top

# Homework: Electric Circuits

## Section 20.11 The Measurement of Current and Voltage

84. Chapter 20, Problem 84
    - (A) 0.00167 Ω
    - (B) 0.0835 Ω
    - (C) 0.333 Ω
    - (D) 0.512 Ω
    - (E) 0.649 Ω

86. Chapter 20, Problem 86
    - (A) 2.6 %
    - (B) 7.7 %
    - (C) 9.8 %
    - (D) 13 %
    - (E) 16 %

88. Chapter 20, Problem 88
    - (A) $I = 2.40 \times 10^{-3}$ A, $R_{coil} = 480$ Ω
    - (B) $I = 4.80 \times 10^{-3}$ A, $R_{coil} = 820$ Ω
    - (C) $I = 3.60 \times 10^{-3}$ A, $R_{coil} = 260$ Ω
    - (D) $I = 8.00 \times 10^{-3}$ A, $R_{coil} = 820$ Ω
    - (E) $I = 6.10 \times 10^{-3}$ A, $R_{coil} = 480$ Ω

## Section 20.12 Capacitors in Series and Parallel

90a. Chapter 20, Problem 90a
   - (A) $8.00 \times 10^{-5}$ C
   - (B) $1.33 \times 10^{-5}$ C
   - (C) $1.80 \times 10^{-4}$ C
   - (D) $3.60 \times 10^{-4}$ C
   - (E) $4.80 \times 10^{-5}$ C

90b. Chapter 20, Problem 90b
   - (A) $1.33 \times 10^{-5}$ C
   - (B) $8.00 \times 10^{-5}$ C
   - (C) $1.80 \times 10^{-4}$ C
   - (D) $4.80 \times 10^{-5}$ C
   - (E) $3.60 \times 10^{-4}$ C

92. Chapter 20, Problem 92
    - (A) 1.0 μF
    - (B) 1.6 μF
    - (C) 2.0 μF
    - (D) 6.4 μF
    - (E) 10.0 μF

94. Chapter 20, Problem 94
    - (A) 5.0 V
    - (B) 12 V
    - (C) 17 V
    - (D) 25 V
    - (E) 31 V

96. Chapter 20, Problem 96
    *The solution is a proof.*

98. Chapter 20, Problem 98
    - (A) 2.32 V
    - (B) 6.03 V
    - (C) 13.6 V
    - (D) 22.6 V
    - (E) 10.0 V

## Section 20.13 RC Circuits

100. Chapter 20, Problem 100
     - (A) $1.4 \times 10^3$ Ω
     - (B) $2.3 \times 10^3$ Ω
     - (C) $6.0 \times 10^3$ Ω
     - (D) $5.2 \times 10^3$ Ω
     - (E) $4.0 \times 10^3$ Ω

*HOMEWORK CHOICES*      **Chapter 20**    *441*

102. Chapter 20, Problem 102
    (A) 0.028 s      (C) 0.15 s      (E) 150 s
    (B) 0.097 s      (D) 1.1 s

104. Chapter 20, Problem 104
    (A) 2.3      (C) 5.6      (E) 8.3
    (B) 4.4      (D) 6.9

## *Additional Problems*

106a. Chapter 20, Problem 106a
    (A) 2.8 $\Omega$      (C) 19 $\Omega$      (E) 91 $\Omega$
    (B) 4.4 $\Omega$      (D) 54 $\Omega$

106b. Chapter 20, Problem 106b
    (A) 3.5 A      (C) 5.9 A      (E) 2.8 A
    (B) 4.0 A      (D) 4.4 A

108. Chapter 20, Problem 108
    (A) 0.73 A, from right to left      (C) 1.73 A, from left to right      (E) 2.06 A, from left to right
    (B) 0.73 A, from left to right      (D) 1.73 A, from right to left

110. Chapter 20, Problem 110
    (A) 37.8 °C      (C) 42.9 °C      (E) 53.4 °C
    (B) 22.6 °C      (D) 28.9 °C

112. Chapter 20, Problem 112
    (A) $5.6 \times 10^4$ s      (C) $4.0 \times 10^4$ s      (E) $3.4 \times 10^4$ s
    (B) $1.7 \times 10^4$ s      (D) $2.8 \times 10^4$ s

114a. Chapter 20, Problem 114a
    (A) 480 W      (C) 1300 W      (E) 5300 W
    (B) 920 W      (D) 1800 W

114b. Chapter 20, Problem 114b
    (A) 5300 W      (C) 1800 W      (E) 1300 W
    (B) 920 W      (D) 480 W

114c. Chapter 20, Problem 114c
    (A) 0.70      (C) 1.4      (E) 2.8
    (B) 1.0      (D) 2.1

116. Chapter 20, Problem 116
    (A) 0.47 $\Omega$, 3.6 $\Omega$      (C) 0.58 $\Omega$, 2.3 $\Omega$      (E) 0.94 $\Omega$, 7.2 $\Omega$
    (B) 0.38 $\Omega$, 2.6 $\Omega$      (D) 0.76 $\Omega$, 5.2 $\Omega$

118. Chapter 20, Problem 118
    (A) 8.26 V
    (B) 8.99 V
    (C) 7.91 V
    (D) 7.84 V
    (E) 8.70 V

120. Chapter 20, Problem 120
    (A) 3
    (B) 4
    (C) 1
    (D) 2
    (E) 6

122. Chapter 20, Problem 122
    (A) 37.8 °C
    (B) 39.5 °C
    (C) 38.2 °C
    (D) 43.7 °C
    (E) 40.1 °C

# Homework Choices
## CHAPTER 21
## Magnetic Forces and Magnetic Fields

*Section 21.1 Magnetic Fields*
*Section 21.2 The Force That a Magnetic Field Exerts on a Moving Charge*

2a. Chapter 21, Problem 2a
  (A) zero
  (B) $2.3 \times 10^{-4}$ N, out of the paper
  (C) $2.3 \times 10^{-4}$ N, into the paper
  (D) $5.7 \times 10^{-5}$ N, out of the paper
  (E) $5.7 \times 10^{-5}$ N, into the paper

2b. Chapter 21, Problem 2b
  (A) zero
  (B) $1.1 \times 10^{-4}$ N, out of the paper
  (C) $1.1 \times 10^{-4}$ N, into the paper
  (D) $2.3 \times 10^{-4}$ N, out of the paper
  (E) $2.3 \times 10^{-4}$ N, into the paper

2c. Chapter 21, Problem 2c
  (A) zero
  (B) $5.7 \times 10^{-5}$ N, out of the paper
  (C) $5.7 \times 10^{-5}$ N, into the paper
  (D) $1.1 \times 10^{-4}$ N, out of the paper
  (E) $1.1 \times 10^{-4}$ N, into the paper

4. Chapter 21, Problem 4
  (A) $2.3 \times 10^{-5}$ C
  (B) $4.6 \times 10^{-5}$ C
  (C) $7.8 \times 10^{-5}$ C
  (D) $8.9 \times 10^{-5}$ C
  (E) $1.0 \times 10^{-4}$ C

6. Chapter 21, Problem 6
  (A) 19.7°
  (B) 21.3°
  (C) 23.1°
  (D) 18.0°
  (E) 17.4°

8a. Chapter 21, Problem 8a
  (A) $2.81 \times 10^{3}$ N
  (B) $1.32 \times 10^{4}$ N
  (C) $4.50 \times 10^{4}$ N
  (D) $3.44 \times 10^{3}$ N
  (E) $1.06 \times 10^{4}$ N

8b. Chapter 21, Problem 8b
  (A) along the $+y$ axis
  (B) along the $+z$ axis
  (C) along the $+x$ axis
  (D) along the $-x$ axis
  (E) along the $-z$ axis

10. Chapter 21, Problem 10
  (A) 0.123 T
  (B) 0.287 T
  (C) 0.350 T
  (D) 0.592 T
  (E) 0.824 T

## Section 21.3 The Motion of a Charged Particle in a Magnetic Field
## Section 21.4 The Mass Spectrometer

12. Chapter 21, Problem 12
    (A) The charge on the ionized helium atom is $+e$.
    (B) The charge on the ionized helium atom is $+2e$.

14a. Chapter 21, Problem 14a
    (A) $5.00 \times 10^6$ m/s
    (B) $4.78 \times 10^6$ m/s
    (C) $9.56 \times 10^6$ m/s
    (D) $1.08 \times 10^7$ m/s
    (E) $1.52 \times 10^7$ m/s

14b. Chapter 21, Problem 14b
    (A) $7.60 \times 10^{-12}$ N
    (B) $6.95 \times 10^{-12}$ N
    (C) $5.77 \times 10^{-12}$ N
    (D) $4.81 \times 10^{-12}$ N
    (E) $3.50 \times 10^{-12}$ N

14c. Chapter 21, Problem 14c
    (A) 0.025 m
    (B) 0.051 m
    (C) 0.102 m
    (D) 0.153 m
    (E) 0.204 m

16. Chapter 21, Problem 16
    (A) $2.31 \times 10^{-4}$ m
    (B) $1.18 \times 10^{-3}$ m
    (C) $1.52 \times 10^{-2}$ m
    (D) $2.14 \times 10^{-1}$ m
    (E) $1.07 \times 10^{-1}$ m

18. Chapter 21, Problem 18
    (A) $v = q^2 EB$
    (B) $v = B/E$
    (C) $v = EB$
    (D) $v = E/B$
    (E) $v = qE/B$

20. Chapter 21, Problem 20
    (A) 0.46 cm
    (B) 0.55 cm
    (C) 0.69 cm
    (D) 0.88 cm
    (E) 1.10 cm

22. Chapter 21, Problem 22
    (A) $9.58 \times 10^6$ C/kg
    (B) $4.79 \times 10^6$ C/kg
    (C) $3.19 \times 10^6$ C/kg
    (D) $1.60 \times 10^6$ C/kg
    (E) $6.8 \times 10^5$ C/kg

24. Chapter 21, Problem 24
    (A) 441
    (B) 484
    (C) 289
    (D) 256
    (E) 361

26. Chapter 21, Problem 26
    (A) $6.06 \times 10^6$ rad/s
    (B) $4.04 \times 10^6$ rad/s
    (C) $2.02 \times 10^6$ rad/s
    (D) $1.01 \times 10^6$ rad/s
    (E) $5.05 \times 10^5$ rad/s

HOMEWORK CHOICES  Chapter 21  445

## Section 21.5 The Force on a Current in a Magnetic Field

28. Chapter 21, Problem 28

|     | left wire | center wire | right wire |
|-----|-----------|-------------|------------|
| (A) | 2.3 N     | 2.3 N       | zero       |
| (B) | zero      | 1.2 N       | zero       |
| (C) | zero      | 1.2 N       | 2.3 N      |
| (D) | 2.3 N     | 2.3 N       | 2.3 N      |
| (E) | 2.3 N     | 1.2 N       | 2.3 N      |

30. Chapter 21, Problem 30
    - (A) 17 A, east to west
    - (B) 25 A, west to east
    - (C) 39 A, east to west
    - (D) 44 A, west to east
    - (E) 49 A, east to west

32a. Chapter 21, Problem 32a

|     | side AB                  | side AC                  | side BC                  |
|-----|--------------------------|--------------------------|--------------------------|
| (A) | zero                     | 24.2 N, into the paper   | 24.2 N, out of the paper |
| (B) | zero                     | 24.2 N, out of the paper | 24.2 N, into the paper   |
| (C) | 24.2 N, into the paper   | 24.2 N, into the paper   | 24.2 N, into the paper   |
| (D) | 24.2 N, out of the paper | 24.2 N, out of the paper | 24.2 N, out of the paper |
| (E) | 24.2 N, out of the paper | zero                     | 24.2 N, into the paper   |

32b. Chapter 21, Problem 32b
   - (A) 12.1 N
   - (B) 24.2 N
   - (C) 48.4 N
   - (D) 72.3 N
   - (E) zero

34. Chapter 21, Problem 34
   - (A) $7.9 \times 10^{-3}$ N
   - (B) $9.2 \times 10^{-3}$ N
   - (C) $1.5 \times 10^{-3}$ N
   - (D) $3.1 \times 10^{-3}$ N
   - (E) $5.5 \times 10^{-3}$ N

36. Chapter 21, Problem 36
   - (A) 0.325 kg
   - (B) 0.748 kg
   - (C) 1.78 kg
   - (D) 1.63 kg
   - (E) 3.19 kg

38a. Chapter 21, Problem 38a
   - (A) 11 A
   - (B) 18 A
   - (C) 24 A
   - (D) 29 A
   - (E) 55 A

38b. Chapter 21, Problem 38b
   **Note**: In the following answers, the rail connected to the positive terminal of the battery is referred to as the *top* rail and the rail connected to the negative terminal is the *bottom* rail.
   - (A) The rod will move away from the battery because this is the direction of the force as determined by the right-hand rule since current flows from the top rail toward the bottom rail and the magnetic field is directed upward.
   - (B) The rod will move toward the battery because this is the direction of the force as determined by the right-hand rule since current flows from the top rail toward the bottom rail and the magnetic field is directed upward.
   - (C) The rod will move away from the battery because this is the direction of the force as determined by the right-hand rule since current flows from the bottom rail toward the top rail and the magnetic field is directed upward.
   - (D) The rod will move toward the battery because this is the direction of the force as determined by the right-hand rule since current flows from the bottom rail toward the top rail and the magnetic field is directed upward.

## Section 21.6 The Torque on a Current-Carrying Coil

40. Chapter 21, Problem 40
    - (A) $7.00 \times 10^{-3}$ N·m
    - (B) $6.14 \times 10^{-3}$ N·m
    - (C) $5.97 \times 10^{-3}$ N·m
    - (D) $4.96 \times 10^{-3}$ N·m
    - (E) $4.19 \times 10^{-3}$ N·m

42a. Chapter 21, Problem 42a
    - (A) 10.8 A·m$^2$
    - (B) 13.4 A·m$^2$
    - (C) 16.2 A·m$^2$
    - (D) 17.0 A·m$^2$
    - (E) 12.0 A·m$^2$

42b. Chapter 21, Problem 42b
    - (A) 26.8 N·m
    - (B) 20.6 N·m
    - (C) 32.3 N·m
    - (D) 24.1 N·m
    - (E) 34.0 N·m

44a. Chapter 21, Problem 44a
    - (A) 28.7°
    - (B) 36.9°
    - (C) 45.0°
    - (D) 53.1°
    - (E) 72.0°

44b. Chapter 21, Problem 44b
    *The solution is a drawing.*

46. Chapter 21, Problem 46
    - (A) $\tau_{max} = 4NIL^2B$
    - (B) $\tau_{max} = \dfrac{NIL^2B}{16}$
    - (C) $\tau_{max} = \dfrac{IL^2B}{16N}$
    - (D) $\tau_{max} = \dfrac{L^2B}{4NI}$
    - (E) $\tau_{max} = \dfrac{4NIB}{L^2}$

48. Chapter 21, Problem 48
    - (A) 1.50
    - (B) 1.33
    - (C) 1.29
    - (D) 1.26
    - (E) 1.13

## Section 21.7 Magnetic Fields Produced by Currents

50. Chapter 21, Problem 50
    - (A) 0.060 m
    - (B) 0.12 m
    - (C) 0.16 m
    - (D) 0.18 m
    - (E) 0.22 m

52. Chapter 21, Problem 52
    - (A) 0.018 m
    - (B) 0.027 m
    - (C) 0.042 m
    - (D) 0.054 m
    - (E) 0.086 m

54. Chapter 21, Problem 54
    - (A) 1.9 A
    - (B) 2.7 A
    - (C) 3.6 A
    - (D) 5.0 A
    - (E) 6.1 A

# HOMEWORK CHOICES — Chapter 21

56a. Chapter 21, Problem 56a
  (A) $2.7 \times 10^{-5}$ T
  (B) $5.3 \times 10^{-5}$ T
  (C) $1.1 \times 10^{-5}$ T
  (D) $4.3 \times 10^{-5}$ T
  (E) $5.3 \times 10^{-6}$ T

56b. Chapter 21, Problem 56b
  (A) $2.7 \times 10^{-5}$ T
  (B) $5.3 \times 10^{-5}$ T
  (C) $1.1 \times 10^{-5}$ T
  (D) $4.3 \times 10^{-5}$ T
  (E) $5.3 \times 10^{-6}$ T

58a. Chapter 21, Problem 58a
  (A) $2.8 \times 10^{-6}$ T, out of the paper
  (B) $2.8 \times 10^{-6}$ T, into the paper
  (C) $1.4 \times 10^{-6}$ T, into the paper
  (D) $1.4 \times 10^{-6}$ T, out of the paper
  (E) $5.6 \times 10^{-6}$ T, into the paper

58b. Chapter 21, Problem 58b
  (A) $2.8 \times 10^{-6}$ T, out of the paper
  (B) $2.8 \times 10^{-6}$ T, into the paper
  (C) $1.4 \times 10^{-6}$ T, into the paper
  (D) $1.4 \times 10^{-6}$ T, out of the paper
  (E) $5.6 \times 10^{-6}$ T, into the paper

60. Chapter 21, Problem 60
  (A) $2.1 \times 10^{-3}$ N
  (B) $5.6 \times 10^{-3}$ N
  (C) $1.5 \times 10^{-4}$ N
  (D) $3.0 \times 10^{-4}$ N
  (E) $5.6 \times 10^{-3}$ N

62. Chapter 21, Problem 62
  (A) $9.8 \times 10^{-8}$ T
  (B) $3.6 \times 10^{-7}$ T
  (C) $2.9 \times 10^{-7}$ T
  (D) $6.5 \times 10^{-8}$ T
  (E) $1.1 \times 10^{-7}$ T

64. Chapter 21, Problem 64
  (A) 570 A
  (B) 450 A
  (C) 320 A
  (D) 210 A
  (E) 130 A

## Section 21.8 Ampere's Law

66. Chapter 21, Problem 66
  *The solution is a proof.*

68. Chapter 21, Problem 68
  *The solution is a proof.*

## Additional Problems

70a. Chapter 21, Problem 70a
  (A) 0.18 T, due west
  (B) 0.20 T, downward (toward the earth)
  (C) 0.20 T, upward (away from the earth)
  (D) 0.11 T, upward (away from the earth)
  (E) 0.11 T, downward (toward the earth)

70b. Chapter 21, Problem 70b
- (A) 0.18 T, due west
- (B) 0.20 T, downward (toward the earth)
- (C) 0.20 T, upward (away from the earth)
- (D) 0.11 T, upward (away from the earth)
- (E) 0.11 T, downward (toward the earth)

72. Chapter 21, Problem 72
- (A) $3.81 \times 10^{-3}$ m
- (B) $4.73 \times 10^{-3}$ m
- (C) $6.37 \times 10^{-3}$ m
- (D) $8.71 \times 10^{-3}$ m
- (E) $9.83 \times 10^{-3}$ m

74. Chapter 21, Problem 74
- (A) 4.9 A
- (B) 6.7 A
- (C) 5.4 A
- (D) 3.2 A
- (E) 4.2 A

76. Chapter 21, Problem 76
- (A) $H = 3.3R$
- (B) $H = 2.1R$
- (C) $H = 1.5R$
- (D) $H = R/2.1$
- (E) $H = R/3.3$

78. Chapter 21, Problem 78
- (A) $3.7 \times 10^{-12}$ N
- (B) $4.1 \times 10^{-12}$ N
- (C) $1.1 \times 10^{-11}$ N
- (D) $1.3 \times 10^{-11}$ N
- (E) $2.3 \times 10^{-11}$ N

80. Chapter 21, Problem 80
- (A) $2.1 \times 10^{-10}$ N, out of the paper
- (B) $1.3 \times 10^{-10}$ N, out of the paper
- (C) $1.0 \times 10^{-10}$ N, out of the paper
- (D) $1.7 \times 10^{-10}$ N, into the paper
- (E) $1.5 \times 10^{-10}$ N, into the paper

82a. Chapter 21, Problem 82a
- (A) 0.749
- (B) 1.63
- (C) 1.05
- (D) 2.11
- (E) 0.857

82b. Chapter 21, Problem 82b
- (A) 0.74
- (B) 1.6
- (C) 2.1
- (D) 1.0
- (E) 0.85

# Homework Choices
# CHAPTER 22
# Electromagnetic Induction

## Section 22.2 Motional Emf

2. Chapter 22, Problem 2
   - (A) 0.065 V
   - (B) 0.077 V
   - (C) 0.086 V
   - (D) 0.049 V
   - (E) 0.037 V

4. Chapter 22, Problem 4
   - (A) 1.1 m/s
   - (B) 1.4 m/s
   - (C) 2.3 m/s
   - (D) 7.5 m/s
   - (E) 15 m/s

6. Chapter 22, Problem 6
   - (A) 18 m
   - (B) 15 m
   - (C) 7.6 m
   - (D) 5.8 m
   - (E) 2.1 m

8. Chapter 22, Problem 8
   - (A) 7.5 m/s
   - (B) 12 m/s
   - (C) 15 m/s
   - (D) 25 m/s
   - (E) 36 m/s

10a. Chapter 22, Problem 10a
   - (A) 0.23 kg
   - (B) 0.54 kg
   - (C) 0.016 kg
   - (D) 0.65 kg
   - (E) 2.8 kg

10b. Chapter 22, Problem 10b
   - (A) +4.2 J
   - (B) +3.4 J
   - (C) −3.4 J
   - (D) +1.8 J
   - (E) −1.8 J

10c. Chapter 22, Problem 10c
   - (A) 7.2 J
   - (B) 6.0 J
   - (C) 3.5 J
   - (D) 1.8 J
   - (E) 3.0 J

## Section 22.3 Magnetic Flux

12. Chapter 22, Problem 12
   - (A) $1.0 \times 10^{-3}$ Wb
   - (B) $2.2 \times 10^{-3}$ Wb
   - (C) $4.1 \times 10^{-3}$ Wb
   - (D) $5.5 \times 10^{-3}$ Wb
   - (E) $7.0 \times 10^{-3}$ Wb

14a. Chapter 22, Problem 14a
   - (A) $3.6 \times 10^{-4}$ Wb
   - (B) $3.2 \times 10^{-4}$ Wb
   - (C) $2.4 \times 10^{-4}$ Wb
   - (D) $2.0 \times 10^{-4}$ Wb
   - (E) $1.2 \times 10^{-4}$ Wb

14b. Chapter 22, Problem 14b
   - (A) $3.6 \times 10^{-4}$ Wb
   - (B) $3.2 \times 10^{-4}$ Wb
   - (C) $2.4 \times 10^{-4}$ Wb
   - (D) $2.0 \times 10^{-4}$ Wb
   - (E) $1.2 \times 10^{-4}$ Wb

14c. Chapter 22, Problem 14c
- (A) $3.6 \times 10^{-4}$ Wb
- (B) $3.2 \times 10^{-4}$ Wb
- (C) $2.4 \times 10^{-4}$ Wb
- (D) $2.0 \times 10^{-4}$ Wb
- (E) $1.2 \times 10^{-4}$ Wb

16. Chapter 22, Problem 16
- (A) $3.2 \times 10^{-3}$ Wb
- (B) $3.9 \times 10^{-3}$ Wb
- (C) $7.7 \times 10^{-3}$ Wb
- (D) $8.4 \times 10^{-3}$ Wb
- (E) $9.1 \times 10^{-3}$ Wb

## Section 22.4 Faraday's Law of Electromagnetic Induction

18. Chapter 22, Problem 18
- (A) 0.48 T
- (B) 0.57 T
- (C) 0.13 T
- (D) 0.22 T
- (E) 0.31 T

20. Chapter 22, Problem 20
- (A) 1.5 m$^2$/s
- (B) 0.65 m$^2$/s
- (C) 1.1 m$^2$/s
- (D) 0.92 m$^2$/s
- (E) 2.1 m$^2$/s

22. Chapter 22, Problem 22
- (A) 0.011 T/s
- (B) 0.088 T/s
- (C) 0.15 T/s
- (D) 0.32 T/s
- (E) 0.45 T/s

24. Chapter 22, Problem 24
- (A) 0.15 V
- (B) 0.27 V
- (C) 0.43 V
- (D) 0.55 V
- (E) 0.68 V

26. Chapter 22, Problem 26
- (A) 0.041 V
- (B) 0.036 V
- (C) 0.032 V
- (D) 0.027 V
- (E) 0.023 V

28. Chapter 22, Problem 28
- (A) 960 rad/s
- (B) 1100 rad/s
- (C) 1300 rad/s
- (D) 1700 rad/s
- (E) 2100 rad/s

HOMEWORK CHOICES  Chapter 22  451

## Section 22.5 Lenz's Law

30. Chapter 22, Problem 30
    Note the following statement that applies to each of the following choices:
    The current, $I$, produces a magnetic field, $B$, which is perpendicular to the coil.
    (A) As $I$ increases, $B$ decreases. The induced field must be directed out of the paper to counteract the decrease in $B$. The current in the coil must be counterclockwise, e. g. it flows from right to left through the resistor.
    (B) As $I$ increases, $B$ increases. The induced field must be directed into the paper to counteract the increase in $B$. The current in the coil must be counterclockwise, e. g. it flows from right to left through the resistor.
    (C) As $I$ increases, $B$ decreases. The induced field must be directed into the paper to counteract the decrease in $B$. The current in the coil must be clockwise, e. g. it flows from left to right through the resistor.
    (D) As $I$ increases, $B$ increases. The induced field must be directed into the paper to counteract the increase in $B$. The current in the coil must be clockwise, e. g. it flows from left to right through the resistor.
    (E) As $I$ increases, $B$ decreases. The induced field must be directed out of paper to counteract the decrease in $B$. The current in the coil must be clockwise, e. g. it flows from left to right through the resistor.

32a. Chapter 22, Problem 32a
    **Note**: *In the following choices* CW = clockwise *and* CCW = counterclockwise.
    (A) CW for $A$ and CW for $B$           (C) CCW for $A$ and CCW for $B$
    (B) CW for $A$ and CCW for $B$          (D) CCW for $A$ and CW for $B$

32b. Chapter 22, Problem 32b
    (A) CW for $A$ and CW for $B$           (C) CCW for $A$ and CCW for $B$
    (B) CW for $A$ and CCW for $B$          (D) CCW for $A$ and CW for $B$

34. Chapter 22, Problem 34
    (A) Since the applied magnetic field is decreasing in time, the flux through the circuit is increasing. An induced magnetic field is generated which is directed into the paper. Therefore, the induced current in the circuit flows clockwise. The lower plate is negative and the upper plate is positive. Thus, the electric field points downward.
    (B) Since the applied magnetic field is decreasing in time, the flux through the circuit is decreasing. An induced magnetic field is generated which is directed out of the paper. Therefore, the induced current in the circuit flows counterclockwise. The lower plate is negative and the upper plate is positive. Thus, the electric field points downward.
    (C) Since the applied magnetic field is decreasing in time, the flux through the circuit is decreasing. An induced magnetic field is generated which is directed into the paper. Therefore, the induced current in the circuit flows counterclockwise. The lower plate is negative and the upper plate is positive. Thus, the electric field points upward.
    (D) Since the applied magnetic field is decreasing in time, the flux through the circuit is decreasing. An induced magnetic field is generated which is directed out of the paper. Therefore, the induced current in the circuit flows counterclockwise. The lower plate is positive and the upper plate is negative. Thus, the electric field points upward.
    (E) No current can be induced in the circuit since there is no complete loop. Therefore, the electric field between the plates will be zero.

36a. Chapter 22, Problem 36a
    (A) at location I, $x \rightarrow y \rightarrow z$ and at location II, $z \rightarrow y \rightarrow x$
    (B) at location I, $z \rightarrow y \rightarrow x$ and at location II, $x \rightarrow y \rightarrow z$
    (C) at location I, $x \rightarrow y \rightarrow z$ and at location II, $x \rightarrow y \rightarrow z$
    (D) at location I, $z \rightarrow y \rightarrow x$ and at location II, $z \rightarrow y \rightarrow x$

# Homework: Electromagnetic Induction

36b. Chapter 22, Problem 36b
    (A) at location I, $x \to y \to z$ and at location II, $z \to y \to x$
    (B) at location I, $z \to y \to x$ and at location II, $x \to y \to z$
    (C) at location I, $x \to y \to z$ and at location II, $x \to y \to z$
    (D) at location I, $z \to y \to x$ and at location II, $z \to y \to x$

## Section 22.7 The Electric Generator

38. Chapter 22, Problem 38
    (A) 0.63 T      (C) 0.11 T      (E) 0.36 T
    (B) 0.49 T      (D) 0.23 T

40. Chapter 22, Problem 40
    (A) 3.0 V      (C) 12 V      (E) 36 V
    (B) 4.0 V      (D) 24 V

42. Chapter 22, Problem 42
    (A) 4.0 Ω      (C) 16 Ω      (E) 40 Ω
    (B) 8.0 Ω      (D) 24 Ω

44. Chapter 22, Problem 44
    (A) 2.0 V      (C) 5.5 V      (E) 14 V
    (B) 4.0 V      (D) 9.4 V

46a. Chapter 22, Problem 46a
    (A) 0.64 A      (C) 4.0 A      (E) 12 A
    (B) 0.80 A      (D) 7.2 A

46b. Chapter 22, Problem 46b
    (A) 8.00 A      (C) 3.60 A      (E) zero
    (B) 6.40 A      (D) 1.60 A

46c. Chapter 22, Problem 46c
    (A) 3.60 A      (C) 1.60 A      (E) 4.40 A
    (B) 8.00 A      (D) 6.40 A

## Section 22.8 Mutual Inductance and Self-Inductance

48. Chapter 22, Problem 48
    (A) 0.69 V      (C) 1.1 V      (E) 1.7 V
    (B) 0.92 V      (D) 1.4 V

50. Chapter 22, Problem 50
    (A) 170 V      (C) 113 V      (E) 21 V
    (B) 130 V      (D) 64 V

HOMEWORK CHOICES

52. Chapter 22, Problem 52
    (A) 0.53 J
    (B) 0.68 J
    (C) 0.15 J
    (D) 0.40 J
    (E) 0.27 J

54. Chapter 22, Problem 54
    (A) $1.28 \times 10^{-4}$ H
    (B) $2.29 \times 10^{-4}$ H
    (C) $2.80 \times 10^{-4}$ H
    (D) $5.29 \times 10^{-4}$ H
    (E) $6.39 \times 10^{-4}$ H

56. Chapter 22, Problem 56
    *The solution is a proof.*

## Section 22.9 Transformers

58. Chapter 22, Problem 58
    (A) 2.8 V
    (B) 5.9 V
    (C) 7.1 V
    (D) 9.2 V
    (E) 13 V

60. Chapter 22, Problem 60
    (A) 441
    (B) 623
    (C) 756
    (D) 1070
    (E) 1310

62. Chapter 22, Problem 62
    (A) 0.27 A
    (B) 0.43 A
    (C) 1.9 A
    (D) 13 A
    (E) 18 A

64. Chapter 22, Problem 64
    (A) 15
    (B) 45
    (C) 90
    (D) 150
    (E) 210

66. Chapter 22, Problem 66
    *The solution is a proof.*

## Additional Problems

68. Chapter 22, Problem 68
    (A) 0.30 T
    (B) 0.65 T
    (C) 0.90 T
    (D) 1.2 T
    (E) 1.6 T

70. Chapter 22, Problem 70
    (A) 0.0025 V
    (B) 0.050 V
    (C) 0.11 V
    (D) 0.22 V
    (E) 0.45 V

72a. Chapter 22, Problem 72a
    (A) a step-up transformer
    (B) a step-down transformer

# Homework: Electromagnetic Induction

72b. Chapter 22, Problem 72b
- (A) 220
- (B) 110
- (C) 100
- (D) 78
- (E) 55

74. Chapter 22, Problem 74
- (A) 0.15 A
- (B) 0.37 A
- (C) 0.73 A
- (D) 0.86 A
- (E) 0.99 A

76a. Chapter 22, Problem 76a
- (A) The motion is retarded because of the induced magnetic field produced by the induced current that flows in the ring. The induced magnetic field points downward.
- (B) The motion is retarded because of the induced magnetic field produced by the induced current that flows in the ring. The induced magnetic field points upward.

76b. Chapter 22, Problem 76b
- (A) The motion is not retarded because of the induced magnetic field produced by the induced current that flows in the ring. The induced magnetic field points downward and accelerates the magnet.
- (B) The motion is not retarded because no induced current can flow in the ring when it is cut through. Without an induced current, there is no induced magnetic field.

78. Chapter 22, Problem 78
- (A) $9.6 \times 10^{-3}$ A
- (B) $7.1 \times 10^{-3}$ A
- (C) $5.9 \times 10^{-3}$ A
- (D) $2.4 \times 10^{-3}$ A
- (E) $1.2 \times 10^{-3}$ A

80. Chapter 22, Problem 80
- (A) $m = \dfrac{\mu_0 \pi R_2^2}{2 N_1 N_2 R_1}$
- (B) $m = \dfrac{\mu_0 \pi R_1 R_2}{2 N_1 N_2}$
- (C) $m = \dfrac{\mu_0 \pi N_1 N_2}{2 R_1 R_2}$
- (D) $m = \dfrac{\mu_0 \pi N_1 N_2 R_2}{2 R_1}$
- (E) $m = \dfrac{\mu_0 \pi N_1 N_2 R_2^2}{2 R_1}$

# Homework Choices
## CHAPTER 23 — Alternating Current Circuits

### Section 23.1 Capacitors and Capacitive Reactance

2. Chapter 23, Problem 2
   - (A) 1.9 V
   - (B) 2.2 V
   - (C) 2.9 V
   - (D) 3.8 V
   - (E) 4.4 V

4. Chapter 23, Problem 4
   - (A) 110 Ω
   - (B) 24 Ω
   - (C) 8.0 Ω
   - (D) 12 Ω
   - (E) 36 Ω

6a. Chapter 23, Problem 6a
   - (A) 1.0 µF
   - (B) 2.4 µF
   - (C) 5.5 µF
   - (D) 6.4 µF
   - (E) 7.8 µF

6b. Chapter 23, Problem 6b
   - (A) $3.6 \times 10^{-4}$ C
   - (B) $9.0 \times 10^{-4}$ C
   - (C) $7.2 \times 10^{-4}$ C
   - (D) $1.8 \times 10^{-5}$ C
   - (E) $1.4 \times 10^{-3}$ C

8. Chapter 23, Problem 8
   - (A) 1/3
   - (B) 1/2
   - (C) 3/2
   - (D) 2/3
   - (E) 5/2

### Section 23.2 Inductors and Inductive Reactance

10. Chapter 23, Problem 10
    - (A) 89 V
    - (B) 75 V
    - (C) 62 V
    - (D) 43 V
    - (E) 23 V

12. Chapter 23, Problem 12
    - (A) 0.22 A
    - (B) 0.44 A
    - (C) 1.1 A
    - (D) 2.3 A
    - (E) 4.5 A

14. Chapter 23, Problem 14
    - (A) $2.09 \times 10^{-2}$ A
    - (B) $4.95 \times 10^{-2}$ A
    - (C) $6.16 \times 10^{-2}$ A
    - (D) $8.41 \times 10^{-2}$ A
    - (E) $1.22 \times 10^{-1}$ A

16. Chapter 23, Problem 16
    - (A) 0.066 H
    - (B) 0.040 H
    - (C) 0.052 H
    - (D) 0.020 H
    - (E) 0.033 H

## Section 23.3 Circuits Containing Resistance, Capacitance, and Inductance

18. Chapter 23, Problem 18
    - (A)  $V_R = 19.9$ V     $V_C = 26.6$ V     $V_L = 13.0$ V
    - (B)  $V_R = 18.5$ V     $V_C = 36.0$ V     $V_L = 29.6$ V
    - (C)  $V_R = 10.5$ V     $V_C = 26.6$ V     $V_L = 36.0$ V
    - (D)  $V_R = 9.0$ V      $V_C = 18.5$ V     $V_L = 30.0$ V
    - (E)  $V_R = 10.5$ V     $V_C = 19.0$ V     $V_L = 29.6$ V

20a. Chapter 23, Problem 20a
   - (A) 10.7 V
   - (B) 8.70 V
   - (C) 12.5 V
   - (D) 6.00 V
   - (E) 14.8 V

20b. Chapter 23, Problem 20b
   - (A) −29.8°
   - (B) −20.4°
   - (C) −15.0°
   - (D) 20.4°
   - (E) 29.8°

22a. Chapter 23, Problem 22a
   - (A) 1.50 A
   - (B) 3.00 A
   - (C) 4.50 A
   - (D) 7.50 A
   - (E) 12.0 A

22b. Chapter 23, Problem 22b
   - (A) 27.0 W
   - (B) 108 W
   - (C) 243 W
   - (D) 675 W
   - (E) 1730 W

22c. Chapter 23, Problem 22c
   - (A) 1.21 mH
   - (B) 4.90 mH
   - (C) 7.96 mH
   - (D) 8.44 mH
   - (E) 9.67 mH

24a. Chapter 23, Problem 24a
   - (A) 0.39 A
   - (B) 0.11 A
   - (C) 0.33 A
   - (D) 0.19 A
   - (E) 0.26 A

24b. Chapter 23, Problem 24b
   - (A) 0.39 A
   - (B) 0.11 A
   - (C) 0.33 A
   - (D) 0.19 A
   - (E) 0.26 A

26. Chapter 23, Problem 26
   - (A) 0.25 Ω
   - (B) 1.0 Ω
   - (C) 8.8 Ω
   - (D) 14 Ω
   - (E) 3.2 Ω

28. Chapter 23, Problem 28
   - (A) 0.382 W
   - (B) 0.451 W
   - (C) 0.651 W
   - (D) 0.839 W
   - (E) 0.967 W

*Section 23.4 Resonance in Electric Circuits*

30. Chapter 23, Problem 30
    - (A) $8.4 \times 10^{-10}$ F
    - (B) $1.9 \times 10^{-9}$ F
    - (C) $2.4 \times 10^{-9}$ F
    - (D) $3.1 \times 10^{-9}$ F
    - (E) $3.8 \times 10^{-9}$ F

32a. Chapter 23, Problem 32a
    - (A) 0.34 A
    - (B) 0.44 A
    - (C) 0.50 A
    - (D) 0.71 A
    - (E) 2.0 A

32b. Chapter 23, Problem 32b
    - (A) 0.34 A
    - (B) 0.44 A
    - (C) 0.50 A
    - (D) 0.71 A
    - (E) 2.0 A

32c. Chapter 23, Problem 32c
    - (A) 0.704 H
    - (B) 0.599 H
    - (C) 0.462 H
    - (D) 0.410 H
    - (E) 0.359 H

34. Chapter 23, Problem 34
    - (A) 0.35
    - (B) 0.59
    - (C) 0.77
    - (D) 1.3
    - (E) 1.7

36. Chapter 23, Problem 36
    - (A) 80.4 Hz
    - (B) 328 Hz
    - (C) 521 Hz
    - (D) 849 Hz
    - (E) 1210 Hz

38. Chapter 23, Problem 38
    - (A) 2
    - (B) $\sqrt{2}$
    - (C) $2\sqrt{2}$
    - (D) $1/\sqrt{2}$
    - (E) 1/2

*Additional Problems*

40. Chapter 23, Problem 40
    - (A) 45 V
    - (B) 35 V
    - (C) 24 V
    - (D) 28 V
    - (E) 38 V

42. Chapter 23, Problem 42
    - (A) 2
    - (B) 4
    - (C) 1/2
    - (D) 1/4
    - (E) 1/8

44. Chapter 23, Problem 44
    - (A) 120 Hz
    - (B) 180 Hz
    - (C) 270 Hz
    - (D) 310 Hz
    - (E) 350 Hz

46a. Chapter 23, Problem 46a
    - (A) The resistor is in series with an inductor.
    - (B) The resistor is in series with a capacitor.

# Homework: Alternating Current Circuits

46b. Chapter 23, Problem 46b
- (A) $R = 129\ \Omega$, $X_L = 243\ \Omega$
- (B) $R = 173\ \Omega$, $X_L = 325\ \Omega$
- (C) $R = 261\ \Omega$, $X_C = 491\ \Omega$
- (D) $R = 325\ \Omega$, $X_C = 306\ \Omega$
- (E) $R = 784\ \Omega$, $X_C = 369\ \Omega$

48. Chapter 23, Problem 48
- (A) $4.5 \times 10^{-4}$ m
- (B) $7.7 \times 10^{-4}$ m
- (C) $2.3 \times 10^{-4}$ m
- (D) $9.8 \times 10^{-4}$ m
- (E) $6.8 \times 10^{-4}$ m

50. Chapter 23, Problem 50
- (A) 2
- (B) 6
- (C) 4
- (D) 5
- (E) 3

# Homework Choices
# CHAPTER 24 — Electromagnetic Waves

## Section 24.1 The Nature of Electromagnetic Waves

2a. Chapter 24, Problem 2a
   (A) 0.781 s
   (B) 1.03 s
   (C) 1.28 s
   (D) 2.06 s
   (E) 2.56 s

2b. Chapter 24, Problem 2b
   (A) 110 s
   (B) 130 s
   (C) 150 s
   (D) 170 s
   (E) 190 s

4. Chapter 24, Problem 4
   (A) $4.97 \times 10^{-11}$ F
   (B) $2.19 \times 10^{-11}$ F
   (C) $2.69 \times 10^{-11}$ F
   (D) $2.97 \times 10^{-11}$ F
   (E) $4.05 \times 10^{-11}$ F

6a. Chapter 24, Problem 6a
   (A) $2.4 \times 10^9$ Hz
   (B) $1.5 \times 10^{10}$ Hz
   (C) $4.8 \times 10^9$ Hz
   (D) $7.5 \times 10^9$ Hz
   (E) $3.6 \times 10^9$ Hz

6b. Chapter 24, Problem 6b
   (A) 0.016 m
   (B) 0.032 m
   (C) 0.048 m
   (D) 0.063 m
   (E) 0.126 m

## Section 24.2 The Electromagnetic Spectrum

8. Chapter 24, Problem 8
   (A) 11.118 m
   (B) 10.214 m
   (C) 9.6352 m
   (D) 8.9995 m
   (E) 8.0838 m

10. Chapter 24, Problem 10
   (A) $5.18 \times 10^{11}$ Hz
   (B) $1.93 \times 10^{12}$ Hz
   (C) $4.17 \times 10^{12}$ Hz
   (D) $9.05 \times 10^{12}$ Hz
   (E) $1.40 \times 10^{13}$ Hz

12a. Chapter 24, Problem 12a
   (A) 606 nm
   (B) 428 nm
   (C) 512 nm
   (D) 473 nm
   (E) 659 nm

12b. Chapter 24, Problem 12b
   (A) 606 nm
   (B) 428 nm
   (C) 512 nm
   (D) 473 nm
   (E) 659 nm

14. Chapter 24, Problem 14
   (A) $7.3 \times 10^9$ Hz
   (B) $1.5 \times 10^{10}$ Hz
   (C) $3.0 \times 10^{10}$ Hz
   (D) $4.5 \times 10^{10}$ Hz
   (E) $6.0 \times 10^{10}$ Hz

## Section 24.3 The Speed of Light

16. Chapter 24, Problem 16
    - (A) 1.18 min
    - (B) 15.8 min
    - (C) 48.0 min
    - (D) 65.6 min
    - (E) 70.9 min

18. Chapter 24, Problem 18
    - (A) 4.2 y
    - (B) 5.9 y
    - (C) 8.8 y
    - (D) 12 y
    - (E) 16 y

20. Chapter 24, Problem 20
    - (A) 140 rev/s
    - (B) 270 rev/s
    - (C) 540 rev/s
    - (D) 1100 rev/s
    - (E) 4300 rev/s

22. Chapter 24, Problem 22
    - (A) $3.2 \times 10^7$ m
    - (B) $1.5 \times 10^7$ m
    - (C) $5.0 \times 10^6$ m
    - (D) $1.6 \times 10^6$ m
    - (E) $3.0 \times 10^8$ m

## Section 24.4 The Energy Carried by Electromagnetic Waves

24. Chapter 24, Problem 24
    - (A) $6.4 \times 10^{-7}$ T
    - (B) $8.8 \times 10^{-7}$ T
    - (C) $1.2 \times 10^{-6}$ T
    - (D) $2.0 \times 10^{-6}$ T
    - (E) $4.6 \times 10^{-6}$ T

26. Chapter 24, Problem 26
    - (A) $4.8 \times 10^{-5}$ J
    - (B) $1.9 \times 10^{-4}$ J
    - (C) $1.5 \times 10^{-4}$ J
    - (D) $1.1 \times 10^{-6}$ J
    - (E) $9.6 \times 10^{-5}$ J

28. Chapter 24, Problem 28
    - (A) $4.2 \times 10^{-5}$ J
    - (B) $6.9 \times 10^{-5}$ J
    - (C) $3.1 \times 10^{-5}$ J
    - (D) $1.8 \times 10^{-5}$ J
    - (E) $9.6 \times 10^{-5}$ J

30. Chapter 24, Problem 30
    - (A) $4.69 \times 10^{-9}$ J
    - (B) $5.37 \times 10^{-9}$ J
    - (C) $6.25 \times 10^{-9}$ J
    - (D) $7.50 \times 10^{-9}$ J
    - (E) $8.33 \times 10^{-9}$ J

32a. Chapter 24, Problem 32a
    - (A) $1.4 \times 10^{-5}$ N
    - (B) $2.7 \times 10^{-5}$ N
    - (C) $5.4 \times 10^{-5}$ N
    - (D) $6.8 \times 10^{-5}$ N
    - (E) $8.3 \times 10^{-5}$ N

32b. Chapter 24, Problem 32b
    - (A) $1.1 \times 10^{-9}$ N
    - (B) $1.7 \times 10^{-9}$ N
    - (C) $2.3 \times 10^{-9}$ N
    - (D) $2.8 \times 10^{-9}$ N
    - (E) $3.3 \times 10^{-9}$ N

# HOMEWORK CHOICES

## Section 24.5 The Doppler Effect and Electromagnetic Waves

34a. Chapter 24, Problem 34a
    (A) zero
    (B) 320 Hz
    (C) 640 Hz
    (D) 960 Hz
    (E) 1200 Hz

34b. Chapter 24, Problem 34b
    (A) The wave that returns to the police car has the greater frequency.
    (B) The wave emitted by the radar gun has the greater frequency.
    (C) Both waves have the sane frequency.

36. Chapter 24, Problem 36
    (A) 1/2
    (B) 3/5
    (C) 4/7
    (D) 2/5
    (E) 1/3

## Section 24.6 Polarization

38. Chapter 24, Problem 38
    (A) 79 %
    (B) 21 %
    (C) 54 %
    (D) 31 %
    (E) 62 %

40a. Chapter 24, Problem 40a
    (A) 0°
    (B) 15°
    (C) 30°
    (D) 45°
    (E) 60°

40b. Chapter 24, Problem 40b
    (A) 0°
    (B) 15°
    (C) 30°
    (D) 45°
    (E) 60°

42a. Chapter 24, Problem 42a
    (A) 314 W/m$^2$
    (B) 611 W/m$^2$
    (C) 681 W/m$^2$
    (D) 392 W/m$^2$
    (E) 911 W/m$^2$

42b. Chapter 24, Problem 42b
    (A) 314 W/m$^2$
    (B) 611 W/m$^2$
    (C) 681 W/m$^2$
    (D) 392 W/m$^2$
    (E) 911 W/m$^2$

44. Chapter 24, Problem 44
    (A) 270 W/m$^2$
    (B) 440 W/m$^2$
    (C) 680 W/m$^2$
    (D) 720 W/m$^2$
    (E) 850 W/m$^2$

## Additional Problems

46. Chapter 24, Problem 46
    (A) 2.93 m to 3.24 m
    (B) 2.44 m to 3.68 m
    (C) 2.32 m to 3.33 m
    (D) 2.19 m to 3.75 m
    (E) 2.78 m to 3.41 m

## Homework: Electromagnetic Waves

48a. Chapter 24, Problem 48a
    (A) $4.64 \times 10^5$ N/C      (C) $8.90 \times 10^5$ N/C      (E) $1.40 \times 10^6$ N/C
    (B) $2.27 \times 10^5$ N/C      (D) $6.81 \times 10^5$ N/C

48b. Chapter 24, Problem 48b
    (A) $4.64 \times 10^{-3}$ T      (C) $8.90 \times 10^{-3}$ T      (E) $1.40 \times 10^{-2}$ T
    (B) $2.27 \times 10^{-3}$ T      (D) $6.81 \times 10^{-3}$ T

50. Chapter 24, Problem 50
    (A) $2.3 \times 10^9$ J      (C) $4.9 \times 10^9$ J      (E) $7.3 \times 10^9$ J
    (B) $3.0 \times 10^9$ J      (D) $5.6 \times 10^9$ J

52. Chapter 24, Problem 52
    (A) 2.71      (C) 4.76      (E) 11.9
    (B) 3.51      (D) 8.37

54. Chapter 24, Problem 54
    (A) 32 m/s      (C) 36 m/s      (E) 34 m/s
    (B) 27 m/s      (D) 29 m/s

56. Chapter 24, Problem 56
    (A) $1.2 \times 10^{-5}$ V      (C) $2.7 \times 10^{-5}$ V      (E) $8.5 \times 10^{-5}$ V
    (B) $2.1 \times 10^{-5}$ V      (D) $5.4 \times 10^{-5}$ V

58. Chapter 24, Problem 58
    (A) $4.2 \times 10^5$ W      (C) $6.6 \times 10^5$ W      (E) $9.1 \times 10^5$ W
    (B) $8.3 \times 10^5$ W      (D) $2.1 \times 10^5$ W

60. Chapter 24, Problem 60
*The solution is a proof.*

**Homework Choices**
# CHAPTER 25
## The Reflection of Light: Mirrors

*Section 25.2 The Reflection of Light*
*Section 25.3 The Formation of Images by a Plane Mirror*

2a. Chapter 25, Problem 2a
- (A) 0.24 m
- (B) 0.85 m
- (C) 0.91 m
- (D) 1.70 m
- (E) 1.82 m

2b. Chapter 25, Problem 2b
- (A) 0.24 m
- (B) 0.85 m
- (C) 0.91 m
- (D) 1.70 m
- (E) 1.82 m

4. Chapter 25, Problem 4

|     | *Image 1* | *Image 2* | *Image 3* |
|-----|-----------|-----------|-----------|
| (A) | $x = 2.0$ m, $y = -1.0$ m | $x = -2.0$ m, $y = 1.0$ m | $x = -2.0$ m, $y = -1.0$ m |
| (B) | $x = 1.0$ m, $y = -2.0$ m | $x = -1.0$ m, $y = 2.0$ m | $x = -1.0$ m, $y = -2.0$ m |
| (C) | $x = 2.0$ m, $y = -2.0$ m | $x = -1.0$ m, $y = 1.0$ m | $x = -2.0$ m, $y = -2.0$ m |
| (D) | $x = 1.0$ m, $y = -1.0$ m | $x = -2.0$ m, $y = 2.0$ m | $x = -1.0$ m, $y = -1.0$ m |
| (E) | $x = 2.0$ m, $y = -1.0$ m | $x = -1.0$ m, $y = 2.0$ m | $x = -1.0$ m, $y = -2.0$ m |

6a. Chapter 25, Problem 6a
- (A) 0.90 m/s
- (B) 1.2 m/s
- (C) 1.8 m/s
- (D) 2.4 m/s
- (E) 2.7 m/s

6b. Chapter 25, Problem 6b
- (A) 0.90 m/s
- (B) 1.2 m/s
- (C) 1.8 m/s
- (D) 2.4 m/s
- (E) 2.7 m/s

8. Chapter 25, Problem 8
- (A) 50°
- (B) 63°
- (C) 75°
- (D) 87°
- (E) 100°

*Section 25.4 Spherical Mirrors*
*Section 25.5 The Formation of Images by Spherical Mirrors*

10a. Chapter 25, Problem 10a
Note: *the following are estimated lengths based upon drawing a ray diagram to scale.*
- (A) 25 cm in front of the mirror
- (B) 15 cm behind the mirror
- (C) 7.5 cm in front of the mirror
- (D) 7.5 cm behind the mirror
- (E) 10 cm in front of the mirror

10b. Chapter 25, Problem 10b
Note: *the following are estimated lengths based upon drawing a ray diagram to scale.*
- (A) 2.5 cm
- (B) 1.5 cm
- (C) 0.5 cm
- (D) 1.0 cm
- (E) 2.0 cm

# 464 Homework: The Reflection of Light: Mirrors

12a. Chapter 25, Problem 12a
    (A)  +68 cm
    (B)  −48 cm
    (C)  +48 cm
    (D)  +24 cm
    (E)  −24 cm

12b. Chapter 25, Problem 12b
*The solution is a drawing.*

14a. Chapter 25, Problem 14a
**Note**: *the following are estimated lengths based upon drawing a ray diagram to scale.*
    (A)  15.0 cm in front of the mirror
    (B)  23.6 cm in front of the mirror
    (C)  7.50 cm behind of the mirror
    (D)  15.0 cm behind the mirror
    (E)  36.0 cm in front of the mirror

14b. Chapter 25, Problem 14b
**Note**: *the following are estimated lengths based upon drawing a ray diagram to scale.*
    (A)  2.14 cm
    (B)  1.00 cm
    (C)  1.25 cm
    (D)  1.50 cm
    (E)  2.00 cm

## Section 25.6 The Mirror Equation and the Magnification Equation

16. Chapter 25, Problem 16
    (A)  +55 cm
    (B)  −55 cm
    (C)  +21 cm
    (D)  −21 cm
    (E)  +31 cm

18a. Chapter 25, Problem 18a
    (A)  −12.0 cm
    (B)  −14.9 cm
    (C)  −17.1 cm
    (D)  −19.4 cm
    (E)  −24.1 cm

18b. Chapter 25, Problem 18b
    (A)  −12.0 cm
    (B)  −14.9 cm
    (C)  −17.1 cm
    (D)  −19.4 cm
    (E)  −24.1 cm

20a. Chapter 25, Problem 20a
    (A)  +23 cm
    (B)  −23 cm
    (C)  +68 cm
    (D)  −68 cm
    (E)  +46 cm

20b. Chapter 25, Problem 20b
    (A)  1.2
    (B)  1.7
    (C)  2.4
    (D)  3.0
    (E)  4.1

22. Chapter 25, Problem 22
    (A)  $3.2 \times 10^4$ m
    (B)  $5.5 \times 10^2$ m
    (C)  $8.2 \times 10^3$ m
    (D)  $2.1 \times 10^4$ m
    (E)  $3.0 \times 10^5$ m

24a. Chapter 25, Problem 24a
    (A)  plane
    (B)  convex
    (C)  concave

# HOMEWORK CHOICES

24b. Chapter 25, Problem 24b
- (A) −2.5 cm
- (B) −4.0 cm
- (C) +5.5 cm
- (D) +3.9 cm
- (E) +3.1 cm

24c. Chapter 25, Problem 24c
- (A) 2.0
- (B) 3.5
- (C) 4.2
- (D) 2.8
- (E) 1.4

24d. Chapter 25, Problem 24d
- (A) upright
- (B) inverted

26a. Chapter 25, Problem 26a

|     | inverted image | upright image |
| --- | --- | --- |
| (A) | $R/4$ | $R/2$ |
| (B) | $R/2$ | $R/4$ |
| (C) | $3R/4$ | $R/2$ |
| (D) | $R/4$ | $3R/4$ |
| (E) | $3R/2$ | $R/4$ |

26b. Chapter 25, Problem 26b
*The solution is a drawing.*

28. Chapter 25, Problem 28
*The solution is a proof.*

30a. Chapter 25, Problem 30a
- (A) $d_o = +3.0 \times 10^1$ cm, $d_i = -1.5 \times 10^1$ cm
- (B) $d_o = +9.0 \times 10^1$ cm, $d_i = +4.5 \times 10^1$ cm
- (C) $d_o = +4.5 \times 10^1$ cm, $d_i = +9.0 \times 10^1$ cm
- (D) $d_o = +1.5 \times 10^1$ cm, $d_i = +6.0 \times 10^1$ cm
- (E) $d_o = +1.5 \times 10^1$ cm, $d_i = -3.0 \times 10^1$ cm

30b. Chapter 25, Problem 30b
- (A) $d_o = +3.0 \times 10^1$ cm, $d_i = -1.5 \times 10^1$ cm
- (B) $d_o = +9.0 \times 10^1$ cm, $d_i = +4.5 \times 10^1$ cm
- (C) $d_o = +4.5 \times 10^1$ cm, $d_i = +9.0 \times 10^1$ cm
- (D) $d_o = +1.5 \times 10^1$ cm, $d_i = +6.0 \times 10^1$ cm
- (E) $d_o = +1.5 \times 10^1$ cm, $d_i = -3.0 \times 10^1$ cm

## *Additional Problems*

32a. Chapter 25, Problem 32a
**Note**: *the following are estimated lengths based upon drawing a ray diagram to scale.*
- (A) 20.0 cm behind the mirror
- (B) 35.0 cm behind the mirror
- (C) 10.0 cm in front of the mirror
- (D) 20.0 cm in front of the mirror
- (E) 15.0 cm behind the mirror

# Homework: The Reflection of Light: Mirrors

32b. Chapter 25, Problem 32b
 (A) 4.0 cm
 (B) 6.0 cm
 (C) 8.0 cm
 (D) 12 cm
 (E) 15 cm

34. Chapter 25, Problem 34
 (A) Only arrow $A$ can be seen by an observer at point $P$.
 (B) Only arrow $B$ can be seen by an observer at point $P$.
 (C) Both arrow $A$ and arrow $B$ can be seen by an observer at point $P$.

36. Chapter 25, Problem 36
 (A) +52 cm
 (B) −43 cm
 (C) +32 cm
 (D) −13 cm
 (E) −6.5 cm

38. Chapter 25, Problem 38
 **Note**: Label each wall as side **A** (left-hand mirror, opposite target), side **B** (top wall), and side **C** (with target). The following choices indicate the directions in which the laser may be fired. These direction are given as angles measured counterclockwise with respect to the $+x$ axis, which is the line drawn from point **P** to the bottom of wall **C**.
 (A) 11.3°, 45.0°, 71.6°, 135°, 162°
 (B) 9.23°, 45.0°, 82.5°, 135°, 158°
 (C) 14.8°, 45.0°, 93.1°, 135°, 176°
 (D) 9.23°, 45.0°, 82.5°, 135°, 162°
 (E) 14.8°, 45.0°, 71.6°, 135°, 158°

40. Chapter 25, Problem 40
 (A) 1.00 m
 (B) 0.259 m
 (C) 0.533 m
 (D) 0.757 m
 (E) 2.00 m

42. Chapter 25, Problem 42
 (A) −1/4
 (B) +1/3
 (C) +1/2
 (D) −1/3
 (E) −1/2

# Homework Choices
## CHAPTER 26

# The Refraction of Light: Lenses and Optical Instruments

### Section 26.1 The Index of Refraction

2. Chapter 26, Problem 2
   - (A) water
   - (B) fused quartz
   - (C) diamond
   - (D) benzene
   - (E) ethyl alcohol

4. Chapter 26, Problem 4
   - (A) 0.7342
   - (B) 0.8347
   - (C) 1.094
   - (D) 0.6127
   - (E) 1.198

6. Chapter 26, Problem 6
   - (A) 0.728
   - (B) 0.800
   - (C) 1.06
   - (D) 1.12
   - (E) 1.25

8. Chapter 26, Problem 8
   - (A) 4.3 cm
   - (B) 4.0 cm
   - (C) 3.7 cm
   - (D) 3.4 cm
   - (E) 3.1 cm

### Section 26.2 Snell's Law and the Refraction of Light

10. Chapter 26, Problem 10
    - (A) 1.25
    - (B) 1.46
    - (C) 1.33
    - (D) 1.87
    - (E) 1.64

12. Chapter 26, Problem 12
    - (A) 55.4°
    - (B) 32.1°
    - (C) 21.7°
    - (D) 6.68°
    - (E) 44.6°

14. Chapter 26, Problem 14
    - (A) 1.6 cm
    - (B) 2.1 cm
    - (C) 3.3 cm
    - (D) 3.9 cm
    - (E) 2.6 cm

16. Chapter 26, Problem 16
    - (A) 87.9 cm
    - (B) 12.1 cm
    - (C) 35.0 cm
    - (D) 177 cm
    - (E) 333 cm

18. Chapter 26, Problem 18
    - (A) 1.6 cm
    - (B) 2.1 cm
    - (C) 1.8 cm
    - (D) 2.3 cm
    - (E) 1.4 cm

20. Chapter 26, Problem 20
    - (A) 2.0 m
    - (B) 1.1 m
    - (C) 2.7 cm
    - (D) 35 cm
    - (E) 50 cm

## 468  Homework: The Refraction of Light: Lenses and Optical Instruments

22. Chapter 26, Problem 22
    - (A) 13.5 cm
    - (B) 16.7 cm
    - (C) 31.7 cm
    - (D) 21.4 cm
    - (E) 46.9 cm

### Section 26.3 Total Internal Reflection

24a. Chapter 26, Problem 24a
    - (A) 1.50
    - (B) 1.27
    - (C) 1.73
    - (D) 1.44
    - (E) 1.62

24b. Chapter 26, Problem 24b
    - (A) 1.50
    - (B) 1.27
    - (C) 1.73
    - (D) 1.44
    - (E) 1.62

26a. Chapter 26, Problem 26a
    - (A) 46.8°
    - (B) 32.1°
    - (C) 38.9°
    - (D) 57.8°
    - (E) 51.1°

26b. Chapter 26, Problem 26b
    - (A) yes, 46.8°
    - (B) yes, 57.8°
    - (C) yes, 38.9°
    - (D) yes, 32.1°
    - (E) no

28. Chapter 26, Problem 28
    - (A) 2.5 m
    - (B) 1.2 m
    - (C) 4.8 m
    - (D) 3.6 m
    - (E) 4.1 m

30a. Chapter 26, Problem 30a
    - (A) *A*
    - (B) *B*
    - (C) *C*

30b. Chapter 26, Problem 30b
    - (A) *A*
    - (B) *B*
    - (C) *C*

32. Chapter 26, Problem 32
    - (A) 1.69
    - (B) 1.22
    - (C) 1.50
    - (D) 1.35
    - (E) 1.96

### Section 26.4 Polarization and the Reflection and Refraction of Light

34a. Chapter 26, Problem 34a
    - (A) 51.90°
    - (B) 52.12°
    - (C) 52.62°
    - (D) 53.12°
    - (E) 53.62°

34b. Chapter 26, Problem 34b
    - (A) 51.90°
    - (B) 52.12°
    - (C) 52.62°
    - (D) 53.12°
    - (E) 53.62°

# HOMEWORK CHOICES Chapter 26

36. Chapter 26, Problem 36
    - (A) 61.79°
    - (B) 43.69°
    - (C) 24.39°
    - (D) 38.21°
    - (E) 34.39°

38a. Chapter 26, Problem 38a
    - (A) 60.0°
    - (B) 75.0°
    - (C) 32.0°
    - (D) 45.0°
    - (E) 58.0°

38b. Chapter 26, Problem 38b
    - (A) 1.33
    - (B) 1.60
    - (C) 1.82
    - (D) 1.53
    - (E) 1.41

40. Chapter 26, Problem 40
    *The solution is a proof.*

## Section 26.5 The Dispersion of Light: Prisms and Rainbows

42. Chapter 26, Problem 42
    - (A) 29.5°
    - (B) 67.5°
    - (C) 60.5°
    - (D) 22.5°
    - (E) 45.0°

44a. Chapter 26, Problem 44a
    - (A) 1.81
    - (B) 1.77
    - (C) 1.73
    - (D) 1.69
    - (E) 1.65

44b. Chapter 26, Problem 44b
    - (A) 1.81
    - (B) 1.77
    - (C) 1.73
    - (D) 1.69
    - (E) 1.65

46. Chapter 26, Problem 46
    - (A) 20.4°
    - (B) 24.8°
    - (C) 28.6°
    - (D) 37.1°
    - (E) 53.9°

## Section 26.6 Lenses
## Section 26.7 The Formation of Images by Lenses
## Section 26.8 The Thin-Lens Equation and the Magnification Equation

48. Chapter 26, Problem 48
    - (A) −6.3 cm
    - (B) −4.2 cm
    - (C) +6.3 cm
    - (D) −8.1 cm
    - (E) +4.2 cm

50. Chapter 26, Problem 50
    - (A) 1.4
    - (B) 4.2
    - (C) 7.8
    - (D) 2.0
    - (E) 2.8

52a. Chapter 26, Problem 52a
- (A) −26 cm
- (B) +26 cm
- (C) −15 cm
- (D) +15 cm
- (E) −21 cm

52b. Chapter 26, Problem 52b
- (A) real
- (B) virtual

54a. Chapter 26, Problems 54a
- (A) +47 cm
- (B) +12 cm
- (C) −12 cm
- (D) −16 cm
- (E) −47 cm

54b. Chapter 26, Problems 54b
- (A) 1.6
- (B) 2.5
- (C) 0.40
- (D) 0.63
- (E) 0.84

54c. Chapter 26, Problems 54c
- (A) real
- (B) virtual

54d. Chapter 26, Problems 54d
- (A) upright
- (B) inverted

54e. Chapter 26, Problems 54e
- (A) enlarged
- (B) reduced

56a. Chapter 26, Problem 56a
- (A) $8.02 \times 10^{-8}$ m$^2$
- (B) $1.15 \times 10^{-7}$ m$^2$
- (C) $3.59 \times 10^{-7}$ m$^2$
- (D) $4.98 \times 10^{-7}$ m$^2$
- (E) $6.74 \times 10^{-7}$ m$^2$

56b. Chapter 26, Problem 56b
- (A) $1.06 \times 10^{5}$ W/m$^2$
- (B) $4.61 \times 10^{5}$ W/m$^2$
- (C) $6.60 \times 10^{4}$ W/m$^2$
- (D) $1.47 \times 10^{5}$ W/m$^2$
- (E) $7.86 \times 10^{5}$ W/m$^2$

58a. Chapter 26, Problem 58a
- (A) toward the lens
- (B) away from the lens

58b. Chapter 26, Problem 58b
- (A) 0.20 m
- (B) 0.25 m
- (C) 0.15 m
- (D) 0.45 m
- (E) 0.30 m

60. Chapter 26, Problem 60
- (A) 12 m
- (B) 24 m
- (C) 36 m
- (D) 48 m
- (E) 60 m

62. Chapter 26, Problem 62
- (A) +27 cm, +98.0 cm
- (B) +35 cm, +90.5 cm
- (C) +42 cm, +78.5 cm
- (D) −63 cm, +62.5 cm
- (E) −27 cm, +98.0 cm

*HOMEWORK CHOICES*

## Section 26.9 Lenses in Combination

64. Chapter 26, Problem 64
    - (A) 0.13 m, to the right of the second lens
    - (B) 0.20 m, to the right of the second lens
    - (C) 0.28 m, to the right of the second lens
    - (D) 0.16 m, to the left of the second lens
    - (E) 0.24 m, to the left of the second lens

66. Chapter 26, Problems 66a and 66b
    **Note**: The solution to part (a) is the drawing of an accurate ray diagram to estimate the following parameters:

    |     | final image distance | overall magnification |
    |-----|----------------------|------------------------|
    | (A) | 4.00 cm to the right | 0.10 |
    | (B) | 12.0 cm to the right | 0.30 |
    | (C) | 12.0 cm to the left  | 0.30 |
    | (D) | 10.4 cm to the left  | 0.52 |
    | (E) | 10.4 cm to the right | 0.52 |

68. Chapter 26, Problem 68
    - (A) 36.0 cm
    - (B) 72.0 cm
    - (C) 160.0 cm
    - (D) 140.0 cm
    - (E) 120.0 cm

70a. Chapter 26, Problem 70a
    - (A) 3.62 cm, to the right of the second lens
    - (B) 4.50 cm, to the left of the second lens
    - (C) 5.22 cm, to the right of the second lens
    - (D) 5.22 cm, to the left of the second lens
    - (E) 4.50 cm, to the right of the second lens

70b. Chapter 26, Problem 70b
    - (A) +2.3
    - (B) −3.3
    - (C) −0.75
    - (D) +0.25
    - (E) −3.0

70c. Chapter 26, Problem 70c
    - (A) virtual
    - (B) real

70d. Chapter 26, Problem 70d
    - (A) inverted
    - (B) upright

70e. Chapter 26, Problem 70e
    - (A) smaller
    - (B) larger

## Section 26.10 The Human Eye

72. Chapter 26, Problem 72
    - (A) 14.6 cm
    - (B) 60.1 cm
    - (C) 42.5 cm
    - (D) 86.9 cm
    - (E) 94.7 cm

74a. Chapter 26, Problem 74a
    - (A) myopic
    - (B) hyperopic

**472** Homework: The Refraction of Light: Lenses and Optical Instruments

74b. Chapter 26, Problem 74b
    (A) 2.1 diopters      (C) 2.6 diopters      (E) 3.0 diopters
    (B) 2.3 diopters      (D) 2.8 diopters

76. Chapter 26, Problem 76
    (A) −5.2 m      (C) −9.2 m      (E) −6.8 m
    (B) +8.1 m      (D) +7.0 m

78a. Chapter 26, Problem 78a
    (A) −2.0 cm      (C) −3.2 cm      (E) −4.5 cm
    (B) −2.5 cm      (D) −3.7 cm

78b. Chapter 26, Problem 78b
    (A) 0.20 m      (C) 0.40 m      (E) 0.80 m
    (B) 0.30 m      (D) 0.50 m

## Section 26.11 Angular Magnification and the Magnifying Glass

80a. Chapter 26, Problem 80a
    (A) 0.014 rad      (C) 0.025 rad      (E) 0.079 rad
    (B) 0.020 rad      (D) 0.063 rad

80b. Chapter 26, Problem 80b
    (A) 0.011 rad      (C) 0.040 rad      (E) 0.089 rad
    (B) 0.036 rad      (D) 0.053 rad

80c. Chapter 26, Problem 80c
    (A) the spectator at the game      (B) the TV viewer

82. Chapter 26, Problem 82
    (A) 88      (C) 29      (E) 12
    (B) 42      (D) 18

84. Chapter 26, Problem 84
    (A) −42 cm      (C) −29 cm      (E) −14 cm
    (B) −35 cm      (D) −22 cm

86. Chapter 26, Problem 86
    (A) 11.5      (C) 15.4      (E) 17.5
    (B) 14.4      (D) 16.7

## Section 26.12 The Compound Microscope

88. Chapter 26, Problem 88
    (A) $1.8 \times 10^{-4}$ rad      (C) $5.2 \times 10^{-3}$ rad      (E) $4.8 \times 10^{-3}$ rad
    (B) $2.1 \times 10^{-3}$ rad      (D) $6.6 \times 10^{-4}$ rad

*HOMEWORK CHOICES* **Chapter 26** 473

90. Chapter 26, Problem 90
    - (A) $7 \times 10^{-2}$ rad
    - (B) $1 \times 10^{-2}$ rad
    - (C) $3 \times 10^{-2}$ rad
    - (D) $6 \times 10^{-3}$ rad
    - (E) $9 \times 10^{-3}$ rad

92. Chapter 26, Problem 92
    - (A) 10.6 cm
    - (B) 9.4 cm
    - (C) 11.0 cm
    - (D) 13.4 cm
    - (E) 12.6 cm

## *Section 26.13 The Telescope*

94. Chapter 26, Problem 94
    - (A) −32, −120
    - (B) −48, −130
    - (C) −30, −180
    - (D) −24, −140
    - (E) −42, −110

96. Chapter 26, Problem 96
    - (A) 2.2 m
    - (B) 0.50 m
    - (C) 1.5 m
    - (D) 1.1 m
    - (E) 2.4 m

98a. Chapter 26, Problem 98a
    - (A) 1.16 m
    - (B) 0.179 m
    - (C) 0.880 m
    - (D) 1.48 m
    - (E) 1.21 m

98b. Chapter 26, Problem 98b
    - (A) 0.0148 m
    - (B) 0.0116 m
    - (C) 0.0088 m
    - (D) 0.0121 m
    - (E) 0.0179 m

100. Chapter 26, Problem 100
    - (A) +23
    - (B) −23
    - (C) +31
    - (D) −31
    - (E) −12

## *Additional Problems*

102. Chapter 26, Problem 102
    - (A) 1.0 cm
    - (B) 3.0 cm
    - (C) 1.5 cm
    - (D) 2.5 cm
    - (E) 2.0 cm

104. Chapter 26, Problem 104
    - (A) 49.81°
    - (B) 40.19°
    - (C) 37.79°
    - (D) 52.62°
    - (E) 65.42°

106. Chapter 26, Problem 106
    - (A) 1.8 m
    - (B) 2.5 m
    - (C) 3.0 m
    - (D) 4.4 m
    - (E) 5.3 m

108. Chapter 26, Problem 108
    - (A) 13.7 m
    - (B) 18.4 m
    - (C) 22.9 m
    - (D) 27.4 m
    - (E) 32.1 m

110. Chapter 26, Problem 110
    (A) 0.30 m
    (B) 0.075 m
    (C) 0.094 m
    (D) 0.22 m
    (E) 0.13 m

112. Chapter 26, Problem 112
    (A) 0.77
    (B) 0.85
    (C) 1.28
    (D) 1.17
    (E) 1.35

114a. Chapter 26, Problem 114a
    (A) −2.2 cm
    (B) −3.0 cm
    (C) −1.8 cm
    (D) +1.0 cm
    (E) +2.5 cm

114b. Chapter 26, Problem 114b
    (A) +5.0 cm
    (B) −3.0 cm
    (C) −6.0 cm
    (D) +8.2 cm
    (E) +7.4 cm

116a. Chapter 26, Problem 116a
    (A) $4.2 \times 10^0$ diopters
    (B) $1.0 \times 10^1$ diopters
    (C) $1.3 \times 10^1$ diopters
    (D) $1.7 \times 10^1$ diopters
    (E) $2.0 \times 10^1$ diopters

116b. Chapter 26, Problem 116b
    (A) 4.0
    (B) 5.6
    (C) 6.9
    (D) 8.6
    (E) 12

118. Chapter 26, Problem 118
    (A) $3.3 \times 10^{-3}$ m
    (B) $6.6 \times 10^{-3}$ m
    (C) $5.0 \times 10^{-3}$ m
    (D) $1.7 \times 10^{-3}$ m
    (E) $3.9 \times 10^{-3}$ m

120. Chapter 26, Problem 120
    (A) 25.0 cm
    (B) 23.2 cm
    (C) 24.1 cm
    (D) 26.9 cm
    (E) 27.4 cm

122. Chapter 26, Problem 122
    (A) 14 cm
    (B) 12 cm
    (C) 9.5 cm
    (D) 9.0 cm
    (E) 8.0 cm

# Homework Choices
## CHAPTER 27
# Interference and the Wave Nature of Light

### Section 27.1 The Principle of Linear Superposition
### Section 27.2 Young's Double-Slit Experiment

2a. Chapter 27, Problem 2a
    (A) constructive     (B) destructive

2b. Chapter 27, Problem 2b
    (A) constructive     (B) destructive

2c. Chapter 27, Problem 2c
    (A) constructive     (B) destructive

2d. Chapter 27, Problem 2d
    (A) constructive     (B) destructive

4. Chapter 27, Problem 4
    (A) 10     (C) 14     (E) 18
    (B) 12     (D) 16

6. Chapter 27, Problem 6
    (A) 391 nm     (C) 587 nm     (E) 727 nm
    (B) 445 nm     (D) 652 nm

8. Chapter 27, Problem 8
    (A) $2.2 \times 10^{-3}$ m     (C) $4.3 \times 10^{-3}$ m     (E) $6.9 \times 10^{-3}$ m
    (B) $1.8 \times 10^{-2}$ m     (D) $8.2 \times 10^{-2}$ m

10. Chapter 27, Problem 10
    (A) 0.289 m, away from the slits     (D) 0.115 m, toward the slits
    (B) 0.148 m, toward the slits     (E) 0.385 m, away from the slits
    (C) 0.231 m, away from the slits

### Section 27.3 Thin-Film Interference

12a. Chapter 27, Problem 12a
    (A) 300 nm     (C) 420 nm     (E) 450 nm
    (B) 150 nm     (D) 210 m

12b. Chapter 27, Problem 12b
    (A) 300 nm     (C) 420 nm     (E) 450 nm
    (B) 150 nm     (D) 210 nm

# Homework: Interference and the Wave Nature of Light

14. Chapter 27, Problem 14
    - (A) red
    - (B) blue

16. Chapter 27, Problem 16
    - (A) 30
    - (B) 20
    - (C) 60
    - (D) 40
    - (E) 50

18. Chapter 27, Problem 18
    - (A) 396 nm
    - (B) 721 nm
    - (C) 587 nm
    - (D) 652 nm
    - (E) 417 nm

20. Chapter 27, Problem 20
    - (A) 0.0105 m
    - (B) 0.0128 m
    - (C) 0.0256 m
    - (D) 0.0384 m
    - (E) 0.0512 m

## Section 27.5 Diffraction

22. Chapter 27, Problem 22
    - (A) $1.9 \times 10^{-4}$ m
    - (B) $2.7 \times 10^{-4}$ m
    - (C) $3.8 \times 10^{-4}$ m
    - (D) $4.6 \times 10^{-4}$ m
    - (E) $5.5 \times 10^{-4}$ m

24. Chapter 27, Problem 24
    - (A) 0.0320 m
    - (B) 0.0640 m
    - (C) 0.0400 m
    - (D) 0.0780 m
    - (E) 0.0390 m

26. Chapter 27, Problem 26
    - (A) 9.85°
    - (B) 11.5°
    - (C) 12.8°
    - (D) 14.2°
    - (E) 15.9°

28. Chapter 27, Problem 28
    - (A) 1.66 cm
    - (B) 2.35 cm
    - (C) 1.41 cm
    - (D) 2.00 cm
    - (E) 17.5 cm

30. Chapter 27, Problem 30
    - (A) 0.037
    - (B) 0.031
    - (C) 0.019
    - (D) 0.013
    - (E) 0.025

## Section 27.6 Resolving Power

32a. Chapter 27, Problem 32a
   - (A) 12 m
   - (B) 16 m
   - (C) $1.0 \times 10^2$ m
   - (D) $2.0 \times 10^2$ m
   - (E) $3.0 \times 10^2$ m

32b. Chapter 27, Problem 32b
   - (A) yes
   - (B) no

34. Chapter 27, Problem 34
    (A) $9.1 \times 10^2$ m
    (B) $2.8 \times 10^3$ m
    (C) $5.6 \times 10^3$ m
    (D) $1.8 \times 10^4$ m
    (E) $3.6 \times 10^4$ m

36a. Chapter 27, Problem 36a
    (A) 7.6 m
    (B) 2.5 m
    (C) 4.2 m
    (D) 8.4 m
    (E) 12 m

36b. Chapter 27, Problem 36b
    (A) 12 m
    (B) 29 m
    (C) 39 m
    (D) 56 m
    (E) 75 m

38. Chapter 27, Problem 38
    (A) 1.4 s
    (B) 1.9 s
    (C) 2.0 s
    (D) 2.5 s
    (E) 3.2 s

40a. Chapter 27, Problem 40a
    (A) 370 m
    (B) 460 m
    (C) 550 m
    (D) 670 m
    (E) 790 m

40b. Chapter 27, Problem 40b
    (A) 370 m
    (B) 460 m
    (C) 550 m
    (D) 670 m
    (E) 790 m

## *Section 27.7 The Diffraction Grating*
## *Section 27.8 Compact Discs and the Use of Interference*

42. Chapter 27, Problem 42
    (A) 27.4°
    (B) 25.1°
    (C) 12.2°
    (D) 18.3°
    (E) 35.0°

44. Chapter 27, Problem 44
    (A) 1440 lines/cm
    (B) 2880 lines/cm
    (C) 1150 lines/cm
    (D) 2300 lines/cm
    (E) 1660 lines/cm

46. Chapter 27, Problem 46
    (A) 1.2
    (B) 1.8
    (C) 2.4
    (D) 0.56
    (E) 0.83

48. Chapter 27, Problem 48
    (A) 4320 lines/cm
    (B) 6540 lines/cm
    (C) 3450 lines/cm
    (D) 6900 lines/cm
    (E) 4980 lines/cm

50a. Chapter 27, Problem 50a
    (A) 4
    (B) 2
    (C) 1
    (D) 0.5
    (E) 0.1

# Homework: Interference and the Wave Nature of Light

50b. Chapter 27, Problem 50b
(A) ($M_A = 2$ and $M_B = 4$) and ($M_A = 3$ and $M_B = 5$)
(B) ($M_A = 2$ and $M_B = 3$) and ($M_A = 4$ and $M_B = 6$)
(C) ($M_A = 2$ and $M_B = 3$) and ($M_A = 5$ and $M_B = 6$)
(D) ($M_A = 2$ and $M_B = 4$) and ($M_A = 5$ and $M_B = 6$)
(E) ($M_A = 2$ and $M_B = 4$) and ($M_A = 3$ and $M_B = 6$)

## Additional Problems

52. Chapter 27, Problem 52
    (A) 24.9°
    (B) 36.2°
    (C) 13.7°
    (D) 17.8°
    (E) 22.9°

54. Chapter 27, Problem 54
    (A) 1.1
    (B) 1.3
    (C) 1.5
    (D) 1.7
    (E) 1.9

56. Chapter 27, Problem 56
    (A) $1.8 \times 10^{-4}$ m
    (B) $1.0 \times 10^{-4}$ m
    (C) $3.6 \times 10^{-4}$ m
    (D) $3.2 \times 10^{-4}$ m
    (E) $2.0 \times 10^{-4}$ m

58. Chapter 27, Problem 58
    (A) 50.9°
    (B) 48.0°
    (C) 53.4°
    (D) 54.0°
    (E) 55.8°

60a. Chapter 27, Problem 60a
    (A) $2.3 \times 10^{-7}$ m
    (B) $1.4 \times 10^{-7}$ m
    (C) $1.0 \times 10^{-7}$ m
    (D) $1.8 \times 10^{-7}$ m
    (E) $2.1 \times 10^{-7}$ m

60b. Chapter 27, Problem 60b
    (A) $2.3 \times 10^{-7}$ m
    (B) $1.4 \times 10^{-7}$ m
    (C) $1.0 \times 10^{-7}$ m
    (D) $1.8 \times 10^{-7}$ m
    (E) $2.1 \times 10^{-7}$ m

62. Chapter 27, Problem 62
    (A) 298 nm
    (B) 326 nm
    (C) 403 nm
    (D) 528 nm
    (E) 609 nm

64a. Chapter 27, Problem 64a
    (A) 172 nm
    (B) 216 nm
    (C) 372 nm
    (D) 486 nm
    (E) 516 nm

64b. Chapter 27, Problem 64b
    (A) 516 nm
    (B) 486 nm
    (C) 372 nm
    (D) 216 nm
    (E) 172 nm

# Homework Choices
# CHAPTER 28  Special Relativity

## *Section 28.3 The Relativity of Time: Time Dilation*

2. Chapter 28, Problem 2
   - (A) 42.40 s
   - (B) 54.48 s
   - (C) 67.18 s
   - (D) 78.05 s
   - (E) 95.18 s

4. Chapter 28, Problem 4
   - (A) 61 h
   - (B) 68 h
   - (C) 72 h
   - (D) 82 h
   - (E) 86 h

6. Chapter 28, Problem 6
   - (A) $3.5 \times 10^{-4}$ s
   - (B) $4.4 \times 10^{-4}$ s
   - (C) $5.3 \times 10^{-4}$ s
   - (D) $6.6 \times 10^{-4}$ s
   - (E) $8.8 \times 10^{-4}$ s

8. Chapter 28, Problem 8
   - (A) 4
   - (B) 8
   - (C) 16
   - (D) 32
   - (E) 64

## *Section 28.4 The Relativity of Length: Length Contraction*

10. Chapter 28, Problem 10
    - (A) 310 m/s
    - (B) 220 m/s
    - (C) 180 m/s
    - (D) 120 m/s
    - (E) 260 m/s

12. Chapter 28, Problem 12
    - (A) 42 years
    - (B) 76 years
    - (C) 190 years
    - (D) 350 years
    - (E) 540 years

14. Chapter 28, Problem 14
    - (A) 3400 m $\times$ 2850 m
    - (B) 1800 m $\times$ 2850 m
    - (C) 4650 m $\times$ 1500 m
    - (D) 3400 m $\times$ 1500 m
    - (E) 4650 m $\times$ 3870 m

16. Chapter 28, Problem 16
    - (A) 37.3°
    - (B) 38.1°
    - (C) 39.6°
    - (D) 40.2°
    - (E) 49.4°

18. Chapter 28, Problem 18
    - (A) 1.06
    - (B) 4.83
    - (C) 8.59
    - (D) 12.6
    - (E) 16.7

## Homework: Special Relativity

### Section 28.5 Relativistic Momentum

20a. Chapter 28, Problem 20a
 (A) $5.1 \times 10^{15}$ kg · m/s
 (B) $6.8 \times 10^{15}$ kg · m/s
 (C) $7.9 \times 10^{15}$ kg · m/s
 (D) $9.7 \times 10^{15}$ kg · m/s
 (E) $1.5 \times 10^{16}$ kg · m/s

20b. Chapter 28, Problem 20b
 (A) $5.1 \times 10^{15}$ kg · m/s
 (B) $6.8 \times 10^{15}$ kg · m/s
 (C) $7.9 \times 10^{15}$ kg · m/s
 (D) $9.7 \times 10^{15}$ kg · m/s
 (E) $1.5 \times 10^{16}$ kg · m/s

22. Chapter 28, Problem 22
 (A) 1.2 m
 (B) 1.0 m
 (C) 1.6 m
 (D) 1.4 m
 (E) 1.8 m

### Section 28.6 The Equivalence of Mass and Energy

24. Chapter 28, Problem 24
 (A) $9.8 \times 10^{-30}$ kg
 (B) $8.7 \times 10^{-30}$ kg
 (C) $6.9 \times 10^{-30}$ kg
 (D) $3.3 \times 10^{-30}$ kg
 (E) $2.7 \times 10^{-30}$ kg

26. Chapter 28, Problem 26
 (A) $0.982c$
 (B) $0.996c$
 (C) $0.975c$
 (D) $0.956c$
 (E) $0.968c$

28. Chapter 28, Problem 28
 (A) $6.9 \times 10^{-13}$ J
 (B) $8.2 \times 10^{-13}$ J
 (C) $9.2 \times 10^{-13}$ J
 (D) $1.7 \times 10^{-13}$ J
 (E) $3.4 \times 10^{-13}$ J

30a. Chapter 28, Problem 30a
 (A) $6.7 \times 10^{5}$ J
 (B) $8.9 \times 10^{5}$ J
 (C) $7.3 \times 10^{5}$ J
 (D) $8.0 \times 10^{5}$ J
 (E) $5.1 \times 10^{5}$ J

30b. Chapter 28, Problem 30b
 (A) $6.1 \times 10^{-12}$ kg
 (B) $1.8 \times 10^{-12}$ kg
 (C) $5.2 \times 10^{-12}$ kg
 (D) $7.4 \times 10^{-12}$ kg
 (E) $2.3 \times 10^{-12}$ kg

32. Chapter 28, Problem 32
 (A) $6.00 \times 10^{-16}$ m
 (B) $1.00 \times 10^{-15}$ m
 (C) $1.40 \times 10^{-15}$ m
 (D) $1.80 \times 10^{-15}$ m
 (E) $2.20 \times 10^{-15}$ m

### Section 28.7 The Relativistic Addition of Velocities

34. Chapter 28, Problem 34
 (A) $0.60c$
 (B) $0.65c$
 (C) $0.70c$
 (D) $0.75c$
 (E) $0.80c$

## HOMEWORK CHOICES

36a. Chapter 28, Problem 36a
- (A) 0.194c
- (B) 0.200c
- (C) 0.800c
- (D) 0.994c
- (E) c

36b. Chapter 28, Problem 36b
- (A) 0.194c
- (B) 0.200c
- (C) 0.800c
- (D) 0.994c
- (E) c

36c. Chapter 28, Problem 36c
- (A) 0.194c
- (B) 0.200c
- (C) 0.800c
- (D) 0.994c
- (E) c

36d. Chapter 28, Problem 36d
- (A) 0.194c
- (B) 0.200c
- (C) 0.800c
- (D) 0.994c
- (E) c

38a. Chapter 28, Problem 38a
- (A) 0.700c
- (B) 0.980c
- (C) 0.989c
- (D) 0.750c
- (E) 0.940c

38b. Chapter 28, Problem 38b
- (A) $1.0 \times 10^{-16}$ kg · m/s
- (B) $1.4 \times 10^{-16}$ kg · m/s
- (C) $1.8 \times 10^{-16}$ kg · m/s
- (D) $2.0 \times 10^{-16}$ kg · m/s
- (E) $2.2 \times 10^{-16}$ kg · m/s

## Additional Problems

40. Chapter 28, Problem 40
- (A) 1.8 breaths/min
- (B) 2.3 breaths/min
- (C) 3.5 breaths/min
- (D) 4.0 breaths/min
- (E) 5.0 breaths/min

42. Chapter 28, Problem 42
- (A) $2.31 \times 10^8$ m/s
- (B) $2.83 \times 10^8$ m/s
- (C) $2.02 \times 10^8$ m/s
- (D) $2.55 \times 10^8$ m/s
- (E) $2.97 \times 10^8$ m/s

44. Chapter 28, Problem 44
- (A) 0.89c
- (B) 0.92c
- (C) 0.94c
- (D) 0.97c
- (E) 0.99c

46a. Chapter 28, Problem 46a
- (A) $3.8 \times 10^{-12}$ J
- (B) $4.2 \times 10^{-12}$ J
- (C) $5.8 \times 10^{-12}$ J
- (D) $7.6 \times 10^{-12}$ J
- (E) $8.4 \times 10^{-12}$ J

46b. Chapter 28, Problem 46b
- (A) 0.9998c
- (B) 0.9960c
- (C) 0.9872c
- (D) 0.9747c
- (E) 0.9501c

# Homework Choices
## CHAPTER 29 — Particles and Waves

### Section 29.3 Photons and the Photoelectric Effect

2a. Chapter 29, Problem 2a
- (A) $7.47 \times 10^{-8}$ m
- (B) $9.33 \times 10^{-7}$ m
- (C) $1.42 \times 10^{-7}$ m
- (D) $1.63 \times 10^{-7}$ m
- (E) $1.80 \times 10^{-7}$ m

2b. Chapter 29, Problem 2b
- (A) $1.84 \times 10^{15}$ Hz
- (B) $1.91 \times 10^{18}$ Hz
- (C) $2.03 \times 10^{13}$ Hz
- (D) $2.14 \times 10^{17}$ Hz
- (E) $4.90 \times 10^{14}$ Hz

2c. Chapter 29, Problem 2c
- (A) infrared
- (B) visible
- (C) ultraviolet
- (D) X-rays
- (E) gamma rays

4. Chapter 29, Problem 4
- (A) 390 nm
- (B) 409 nm
- (C) 545 nm
- (D) 599 nm
- (E) 663 nm

6a. Chapter 29, Problem 6a
- (A) $1.0 \times 10^{-19}$ J
- (B) $1.6 \times 10^{-19}$ J
- (C) $2.0 \times 10^{-19}$ J
- (D) $2.7 \times 10^{-19}$ J
- (E) $5.2 \times 10^{-19}$ J

6b. Chapter 29, Problem 6b
- (A) 380 nm to 545 nm
- (B) 380 nm to 480 nm
- (C) 480 nm to 750 nm
- (D) 545 nm to 750 nm
- (E) 610 nm to 750 nm

8. Chapter 29, Problem 8
- (A) 111
- (B) 138
- (C) 333
- (D) 528
- (E) 723

10. Chapter 29, Problem 10
- (A) potassium
- (B) calcium
- (C) uranium
- (D) aluminum
- (E) gold

12. Chapter 29, Problem 12
- (A) $7.00 \times 10^{-12}$ m
- (B) $9.56 \times 10^{-12}$ m
- (C) $2.63 \times 10^{-11}$ m
- (D) $4.40 \times 10^{-12}$ m
- (E) $9.05 \times 10^{-11}$ m

14a. Chapter 29, Problem 14a
- (A) 4020 N/C
- (B) 6020 N/C
- (C) 7760 N/C
- (D) 9780 N/C
- (E) 13 800 N/C

14b. Chapter 29, Problem 14b
- (A) $1.34 \times 10^{-5}$ T
- (B) $2.59 \times 10^{-5}$ T
- (C) $2.01 \times 10^{-5}$ T
- (D) $3.26 \times 10^{-5}$ T
- (E) $4.60 \times 10^{-5}$ T

## *Section 29.4 The Momentum of a Photon and the Compton Effect*

16. Chapter 29, Problem 16
- (A) 0.52
- (B) 0.76
- (C) 1.3
- (D) 1.6
- (E) 1.9

18a. Chapter 29, Problem 18a
- (A) $4.86 \times 10^{-12}$ m
- (B) $4.01 \times 10^{-12}$ m
- (C) $6.39 \times 10^{-12}$ m
- (D) $5.52 \times 10^{-12}$ m
- (E) $3.03 \times 10^{-12}$ m

18b. Chapter 29, Problem 18b
- (A) $1.29 \times 10^{-13}$ m
- (B) $1.98 \times 10^{-13}$ m
- (C) $2.97 \times 10^{-13}$ m
- (D) $3.26 \times 10^{-13}$ m
- (E) $4.20 \times 10^{-13}$ m

20a. Chapter 29, Problem 20a
- (A) 0.1209 nm
- (B) 0.1367 nm
- (C) 0.1583 nm
- (D) 0.1667 nm
- (E) 0.1819 nm

20b. Chapter 29, Problem 20b
- (A) $1.064 \times 10^{-15}$ J
- (B) $1.092 \times 10^{-15}$ J
- (C) $2.203 \times 10^{-15}$ J
- (D) $2.336 \times 10^{-15}$ J
- (E) $2.932 \times 10^{-15}$ J

20c. Chapter 29, Problem 20c
- (A) $2.932 \times 10^{-15}$ J
- (B) $2.336 \times 10^{-15}$ J
- (C) $1.092 \times 10^{-15}$ J
- (D) $1.064 \times 10^{-15}$ J
- (E) $2.203 \times 10^{-15}$ J

20d. Chapter 29, Problem 20d
- (A) $2.8 \times 10^{-17}$ J
- (B) $2.0 \times 10^{-17}$ J
- (C) $1.3 \times 10^{-17}$ J
- (D) $4.3 \times 10^{-17}$ J
- (E) $3.3 \times 10^{-17}$ J

22a. Chapter 29, Problem 22a
- (A) $5.0 \times 10^{-9}$ N
- (B) $2.1 \times 10^{-8}$ N
- (C) $6.5 \times 10^{-9}$ N
- (D) $3.4 \times 10^{-8}$ N
- (E) $1.0 \times 10^{-8}$ N

22b. Chapter 29, Problem 22b
- (A) $5.0 \times 10^{-9}$ N
- (B) $2.1 \times 10^{-8}$ N
- (C) $6.5 \times 10^{-9}$ N
- (D) $3.4 \times 10^{-8}$ N
- (E) $1.0 \times 10^{-8}$ N

## Section 29.5 The de Broglie Wavelength and the Wave Nature of Matter

24. Chapter 29, Problem 24
    - (A) $1.19 \times 10^3$ m/s
    - (B) $1.30 \times 10^3$ m/s
    - (C) $1.41 \times 10^3$ m/s
    - (D) $1.52 \times 10^3$ m/s
    - (E) $1.63 \times 10^3$ m/s

26a. Chapter 29, Problem 26a
    - (A) $8.0 \times 10^{-8}$ m
    - (B) $1.0 \times 10^{-7}$ m
    - (C) $2.0 \times 10^{-7}$ m
    - (D) $2.5 \times 10^{-7}$ m
    - (E) $4.5 \times 10^{-7}$ m

26b. Chapter 29, Problem 26b
    - (A) $8.4 \times 10^{-8}$ m
    - (B) $7.5 \times 10^{-9}$ m
    - (C) $2.0 \times 10^{-10}$ m
    - (D) $3.5 \times 10^{-10}$ m
    - (E) $5.6 \times 10^{-10}$ m

28. Chapter 29, Problem 28
    - (A) $1.42 \times 10^{-6}$ rad
    - (B) $1.58 \times 10^{-6}$ rad
    - (C) $1.73 \times 10^{-6}$ rad
    - (D) $1.99 \times 10^{-6}$ rad
    - (E) $2.07 \times 10^{-6}$ rad

30a. Chapter 29, Problem 30a
    - (A) $8.7 \times 10^{-12}$ m
    - (B) $3.4 \times 10^{-12}$ m
    - (C) $5.6 \times 10^{-12}$ m
    - (D) $2.8 \times 10^{-12}$ m
    - (E) $4.6 \times 10^{-12}$ m

30b. Chapter 29, Problem 30b
    - (A) $1.2 \times 10^{-22}$ kg · m/s
    - (B) $3.4 \times 10^{-22}$ kg · m/s
    - (C) $5.6 \times 10^{-22}$ kg · m/s
    - (D) $7.2 \times 10^{-22}$ kg · m/s
    - (E) $8.9 \times 10^{-22}$ kg · m/s

30c. Chapter 29, Problem 30c
    - (A) $8.3 \times 10^{-15}$ J
    - (B) $7.7 \times 10^{-15}$ J
    - (C) $5.1 \times 10^{-15}$ J
    - (D) $4.0 \times 10^{-15}$ J
    - (E) $2.4 \times 10^{-15}$ J

32. Chapter 29, Problem 32
    - (A) $7.3 \times 10^1$
    - (B) $6.0 \times 10^1$
    - (C) $5.2 \times 10^1$
    - (D) $2.8 \times 10^1$
    - (E) $4.0 \times 10^1$

## Section 29.6 The Heisenberg Uncertainty Principle

34. Chapter 29, Problem 34
    - (A) $\Delta y_{electron} = 1.0 \times 10^{-4}$ m, $\Delta y_{ball} = 1.7 \times 10^{-32}$ m
    - (B) $\Delta y_{electron} = 6.2 \times 10^{-5}$ m, $\Delta y_{ball} = 1.2 \times 10^{-33}$ m
    - (C) $\Delta y_{electron} = 5.0 \times 10^{-3}$ m, $\Delta y_{ball} = 1.0 \times 10^{-32}$ m
    - (D) $\Delta y_{electron} = 1.6 \times 10^{-3}$ m, $\Delta y_{ball} = 3.3 \times 10^{-33}$ m
    - (E) $\Delta y_{electron} = 3.2 \times 10^{-3}$ m, $\Delta y_{ball} = 7.1 \times 10^{-33}$ m

*HOMEWORK CHOICES*      **Chapter 29**

36. Chapter 29, Problem 36
    - (A) 37 J·s
    - (B) 63 J·s
    - (C) 74 J·s
    - (D) 130 J·s
    - (E) 170 J·s

38. Chapter 29, Problem 38
    - (A) $3.7 \times 10^4$ m/s
    - (B) $4.9 \times 10^4$ m/s
    - (C) $5.7 \times 10^4$ m/s
    - (D) $7.2 \times 10^4$ m/s
    - (E) $8.1 \times 10^4$ m/s

## *Additional Problems*

40. Chapter 29, Problem 40
    - (A) The $3.3 \times 10^{-16}$ J photon has a frequency in the ultraviolet region; and the other photon has a frequency in the radio region.
    - (B) The $3.3 \times 10^{-16}$ J photon has a frequency in the gamma ray region; and the other photon has a frequency in the infrared region.
    - (C) The $3.3 \times 10^{-16}$ J photon has a frequency in the gamma ray region; and the other photon has a frequency in the visible region.
    - (D) The $3.3 \times 10^{-16}$ J photon has a frequency in the X-ray region; and the other photon has a frequency in the infrared region.
    - (E) The $3.3 \times 10^{-16}$ J photon has a frequency in the X-ray region; and the other photon has a frequency in the visible region.

42. Chapter 29, Problem 42
    - (A) 1830
    - (B) 1520
    - (C) 1310
    - (D) 1190
    - (E) 1050

44. Chapter 29, Problem 44
    - (A) 4.00 eV
    - (B) 2.10 eV
    - (C) 4.29 eV
    - (D) 2.68 eV
    - (E) 3.54 eV

46. Chapter 29, Problem 46
    - (A) $6.59 \times 10^5$ m/s
    - (B) $8.00 \times 10^5$ m/s
    - (C) $9.32 \times 10^5$ m/s
    - (D) $7.12 \times 10^4$ m/s
    - (E) $4.98 \times 10^4$ m/s

48. Chapter 29, Problem 48
    - (A) $3.09 \times 10^{-10}$ m
    - (B) $2.18 \times 10^{-10}$ m
    - (C) $1.59 \times 10^{-10}$ m
    - (D) $1.13 \times 10^{-10}$ m
    - (E) $9.12 \times 10^{-11}$ m

# Homework Choices
## CHAPTER 30 — The Nature of the Atom

*Section 30.1 Rutherford Scattering and the Nuclear Atom*

2a. Chapter 30, Problem 2a
   (A) 12 000 nuclei     (C) 25 000 nuclei     (E) 50 000 nuclei
   (B) 19 000 nuclei     (D) 32 000 nuclei

2b. Chapter 30, Problem 2b
   (A) 0.25 km     (C) 0.60 km     (E) 1.0 km
   (B) 0.40 km     (D) 0.75 km

4. Chapter 30, Problem 4
   (A) $5.0 \times 10^{13}$     (C) $1.0 \times 10^{14}$     (E) $3.0 \times 10^{14}$
   (B) $7.5 \times 10^{13}$     (D) $1.5 \times 10^{14}$

6. Chapter 30, Problem 6
   (A) $1.7 \times 10^{-14}$ m     (C) $4.9 \times 10^{-14}$ m     (E) $7.3 \times 10^{-14}$ m
   (B) $3.8 \times 10^{-14}$ m     (D) $6.1 \times 10^{-14}$ m

*Section 30.2 Line Spectra*
*Section 30.3 The Bohr Model of the Hydrogen Atom*

8. Chapter 30, Problem 8
   (A) 2     (C) 4     (E) 6
   (B) 3     (D) 5

10a. Chapter 30, Problem 10a
   (A) 3     (C) 9     (E) 16
   (B) 4     (D) 12

10b. Chapter 30, Problem 10b
   (A) 0.25     (C) 0.45     (E) 0.70
   (B) 0.33     (D) 0.58

12. Chapter 30, Problem 12
   (A) 13.6 eV     (C) 30.6 eV     (E) 27.2 eV
   (B) 54.4 eV     (D) 61.2 eV

# HOMEWORK CHOICES  Chapter 30

14. Chapter 30, Problem 14

|     | $n_{He}$ | $n_{Li}$ | Energy |
|-----|----------|----------|--------|
| (A) | 2        | 3        | −13.6 eV |
|     | 3        | 4        | −5.62 eV |
|     | 4        | 5        | −1.88 eV |
| (B) | 2        | 2        | −13.6 eV |
|     | 4        | 4        | −3.40 eV |
|     | 8        | 8        | −1.51 eV |
| (C) | 2        | 3        | −13.6 eV |
|     | 4        | 6        | −3.40 eV |
|     | 6        | 9        | −1.51 eV |
| (D) | 2        | 3        | −13.6 eV |
|     | 3        | 6        | −5.62 eV |
|     | 4        | 9        | −1.88 eV |
| (E) | 3        | 2        | −13.6 eV |
|     | 6        | 4        | −2.76 eV |
|     | 9        | 6        | −1.25 eV |

16. Chapter 30, Problem 16

|     | kinetic energy | electric potential energy |
|-----|----------------|---------------------------|
| (A) | +4.90 eV       | −9.80 eV                  |
| (B) | −2.45 eV       | −2.45 eV                  |
| (C) | −4.90 eV       | +9.80 eV                  |
| (D) | −7.35 eV       | +2.45 eV                  |
| (E) | +2.45 eV       | −7.35 eV                  |

18. Chapter 30, Problem 18
    - (A) $1.47 \times 10^{-10}$ m
    - (B) $4.90 \times 10^{-11}$ m
    - (C) $4.41 \times 10^{-10}$ m
    - (D) $2.20 \times 10^{-10}$ m
    - (E) $2.94 \times 10^{-11}$ m

20. Chapter 30, Problem 20
    - (A) 2180 lines/cm
    - (B) 3920 lines/cm
    - (C) 2550 lines/cm
    - (D) 4860 lines/cm
    - (E) 3240 lines/cm

## Section 30.5 The Quantum Mechanical Picture of the Hydrogen Atom

22. Chapter 30, Problem 22
    - (A) 2.26 eV
    - (B) 0.442 eV
    - (C) 0.544 eV
    - (D) 0.378 eV
    - (E) 0.367 eV

24. Chapter 30, Problem 24
    *The solution is a table of quantum numbers.*

26. Chapter 30, Problem 26
    - (A) 3.000
    - (B) 1.500
    - (C) 1.225
    - (D) 1.414
    - (E) 1.732

488  Homework: The Nature of the Atom

28a. Chapter 30, Problem 28a

|     | Bohr model | quantum mechanics |
|-----|------------|-------------------|
| (A) | zero | $L = \dfrac{h}{2\pi}$ |
| (B) | $L = \dfrac{h}{2\pi}$ | zero |
| (C) | zero | $L = \dfrac{\sqrt{2}h}{2\pi}$ |
| (D) | $L = \dfrac{\sqrt{2}h}{2\pi}$ | zero |
| (E) | $L = \dfrac{h}{2\pi}$ | $L = \dfrac{\sqrt{3}h}{2\pi}$ |

28b. Chapter 30, Problem 28b

|     | Bohr model | quantum mechanics |
|-----|------------|-------------------|
| (A) | zero | zero, $L = \dfrac{h}{2\pi}$, $L = \dfrac{\sqrt{2}h}{2\pi}$ |
| (B) | zero | zero, $L = \dfrac{\sqrt{2}h}{2\pi}$, $L = \dfrac{\sqrt{3}h}{2\pi}$ |
| (C) | $L = \dfrac{\sqrt{2}h}{2\pi}$ | zero, $L = \dfrac{\sqrt{2}h}{2\pi}$, $L = \dfrac{\sqrt{6}h}{2\pi}$ |
| (D) | $L = \dfrac{3h}{2\pi}$ | zero, $L = \dfrac{\sqrt{2}h}{2\pi}$, $L = \dfrac{\sqrt{6}h}{2\pi}$ |
| (E) | $L = \dfrac{h}{2\pi}$ | zero, $L = \dfrac{\sqrt{2}h}{2\pi}$, $L = \dfrac{\sqrt{3}h}{2\pi}$ |

## Section 30.6 The Pauli Exclusion Principle and the Periodic Table of the Elements

30. Chapter 30, Problem 30
(A) $4s^2\, 4p^3\, 3d^{10}$
(B) $1s^2\, 2s^2\, 3s^2\, 3p^6\, 4s^2\, 4p^6\, 3d^5$
(C) $1s^2\, 2s^2\, 2p^6\, 3s^2\, 3p^6\, 4s^2\, 3d^5$
(D) $1s^2\, 2s^2\, 2p^8\, 3s^2\, 3p^6\, 4d^{10}$
(E) $1s^2\, 2s^2\, 3s^2\, 3p^6\, 4p^8\, 4d^5$

32a. Chapter 30, Problem 32a
(A) 12
(B) 16
(C) 24
(D) 32
(E) 32

32b. Chapter 30, Problem 32b
(A) 28
(B) 50
(C) 56
(D) 44
(E) 22

34. Chapter 30, Problem 34
(A) barium (Ba)
(B) calcium (Ca)
(C) potassium (K)
(D) rubidium (Rb)
(E) strontium (Sr)

# HOMEWORK CHOICES

## Section 30.7 X-rays

36. Chapter 30, Problem 36
    - (A) $3.2 \times 10^{-11}$ m
    - (B) $2.4 \times 10^{-11}$ m
    - (C) $1.8 \times 10^{-11}$ m
    - (D) $4.6 \times 10^{-11}$ m
    - (E) $5.0 \times 10^{-11}$ m

38. Chapter 30, Problem 38
    - (A) $1.07 \times 10^{-14}$ J
    - (B) $1.99 \times 10^{-15}$ J
    - (C) $1.42 \times 10^{-14}$ J
    - (D) $7.13 \times 10^{-15}$ J
    - (E) $3.12 \times 10^{-14}$ J

40. Chapter 30, Problem 40
    - (A) $8.23 \times 10^{-9}$ m
    - (B) $1.55 \times 10^{-10}$ m
    - (C) $2.17 \times 10^{-10}$ m
    - (D) $2.88 \times 10^{-10}$ m
    - (E) $1.20 \times 10^{-10}$ m

## Section 30.8 The Laser

42. Chapter 30, Problem 42
    - (A) $3.09 \times 10^{-18}$ J
    - (B) $5.12 \times 10^{-19}$ J
    - (C) $7.18 \times 10^{-19}$ J
    - (D) $1.03 \times 10^{-18}$ J
    - (E) $8.04 \times 10^{-19}$ J

44. Chapter 30, Problem 44
    - (A) 110
    - (B) 79
    - (C) 54
    - (D) 35
    - (E) 19

46. Chapter 30, Problem 46
    - (A) 21 days
    - (B) 16 days
    - (C) 11 days
    - (D) 7.6 days
    - (E) 5.4 days

## Additional Problems

48. Chapter 30, Problem 48
    - (A) $9.20 \times 10^{3}$ V
    - (B) $1.47 \times 10^{4}$ V
    - (C) $1.07 \times 10^{4}$ V
    - (D) $1.30 \times 10^{4}$ V
    - (E) $1.24 \times 10^{4}$ V

50. Chapter 30, Problem 50
    - (A) $4.5 \times 10^{47}$
    - (B) $2.2 \times 10^{47}$
    - (C) $9.7 \times 10^{47}$
    - (D) $7.1 \times 10^{47}$
    - (E) $3.9 \times 10^{47}$

52. Chapter 30, Problem 52
    *The solution is a table of quantum numbers.*

54. Chapter 30, Problem 54
    - (A) 30.3 nm
    - (B) 22.8 nm
    - (C) 25.9 nm
    - (D) 20.3 nm
    - (E) 10.1 nm

# Homework: The Nature of the Atom

56. Chapter 30, Problem 56

(A) $KE = \dfrac{n^2 h^2}{8mL^2}$ where n = 1, 2, 3, ...

(B) $KE = \dfrac{nh^2}{4\pi mL^2}$ where n = 1, 2, 3, ...

(C) $KE = \dfrac{2\pi nh^2}{mL^2}$ where n = 0, 1, 2, ...

(D) $KE = \dfrac{2n^2 h^2}{m\sqrt{L}}$ where n = 1, 2, 3, ...

(E) $KE = \dfrac{2\pi n^2 h^2}{mL^4}$ where n = 1, 2, 3, ...

58a. Chapter 30, Problem 58a

(A) $T_n = \dfrac{8\pi^2 n^3 h^3}{mke^4 Z^2}$

(B) $T_n = \dfrac{n^2 h^3}{8\pi^2 mk^2 e^2 Z^2}$

(C) $T_n = \dfrac{2\pi nh^3}{mk^2 e^4 Z^2}$

(D) $T_n = \dfrac{n^3 h^3}{4\pi^2 mk^2 e^4 Z^2}$

(E) $T_n = \dfrac{4\pi^2 n^2 h^2}{mk^2 e^4 Z^2}$

58b. Chapter 30, Problem 58b

(A) $3.06 \times 10^{-16}$ s
(B) $1.53 \times 10^{-16}$ s
(C) $2.44 \times 10^{-15}$ s
(D) $1.22 \times 10^{-15}$ s
(E) $4.74 \times 10^{-15}$ s

58c. Chapter 30, Problem 58c

(A) $3.06 \times 10^{-16}$ s
(B) $1.53 \times 10^{-16}$ s
(C) $2.44 \times 10^{-15}$ s
(D) $1.22 \times 10^{-15}$ s
(E) $4.74 \times 10^{-15}$ s

# Homework Choices
# CHAPTER 31
# Nuclear Physics and Radioactivity

*Section 31.1 Nuclear Structure*
*Section 31.2 The Strong Nuclear Force and the Stability of the Nucleus*

2. Chapter 31, Problem 2
   - (A) $8.3 \times 10^{-15}$ m
   - (B) $4.4 \times 10^{-15}$ m
   - (C) $5.8 \times 10^{-14}$ m
   - (D) $2.9 \times 10^{-14}$ m
   - (E) $2.2 \times 10^{-14}$ m

4a. Chapter 31, Problem 4a
   - (A) $+9.08 \times 10^{-17}$ C
   - (B) $+8.51 \times 10^{-17}$ C
   - (C) $+4.82 \times 10^{-17}$ C
   - (D) $+3.32 \times 10^{-17}$ C
   - (E) $+1.31 \times 10^{-17}$ C

4b. Chapter 31, Problem 4b
   - (A) 82
   - (B) 126
   - (C) 164
   - (D) 208
   - (E) 104

4c. Chapter 31, Problem 4c
   - (A) 82
   - (B) 126
   - (C) 164
   - (D) 208
   - (E) 104

4d. Chapter 31, Problem 4d
   - (A) $7.1 \times 10^{-15}$ m
   - (B) $1.7 \times 10^{-14}$ m
   - (C) $5.6 \times 10^{-15}$ m
   - (D) $3.4 \times 10^{-15}$ m
   - (E) $3.2 \times 10^{-15}$ m

4e. Chapter 31, Problem 4e
   - (A) $2.0 \times 10^{17}$ kg/m$^3$
   - (B) $2.3 \times 10^{17}$ kg/m$^3$
   - (C) $2.6 \times 10^{17}$ kg/m$^3$
   - (D) $3.0 \times 10^{17}$ kg/m$^3$
   - (E) $3.4 \times 10^{17}$ kg/m$^3$

6. Chapter 31, Problem 6
   - (A) 4.7 N
   - (B) 0.20 N
   - (C) 1.2 N
   - (D) 0.40 N
   - (E) 2.4 N

8. Chapter 31, Problem 8
   - (A) 1.14
   - (B) 1.22
   - (C) 1.31
   - (D) 1.45
   - (E) 1.50

10a. Chapter 31, Problem 10a
   - (A) $4.6 \times 10^{17}$ kg/m$^3$
   - (B) $6.6 \times 10^{17}$ kg/m$^3$
   - (C) $1.2 \times 10^{17}$ kg/m$^3$
   - (D) $2.3 \times 10^{17}$ kg/m$^3$
   - (E) $3.9 \times 10^{17}$ kg/m$^3$

10b. Chapter 31, Problem 10b
   - (A) $1.2 \times 10^{10}$ kg
   - (B) $2.3 \times 10^{10}$ kg
   - (C) $4.6 \times 10^{10}$ kg
   - (D) $5.9 \times 10^{10}$ kg
   - (E) $6.8 \times 10^{10}$ kg

# Homework: Nuclear Physics and Radioactivity

10c. Chapter 31, Problem 10c
- (A) 3
- (B) 12
- (C) 80
- (D) 150
- (E) 200

## Section 31.3 The Mass Defect of the Nucleus and Nuclear Binding Energy

12. Chapter 31, Problem 12
    - (A) 1.71 MeV/nucleon
    - (B) 2.07 MeV/nucleon
    - (C) 4.17 MeV/nucleon
    - (D) 5.60 MeV/nucleon
    - (E) 7.90 MeV/nucleon

14. Chapter 31, Problem 14
    - (A) $1.5 \times 10^{16}$ kg
    - (B) $2.9 \times 10^{16}$ kg
    - (C) $4.4 \times 10^{16}$ kg
    - (D) $5.9 \times 10^{16}$ kg
    - (E) $7.4 \times 10^{16}$ kg

16a. Chapter 31, Problem 16a
    - (A) 11.61 MeV
    - (B) 8.64 MeV
    - (C) 9.61 MeV
    - (D) 10.55 MeV
    - (E) 7.55 MeV

16b. Chapter 31, Problem 16b
    - (A) 11.61 MeV
    - (B) 8.64 MeV
    - (C) 9.61 MeV
    - (D) 10.55 MeV
    - (E) 7.55 MeV

16c. Chapter 31, Problem 16c
    - (A) proton
    - (B) neutron

## Section 31.4 Radioactivity

18a. Chapter 31, Problem 18a
    - (A) $^{212}_{84}\text{Po} \rightarrow ^{208}_{80}\text{Hg} + ^{4}_{2}\text{He}$
    - (B) $^{212}_{84}\text{Po} \rightarrow ^{212}_{80}\text{Bi} + ^{4}_{2}\text{He}$
    - (C) $^{212}_{84}\text{Po} \rightarrow ^{208}_{82}\text{Pb} + ^{4}_{2}\text{He}$
    - (D) $^{212}_{84}\text{Po} \rightarrow ^{216}_{88}\text{Ra} + ^{4}_{2}\text{He}$
    - (E) $^{212}_{84}\text{Po} \rightarrow ^{204}_{81}\text{Tl} + ^{4}_{2}\text{He}$

18b. Chapter 31, Problem 18b
    - (A) $^{232}_{92}\text{U} \rightarrow ^{228}_{88}\text{Ra} + ^{4}_{2}\text{He}$
    - (B) $^{232}_{92}\text{U} \rightarrow ^{232}_{90}\text{Th} + ^{4}_{2}\text{He}$
    - (C) $^{232}_{92}\text{U} \rightarrow ^{224}_{96}\text{Cm} + ^{4}_{2}\text{He}$
    - (D) $^{232}_{92}\text{U} \rightarrow ^{236}_{94}\text{Pu} + ^{4}_{2}\text{He}$
    - (E) $^{232}_{92}\text{U} \rightarrow ^{228}_{90}\text{Th} + ^{4}_{2}\text{He}$

20a. Chapter 31, Problem 20a
    - (A) $^{14}_{6}\text{C} \rightarrow ^{14}_{7}\text{N} + ^{0}_{-1}e$
    - (B) $^{14}_{6}\text{C} \rightarrow ^{13}_{7}\text{N} + ^{0}_{-1}e$
    - (C) $^{14}_{6}\text{C} \rightarrow ^{13}_{6}\text{C} + ^{0}_{-1}e$
    - (D) $^{14}_{6}\text{C} \rightarrow ^{15}_{6}\text{C} + ^{0}_{-1}e$
    - (E) $^{14}_{6}\text{C} \rightarrow ^{14}_{5}\text{B} + ^{0}_{-1}e$

## HOMEWORK CHOICES

20b. Chapter 31, Problem 20b
- (A) $^{212}_{82}Pb \rightarrow {}^{212}_{81}Tl + {}^{0}_{-1}e$
- (B) $^{212}_{82}Pb \rightarrow {}^{213}_{82}Pb + {}^{0}_{-1}e$
- (C) $^{212}_{82}Pb \rightarrow {}^{212}_{83}Bi + {}^{0}_{-1}e$
- (D) $^{212}_{82}Pb \rightarrow {}^{213}_{81}Tl + {}^{0}_{-1}e$
- (E) $^{212}_{82}Pb \rightarrow {}^{211}_{83}Bi + {}^{0}_{-1}e$

22a. Chapter 31, Problem 22a
- (A) $^{14}_{6}C \rightarrow {}^{14}_{5}B + {}^{0}_{-1}e$
- (B) $^{14}_{6}C \rightarrow {}^{15}_{6}C + {}^{0}_{-1}e$
- (C) $^{14}_{6}C \rightarrow {}^{13}_{6}C + {}^{0}_{-1}e$
- (D) $^{14}_{6}C \rightarrow {}^{13}_{7}N + {}^{0}_{-1}e$
- (E) $^{14}_{6}C \rightarrow {}^{14}_{7}N + {}^{0}_{-1}e$

22b. Chapter 31, Problem 22b
- (A) 0.156 MeV
- (B) 0.834 MeV
- (C) 0.226 MeV
- (D) 0.328 MeV
- (E) 0.103 MeV

24. Chapter 31, Problem 24
- (A) 0.10 MeV
- (B) 0.16 MeV
- (C) 0.43 MeV
- (D) 0.81 MeV
- (E) 0.64 MeV

26a. Chapter 31, Problem 26a
- (A) 2
- (B) 3
- (C) 4
- (D) 5
- (E) 6

26b. Chapter 31, Problem 26b
- (A) 1
- (B) 2
- (C) 3
- (D) 4
- (E) 5

28. Chapter 31, Problem 28
- (A) $KE_{Th}$ = 0.039 MeV, $KE_\alpha$ = 8.4 MeV
- (B) $KE_{Th}$ = 0.089 MeV, $KE_\alpha$ = 6.6 MeV
- (C) $KE_{Th}$ = 0.072 MeV, $KE_\alpha$ = 4.2 MeV
- (D) $KE_{Th}$ = 0.042 MeV, $KE_\alpha$ = 7.2 MeV
- (E) $KE_{Th}$ = 0.060 MeV, $KE_\alpha$ = 5.2 MeV

30. Chapter 31, Problem 30
- (A) $1.89 \times 10^3$ m/s
- (B) $2.84 \times 10^3$ m/s
- (C) $5.67 \times 10^3$ m/s
- (D) $8.52 \times 10^3$ m/s
- (E) $1.13 \times 10^4$ m/s

### Section 31.6 Radioactive Decay and Activity

32. Chapter 31, Problem 32
- (A) $1.23 \times 10^{-7}$ s
- (B) $2.40 \times 10^{-7}$ s
- (C) $3.98 \times 10^{-7}$ s
- (D) $5.62 \times 10^{-7}$ s
- (E) $6.77 \times 10^{-7}$ s

34. Chapter 31, Problem 34
- (A) 2.59 %
- (B) 3.66 %
- (C) 5.17 %
- (D) 6.34 %
- (E) 7.53 %

## Homework: Nuclear Physics and Radioactivity

36. Chapter 31, Problem 36
    - (A) $2.1 \times 10^{-11}$ kg
    - (B) $5.5 \times 10^{-11}$ kg
    - (C) $1.0 \times 10^{-11}$ kg
    - (D) $3.8 \times 10^{-11}$ kg
    - (E) $8.2 \times 10^{-11}$ kg

38. Chapter 31, Problem 38
    - (A) $1.3 \times 10^{14}$ Bq
    - (B) $1.0 \times 10^{14}$ Bq
    - (C) $6.1 \times 10^{13}$ Bq
    - (D) $3.6 \times 10^{13}$ Bq
    - (E) $2.1 \times 10^{13}$ Bq

40. Chapter 31, Problem 40
    - (A) 0.00418
    - (B) 0.00591
    - (C) 0.00887
    - (D) 0.00913
    - (E) 0.01023

42. Chapter 31, Problem 42
    - (A) 57 C°
    - (B) 61 C°
    - (C) 65 C°
    - (D) 77 C°
    - (E) 81 C°

## Section 31.7 Radioactive Dating
## Section 31.8 Radioactive Decay Series

44a. Chapter 31, Problem 44a
   - (A) 0.999
   - (B) 0.992
   - (C) 0.984
   - (D) 0.968
   - (E) 0.938

44b. Chapter 31, Problem 44b
   - (A) $6.51 \times 10^{-10}$
   - (B) $1.36 \times 10^{-9}$
   - (C) $4.99 \times 10^{-8}$
   - (D) $1.05 \times 10^{-7}$
   - (E) $2.22 \times 10^{-6}$

44c. Chapter 31, Problem 44c
   - (A) 0.410
   - (B) 0.595
   - (C) 0.614
   - (D) 0.755
   - (E) 0.838

46. Chapter 31, Problem 46
    - (A) 1200 years
    - (B) 1500 years
    - (C) 1900 years
    - (D) 2200 years
    - (E) 2600 years

48. Chapter 31, Problem 48
    *The solution is a plot.*

## Additional Problems

50. Chapter 31, Problem 50
    - (A) 235.0 MeV
    - (B) 143.0 MeV
    - (C) 215.0 MeV
    - (D) 153.0 MeV
    - (E) 225.0 MeV

*HOMEWORK CHOICES*  **Chapter 31**

52. Chapter 31, Problem 52
    - (A) 1.06 MeV
    - (B) 4.87 MeV
    - (C) 3.22 MeV
    - (D) 2.39 MeV
    - (E) 6.31 MeV

54a. Chapter 31, Problem 54a
    - (A) $^{238}_{94}$Pu
    - (B) $^{238}_{92}$U
    - (C) $^{240}_{90}$Th
    - (D) $^{240}_{98}$Cf
    - (E) $^{236}_{96}$Cm

54b. Chapter 31, Problem 54b
    - (A) $^{24}_{10}$Ne
    - (B) $^{28}_{10}$Ne
    - (C) $^{24}_{12}$Mg
    - (D) $^{28}_{12}$Mg
    - (E) $^{23}_{13}$Al

54c. Chapter 31, Problem 54c
    - (A) $^{13}_{8}$O
    - (B) $^{13}_{6}$C
    - (C) $^{14}_{6}$C
    - (D) $^{12}_{7}$N
    - (E) $^{14}_{7}$N

56. Chapter 31, Problem 56
    - (A) 2.017 330 u
    - (B) 1.512 998 u
    - (C) 2.521 625 u
    - (D) 3.025 995 u
    - (E) 1.008 665 u

58. Chapter 31, Problem 58
    - (A) 0.647
    - (B) 0.695
    - (C) 0.771
    - (D) 0.838
    - (E) 0.942

60. Chapter 31, Problem 60
    - (A) 9.37 days
    - (B) 8.43 days
    - (C) 6.24 days
    - (D) 5.82 days
    - (E) 3.37 days

# Homework Choices
# CHAPTER 32
# Ionizing Radiation, Nuclear Energy, and Elementary Particles

## Section 32.1 Biological Effects of Ionizing Radiation

2. Chapter 32, Problem 2
   - (A) $1.5 \times 10^{-6}$ J
   - (B) $2.0 \times 10^{-6}$ J
   - (C) $2.4 \times 10^{-6}$ J
   - (D) $3.1 \times 10^{-6}$ J
   - (E) $3.7 \times 10^{-6}$ J

4. Chapter 32, Problem 4
   - (A) 25 rad
   - (B) 50 rad
   - (C) 100 rad
   - (D) 250 rad
   - (E) 500 rad

6. Chapter 32, Problem 6
   - (A) 40 mrem
   - (B) 80 mrem
   - (C) 100 mrem
   - (D) 50 mrem
   - (E) 60 mrem

8. Chapter 32, Problem 8
   - (A) $2.8 \times 10^{13}$
   - (B) $3.4 \times 10^{13}$
   - (C) $4.8 \times 10^{13}$
   - (D) $6.2 \times 10^{13}$
   - (E) $7.9 \times 10^{13}$

10. Chapter 32, Problem 10
    - (A) $3.2 \times 10^{11}$ s$^{-1}$
    - (B) $4.4 \times 10^{11}$ s$^{-1}$
    - (C) $1.3 \times 10^{11}$ s$^{-1}$
    - (D) $2.0 \times 10^{11}$ s$^{-1}$
    - (E) $7.9 \times 10^{11}$ s$^{-1}$

## Section 32.2 Induced Nuclear Reactions

12. Chapter 32, Problem 12
    - (A) 31
    - (B) 27
    - (C) 30
    - (D) 28
    - (E) 26

14a. Chapter 32, Problem 14a
   - (A) $^{27}_{13}$Al (n, p) $^{27}_{12}$Mg
   - (B) $^{27}_{13}$Al (n, $\alpha$) $^{27}_{12}$Mg
   - (C) $^{27}_{13}$Al (p, n) $^{27}_{12}$Mg
   - (D) $^{27}_{13}$Al ($\alpha$, n) $^{27}_{12}$Mg
   - (E) $^{27}_{13}$Al ($\alpha$, $\beta$) $^{27}_{12}$Mg

14b. Chapter 32, Problem 14b
   - (A) $^{40}_{18}$Ar (p, $\alpha$)$^{43}_{19}$K
   - (B) $^{40}_{18}$Ar (n, $\alpha$)$^{43}_{19}$K
   - (C) $^{40}_{18}$Ar ($\alpha$, n)$^{43}_{19}$K
   - (D) $^{40}_{18}$Ar ($\alpha$, p)$^{43}_{19}$K
   - (E) $^{40}_{18}$Ar ($\alpha$, $\beta$)$^{43}_{19}$K

16. Chapter 32, Problem 16
    - (A) nickel, $A = 61$, $Z = 28$
    - (B) germanium, $A = 62$, $Z = 32$
    - (C) copper, $A = 64$, $Z = 29$
    - (D) gallium, $A = 62$, $Z = 31$
    - (E) zinc, $A = 64$, $Z = 30$

*HOMEWORK CHOICES*

## Section 32.3 Nuclear Fission
## Section 32.4 Nuclear Reactions

18. Chapter 32, Problem 18
    - (A) 2 neutrons
    - (B) 2 protons
    - (C) 3 neutrons
    - (D) 3 protons
    - (E) 4 neutrons

20. Chapter 32, Problem 20
    - (A) $2.0 \times 10^{-3}$ kg
    - (B) $1.3 \times 10^{-3}$ kg
    - (C) $5.4 \times 10^{-3}$ kg
    - (D) $3.6 \times 10^{-3}$ kg
    - (E) $4.0 \times 10^{-3}$ kg

22. Chapter 32, Problem 22
    - (A) 38 MeV
    - (B) 79 MeV
    - (C) 139 MeV
    - (D) 173 MeV
    - (E) 208 MeV

24. Chapter 32, Problem 24
    - (A) 16
    - (B) 22
    - (C) 28
    - (D) 35
    - (E) 41

26. Chapter 32, Problem 26
    - (A) 140 MeV
    - (B) 160 MeV
    - (C) 12 MeV
    - (D) 91 MeV
    - (E) 70 MeV

28a. Chapter 32, Problem 28a
    - (A) $5.1 \times 10^{24}$
    - (B) $6.3 \times 10^{24}$
    - (C) $2.7 \times 10^{24}$
    - (D) $3.1 \times 10^{24}$
    - (E) $4.0 \times 10^{24}$

28b. Chapter 32, Problem 28b
    - (A) 0.30 kg
    - (B) 0.72 kg
    - (C) 1.2 kg
    - (D) 2.1 kg
    - (E) 2.8 kg

28c. Chapter 32, Problem 28c
    - (A) $2.8 \times 10^{-3}$ kg
    - (B) $1.1 \times 10^{-3}$ kg
    - (C) $5.4 \times 10^{-3}$ kg
    - (D) $9.3 \times 10^{-3}$ kg
    - (E) $7.1 \times 10^{-3}$ kg

## Section 32.5 Nuclear Fusion

30. Chapter 32, Problem 30
    - (A) 1.0 MeV
    - (B) 3.3 MeV
    - (C) 3.6 MeV
    - (D) 4.0 MeV
    - (E) 5.4 MeV

32. Chapter 32, Problem 32
    - (A) 1.0 MeV
    - (B) 2.2 MeV
    - (C) 5.8 MeV
    - (D) 9.4 MeV
    - (E) 18 MeV

**498  Homework: Ionizing Radiation, Nuclear Energy, and Elementary Particles**

34a. Chapter 32, Problem 34a
 (A)  $1.5 \times 10^{21}$
 (B)  $2.2 \times 10^{21}$
 (C)  $1.0 \times 10^{22}$
 (D)  $4.0 \times 10^{23}$
 (E)  $6.7 \times 10^{25}$

34b. Chapter 32, Problem 34b
 (A)  $4.3 \times 10^{9}$ kg
 (B)  $8.4 \times 10^{8}$ kg
 (C)  $6.8 \times 10^{9}$ kg
 (D)  $9.3 \times 10^{8}$ kg
 (E)  $8.1 \times 10^{9}$ kg

## *Section 32.6 Elementary Particles*

36. Chapter 32, Problem 36
 (A)  u, d, d
 (B)  t, b, b
 (C)  $\bar{d}, \bar{u}, \bar{u}$
 (D)  $\bar{u}, \bar{d}, \bar{d}$
 (E)  $\bar{t}, \bar{b}, \bar{b}$

38a. Chapter 32, Problem 38a
 (A)  0.221 MeV
 (B)  0.319 MeV
 (C)  0.429 MeV
 (D)  0.513 MeV
 (E)  0.667 MeV

38b. Chapter 32, Problem 38b
 (A)  $2.43 \times 10^{-12}$ m
 (B)  $2.71 \times 10^{-12}$ m
 (C)  $2.96 \times 10^{-12}$ m
 (D)  $3.07 \times 10^{-12}$ m
 (E)  $5.33 \times 10^{-12}$ m

38c. Chapter 32, Problem 38c
 (A)  $2.73 \times 10^{-22}$ kg · m/s
 (B)  $3.64 \times 10^{-22}$ kg · m/s
 (C)  $2.16 \times 10^{-22}$ kg · m/s
 (D)  $2.24 \times 10^{-22}$ kg · m/s
 (E)  $1.24 \times 10^{-22}$ kg · m/s

40. Chapter 32, Problem 40
 (A)  116 MeV
 (B)  82.0 MeV
 (C)  336 MeV
 (D)  214 MeV
 (E)  418 MeV

42. Chapter 32, Problem 42
 (A)  0.83 MeV
 (B)  0.66 MeV
 (C)  0.43 MeV
 (D)  0.26 MeV
 (E)  0.18 MeV

## *Additional Problems*

44. Chapter 32, Problem 44
 (A)  178 MeV
 (B)  675 MeV
 (C)  815 MeV
 (D)  919 MeV
 (E)  1750 MeV

46. Chapter 32, Problem 46
 (A)  138.2567 u
 (B)  198.3242 u
 (C)  232.7851 u
 (D)  254.3621 u
 (E)  332.5325 u

*HOMEWORK CHOICES*  **Chapter 32**

48. Chapter 32, Problem 48
    - (A) $7.5 \times 10^{-3}$ C°
    - (B) $6.2 \times 10^{-3}$ C°
    - (C) $4.2 \times 10^{-3}$ C°
    - (D) $3.1 \times 10^{-3}$ C°
    - (E) $1.8 \times 10^{-3}$ C°

50. Chapter 32, Problem 50
    - (A) $7.4 \times 10^{2}$ kg/s
    - (B) $4.4 \times 10^{3}$ kg/s
    - (C) $8.1 \times 10^{3}$ kg/s
    - (D) $1.8 \times 10^{4}$ kg/s
    - (E) $2.9 \times 10^{4}$ kg/s

52. Chapter 32, Problem 52
    - (A) 32 eV
    - (B) 66 eV
    - (C) 14 eV
    - (D) 54 eV
    - (E) 26 eV

# ANSWERS — Homework Choices

## Chapter 1
### Introduction and Mathematical Concepts

| | | | | | | | | | | | |
|---|---|---|---|---|---|---|---|---|---|---|---|
| 2. | B | 18. | C | 30a. | D | 40a. | A | 50. | C | 64. | B |
| 4. | D | 20. | proof | 30b. | E | 40b. | B | 52. | D | 66a. | E |
| 6. | C | 22a. | D | 32a. | B | 42a. | D | 54. | A | 66b. | D |
| 8. | B | 22b. | E | 32b. | A | 42b. | E | 56. | E | | |
| 10. | A | 24. | C | 34. | D | 44. | A | 58. | C | | |
| 12. | A | 26a. | E | 36a. | C | 46a. | C | 60. | B | | |
| 14. | E | 26b. | C | 36b. | C | 46b. | E | 62a. | A | | |
| 16. | A | 28. | B | 38. | D | 48. | B | 62b. | D | | |

## Chapter 2
### Kinematics in One Dimension

| | | | | | | | | | | | |
|---|---|---|---|---|---|---|---|---|---|---|---|
| 2a. | A | 18. | D | 32b. | B | 46. | C | 62. | A | 78. | A |
| 2b. | D | 20. | B | 34. | D | 48. | E | 64a. | C | 80. | D |
| 4. | E | 22a. | E | 36a. | B | 50. | C | 64b. | E | 82. | B |
| 6. | E | 22b. | B | 36b. | A | 52. | C | 66. | graph | 84. | E |
| 8. | B | 24. | C | 38. | A | 54. | D | 68. | C | 86. | E |
| 10. | D | 26. | C | 40a. | C | 56a. | E | 70. | D | | |
| 12. | C | 28. | C | 40b. | D | 56b. | E | 72. | A | | |
| 14. | A | 30. | A | 42. | D | 58. | A | 74. | B | | |
| 16. | B | 32a. | E | 44. | B | 60. | B | 76. | D | | |

## Chapter 3
### Kinematics in Two Dimensions

| | | | | | | | | | | | |
|---|---|---|---|---|---|---|---|---|---|---|---|
| 2. | E | 16. | B | 30. | C | 46. | D | 62. | A | 72. | C |
| 4. | D | 18a. | D | 32. | A | 48. | B | 64a. | C | 74. | A |
| 6. | C | 18b. | E | 34. | D | 50. | C | 64b. | E | 76. | E |
| 8. | A | 20. | A | 36. | B | 52. | D | 66. | B | | |
| 10a. | D | 22. | E | 38. | E | 54. | D | 68a. | D | | |
| 10b. | A | 24. | C | 40. | E | 56. | C | 68b. | A | | |
| 12. | B | 26. | A | 42. | C | 58. | E | 70a. | E | | |
| 14. | C | 28. | B | 44. | A | 60. | B | 70b. | B | | |

## Chapter 4
### Forces and Newton's Laws of Motion

| | | | | | | | | | | | |
|---|---|---|---|---|---|---|---|---|---|---|---|
| 2. | E | 12. | A | 22. | B | 32. | B | 38. | D | 44c. | B |
| 4. | B | 14. | E | 24. | D | 34. | B | 40. | A | 46. | D |
| 6. | C | 16. | D | 26. | E | 36a. | D | 42. | C | 48. | C |
| 8. | A | 18. | D | 28. | A | 36b. | A | 44a. | E | 50a. | D |
| 10. | C | 20. | E | 30. | C | 36c. | B | 44b. | A | 50b. | A |

## Chapter 4 (continued)

| | | | | | | | | | | | |
|---|---|---|---|---|---|---|---|---|---|---|---|
| 52. | E | 66. | C | 78. | A | 88b. | A | 96. | D | 110. | C |
| 54. | C | 68a. | C | 80. | A | 88c. | D | 98. | B | 112. | E |
| 56. | A | 68b. | A | 82a. | B | 90a. | E | 100. | C | 114a. | B |
| 58. | B | 70. | E | 82b. | A | 90b. | A | 102. | B | 114b. | E |
| 60. | B | 72a. | D | 84. | E | 92a. | C | 104. | C | 114c. | D |
| 62. | E | 72b. | D | 86a. | C | 92b. | E | 106. | D | 116a. | B |
| 64a. | B | 74. | E | 86b. | B | 94a. | D | 108a. | C | 116b. | B |
| 64b. | E | 76. | C | 88.a. | D | 94b. | A | 108b. | B | | |

## Chapter 5
## Dynamics of Uniform Circular Motion

| | | | | | | | | | | | |
|---|---|---|---|---|---|---|---|---|---|---|---|
| 2. | A | 14a. | A | 24b. | E | 36a. | A | 46a. | D | 56a. | B |
| 4. | D | 14b. | E | 26. | C | 36b. | B | 46b. | A | 56b. | D |
| 6. | B | 16. | C | 28. | D | 36c. | B | 46c. | B | 58a. | C |
| 8a. | D | 18. | B | 30. | D | 38. | C | 48. | E | 58b. | A |
| 8b. | B | 20. | C | 32a. | A | 40. | D | 50. | C | 58c. | E |
| 10. | A | 22. | E | 32b. | E | 42. | E | 52. | C | | |
| 12. | E | 24a. | B | 34. | D | 44. | C | 54. | A | | |

## Chapter 6
## Work and Energy

| | | | | | | | | | | | |
|---|---|---|---|---|---|---|---|---|---|---|---|
| 2. | B | 16.a. | E | 30a. | A | 48. | B | 66b. | E | 76. | B |
| 4a. | C | 16b. | D | 30b. | B | 50. | E | 68a. | B | 78. | A |
| 4b. | D | 18. | E | 32. | C | 52. | A | 68b. | A | 80. | C |
| 6. | C | 20a. | E | 34. | A | 54. | D | 68c. | E | 82. | C |
| 8a. | A | 20b. | E | 36. | D | 56. | C | 70a. | D | 84. | D |
| 8b. | A | 22. | B | 38. | D | 58. | B | 70b. | A | | |
| 10a. | B | 24. | C | 40. | A | 60. | E | 72a. | E | | |
| 10b. | C | 26. | D | 42. | E | 62. | A | 72b. | D | | |
| 12. | B | 28a. | C | 44. | C | 64. | D | 74a. | E | | |
| 14. | A | 28b. | D | 46. | B | 66a. | C | 74b. | D | | |

## Chapter 7
## Impulse and Momentum

| | | | | | | | | | | | |
|---|---|---|---|---|---|---|---|---|---|---|---|
| 2. | D | 12. | A | 24. | E | 34b. | D | 44. | A | 52b. | E |
| 4a. | A | 14. | D | 26. | D | 36. | A | 46. | C | 54. | D |
| 4b. | C | 16. | B | 28. | B | 38. | B | 48a. | D | 56a. | C |
| 6a. | C | 18. | C | 30. | E | 40. | A | 48b. | A | 56b. | C |
| 6b. | E | 20a. | A | 32a. | C | 42a. | B | 50a. | D | 58. | E |
| 8. | E | 20b. | D | 32b. | B | 42b. | B | 50b. | E | 60. | B |
| 10. | B | 22. | C | 34a. | E | 42c. | B | 52a. | A | | |

## Chapter 8
### Rotational Kinematics

| | | | | | | | | | | | |
|---|---|---|---|---|---|---|---|---|---|---|---|
| 2a. | A | 16b. | D | 30. | C | 42. | E | 54b. | E | 66a. | C |
| 2b. | D | 16c. | B | 32a. | A | 44a. | A | 56. | B | 66b. | D |
| 4. | C | 18a. | B | 32b. | B | 44b. | A | 58a. | B | 68. | E |
| 6. | B | 18b. | D | 34a. | B | 46a. | B | 58b. | A | 70. | C |
| 8. | A | 20. | E | 34b. | D | 46b. | C | 60. | C | 72. | B |
| 10. | A | 22. | E | 36. | A | 48. | A | 62a. | A | 74. | E |
| 12. | E | 24. | C | 38a. | D | 50. | D | 62b. | E | | |
| 14. | C | 26. | E | 38b. | D | 52. | C | 64a. | C | | |
| 16a. | E | 28. | D | 40. | C | 54a. | B | 64b. | C | | |

## Chapter 9
### Rotational Dynamics

| | | | | | | | | | | | |
|---|---|---|---|---|---|---|---|---|---|---|---|
| 2. | D | 16. | A | 24b. | C | 36. | E | 50. | C | 66. | D |
| 4. | D | 18a. | E | 26. | E | 38. | A | 52. | B | 68a. | D |
| 6. | B | 18b. | E | 28. | D | 40. | A | 54. | E | 68b. | E |
| 8a. | C | 20a. | D | 30a. | B | 42. | A | 56. | B | 70. | D |
| 8b. | C | 20b. | A | 30b. | E | 44. | D | 58. | C | 72a. | C |
| 8c. | C | 22a. | C | 30c. | A | 46. | B | 60a. | A | 72b. | B |
| 10. | A | 22b. | A | 32a. | D | 48a. | C | 60b. | D | 74. | C |
| 12. | E | 22c. | C | 32b. | B | 48b. | B | 62. | E | 76. | E |
| 14. | B | 24a. | B | 34. | D | 48c. | D | 64. | A | | |

## Chapter 10
### Elasticity and Simple Harmonic Motion

| | | | | | | | | | | | |
|---|---|---|---|---|---|---|---|---|---|---|---|
| 2. | A | 18. | E | 36. | D | 44. | D | 60. | E | 80. | A |
| 4a. | D | 20. | B | 38a. | A | 46. | D | 62. | B | 82. | C |
| 4b. | E | 22. | C | 38b. | A | 48. | B | 64. | C | 84. | D |
| 6. | D | 24. | A | 38c. | C | 50. | D | 66. | C | 86. | D |
| 8. | B | 26. | C | 38d. | E | 52. | E | 68. | A | 88a. | E |
| 10. | B | 28. | C | 38e. | B | 54. | C | 70. | D | 88b. | D |
| 12. | D | 30. | B | 40. | E | 56a. | B | 72. | E | | |
| 14a. | A | 32. | A | 42a. | C | 56b. | A | 74. | E | | |
| 14b. | C | 34a. | B | 42b. | A | 58a. | D | 76. | A | | |
| 16. | E | 34b. | E | 42c. | C | 58b. | B | 78. | B | | |

## Chapter 11
### Fluids

| | | | | | | | | | | | |
|---|---|---|---|---|---|---|---|---|---|---|---|
| 2a. | B | 12a. | C | 20. | B | 32a. | C | 42. | D | 54. | B |
| 2b. | A | 12b. | D | 22. | E | 32b. | C | 44. | C | 56a. | A |
| 4. | E | 14. | E | 24. | C | 34. | B | 46. | D | 56b. | E |
| 6. | B | 16a. | A | 26. | B | 36. | A | 48. | E | 58. | C |
| 8. | D | 16b. | A | 28. | D | 38. | E | 50. | A | 60. | D |
| 10. | C | 18. | D | 30. | E | 40. | A | 52. | B | 62a. | A |

## Chapter 11 (continued)

| | | | | | |
|---|---|---|---|---|---|
| 62b. A | 66b. A | 72b. D | 82. A | 88. E | 98. C |
| 64a. B | 68. B | 74. proof | 84a. C | 90. E | 100. A |
| 64b. E | 70a. E | 76. E | 84b. D | 92. D | 102. E |
| 64b. B | 70b. C | 78. A | 86a. D | 94. B | |
| 66a. D | 72a. A | 80. B | 86b. D | 96. C | |

## Chapter 12
### Temperature and Heat

| | | | | | |
|---|---|---|---|---|---|
| 2a. C | 12b. A | 30. A | 46b. E | 66. A | 86. E |
| 2b. B | 14. E | 32. D | 48. B | 68. A | 88. C |
| 4a. A | 16. D | 34. B | 50. C | 70. D | 90. A |
| 4b. E | 18. B | 36. C | 52. E | 72. E | 92. C |
| 6a. B | 20. A | 38a. E | 54. C | 74. D | 94. B |
| 6b. D | 22. D | 38b. D | 56. D | 76. C | 96. A |
| 8. C | 24. B | 40. D | 58. D | 78. B | 98. A |
| 10a. D | 26a. C | 42. E | 60. A | 80. B | 100. E |
| 10b. B | 26b. E | 44. C | 62. E | 82. D | |
| 12a. C | 28. A | 46a. A | 64. B | 84. A | |

## Chapter 13
### The Transfer of Heat

| | | | | | |
|---|---|---|---|---|---|
| 2. D | 10a. C | 16. A | 24. A | 32. E | 42. proof |
| 4a. A | 10b. B | 18. E | 26a. C | 34. B | |
| 4b. C | 12. E | 20. D | 26b. B | 36. A | |
| 6. C | 14a. D | 22a. B | 28. D | 38. D | |
| 8. B | 14b. C | 22b. E | 30. B | 40. D | |

## Chapter 14
### The Ideal Gas Law and Kinetic Theory

| | | | | | |
|---|---|---|---|---|---|
| 2a. E | 10. B | 24. B | 34. C | 46b. B | 58. D |
| 2b. E | 12. A | 26a. D | 36. E | 46c. B | 60a. C |
| 4. B | 14. C | 26b. A | 38. E | 48. D | 60b. C |
| 6a. C | 16. C | 28. E | 40. B | 50. A | 60c. D |
| 6b. A | 18. A | 30. E | 42. A | 52. C | 62. B |
| 8a. C | 20. D | 30. D | 44. D | 54. E | |
| 8b. D | 22. B | 32. C | 46a. C | 56. A | |

## Chapter 15
### Thermodynamics

| | | | | | |
|---|---|---|---|---|---|
| 2a. A | 4a. C | 6a. E | 8b. D | 12. C | 14c. A |
| 2b. D | 4b. B | 6b. E | 10a. E | 14a. D | 16a. B |
| 2c. C | 4c. B | 8a. A | 10b. A | 14b. E | 16b. A |

## Chapter 15 (continued)

| | | | | | |
|---|---|---|---|---|---|
| 18. B | 34b. A | 46b. A | 62. E | 78a. C | 90b. B |
| 20. D | 36a. D | 48. E | 64. A | 78b. B | 92. E |
| 22. C | 36b. E | 50. B | 66. C | 78c. B | 94. D |
| 24. B | 36c. D | 52. D | 68. D | 80a. E | 96. C |
| 26. E | 38. E | 54a. E | 70a. A | 80b. A | 98. D |
| 28. A | 40a. B | 54b. C | 70b. D | 82. E | 100. E |
| 30a. C | 40b. D | 56. D | 70c. B | 84. A | |
| 30b. B | 42. D | 58a. C | 72. C | 86. D | |
| 32. C | 44. B | 58b. A | 74. B | 88. A | |
| 34a. C | 46a. C | 60. B | 76. A | 90a. B | |

## Chapter 16
## Wave Motion

| | | | | | |
|---|---|---|---|---|---|
| 2. C | 18. A | 34. D | 50b. C | 70. E | 90. A |
| 4. C | 20a. E | 36. D | 52. C | 72. D | 92. D |
| 6. B | 20b. E | 38a. E | 54. E | 74. B | 94. E |
| 8a. C | 22. C | 38b. A | 56. A | 76. C | 96. B |
| 8b. D | 24a. B | 40. B | 58. B | 78. D | 98. B |
| 10. A | 24b. B | 42. A | 60. D | 80. E | 100. C |
| 12a. E | 26. D | 44. C | 62. E | 82. A | 102. A |
| 12b. E | 28. A | 46. A | 64. B | 84. B | 104. B |
| 14. B | 30. A | 48. D | 66. B | 86. C | 106. A |
| 16. D | 32. B | 50a. C | 68. D | 88. E | 108. C |

## Chapter 17
## The Principle of Linear Superposition and Interference Phenomena

| | | | | | |
|---|---|---|---|---|---|
| 2. draw | 16. E | 30a. A | 38a. D | 50. C | 58a. A |
| 4. C | 18. B | 30b. E | 38b. B | 52a. B | 58b. E |
| 6. B | 20. C | 30c. C | 40. E | 52b. B | 58c. C |
| 8. D | 22. E | 32. D | 42. C | 52c. E | 58d. A |
| 10. A | 24. A | 34. B | 44. A | 54. D | 60a. E |
| 12. A | 26. C | 36a. D | 46. A | 56a. C | 60b. C |
| 14. D | 28. B | 36b. B | 48. E | 56b. D | |

## Chapter 18
## Electric Forces and Electric Fields

| | | | | | |
|---|---|---|---|---|---|
| 2. D | 18. B | 32. A | 44a. C | 52b. D | 62. D |
| 4. A | 20a. D | 34a. C | 44b. B | 52c. B | 64a. C |
| 6. E | 20b. B | 34b. D | 46. E | 54a. A | 64b. A |
| 8. A | 22. C | 36a. C | 48. A | 54b. C | 66. D |
| 10. B | 24. E | 36b. A | 50a. C | 54c. E | 68a. B |
| 12. C | 26. E | 38. D | 50b. A | 56. proof | 68b. E |
| 14. E | 28. draw | 40. B | 50c. E | 58. draw | 70. D |
| 16. C | 30. A | 42. B | 52a. C | 60. D | |

# HOMEWORK CHOICES

## Chapter 19
### Electric Potential Energy and the Electric Potential

| | | | | | | | | | | | |
|---|---|---|---|---|---|---|---|---|---|---|---|
| 2a. | D | 12. | E | 24. | A | 36b. | B | 48. | E | 58a. | B |
| 2b. | C | 14. | B | 26. | A | 36c. | D | 50. | C | 58b. | A |
| 4a. | D | 16. | A | 28. | D | 38. | C | 52. | A | 60. | E |
| 4b. | A | 18a. | B | 30. | E | 40. | A | 54. | E | 62. | B |
| 6. | C | 18b. | B | 32. | C | 42. | E | 56a. | D | 64. | C |
| 8. | E | 20. | C | 34. | B | 44. | A | 56b. | D | 66a. | D |
| 10. | D | 22. | E | 36a. | E | 46. | C | 56c. | A | 66b. | B |

## Chapter 20
### Electric Circuits

| | | | | | | | | | | | |
|---|---|---|---|---|---|---|---|---|---|---|---|
| 2. | E | 28. | A | 46. | C | 66a. | A | 86. | B | 108. | A |
| 4. | A | 30. | A | 48a. | D | 66b. | E | 88. | D | 110. | A |
| 6. | B | 32. | C | 48b. | B | 68. | C | 90a. | D | 112. | E |
| 8. | B | 34. | D | 50. | B | 70. | D | 90b. | B | 114a. | C |
| 10. | A | 36. | D | 52. | E | 72. | A | 92. | C | 114b. | D |
| 12. | C | 38a. | E | 54. | B | 74. | E | 94. | D | 114c. | C |
| 14. | D | 38b. | A | 56a. | E | 76a. | A | 96. | proof | 116. | D |
| 16. | D | 40. | C | 56b. | E | 76b. | E | 98. | E | 118. | B |
| 18. | E | 42a. | B | 58. | C | 78a. | D | 100. | E | 120. | A |
| 20. | E | 42b. | C | 60. | D | 78b. | proof | 102. | C | 122. | B |
| 22. | B | 44a. | C | 62. | D | 80. | C | 104. | D | | |
| 24. | C | 44b. | E | 64a. | B | 82. | A | 106a. | B | | |
| 26. | B | 44c. | A | 64b. | A | 84. | B | 106b. | E | | |

## Chapter 21
### Electric Circuits

| | | | | | | | | | | | |
|---|---|---|---|---|---|---|---|---|---|---|---|
| 2a. | E | 14a. | D | 28. | D | 42a. | B | 56a. | D | 70a. | D |
| 2b. | C | 14b. | A | 30. | C | 42b. | D | 56b. | B | 70b. | E |
| 2c. | C | 14c. | C | 32a. | A | 44a. | D | 58a. | A | 72. | D |
| 4. | D | 16. | C | 32b. | E | 44b. | draw | 58b. | B | 74. | E |
| 6. | A | 18. | D | 34. | B | 46. | C | 60. | C | 76. | B |
| 8a. | E | 20. | A | 36. | A | 48. | E | 62. | E | 78. | A |
| 8b. | B | 22. | E | 38a. | A | 50. | B | 64. | C | 80. | B |
| 10. | B | 24. | B | 38b. | A | 52. | C | 66. | proof | 82a. | C |
| 12. | B | 26. | D | 40. | E | 54. | E | 68. | proof | 82b. | A |

## Chapter 22
### Electromagnetic Induction

| | | | | | | | | | | | |
|---|---|---|---|---|---|---|---|---|---|---|---|
| 2. | A | 10b. | E | 14c. | D | 24. | C | 32b. | D | 40. | E |
| 4. | C | 10c. | D | 16. | C | 26. | A | 34. | D | 42. | C |
| 6. | B | 12. | B | 18. | E | 28. | E | 36a. | A | 44. | A |
| 8. | D | 14a. | E | 20. | A | 30. | B | 36b. | B | 46a. | B |
| 10a. | A | 14b. | B | 22. | C | 32a. | B | 38. | C | 46b. | A |

## Chapter 22 (continued)

| | | | | | | | | | |
|---|---|---|---|---|---|---|---|---|---|
| 46c. | E | 54. | C | 62. | B | 70. | B | 76a. | B |
| 48. | D | 56. | proof | 64. | D | 72a. | A | 76b. | B |
| 50. | A | 58. | D | 66. | proof | 72b. | E | 78. | D |
| 52. | E | 60. | C | 68. | A | 74. | D | 80. | E |

## Chapter 23
### Alternating Current Circuits

| | | | | | | | | | | | |
|---|---|---|---|---|---|---|---|---|---|---|---|
| 2. | A | 12. | E | 22a. | B | 28. | C | 36. | C | 46b. | B |
| 4. | D | 14. | C | 22b. | B | 30. | D | 38. | D | 48. | D |
| 6a. | D | 16. | D | 22c. | C | 32a. | C | 40. | E | 50. | E |
| 6b. | B | 18. | E | 24a. | E | 32b. | A | 42. | B | | |
| 8. | C | 20a. | A | 24b. | B | 32c. | A | 44. | D | | |
| 10. | B | 20b. | A | 26. | C | 34. | E | 46a. | A | | |

## Chapter 24
### Electromagnetic Waves

| | | | | | | | | | | | |
|---|---|---|---|---|---|---|---|---|---|---|---|
| 2a. | C | 12a. | D | 24. | E | 34b. | A | 44. | E | 56. | E |
| 2b. | E | 12b. | A | 26. | A | 36. | B | 46. | E | 58. | B |
| 4. | B | 14. | B | 28. | D | 38. | E | 48a. | D | 60. | proof |
| 6a. | A | 16. | E | 30. | C | 40a. | A | 48b. | B | | |
| 6b. | D | 18. | C | 32a. | E | 40b. | D | 50. | D | | |
| 8. | A | 20. | C | 32b. | B | 42a. | B | 52. | D | | |
| 10. | B | 22. | D | 34a. | C | 42b. | C | 54. | A | | |

## Chapter 25
### The Reflection of Light: Mirrors

| | | | | | | | | | | | |
|---|---|---|---|---|---|---|---|---|---|---|---|
| 2a. | C | 10a. | C | 16. | E | 24a. | C | 28. | proof | 36. | D |
| 2b. | B | 10b. | D | 18a. | B | 24b. | E | 30a. | B | 38. | A |
| 4. | A | 12a. | D | 18b. | D | 24c. | D | 30b. | E | 40. | C |
| 6a. | C | 12b. | draw | 20a. | E | 24d. | A | 32a. | A | 42. | E |
| 6b. | B | 14a. | B | 20b. | C | 26a. | D | 32b. | B | | |
| 8. | E | 14b. | A | 22. | A | 26b. | draw | 34. | A | | |

## Chapter 26
### The Refraction of Light: Lenses and Optical Instruments

| | | | | | | | | | | | |
|---|---|---|---|---|---|---|---|---|---|---|---|
| 2. | E | 14. | D | 24b. | B | 32. | D | 40. | proof | 50. | E |
| 4. | E | 16. | B | 26a. | C | 34a. | D | 42. | A | 52a. | C |
| 6. | B | 18. | B | 26b. | E | 34b. | C | 44a. | C | 52b. | B |
| 8. | A | 20. | C | 28. | A | 36. | E | 44b. | D | 54a. | C |
| 10. | E | 22. | D | 30a. | B | 38a. | E | 46. | A | 54b. | D |
| 12. | C | 24a. | A | 30b. | A | 38b. | B | 48. | D | 54c. | B |

## Chapter 26 *(continued)*

| | | | | | | | | | |
|---|---|---|---|---|---|---|---|---|---|
| 54d. | A | 68. | C | 78a. | E | 92. | A | 110. | E |
| 54e. | B | 70a. | E | 78b. | D | 94. | A | 112. | B |
| 56a. | E | 70b. | C | 80a. | C | 96. | D | 144a. | A |
| 56b. | E | 70c. | B | 80b. | C | 98a. | D | 114b. | D |
| 58a. | B | 70d. | A | 80c. | B | 98b. | E | 116a. | E |
| 58b. | C | 70e. | A | 82. | D | 100. | D | 116b. | B |
| 60. | A | 72. | D | 84. | A | 102. | B | 118. | A |
| 62. | B | 74a. | B | 86. | C | 104. | C | 120. | D |
| 64. | A | 74b. | D | 88. | E | 106. | C | 122. | E |
| 66. | D | 76. | C | 90. | E | 108. | A | | |

## Chapter 27
### Interference and the Wave Nature of Light

| | | | | | | | | | | | | | |
|---|---|---|---|---|---|---|---|---|---|---|---|---|---|
| 2a. | A | 10. | D | 22. | C | 34. | C | 44. | E | 56. | B |
| 2b. | B | 12a. | B | 24. | E | 36a. | A | 46. | A | 58. | E |
| 2c. | B | 12b. | D | 26. | B | 36b. | D | 48. | B | 60a. | C |
| 2d. | A | 14. | A | 28. | A | 38. | E | 50a. | B | 60b. | E |
| 4. | D | 16. | E | 30. | D | 40a. | C | 50b. | D | 62. | C |
| 6. | C | 18. | E | 32a. | E | 40b. | A | 52. | D | 64. | D |
| 8. | C | 20. | C | 32b. | B | 42. | D | 54. | C | | |

## Chapter 28
### Special Relativity

| | | | | | | | | | | | |
|---|---|---|---|---|---|---|---|---|---|---|---|
| 2. | B | 12. | E | 20b. | D | 30a. | A | 36b. | D | 40. | A |
| 4. | C | 14. | D | 22. | B | 30b. | D | 36c. | B | 42. | B |
| 6. | B | 16. | D | 24. | B | 32. | C | 36d. | A | 44. | C |
| 8. | C | 18. | C | 26. | E | 34. | E | 38a. | E | 46a. | C |
| 10. | A | 20a. | A | 28. | D | 36a. | E | 38b. | C | 46b. | A |

## Chapter 29
### Particles and Waves

| | | | | | | | | | | | |
|---|---|---|---|---|---|---|---|---|---|---|---|
| 2a. | D | 8. | B | 18a. | A | 22a. | E | 30a. | C | 38. | D |
| 2b. | A | 10. | E | 18b. | D | 22b. | A | 30b. | A | 40. | D |
| 2c. | C | 12. | B | 20a. | E | 24. | C | 30c. | B | 42. | A |
| 4. | C | 14a. | C | 20b. | B | 26a. | D | 32. | E | 44. | B |
| 6a. | B | 14b. | B | 20c. | D | 26b. | E | 34. | B | 46. | C |
| 6b. | D | 16. | E | 20d. | A | 28. | C | 36. | A | 48. | A |

## Chapter 30
### The Nature of the Atom

| | | | | | | | | | | | |
|---|---|---|---|---|---|---|---|---|---|---|---|
| 2a. | C | 10b. | A | 22. | D | 32a. | D | 42. | D | 54. | B |
| 2b. | B | 12. | B | 24. | table | 32b. | B | 44. | E | 56. | A |
| 4. | D | 14. | C | 26. | E | 34. | E | 46. | A | 58a. | D |
| 6. | E | 16. | A | 28a. | B | 36. | A | 48. | C | 58b. | B |
| 8. | D | 18. | C | 28b. | D | 38. | A | 50. | C | 58c. | D |
| 10a. | E | 20. | A | 30. | C | 40. | B | 52. | table | | |

## Chapter 31
### Nuclear Physics and Radioactivity

| | | | | | | | | | | | |
|---|---|---|---|---|---|---|---|---|---|---|---|
| 2. | B | 10a. | D | 18a. | C | 26b. | A | 42. | B | 54a. | B |
| 4a. | E | 10b. | A | 18b. | E | 28. | C | 44a. | A | 54b. | C |
| 4b. | B | 10c. | C | 20a. | A | 30. | C | 44b. | B | 54c. | B |
| 4c. | D | 12. | E | 20b. | C | 32. | D | 44c. | D | 56. | A |
| 4d. | A | 14. | B | 22a. | E | 34. | E | 46. | D | 58. | D |
| 4e. | B | 16a. | D | 22b. | A | 36. | A | 48. | plot | 60. | A |
| 6. | C | 16b. | E | 24. | D | 38. | E | 50. | E | | |
| 8. | A | 16c. | B | 26a. | C | 40. | D | 52. | B | | |

## Chapter 32
### Ionizing Radiation, Nuclear Energy, and Elementary Particles

| | | | | | | | | | | | |
|---|---|---|---|---|---|---|---|---|---|---|---|
| 2. | E | 14a. | A | 24. | E | 32. | B | 38c. | A | 50. | D |
| 4. | B | 14b. | D | 26. | B | 34a. | C | 40. | A | 52. | A |
| 6. | C | 16. | E | 28a. | D | 34b. | E | 42. | E | | |
| 8. | D | 18. | A | 28b. | C | 36. | C | 44. | C | | |
| 10. | B | 20. | A | 28c. | B | 38a. | D | 46. | C | | |
| 12. | C | 22. | D | 30. | B | 38b. | A | 48. | E | | |

# NOTES

# NOTES